# A First Course
## *in*
# Abstract Algebra

## Rings, Groups, and Fields

Third Edition

# A First Course
## *in*
# Abstract Algebra
## Rings, Groups, and Fields
### Third Edition

## Marlow Anderson
Colorado College
Colorado Springs, USA

## Todd Feil
Denison University
Granville, Ohio, USA

CRC Press
Taylor & Francis Group
Boca Raton   London   New York

CRC Press is an imprint of the
Taylor & Francis Group, an **informa** business

A CHAPMAN & HALL BOOK

CRC Press
Taylor & Francis Group
6000 Broken Sound Parkway NW, Suite 300
Boca Raton, FL 33487-2742

© 2015 by Taylor & Francis Group, LLC
CRC Press is an imprint of Taylor & Francis Group, an Informa business

No claim to original U.S. Government works

Printed on acid-free paper
Version Date: 20141001

International Standard Book Number-13: 978-1-4822-4552-3 (Pack - Book and Ebook)

**Visit the Taylor & Francis Web site at**
**http://www.taylorandfrancis.com**

**and the CRC Press Web site at**
**http://www.crcpress.com**

# Contents

# II   Rings, Domains, and Fields                              55

## 6   Rings                                                    57

## 7   Subrings and Unity                                       69

## 8   Integral Domains and Fields                              81

## 9   Ideals                                                   97

## 10   Polynomials over a Field                                109

**Section II in a Nutshell**                                   120

# III   Ring Homomorphisms and Ideals                          123

## 11   Ring Homomorphisms                                      125

# *Preface*

Traditionally, a first course in abstract algebra introduces groups, rings, and fields, in that order. In contrast, we have chosen to develop ring theory first, in order to draw upon the student's familiarity with integers and with polynomials, which we use as the motivating examples for studying rings.

This approach has worked well for us in motivating students in the study of abstract algebra and in showing them the power of abstraction. Our students have found the process of abstraction easier to understand, when they have more familiar examples to base it upon. We introduce groups later on, again by first looking at concrete examples, in this case symmetries of figures in the plane and space and permutations. By this time students are more experienced, and they handle the abstraction much more easily. Indeed, these parts of the text move quite quickly, which initially surprised (and pleased) the authors.

This is the third edition of this text and significant changes have been made to the second edition. Both comments from adopters and our own experiences have prompted this. The biggest change is in moving the section on Unique Factorization (now Section VII) further back in the book reflecting that for many teaching a first course, this topic is optional. Doing so necessitated reorganizing the introduction of ideals. Now Sections I, II, and III form the core material on rings, integral domains, and fields.

Sections IV and V contain the basic group theory material. We have compressed the motivating examples of symmetries of the plane and of space into one chapter, following it with a more detailed treatment of permutations, our other motivating example for abstract groups. Section VI introduces more topics in group theory including new chapters on the Sylow theorems. Sections VIII, IX, and X remain sections on Constructibility, Vector Spaces and Field Extensions, and Galois Theory; the latter contains much edited material and many new exercises.

The diagram below roughly indicates the dependency of the large sections.

$$
\begin{array}{c}
\text{VII} \\
\uparrow \\
\text{I} \longrightarrow \text{II} \longrightarrow \text{III} \longrightarrow \text{IV} \longrightarrow \text{V} \longrightarrow \text{VI} \\
\downarrow \qquad\qquad\searrow \qquad\qquad\qquad\downarrow \\
\text{VIII} \qquad\qquad\qquad \text{IX} \longrightarrow \text{X}
\end{array}
$$

## Descriptions of Sections

Section I (*Numbers, Polynomials, and Factoring*) introduces the integers $\mathbb{Z}$, and the polynomials $\mathbb{Q}[x]$ over the rationals. In both cases we emphasize the idea of factoring into irreducibles, pointing out the structural similarities. We also introduce the rings of integers modulo $n$ in this section. Induction, the most important proof technique used in the early part of this text, is introduced in Chapter 1.

In Section II (*Rings, Domains, and Fields*) we define a ring as the abstract concept encompassing our specific examples from Section I. We define integral domains and fields and then look at polynomials over an arbitrary field. We make the point that the important properties of $\mathbb{Q}[x]$ are really due to the fact that we have coefficients from a field; this gives students a nice example of the power of abstraction. An introduction to complex numbers is given. We also introduce ideals in this section.

Section III (*Ring Homomorphisms and Ideals*) has as its main goal the proof of the Fundamental Isomorphism Theorem. Section III also includes a chapter about the connection between maximal ideals and fields, and prime ideals and domains. There is also an optional chapter on the Chinese Remainder Theorem.

Section IV (*Groups*) begins with two chapters on concrete examples motivating abstract groups: symmetries of geometric figures (in the plane and in space) and permutations. We then define abstract groups and group isomorphisms and consider subgroups and cyclic groups.

Section V (*Group Homomorphisms*) defines group homomorphisms with one of its goals the Fundamental Isomorphism Theorem for groups. Cayley's and Lagrange's theorems are presented in this section.

Section VI (*Topics from Group Theory*) explores three topics: the alternating group, the Sylow theorems, and solvable groups. The latter is needed in Section X. We use groups acting on sets to prove the Sylow theorems, which provides not only a very accessible method of proof but also experience with permutations from a slightly different perspective.

In Section VII (*Unique Factorization*) we explore more general contexts in which unique factorization is possible. Chapter 33 concludes with the theorem that every principal ideal domain is a unique factorization domain. In the interest of time, many instructors may wish to skip the last two chapters of this section.

Section VIII (*Constructibility Problems*) is an optional section that is a great example of the power of abstract algebra. In it, we show that the three Greek constructibility problems using a compass and straightedge are impossible. This section does not use Kronecker's Theorem and is very computational in flavor. It does not depend on knowing any group theory and can be taught immediately after Section III, if the instructor wishes to delay the introduction of groups.

We revisit the impossibility proofs in Section IX (*Vector Spaces and Field Extensions*), where we give enough vector space theory to introduce students to the theory of algebraic field extensions. Seeing the impossibility proofs again, in a more abstract context, emphasizes the power of abstract field theory.

Section X develops *Galois Theory* with the goal of showing the impossibility of solving the quintic with radicals. This section depends heavily on Section IX, as well as Chapters 27 and 30.

Each chapter includes Quick Exercises, which are intended to be done by the student as the text is read (or perhaps on the second reading) to make sure the topic just covered is understood. Quick Exercises are typically straightforward, rather short verifications of facts just stated that act to reinforce the information just presented. They also act as an early warning to the student that something basic was missed. We often use some of them as a basis for in-class discussion. The exercises following each chapter begin with the Warm-up Exercises, which test fundamental comprehension and should be done by all students. These are followed by the regular exercises, which include both computational problems and supply-the-proof problems. Answers to most odd-numbered exercises that do not require proof are given in the Hints and Answers section. Hints, of varying depth, to odd-numbered proof problems are also given there.

Historical remarks follow many of the chapters. For the most part we try to make use of the history of algebra to make certain pedagogical points. We find that students enjoy finding out a bit about the history of the subject, and how long it took for some of the concepts of abstract algebra to evolve. We've relied on such authors as Boyer & Merzbach, Eves, Burton, Kline, and Katz for this material.

We find that in a first (or second) course, students lose track of the forest, getting bogged down in the details of new material. With this in mind, we've ended each section with a short synopsis that we've called a "Nutshell" in which we've laid out the important definitions and theorems developed in that section, sometimes in an order slightly altered from the text. It's a way for the student to organize their thoughts on the material and see what the major points were, in case they missed them the first time through.

We include an appendix entitled "Guide to Notation," which provides a list of mathematical notations used in the book, and citations to where they are introduced in the text. We group the notations together conceptually. There is also a complete index, which will enable readers to find theorems, definitions, and biographical citations easily in the text.

## Notes for the Instructor

There is more material here than can be used in one semester. Those teaching a one-semester course may choose among various topics they might wish to include. There is sufficient material in this text for a two-semester course, probably more, in most cases.

The suggested track through the text for a one-semester course is to tackle Sections I through V (except possibly Chapter 16). We consider these chapters as the core ideas in the text. If time permits, one could include topics from Section VI, VII, VIII, or a combination, as to your taste.

A second semester could pick up wherever the first semester left off with the goal of completing Section X on Galois Theory, or else finish the remaining topics in the text.

We assume that the students using the text have had the usual calculus sequence; this is mostly an assumption of a little mathematical maturity, since we only occasionally make any real use of the calculus. We do not assume any familiarity with linear algebra, although it would be helpful. We regularly use the multiplication of $2 \times 2$ matrices, mostly as an example of a non-commutative operation; we find that a short in-class discussion of this (perhaps supplemented with some of our exercises) is sufficient even for students who've never seen matrices before. We make heavy use of complex numbers in the text but do not assume any prior acquaintance with them; our introduction to them in Chapters 8 and 10 should be quite adequate.

Instructors may contact the publisher for an Instructor's Manual which contains answers to all of the exercises, including in-depth outlines for proof problems.

## Acknowledgments

Each of us has spent over thirty-five years studying, reading, teaching and doing scholarship in abstract algebra, and thinking about the undergraduate course in this subject. Over the course of that time we have been influenced importantly by many textbooks on the subject. These books have inspired us to think about what topics to include, the order in which they might be presented, and techniques of proof and presentation. Both of us took courses from I. N. Herstein's text *Topics in Algebra*. We've had many occasions to look into the classic works by Van der Waerden and Jacobson. We've taught and learned from texts by such authors as Goldstein, Fraleigh, Burton, and Gallian. Burton's text on elementary number theory and the Maxfields' little book on Galois theory have also been important. We offer our thanks here to those mentioned, and the many influences on our approach we cannot trace.

We owe considerable gratitude to the many people who have helped us in writing this book. We have discussed this book informally with many of our colleagues, immeasurably shaping the final product. Special mention should be made of Mike Westmoreland and

Robin Wilson, who gave parts of the original manuscript a close reading. We're grateful to our respective institutions (The Colorado College and Denison University), which provided financial support to this project at various crucial stages. We're also grateful to the readers of the first two editions, who have made helpful suggestions, and pointed out errors both typographical and mathematical.

A big thank you goes to Sunil Nair from Chapman & Hall/CRC for encouraging us to undertake this third edition. His support of this project is very much appreciated.

Most of all, we thank Audrey and Robin, for their moral support that was, as always, indispensable.

We take the only reasonable position regarding errors that may remain in the text, both typographical and otherwise, and blame each other.

Marlow Anderson
Mathematics Department
The Colorado College
Colorado Springs, CO
manderson@coloradocollege.edu

Todd Feil
Department of Mathematics and Computer Science
Denison University
Granville, OH
feil@denison.edu

# Part I

# Numbers, Polynomials, and Factoring

# Chapter 1

## The Natural Numbers

All mathematics begins with counting. This is the process of putting the set of objects to be counted in one-to-one correspondence with the first several **natural numbers** (or **counting numbers**):

$$1, 2, 3, 4, 5, \cdots.$$

We denote by $\mathbb{N}$ the infinite set consisting of all these numbers. Amazingly, despite the antiquity of its study, humankind has barely begun to understand the algebra of this set. This introduction is intended to provide you with a fund of examples and principles that we will generalize in later chapters.

---

## 1.1 Operations on the Natural Numbers

We encounter no trouble as long as we restrict ourselves to *adding* natural numbers, because more natural numbers result. Accordingly we say our set is *closed under addition*. However, consider what happens when we attempt to *subtract* a natural number $a$ from $b$, or, equivalently, we seek a solution to the equation $a + x = b$ in the unknown $x$. We discover that our set of natural numbers is inadequate to the task. This naturally leads to the set of all **integers**, which we denote by $\mathbb{Z}$ (for 'Zahlen,' in German):

$$\cdots - 3, -2, -1, 0, 1, 2, 3, \cdots.$$

This is the smallest set of numbers containing $\mathbb{N}$ and closed under subtraction.

It is easy to make sense of *multiplication* in $\mathbb{N}$, by viewing it as repeated addition:

$$na = \underbrace{a + a + \cdots + a}_{n \text{ times}}.$$

This operation is easily extended to $\mathbb{Z}$ by using the sign conventions with which you are probably familiar. Why minus multiplied by minus needs to be plus is something you might reflect on now. We will return to this question in a more general context later.

We now have a whole new class of equations, many of which lack solutions: $ax = b$. This leads to *division*, and to the **rational numbers** $\mathbb{Q}$, which are precisely the quotients of one integer by another. The reason why we don't allow division by 0 is because if we let $a = 0$ and $b \neq 0$ in the equation above, we obtain $0 = 0x = b \neq 0$. Why $0x = 0$ is another question you might reflect on now – we will return to this later too.

But to address the algebra of $\mathbb{Q}$ takes us too far afield from our present subject. For the present we shall be more than satisfied in considering $\mathbb{Z}$ and its operations.

## 1.2  Well Ordering and Mathematical Induction

A fundamental property of $\mathbb{N}$ (which has a profound influence on the algebra of $\mathbb{Z}$) is that this set is **well ordered**, a property that we state formally as follows, and which we shall accept as an axiom about $\mathbb{N}$:

**The Well-ordering Principle** *Every non-empty subset of $\mathbb{N}$ has a least element.*

For any subset of $\mathbb{N}$ that we might specify by listing the elements, this seems obvious, but the principle applies even to sets that are more indirectly defined. For example, consider the set of all natural numbers expressible as $12x + 28y$, where $x$ and $y$ are allowed to be any integers. The extent of this set is not evident from the definition. Yet the Well-ordering Principle applies and thus there is a smallest natural number expressible in this way. We shall meet this example again, when we prove something called the GCD identity in the next chapter. (See also Exercise 9.)

Suppose we wish to apply the Well-ordering Principle to a particular subset $X$ of $\mathbb{N}$. We may then consider a sequence of yes/no questions of the following form:

$$\text{Is } 1 \in X?$$
$$\text{Is } 2 \in X?$$
$$\vdots$$

Because $X$ is non-empty, sooner or later one of these questions must be answered yes. The first such occurrence gives the least element of $X$. Of course, such questions might not be easily answerable in practice. But nevertheless, the Well-ordering Principle asserts the existence of this least element, without identifying it explicitly.

The Well-ordering Principle allows us to prove one of the most powerful techniques of proof that you will encounter in this book. (See Theorem 1.1 later in this chapter.) This is the *Principle of Mathematical Induction*:

**Principle of Mathematical Induction** *Suppose $X$ is a subset of $\mathbb{N}$ that satisfies the following two criteria:*

1. *$1 \in X$, and*

2. *If $k \in X$ for all $k < n$, then $n \in X$.*

*Then $X = \mathbb{N}$.*

The Principle of Mathematical Induction is used to prove that certain sets $X$ equal the entire set $\mathbb{N}$. In practice, the set $X$ will usually be "the set of all natural numbers with property such-and-such." To apply it we must check two things:

1. The *base case*: That the least element of $\mathbb{N}$ belongs to $X$, and

2. The *bootstrap*: A general statement which asserts that a natural number belongs to $X$ whenever all its predecessors do.

You should find the Principle of Mathematical Induction plausible because successively applying the bootstrap allows you to conclude that

$$2 \in X, \; 3 \in X, \; 4 \in X, \; \cdots.$$

When checking the bootstrap, we assume that all predecessors of $n$ belong to $X$ and must infer that $n$ belongs to $X$. In practice we often need only that certain predecessors of $n$ belong to $X$. For instance, many times we will need only that $n-1$ belongs to $X$. Indeed, the form of induction you have used before probably assumed only that $n-1$ was in $X$, instead of all $k < n$. It turns out that the version you learned before and the version we will be using are equivalent, although they don't appear to be at first glance. We will find the version given above of more use. (See Exercise 17.)

Before proving the Principle of Mathematical Induction itself, let us look at some simple examples of its use.

## Example 1.1

A finite set with $n$ elements has exactly $2^n$ subsets.

**Proof by Induction**: Let $X$ be the set of those positive integers for which this is true. We first check that $1 \in X$. But a set with exactly one element has two subsets, namely, the empty set $\emptyset$ and the set itself. This is $2 = 2^1$ subsets, as required.

Now suppose that $n > 1$, and $k \in X$ for all $k < n$. We must prove that $n \in X$. Suppose then that $S$ is a set with $n$ elements; we must show that $S$ has $2^n$ subsets. Because $S$ has at least one element, choose one of them and call it $s$. Now every subset of $S$ either contains $s$ or it doesn't. Those subsets that don't contain $s$ are precisely the subsets of $S \backslash \{s\} = \{x \in S : x \neq s\}$. But this latter set has $n-1$ elements, and so by our assumption that $n - 1 \in X$ we know that $S \backslash \{s\}$ has $2^{n-1}$ subsets. Now those subsets of $S$ that *do* contain $s$ are of the form $A \cup \{s\}$, where $A$ is a subset of $S \backslash \{s\}$. There are also $2^{n-1}$ of these subsets. Thus, there are $2^{n-1} + 2^{n-1} = 2^n$ subsets of $S$ altogether. In other words, $n \in X$. Thus, by the Principle of Mathematical Induction, any finite set with $n$ elements has exactly $2^n$ subsets. $\qquad\square$

Notice that this formula for counting subsets also works for a set with zero elements because the empty set has exactly one subset (namely, itself). We could have easily incorporated this fact into the proof above by starting at $n = 0$ instead. This amounts to saying that the set $\{0, 1, 2, \cdots\}$ is well ordered too. In the future we will feel free to start an induction proof at any convenient point, whether that happens to be $n = 1$ or $n = 0$. (We can also start induction at, say, $n = 2$, but in such a case remember that we would then have proved only that $X = \mathbb{N} \backslash \{1\}$.)

## Example 1.2

The sum of the first $n$ positive odd integers is $n^2$. That is,

$$1 + 3 + 5 + \cdots + (2n - 1) = n^2, \text{ for } n \geq 1.$$

**Proof by Induction**: In this proof we proceed slightly less formally than before and suppress explicit mention of the set $X$.

▷ **Quick Exercise.** What is the set $X$ in this proof? ◁

Because $2 \cdot 1 - 1 = 1^2$, our formula certainly holds for $n = 1$. We now assume that the formula holds for $k < n$ and show that it holds for $n$. But then, putting $k = n - 1$, we have

$$1 + 3 + 5 + \cdots + (2(n-1) - 1) = (n-1)^2.$$

Thus,

$$
\begin{aligned}
1 + 3 + 5 + \cdots + (2(n-1) - 1) + (2n - 1) &= \\
(n-1)^2 + (2n-1) &= n^2,
\end{aligned}
$$

which shows that the formula holds for $n$. Thus, by the Principle of Mathematical Induction, the formula holds for *all* $n$. $\qquad\square$

Students new to mathematical induction often feel that in verifying (2) they are assuming exactly what they are required to prove. This feeling arises from a misunderstanding of the fact that (2) is an *implication*: that is, a statement of the form $p \Rightarrow q$. To prove such a statement we must assume $p$, and then derive $q$. Indeed, assuming that $k$ is in $X$ for all $k < n$ is often referred to as **the induction hypothesis**.

Mention should also be made of the fact that mathematical induction is a deductive method of proof and so should not be confused with the notion of inductive reasoning discussed by philosophers. The latter involves inferring likely general principles from particular cases. This sort of reasoning has an important role in mathematics, and we hope you will apply it to make conjectures regarding the more general principles that lie behind many of the particular examples which we will discuss. However, for a mathematician an inductive inference of this sort does not end the story. What is next required is a deductive proof that the conjecture (which might have been verified in particular instances) is always true.

## 1.3   The Fibonacci Sequence

To provide us with another example of proof by induction, we consider a famous sequence of integers, called the **Fibonacci sequence** in honor of the medieval mathematician who first described it. The first several terms are

$$1, \; 1, \; 2, \; 3, \; 5, \; 8, \; 13, \; \cdots.$$

You might have already detected the pattern: A typical element of the sequence is the sum of its two immediate predecessors. This means that we can *inductively* define the sequence by setting

$$\begin{aligned}
a_1 &= 1, \\
a_2 &= 1, \text{ and} \\
a_{n+2} &= a_{n+1} + a_n, \text{ for } n \geq 1.
\end{aligned}$$

This sort of inductive or *recursive* definition of a sequence is often very useful. However, it would still be desirable to have an explicit formula for $a_n$ in terms of $n$. It turns out that the following surprising formula does the job:

$$a_n = \frac{(1 + \sqrt{5})^n - (1 - \sqrt{5})^n}{2^n \sqrt{5}}.$$

At first (or even second) glance, it does not even seem clear that this formula gives integer values, much less the particular integers that make up the Fibonacci sequence. You will prove this formula in Exercise 13. The proof uses the Principle of Mathematical Induction, because the Fibonacci sequence is defined in terms of its two immediate predecessors. We now prove another simpler fact about the Fibonacci sequence:

**Example 1.3**

$a_{n+1} \leq 2a_n$, for all $n \geq 1$.

**Proof by Induction:**   This is trivially true when $n = 1$. In the argument which follows we rely on two successive true instances of our formula—as might be expected because the Fibonacci sequence is defined in terms of two successive terms. Consequently, you should check that the inequality holds when $n = 2$.

▷ **Quick Exercise.** Verify that the inequality $a_{n+1} \leq 2a_n$ holds for $n = 1$ and $n = 2$. ◁

We now assume that $a_{k+1} \leq 2a_k$ for all $k < n$, where $n > 2$. We must show that this inequality holds for $k = n$. Now

$$a_{n+1} = a_n + a_{n-1} \leq 2a_{n-1} + 2a_{n-2},$$

where we have applied the induction hypothesis for both $k = n - 1$ and $k = n - 2$. But because $a_{n-1} + a_{n-2} = a_n$, we have $a_{n+1} \leq 2a_n$, as required. □

## 1.4   Well Ordering Implies Mathematical Induction

We now prove the Principle of Mathematical Induction, using the Well-ordering Principle.

**Theorem 1.1** *The Well-ordering Principle implies the Principle of Mathematical Induction.*

**Proof**:   Suppose that $X$ is a subset of $\mathbb{N}$ satisfying both (1) and (2). Our strategy for showing that $X = \mathbb{N}$ is 'reductio ad absurdum' (or 'proof by contradiction'): We assume the contrary and derive a contradiction.

In this case we assume that $X$ is a proper subset of $\mathbb{N}$, and so $Y = \mathbb{N} \backslash X$ is a non-empty subset of $\mathbb{N}$. By the Well-ordering Principle, $Y$ possesses a least element $m$. Clearly $m \neq 1$ by (1). All natural numbers $k < m$ belong to $X$ because $m$ is the *least* element of $Y$. However, by (2) we conclude that $m \in X$. But now we have concluded that $m \in X$ and $m \notin X$; this is clearly a contradiction. Our assumption that $X$ is a proper subset of $\mathbb{N}$ must have been false. Hence, $X = \mathbb{N}$. □

The converse of this theorem is also true (see Exercise 16).

## 1.5   The Axiomatic Method

Our careful proof of the Principle of Mathematical Induction from the Well-ordering Principle is part of a general program we are beginning in this chapter. We wish eventually to base our analysis of the arithmetic of the integers on as few assumptions as possible. This will be an example of the *axiomatic method* in mathematics. By making our assumptions clear and our proofs careful, we will be able to accept with confidence the truth of statements about the integers which we will prove later, even if the statements themselves are not obviously true. We eventually will also apply the axiomatic method to many algebraic systems other than the integers.

The first extended example of an axiomatic approach to mathematics appears in *The Elements* of Euclid, who was a Greek mathematician living circa 300 B.C. In his book he developed much of ordinary plane geometry by means of a careful logical string of theorems, based on only five axioms and some definitions. The logical structure of Euclid's book is a model of mathematical economy and elegance. So much mathematics is inferred from so few underlying assumptions!

Note of course that we must accept *some* statements without proof (and we call these statements axioms)—for otherwise we'd be led into circular reasoning or an infinite regress.

One cost of the axiomatic method is that we must sometimes prove a statement that already seems 'obvious'. But if we are to be true to the axiomatic method, a statement we believe to be true must either be proved, or else added to our list of axioms. And for reasons of logical economy and elegance, we wish to rely on as few axioms as possible.

Unfortunately, we are not yet in a position to proceed in a completely axiomatic way. We shall accept the Well-ordering Principle as an axiom about the natural numbers. But in addition, we shall accept as given facts your understanding of the elementary arithmetic in $\mathbb{Z}$: that is, addition, subtraction, and multiplication. In Chapter 6, we will finally be able to enumerate carefully the abstract properties which make this arithmetic work. The role of $\mathbb{Z}$ as a familiar, motivating example will be crucial.

The status of division in the integers is quite different. It is considerably trickier (because it is not always possible). We will examine this carefully in the next chapter.

---

## Chapter Summary

In this chapter we introduced the natural numbers $\mathbb{N}$ and emphasized the following facts about this set:

- $\mathbb{N}$ is closed under addition;

- Multiplication in $\mathbb{N}$ can be defined in terms of addition, and under this definition $\mathbb{N}$ is closed under multiplication;

- The *Well-ordering Principle* holds for $\mathbb{N}$.

We then used the Well-ordering Principle to prove the *Principle of Mathematical Induction* and provided examples of its use.

We also introduced the set $\mathbb{Z}$ of all integers, as the smallest set of numbers containing $\mathbb{N}$ that is closed under subtraction.

---

## Warm-up Exercises

a. Explain the arithmetic advantages of $\mathbb{Z}$, as compared with $\mathbb{N}$. How about $\mathbb{Q}$, as compared with $\mathbb{Z}$?

b. Why isn't $\mathbb{Z}$ well ordered? Why isn't $\mathbb{Q}$ well ordered? Why isn't the set of all rational numbers $x$ with $0 \leq x \leq 1$ well ordered?

c. Suppose we have an infinite row of dominoes, set up on end. What sort of induction argument would convince us that knocking down the first domino will knock them all down?

d. Explain why any finite subset of $\mathbb{Q}$ is well ordered.

## Exercises

1. Prove using mathematical induction that for all positive integers $n$,
$$1 + 2 + 3 + \cdots + n = \frac{n(n+1)}{2}.$$

2. Prove using mathematical induction that for all positive integers $n$,
$$1^2 + 2^2 + 3^2 + \cdots + n^2 = \frac{n(2n+1)(n+1)}{6}.$$

3. You probably recall from your previous mathematical work the **triangle inequality**: for any real numbers $x$ and $y$,
$$|x + y| \leq |x| + |y|.$$
   Accept this as given (or see a calculus text to recall how it is proved). Generalize the triangle inequality, by proving that
$$|x_1 + x_2 + \cdots + x_n| \leq |x_1| + |x_2| + \cdots + |x_n|,$$
   for any positive integer $n$.

4. Given a positive integer $n$, recall that $n! = 1 \cdot 2 \cdot 3 \cdots n$ (this is read as $n$ **factorial**). Provide an inductive definition for $n!$. (It is customary to actually start this definition at $n = 0$, setting $0! = 1$.)

5. Prove that $2^n < n!$ for all $n \geq 4$.

6. Prove that for all positive integers $n$,
$$1^3 + 2^3 + \cdots + n^3 = \left( \frac{n(n+1)}{2} \right)^2.$$

7. Prove the familiar **geometric progression** formula. Namely, suppose that $a$ and $r$ are real numbers with $r \neq 1$. Then show that
$$a + ar + ar^2 + \cdots + ar^{n-1} = \frac{a - ar^n}{1 - r}.$$

8. Prove that for all positive integers $n$,
$$\frac{1}{1 \cdot 2} + \frac{1}{2 \cdot 3} + \cdots + \frac{1}{n(n+1)} = \frac{n}{n+1}.$$

9. By trial and error, try to find the smallest positive integer expressible as $12x + 28y$, where $x$ and $y$ are allowed to be any integers.

10. A **complete graph** is a collection of $n$ points, each of which is connected to each other point. The complete graphs on 3, 4, and 5 points are illustrated below:

   Use mathematical induction to prove that the complete graph on $n$ points has exactly $n(n-1)/2$ lines.

11. Consider the sequence $\{a_n\}$ defined inductively as follows:

$$a_1 = a_2 = 1, \quad a_{n+2} = 2a_{n+1} - a_n.$$

Use mathematical induction to prove that $a_n = 1$, for all natural numbers $n$.

12. Consider the sequence $\{a_n\}$ defined inductively as follows:

$$a_1 = 5, \quad a_2 = 7, \quad a_{n+2} = 3a_{n+1} - 2a_n.$$

Use mathematical induction to prove that $a_n = 3 + 2^n$, for all natural numbers $n$.

13. Consider the Fibonacci sequence $\{a_n\}$.

(a) Use mathematical induction to prove that

$$a_{n+1}a_{n-1} = (a_n)^2 + (-1)^n.$$

(b) Use mathematical induction to prove that

$$a_n = \frac{(1+\sqrt{5})^n - (1-\sqrt{5})^n}{2^n\sqrt{5}}.$$

14. In this problem you will prove some results about the **binomial coefficients**, using induction. Recall that

$$\binom{n}{k} = \frac{n!}{(n-k)!k!},$$

where $n$ is a positive integer, and $0 \le k \le n$.

(a) Prove that

$$\binom{n}{k} = \binom{n-1}{k} + \binom{n-1}{k-1},$$

$n \ge 2$ and $k < n$. *Hint:* You do not need induction to prove this. Bear in mind that $0! = 1$.

(b) Verify that $\binom{n}{0} = 1$ and $\binom{n}{n} = 1$. Use these facts, together with part (a), to prove by induction on $n$ that $\binom{n}{k}$ is an integer, for all $k$ with $0 \le k \le n$. (*Note:* You may have encountered $\binom{n}{k}$ as the count of the number of $k$ element subsets of a set of $n$ objects; it follows from this that $\binom{n}{k}$ is an integer. What we are asking for here is an inductive proof based on algebra.)

(c) Use part (a) and induction to prove the **Binomial Theorem**: For non-negative $n$ and variables $x, y$,

$$(x+y)^n = \sum_{k=0}^{n} \binom{n}{k} x^{n-k} y^k.$$

15. Criticize the following 'proof' showing that all cows are the same color.

It suffices to show that any herd of $n$ cows has the same color. If the herd has but one cow, then trivially all the cows in the herd have the same color. Now suppose that we have a herd of $n$ cows and $n > 1$. Pick out a cow and remove it from the herd, leaving $n-1$ cows; by the induction hypothesis these cows all have the same color. Now put the cow back and remove another cow. (We can do so because $n > 1$.) The remaining $n-1$ again must all be the same color. Hence, the first cow selected and the second cow selected have the same color as those not selected, and so the entire herd of $n$ cows has the same color.

16. Prove the converse of Theorem 1.1; that is, prove that the Principle of Mathematical Induction implies the Well-ordering Principle. (This shows that these two principles are logically equivalent, and so from an axiomatic point of view it doesn't matter which we assume is an axiom for the natural numbers.)

17. The *Strong* Principle of Mathematical Induction asserts the following. Suppose that $X$ is a subset of $\mathbb{N}$ that satisfies the following two criteria:

    (a) $1 \in X$, and

    (b) If $n > 1$ and $n - 1 \in X$, then $n \in X$.

    Then $X = \mathbb{N}$. Prove that the Principle of Mathematical Induction holds if and only if the Strong Principle of Mathematical Induction does.

# Chapter 2

## The Integers

In this chapter we analyze how multiplication works in the integers $\mathbb{Z}$, and in particular when division is possible. This is more interesting than asking how multiplication works in the rational numbers $\mathbb{Q}$, where division is always possible (except for division by zero).

We all learned at a very young age that we can always divide one integer by another non-zero integer, as long as we allow for a remainder. For example, $326 \div 21$ gives quotient 15 with remainder 11. The actual computation used to produce this result is our usual long division. Note that the division process halts when we arrive at a number less than the divisor. In this case 11 is less than 21, and so our division process stops. We can record the result of this calculation succinctly as

$$326 = (21)(15) + 11, \text{ where } 0 \leq 11 < 21.$$

## 2.1 The Division Theorem

The following important theorem describes this situation formally. This is the first of many examples in this book of an *existence and uniqueness theorem*: We assert that something exists, and that there is only one such. Both assertions must be proved. We will use induction for the existence proof.

**Theorem 2.1    Division Theorem for $\mathbb{Z}$**    *Let $a, b \in \mathbb{Z}$, with $a \neq 0$. Then there exist unique integers $q$ and $r$ (called the* quotient *and* remainder, *respectively), with $0 \leq r < |a|$, such that $b = aq + r$.*

**Proof**:    We first prove the theorem in case $a > 0$ and $b \geq 0$. To show the existence of $q$ and $r$ in this case, we use induction on $b$.

We must first establish the base case for the induction. You might expect us to check that the theorem holds in case $b = 0$ (the smallest possible value for $b$). But actually, we can establish the theorem for all $b$ where $b < a$; for in this case the quotient is 0 and the remainder is $b$. That is, $b = a \cdot 0 + b$.

We may now assume that $b \geq a$. Our induction hypothesis is that there exist a quotient and remainder whenever we attempt to divide an integer $c < b$ by $a$. So let $c = b - a$. Since $c < b$ we have by the induction hypothesis that $c = aq' + r$, where $0 \leq r < a$. But then

$$b = aq' + r + a = a(q' + 1) + r, \text{ where } 0 \leq r < a.$$

We therefore have a quotient $q = q' + 1$ and a remainder $r$.

We now consider the general case, where $b$ is any integer, and $a$ is any non-zero integer. We apply what we have already proved to the integers $|b|$ and $|a|$ to obtain unique integers

$q'$ and $r'$ so that $|b| = q'|a| + r'$, with $r' < |a|$. We now obtain the quotient and remainders required, depending on the signs of $a$ and $b$, in the following three cases:

Case (i): Suppose that $a < 0$ and $b \leq 0$. Then let $q = q' + 1$ and $r = -a - r'$. Note first that $0 \leq r < |a|$. Now

$$aq + r = a(q' + 1) + -a - r' = aq' + a - a - r'$$
$$= aq' - r' = -(|a|q' + r') = -|b| = b,$$

as required.

You can now check the remaining two cases:

Case (ii): If $a < 0$ and $b \geq 0$, then let $q = -q'$ and $r = r'$.

Case (iii): If $a > 0$ and $b \leq 0$, then let $q = -q' - 1$ and $r = a - r'$.

▷ **Quick Exercise.**   Verify that the quotients and remainders specified in Cases (ii) and (iii) actually work. ◁

Now we prove the uniqueness of $q$ and $r$. Our strategy is to assume that we have two potentially different quotient-remainder pairs, and then show that the different pairs are actually the same. So, suppose that $b = aq + r = aq' + r'$, where $0 \leq r < |a|$ and $0 \leq r' < |a|$. We hope that $q = q'$ and $r = r'$.

Since $aq + r = aq' + r'$, we have that $a(q - q') = r' - r$. Now $|r' - r| < |a|$, and so $|a||q - q'| = |r' - r| < |a|$. Hence, $|q - q'| < 1$. Thus, $q - q'$ is an integer whose absolute value is less than 1, and so $q - q' = 0$. That is, $q = q'$. But then $r' - r = a \cdot 0 = 0$ and so $r' = r$, proving uniqueness.                                                                                                    □

You should exercise some care in applying the Division Theorem with negative integers. The fact that the remainder must be positive leads to some answers that may be surprising.

**Example 2.1**

For example, while 326 divided by 21 gives quotient 15 and remainder 11, $-326$ divided by 21 gives quotient $-16$ and remainder 10, and $-326$ divided by $-21$ gives quotient 16 and remainder 10.

We say an integer $a$ **divides** an integer $b$ if $b = aq$ for some integer $q$. In this case, we say $a$ is a **factor** of $b$, and write $a|b$. In the context of the Division Theorem, $a|b$ means that the remainder obtained is 0.

**Example 2.2**

Thus, $-6|126$, because $126 = (-6)(-21)$. Note that *any* integer divides 0, because $0 = (a)(0)$.

---

## 2.2   The Greatest Common Divisor

In practice, it may be *very* difficult to find the factors of a given integer, if it is large. However, it turns out to be relatively easy to determine whether two given integers have a common factor. To understand this, we must introduce the notion of greatest common divisor: Given two integers $a$ and $b$ (not both zero), then the integer $d$ is the **greatest**

**common divisor** (gcd) of $a$ and $b$ if $d$ divides both $a$ and $b$, and it is the largest positive integer that does. We will often write $\gcd(a, b) = d$ to express this relationship.

For example $6 = \gcd(42, -30)$, as you can check directly by computing all possible common divisors, and picking out the largest one. Because all integers divide 0, we have not allowed ourselves to consider the meaningless expression $\gcd(0, 0)$. However, if $a \neq 0$, it does make sense to consider $\gcd(a, 0)$.

▷ **Quick Exercise.** Argue that for all $a \neq 0$, $\gcd(a, 0) = |a|$. ◁

But why should an arbitrary pair of integers (not both zero) have a gcd? That is, does the definition we have of gcd really make sense? Note that if $c > 0$ and $c|a$ and $c|b$, then $c \leq |a|$ and $c \leq |b|$. This means that there are only finitely many positive integers that could possibly be the gcd of $a$ and $b$, and because 1 *does* divide both $a$ and $b$, $a$ and $b$ do have at least one common divisor. This means that the gcd of any pair of integers exists (and is unique).

To actually determine $\gcd(a, b)$ we would rather not check all the possibilities less than $|a|$ and $|b|$. Fortunately, we don't have to, because there is an algorithm that determines the gcd quite efficiently. This first appears as Proposition 2 of Book 7 of Euclid's *Elements* and depends on repeated applications of the Division Theorem; we call it **Euclid's Algorithm**. We present the algorithm below but first need the following lemma:

**Lemma 2.2** *Suppose that $a, b, q, r$ are integers and $b = aq + r$. Then $\gcd(b, a) = \gcd(a, r)$.*

**Proof:** To show this, we need only check that every common divisor of $b$ and $a$ is a common divisor of $a$ and $r$, and vice versa, for then the greatest element of this set will be both $\gcd(b, a)$ and $\gcd(a, r)$. But if $d|a$ and $d|b$ then $d|r$, because $r = b - aq$. Conversely, if $d|a$ and $d|r$, then $d|b$, because $b = aq + r$. □

We will now give an example of Euclid's Algorithm, before describing it formally below. This example should make the role of the lemma clear.

**Example 2.3**

Suppose we wish to determine the gcd of 285 and 255. If we successively apply the Division Theorem until we reach a remainder of 0, we obtain the following:

$$\begin{aligned}
285 &= 255 \cdot 1 + 30 \\
255 &= 30 \cdot 8 + 15 \\
30 &= 15 \cdot 2 + 0
\end{aligned}$$

By the lemma we have that

$$\gcd(285, 255) = \gcd(255, 30) = \gcd(30, 15) = \gcd(15, 0),$$

and by the Quick Exercise above, this last is equal to 15.

Explicitly, to compute the gcd of $b$ and $a$ using Euclid's Algorithm, where $|b| \geq |a|$, we proceed inductively as follows. First, set $b_0 = b, a_0 = a$, and let $q_0$ and $r_0$ be the quotient and remainder that result when $b_0$ is divided by $a_0$. Then, for $n \geq 0$, let $b_n = a_{n-1}$ and $a_n = r_{n-1}$, and let $q_n$ and $r_n$ be the quotient and remainder that result when $b_n$ is divided by $a_n$. We then continue until $r_n = 0$, and claim that $r_{n-1} = \gcd(b, a)$. Setting aside for a moment the important question of why $r_n$ need ever reach 0, the general form of the algorithm looks like this:

$$
\begin{aligned}
b_0 &= a_0 q_0 + r_0 \\
b_1 &= a_1 q_1 + r_1 \\
&\ \ \vdots \\
b_{n-1} &= a_{n-1} q_{n-1} + r_{n-1} \\
b_n &= a_n q_n + 0
\end{aligned}
$$

We can now formally show that Euclid's Algorithm does indeed compute $\gcd(b, a)$:

**Theorem 2.3** *Euclid's Algorithm computes* $\gcd(b, a)$.

**Proof**:    Using the general form for Euclid's Algorithm above, the lemma says that

$$
\begin{aligned}
\gcd(b, a) = \gcd(b_0, a_0) &= \\
\gcd(a_0, r_0) = \gcd(b_1, a_1) &= \\
\gcd(a_1, r_1) = \cdots &= \\
\gcd(a_{n-1}, r_{n-1}) = \gcd(b_n, a_n) &= \\
\gcd(a_n, 0) = a_n = r_{n-1}. &
\end{aligned}
$$

It remains only to understand why this algorithm halts. That is, why must some remainder $r_n = 0$? But $a_{i+1} = r_i < |a_i| = r_{i-1}$. Thus, the $r_i$'s form a strictly decreasing sequence of non-negative integers. By the Well-ordering Principle, such a sequence is necessarily finite. This means that $r_n = 0$ for some $n$.      $\square$

We have thus proved that after finitely many steps Euclid's Algorithm will produce the gcd of any pair of integers. In fact, this algorithm reaches the gcd quite rapidly, in a sense we cannot make precise here. It is certainly much more rapid than considering all possible common factors case by case.

---

## 2.3   The GCD Identity

In the equations describing Euclid's Algorithm above, we can start with the bottom equation $b_{n-1} = a_{n-1} q_{n-1} + r_{n-1}$ and solve this for $\gcd(b, a) = r_{n-1}$ in terms of $b_{n-1}$ and $a_{n-1}$. Plugging this into the previous equation, we can express $\gcd(b, a)$ in terms of $b_{n-2}$ and $a_{n-2}$. Repeating this process, we can eventually obtain an equation of the form $\gcd(b, a) = ax + by$, where $x$ and $y$ are integers. That is, $\gcd(b, a)$ can be expressed as a **linear combination** of $a$ and $b$. (Here the coefficients of the linear combination are integers $x$ and $y$; we will use this terminology in a more general context later.)

**Example 2.4**

In the case of 285 and 255 we have the following:

$$
\begin{aligned}
15 &= 255 - 30(8) \\
&= 255 - (285 - 255 \cdot 1)(8) \\
&= 255(9) + 285(-8)
\end{aligned}
$$

This important observation we state formally:

**Theorem 2.4     The GCD identity for integers**     *Given integers a and b (not both zero), there exist integers x and y for which* $\gcd(b, a) = ax + by$.

▷ **Quick Exercise.**     Try using Euclid's Algorithm to compute

$$\gcd(120, 27),$$

and then express this gcd as a linear combination of 120 and 27.  ◁

What we have described above is a *constructive* (or *algorithmic*) approach to expressing the gcd of two integers as a linear combination of them. We will now describe an alternative proof of the GCD identity, which shows the existence of the linear combination, without giving us an explicit recipe for finding it. This sort of proof is inherently more abstract than the constructive proof, but we are able to conclude a bit more about the gcd from it. We will also find it valuable when we generalize these notions to more general algebraic structures than the integers.

**Existential proof of the GCD identity**:     We begin by considering the set of all linear combinations of the integers $a$ and $b$. That is, consider the set

$$S = \{ax + by : x, y \in \mathbb{Z}\}.$$

This is obviously an infinite subset of $\mathbb{Z}$. If the GCD identity is to be true, then the gcd of $a$ and $b$ belongs to this set. But which element is it? By the Well-ordering Principle, $S$ contains a unique smallest positive element which we will call $d$.

▷ **Quick Exercise.**     To apply the Well-ordering Principle, the set $S$ must contain at least one positive element. Why is this true?  ◁

Since $d \in S$, we can write it as $d = ax_0 + by_0$, for some particular integers $x_0$ and $y_0$. We claim that $d$ is the gcd of $a$ and $b$.

To prove this, we must first check that $d$ is a common divisor, that is, that it divides both $a$ and $b$. If we apply the Division Theorem 2.1 to $d$ and $a$, we obtain $a = dq + r$. We must show that $r$ is zero. But

$$r = a - dq = a - (ax_0 + by_0)q = a(1 - qx_0) + b(-qy_0),$$

and so $r \in S$. Because $0 \leq r < d$, and $d$ is the smallest positive element of $S$, $r = 0$, as required. A similar argument shows that $d|b$ too.

Now suppose that $c > 0$ and $c|a$ and $c|b$. Then $a = nc$ and $b = mc$. But then $ax + by = ncx + mcy = c(nx + my)$, and so $c$ divides any linear combination of $a$ and $b$. Thus, $c$ divides $d$. But because $c$ and $d$ are both positive, $c \leq d$. That is, $d$ is the gcd of $a$ and $b$.     □

**Example 2.5**

> Thus, the gcd of 12 and 28 is 4, because $4 = 12 \cdot (-2) + 28(1)$ is the smallest positive integer expressible as a linear combination of 12 and 28. We referred to this example when introducing the Well-ordering Principle in the previous chapter; see Exercise 1.9.

We conclude from this proof the following:

**Corollary 2.5** *The gcd of two integers (not both zero) is the least positive linear combination of them.*

## 2.4 The Fundamental Theorem of Arithmetic

We are now ready to tackle the main business of this chapter: Proving that every non-zero integer can be factored uniquely as a product of integers that cannot be further factored. This theorem's importance is emphasized by the fact that it is usually known as the *Fundamental Theorem of Arithmetic*. It first appears (in essence) as Proposition 14 of Book 9 in Euclid's *Elements*.

We first need a formal definition. An integer $p$ (other than $\pm 1$) is **irreducible** if whenever $p = ab$, then $a$ or $b$ is $\pm 1$. We are thus allowing the always possible 'trivial' factorizations $p = (1)(p) = (-1)(-p)$. We are not allowing $\pm 1$ to be irreducible because it would unnecessarily complicate the formal statement of the Fundamental Theorem of Arithmetic that we make below. Because $0 = (a)(0)$ for any integer $a$, it is clear that $0$ is not irreducible. Finally, notice that if $p$ is irreducible, then so is $-p$. This means that in the arguments that follow we can often assume that $p$ is positive.

The positive integers that are irreducible form a familiar list:

$$2, \quad 3, \quad 5, \quad 7, \quad 11, \quad 13, \quad 17, \cdots.$$

You are undoubtedly familiar with these numbers, under the name *prime* integers, and it may seem perverse for us to call them 'irreducible'. But this temporary perversity now will allow us to be consistent with the more general terminology we'll use later.

We reserve the term 'prime' for another definition: An integer $p$ (other than $0$ and $\pm 1$) is **prime** if, whenever $p$ divides $ab$, then either $p$ divides $a$ or $p$ divides $b$. (Notice that when we say 'or' here, we mean one or the other or both. This is what logicians call the *inclusive* 'or', and is the sense of this word that we will always use.)

**Example 2.6**

> For instance, we know that $2$ is a prime integer. For if $2|ab$, then $ab$ is even. But a product of integers is even exactly if at least one of the factors is even, and so $2|a$ or $2|b$.

The prime property generalizes to more than two factors:

**Theorem 2.6** *If $p$ is prime and $p|a_1 a_2 \cdots a_n$, then $p|a_i$ for some $i$.*

**Proof:**   This is Exercise 5. Prove it using induction on $n$.   $\square$

For the integers, the ideas of primeness and irreducibility coincide. This is the content of the next theorem.

**Theorem 2.7** *An integer is prime if and only if it is irreducible.*

**Proof:**   This theorem asserts that the concepts of primeness and irreducibility are equivalent for integers. This amounts to two implications which must be proved: primeness implies irreducibility, and the converse statement that irreducibility implies primeness.

Suppose first that $p$ is prime. To show that it is irreducible, suppose that $p$ has been factored: $p = ab$. Then $p|ab$, and so (without loss of generality) $p|a$. Thus, $a = px$, and so $p = pxb$. But then $1 = xb$, and so both $x$ and $b$ can only be $\pm 1$. This shows that the factorization $p = ab$ is trivial, as required.

Conversely, suppose that $p$ is irreducible, and $p|ab$. We will suppose also that $p$ does not divide $a$. We thus must prove that $p$ does divide $b$. Suppose that $d$ is a positive common divisor of $p$ and $a$. Then, because $p$ is irreducible, $d$ must be $p$ or 1. Because $p$ doesn't divide $a$, it must be that $\gcd(p, a) = 1$. So by the GCD identity 2.4, there exist $x$ and $y$ with $1 = ax + py$. But then $b = abx + bpy$, and because $p$ clearly divides both $abx$ and $bpy$ it thus divides $b$, as required. $\qquad\square$

Again, it may seem strange to have both of the terms 'prime' and 'irreducible', because for $\mathbb{Z}$ we have proved that they amount to the same thing. But we will later discover more general contexts where these concepts are distinct.

We now prove half of the Fundamental Theorem of Arithmetic:

**Theorem 2.8** *Every non-zero integer (other than $\pm 1$) is either irreducible or a product of irreducibles.*

**Proof:** Let $n$ be an integer other than $\pm 1$, which we may as well suppose is positive. We proceed by induction on $n$. We know that $n \neq 1$, and if $n = 2$, then it is irreducible. Now suppose the theorem holds true for all $m < n$. If $n$ is irreducible already, we are done. If not, then $n = bc$, where, without loss of generality, both factors are positive and greater than 1. But then by the induction hypothesis both $b$ and $c$ can be factored as a product of irreducibles, and thus so can their product $n$. $\qquad\square$

For example, we can factor the integer 120 as $2 \cdot 2 \cdot 2 \cdot 3 \cdot 5$. Now $(-5) \cdot 2 \cdot (-2) \cdot 3 \cdot 2$ is a distinct factorization of 120 into irreducibles, but it is clearly essentially the same, where we disregard order and factors of $-1$. The uniqueness half of the Fundamental Theorem of Arithmetic asserts that all distinct factorizations into irreducibles of a given integer are essentially the same, in this sense. To prove this we use the fact that irreducible integers are prime.

**Theorem 2.9    Unique Factorization Theorem for Integers**
*If an integer $x = a_1 a_2 \cdots a_n = b_1 b_2 \cdots b_m$ where the $a_i$ and $b_j$ are all irreducible, then $n = m$ and the $b_j$ may be rearranged so that $a_i = \pm b_i$, for $i = 1, 2, \cdots, n$.*

**Proof:** We use induction on $n$. If $n = 1$, the theorem follows easily.

▷ **Quick Exercise.** Check this. ◁

So we assume $n > 1$. By the primeness property of the irreducible $a_1$, $a_1$ divides one of the $b_j$. By renumbering the $b_j$ if necessary, we may assume $a_1$ divides $b_1$. So, because $b_1$ is irreducible, $a_1 = \pm b_1$. Therefore, by dividing both sides by $a_1$, we have

$$a_2 a_3 \cdots a_n = \pm b_2 \cdots b_m.$$

(Because $b_2$ is irreducible, so is $-b_2$, and we consider $\pm b_2$ as an irreducible factor.) We now have two factorizations into irreducibles, and the number of $a_i$ factors is $n - 1$. So by the induction hypothesis $n - 1 = m - 1$, and by renumbering the $b_j$ as necessary, $a_i = \pm b_i$ for $i = 1, 2, \cdots, n$. This proves the theorem. $\qquad\square$

## 2.5    A Geometric Interpretation

As we have indicated already, both Euclid's Algorithm and the Fundamental Theorem of Arithmetic have their origins in the work of the Greek geometer Euclid. It is important to note that Euclid viewed both of these theorems as *geometric* statements about line segments.

To understand this requires a definition: A line segment $AB$ **measures** a line segment $CD$, if there is a positive integer $n$, so that we can use a compass to lay exactly $n$ copies of $AB$ next to one another, to make up the segment $CD$. In modern language, we would say that the length of $CD$ is $n$ times that of $AB$, but this notion of *length* was foreign to Euclid.

Euclid's Algorithm can now be viewed in the following geometric way: Given two line segments $AB$ and $CD$, can we find a line segment $EF$, which measures both $AB$ and $CD$? In the diagram below, we see by example how Euclid's Algorithm accomplishes this.

We can recapitulate this geometry in algebraic form, which makes the connection with Euclid's Algorithm clear:

$$AB = 3CD + E_1F_1,$$

$$CD = 1E_1F_1 + E_2F_2,$$

$$E_1F_1 = 2E_2F_2.$$

Thus, $AB$ and $CD$ are both measured by $E_2F_2$. In fact, $CD = 3E_2F_2$ and $AB = 11E_2F_2$. In modern language, we would say that the ratio of the length of $AB$ to the length of $CD$ is $11/3$. Note that in this context Euclid's Algorithm halts only in case this ratio of lengths is a *rational number* (that is, a ratio of integers). In fact, it is possible to prove that the ratio of the diagonal of a square to one of the sides is irrational, by showing that in this case Euclid's Algorithm never halts (see Exercises 14 and 15).

Euclid's proposition that is closest to the Fundamental Theorem of Arithmetic says that *if a number be the least that is measured by prime numbers, it will not be measured by any other prime number except those originally measuring it.* This seems much more obscure than our statement, in part because of the geometric language that Euclid uses. Euclid's proposition does assert that if a number is a product of certain primes, it is then not divisible by any other prime, which certainly follows from the Fundamental Theorem, and is indeed the most important idea contained in our theorem. However, Euclid lacked both our flexible notation, and the precisely formulated tool of Mathematical Induction, to make his statement clearer and more modern. It wasn't until the eighteenth century, with such mathematicians as Euler and Legendre, that a modern statement was possible, and a careful proof in modern form did not appear until the work of Gauss, in the early 19th century.

---

## Chapter Summary

In this chapter we examined division and factorization in $\mathbb{Z}$. We proved the *Division Theorem* by induction and then used it to obtain *Euclid's Algorithm* and the *GCD identity*. We defined the notions of *primeness* and *irreducibility* and showed that they are equivalent. We then proved the *Fundamental Theorem of Arithmetic*, which asserts that all integers other than $0, 1, -1$ are irreducible or can be factored uniquely into a product of irreducibles.

## Warm-up Exercises

a. Find the quotient and remainder, as guaranteed by the Division Theorem 2.1, for 13 and $-120$, $-13$ and 120, and $-13$ and $-120$.

b. What are the possible remainders when you divide by 3, using the Division Theorem 2.1? Choose one such remainder, and make a list describing all integers that give this remainder, when dividing by 3.

c. What are the possible answers to $\gcd(a, p)$, where $p$ is prime, and $a$ is an arbitrary integer?

d. Let $m$ be a fixed integer. Describe succinctly the integers $a$ where

$$\gcd(a, m) = m.$$

e. Give the prime factorizations of 92, 100, 101, 102, 502, and 1002.

f. Suppose that we have two line segments. One has length $11/6$ units, and the other has length $29/15$. What length is the longest segment that measures both?

g. We proved the GCD identity 2.4 twice. Explain the different approaches of the two proofs to finding the appropriate linear combination. Which is easier to describe in words? Which is computationally more practical?

## Exercises

1. (a) Find the greatest common divisor of 34 and 21, using Euclid's Algorithm. Then express this gcd as a linear combination of 34 and 21.

   (b) Now do the same for 2424 and 772.

   (c) Do the same for 2007 and 203.

   (d) Do the same for 3604 and 4770.

2. (a) Prove that $\gcd(a, b)$ divides $a - b$. This sometimes provides a short cut in finding gcds.

   (b) Use this to find $\gcd(1962, 1965)$.

   (c) Now find $\gcd(1961, 1965)$.

   (d) Find the gcds in Exercise 1 using this short cut.

3. Prove that the set of all linear combinations of $a$ and $b$ are precisely the multiples of $\gcd(a, b)$.

4. Two numbers are said to be **relatively prime** if their gcd is 1. Prove that $a$ and $b$ are relatively prime if and only if every integer can be written as a linear combination of $a$ and $b$.

5. Prove Theorem 2.6. That is, use induction to prove that if the prime $p$ divides $a_1 a_2 \cdots a_n$, then $p$ divides $a_i$, for some $i$.

6. Suppose that $a$ and $b$ are positive integers. If $a + b$ is prime, prove that $\gcd(a, b) = 1$.

7. (a) A natural number greater than 1 that is not prime is called **composite**. Show that for any $n$, there is a run of $n$ consecutive composite numbers. *Hint:* Think factorial.

   (b) Therefore, there is a string of 5 consecutive composite numbers starting where?

8. Prove that two consecutive members of the Fibonacci sequence are relatively prime.

9. Notice that $\gcd(30, 50) = 5\gcd(6, 10) = 5 \cdot 2$. In fact, this is always true; prove that if $a \neq 0$, then $\gcd(ab, ac) = a \cdot \gcd(b, c)$.

10. Suppose that two integers $a$ and $b$ have been factored into primes as follows:

$$a = p_1^{n_1} p_2^{n_2} \cdots p_r^{n_r}$$

and

$$b = p_1^{m_1} p_2^{m_2} \cdots p_r^{m_r},$$

where the $p_i$'s are primes, and the exponents $m_i$ and $n_i$ are non-negative integers. It is the case that

$$\gcd(a, b) = p_1^{s_1} p_2^{s_2} \cdots p_r^{s_r},$$

where $s_i$ is the smaller of $n_i$ and $m_i$. Show this with $a = 360 = 2^3 3^2 5$ and $b = 900 = 2^2 3^2 5^2$. Now prove this fact in general.

11. The **least common multiple** of natural numbers $a$ and $b$ is the smallest positive common multiple of $a$ and $b$. That is, if $m$ is the least common multiple of $a$ and $b$, then $a|m$ and $b|m$, and if $a|n$ and $b|n$ then $n \geq m$. We will write $\text{lcm}(a, b)$ for the least common multiple of $a$ and $b$. Find $\text{lcm}(20, 114)$ and $\text{lcm}(14, 45)$. Can you find a formula for the lcm of the type given for the gcd in the previous exercise?

12. Show that if $\gcd(a, b) = 1$, then $\text{lcm}(a, b) = ab$. In general, show that

$$\text{lcm}(a, b) = \frac{ab}{\gcd(a, b)}.$$

13. Prove that if $m$ is a common multiple of both $a$ and $b$, then $\text{lcm}(a, b)|m$.

14. Prove that $\sqrt{2}$ is irrational.

15. This problem outlines another proof that $\sqrt{2}$ is irrational. We show that Euclid's Algorithm never halts if applied to a diagonal $d$ and side $s$ of a square. The first step of the algorithm yields

$$d = 1 \cdot s + r,$$

as shown in the picture below:

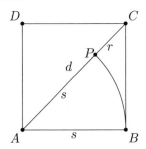

(a) Now find the point $E$ by intersecting the side $CD$ with the perpendicular to the diagonal $AC$ at $P$. It is obvious that the length of segment $EC$ is $\sqrt{2}r$. (Why?) Now prove that the length of segment $DE$ is $r$, by showing that the triangle $DEP$ is an isosceles triangle.

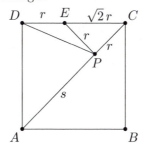

Why does this mean that the next step in Euclid's Algorithm yields

$$s = 2r + (\sqrt{2} - 1)r?$$

(b) Argue that the next step of the algorithm yields

$$r = 2(\sqrt{2} - 1)r + (\sqrt{2} - 1)^2 r.$$

Conclude that this algorithm never halts, and so there is no common measure for the diagonal and side of the square.

16. State Euclid's version of the Fundamental Theorem of Arithmetic in modern language, and prove it carefully as a Corollary of the Fundamental Theorem.

17. (a) As with many algorithms, one can easily write a recursive version of Euclid's Algorithm. This version is for nonnegative $a$ and $b$. (The symbol $\leftarrow$ is the assignment symbol and $a \bmod b$ is the remainder when dividing $a$ by $b$.)

    **function** $gcd(a, b)$;
        **if** $b = 0$ **then** gcd $\leftarrow a$ **else** $gcd \leftarrow gcd(b, a \bmod b)$
    **endfunction**.

    Try this version on 2424 and 772 and a couple of other pairs of your choice.

    (b) One can also write a recursive extended gcd algorithm that returns the linear combination guaranteed by the GCD identity. This procedure again assumes that both $a$ and $b$ are non-negative. When the initial call returns, $g$ will be the gcd of $a$ and $b$ and $g = ax + by$.

    **procedure** $extgcd(a, b, g, x, y)$;
        **if** $b = 0$ **then**
            $g \leftarrow a$; $x \leftarrow 1$; $y \leftarrow 0$;
        **else**
            $extgcd(b, a \bmod b, g, x, y)$;
            $temp \leftarrow y$;
            $y \leftarrow x - \lfloor a/b \rfloor y$;
            $x \leftarrow temp$;
    **endprocedure**.

    (Here, $\lfloor x \rfloor$ is the *floor* function. That is, $\lfloor x \rfloor$ = the greatest integer less than or equal to $x$.) Try this procedure on 285 and 255, then 2424 and 772, and a pair of your choice.

18.  (a) Show that in Euclid's Algorithm, the remainders are at least halved after two steps. That is, $r_{i+2} < 1/2\, r_i$.

   (b) Use part (a) to find the maximum number of steps required for Euclid's Algorithm. (Figure this in terms of the maximum of $a$ and $b$.)

19. Recall from Exercise 1.14 the definition of the binomial coefficient $\binom{n}{k}$. Suppose that $p$ is a positive prime integer, and $k$ is an integer with $1 \le k \le p - 1$. Prove that $p$ divides binomial coefficient $\binom{p}{k}$.

# Chapter 3

## *Modular Arithmetic*

In this chapter we look again at the content of the Division Theorem 2.1, only this time placing our primary interest on the remainders obtained. By adopting a slightly more abstract point of view, we will be able to obtain some new insight into the arithmetic of $\mathbb{Z}$.

### 3.1 Residue Classes

For any positive integer $m$ and integer $a$, the **residue of $a$ modulo $m$** is the remainder one obtains when dividing $a$ by $m$ in the Division Theorem. (We will frequently write 'mod $m$' for 'modulo $m$'.)

**Example 3.1**

    The residue of 8 (mod 5) is 3. The residue of $-22$ (mod 6) is 2.

Of course, many integers have the same residue (mod $m$). Given an integer $a$, the set of all integers with the same residue (mod $m$) as $a$ is called the **residue class (mod $m$) of** $a$ and denoted $[a]_m$.

**Example 3.2**

    For instance,
$$[3]_5 = \{\ldots, -12, -7, -2, 3, 8, \ldots\},$$
    and
$$[-22]_6 = \{\ldots, -22, -16, -10, -4, 2, \ldots\}.$$

If $[a]_m = [b]_m$ we say that $a$ and $b$ are **congruent modulo** $m$, and write $a \equiv b \,(\text{mod}\, m)$. We simplify this notation to $a \equiv b$, if it is clear what **modulus** $m$ is being used.

Our intention in this chapter is to define addition and multiplication on these residue classes to give us interesting new number systems. Before doing this we will explore more about the classes themselves.

Notice that $[3]_5$ consists of the list of every fifth integer, which includes 3. That is,

$$[3]_5 = \{\ldots,\ 3 + (-3)5,\ 3 + (-2)5,\ 3 + (-1)5,\ 3 + (0)5,\ 3 + (1)5, \ldots\}.$$

And similarly, $[-22]_6$ consists of the list of every sixth integer, which includes $-22$. Our first theorem asserts that this is always true.

**Theorem 3.1** $[a]_m = \{a + km : k \in \mathbb{Z}\}.$

**Proof:**   We must show that two infinite sets are in fact equal. Our strategy is to show that each of these sets is a subset of the other. For that purpose, suppose that

$$x \in \{a + km : k \in \mathbb{Z}\}.$$

Then $x = a + k_0 m$ for some $k_0 \in \mathbb{Z}$. Suppose the residue (mod $m$) of $a$ is $r$. That is, when we divide $a$ by $m$, we have remainder $r$. But then $a = qm + r$, where $0 \leq r < m$ and $q$ is some integer. Then

$$x = a + k_0 m = qm + r + k_0 m = (k_0 + q)m + r.$$

But this means that the residue of $x$ modulo $m$ is $r$, and so $x \in [a]_m$. Thus,

$$\{a + km : k \in \mathbb{Z}\} \subseteq [a]_m.$$

Now let $x \in [a]_m$. In other words, $x$ has the same residue (mod $m$) as $a$. Suppose that the common residue of $x$ and $a$ modulo $m$ is $r$, and so $x = q_1 m + r$ and $a = q_2 m + r$. Then $r = a - q_2 m$ and so

$$x = q_1 m + a - q_2 m = (q_1 - q_2)m + a.$$

That is, $x \in \{a + km : k \in \mathbb{Z}\}$, proving the theorem.          □

As our examples above suggest, this theorem says that elements in a given residue class (mod $m$) occur exactly once every $m$ integers. So, if $x \in [a]_m$, the next larger element in $[a]_m$ is $x + m$. Hence, any $m$ consecutive integers will contain exactly one element of $[a]_m$. Thus, there are exactly $m$ residue classes (mod $m$), and we can choose representatives from each class simply by picking any set of $m$ consecutive integers. For example, we could certainly choose the $m$ integers $0, 1, 2, \cdots, m - 1$ (which are of course exactly the possible remainders from division by $m$). Indeed, with this conventional and convenient choice of representatives we can specify the $m$ distinct residue classes as $[0], [1], \ldots, [m-1]$. These $m$ residue classes then **partition** the integers, meaning that each integer belongs to exactly one of these classes, and if distinct classes intersect, they are in fact equal. Alternatively, this means that

$$\mathbb{Z} = [0] \cup [1] \cup [2] \cup \cdots \cup [m - 1],$$

and the sets in this union are disjoint from one another pairwise.

In particular, we have that

$$
\begin{aligned}
\mathbb{Z} &= [0]_4 \cup [1]_4 \cup [2]_4 \cup [3]_4 \\
&= \{\ldots, -4, 0, 4, 8, \ldots\} \cup \{\ldots, -3, 1, 5, 9, \ldots\} \cup \\
&\quad \{\ldots, -2, 2, 6, 10, \ldots\} \cup \{\ldots, -1, 3, 7, 11, \ldots\}.
\end{aligned}
$$

The next theorem provides a very useful way of determining when two integers are in the same residue class. Indeed, we will use this characterization more often than the definition itself.

**Theorem 3.2** *Two integers, $x$ and $y$, have the same residue (mod $m$) if and only if $x - y = km$ for some integer $k$.*

**Proof:**   First, suppose $x \equiv y \pmod{m}$. Then $x = k_1 m + r$, and $y = k_2 m + r$ for some integers $k_1$ and $k_2$ and $0 \leq r < m$. But then $x - y = (k_1 - k_2)m$.

Conversely, suppose $x - y = km$, for some integer $k$ with $x = k_1 m + r_1$ and $y = k_2 m + r_2$, where $0 \leq r_1 < m$ and $0 \leq r_2 < m$. Then

$$km = x - y = (k_1 - k_2)m + r_1 - r_2,$$

which implies that $r_1 - r_2 = (k - k_1 + k_2)m$. Now, because $r_1$ and $r_2$ are both non-negative integers less than $m$, the distance between them is less than $m$. That is, $|r_1 - r_2| < m$. So, $-m < r_1 - r_2 < m$. But we have just shown that $r_1 - r_2$ is an integer multiple of $m$. Hence, that multiple is 0. Therefore, $r_1 - r_2 = 0$ or $r_1 = r_2$. □

**Example 3.3**

We have $[18]_7 = [-38]_7$ because $18 - (-38) = 56 = (7)(8)$.

We now consider the set of all residue classes modulo $m$. We denote this set by $\mathbb{Z}_m$. That is,

$$\mathbb{Z}_m = \{[0], [1], [2], \cdots, [m-1]\}.$$

Be careful to note that we are considering here a *set of sets*: Each element of the finite set $\mathbb{Z}_m$ is in fact an infinite set of the form $[k]$. While this construct seems abstract, you should take heart from the fact that for the most part, we can focus our attention on particular representatives of the residue classes, rather than on the entire set.

---

## 3.2 Arithmetic on the Residue Classes

We are now ready to define an 'arithmetic' on $\mathbb{Z}_m$ which is directly analogous to (and indeed inherited from) the arithmetic on $\mathbb{Z}$. By an 'arithmetic' we mean operations on $\mathbb{Z}_m$ that we call addition and multiplication.

To *add* two elements of $\mathbb{Z}_m$ (that is, two mod $m$ residue classes) simply take a representative from each class. The sum of the two residue classes is defined to be the residue class of their sum. For instance, to add $[3]_5$ and $[4]_5$, we pick, say, $8 \in [3]_5$ and $4 \in [4]_5$. But $[8+4]_5 = [2]_5$, and so $[3]_5 + [4]_5 = [2]_5$. Note that any other choice of representatives would also yield $[2]_5$.

▷ **Quick Exercise.** Try some other representatives of these two residue classes, and see that the same sum is obtained. ◁

It is vitally important that this definition be *independent* of representatives chosen, for otherwise it would be ambiguous and consequently not of much use. We will shortly prove that this independence of representatives in fact holds. Before we do so, we first observe that we can define *multiplication* on $\mathbb{Z}_m$ in a similar way.

More succinctly, the definition of the operations on $\mathbb{Z}_m$ are:

$$\begin{aligned} [a]_m + [b]_m &= [a+b]_m \\ [a]_m \cdot [b]_m &= [a \cdot b]_m. \end{aligned}$$

Thus, $[4]_5[3]_5 = [12]_5 = [2]_5$.

▷ **Quick Exercise.** Try some other representatives of these two residue classes, and see that the same product is obtained. ◁

We now check to see that these definitions are well defined. That is, if one picks different representatives from the residue classes, the result should be the same. You have seen that this worked in the example above for addition and multiplication in $\mathbb{Z}_5$ (at least for the representatives you tried).

**Proof that operations are well defined:**    To show addition on $\mathbb{Z}_m$ is well defined, consider $[a]$ and $[b]$. We pick two representatives from $[a]$, say $x$ and $y$, and two representatives from $[b]$, say $r$ and $s$. Now we must show that $[x + r] = [y + s]$. But $x, y \in [a]$ implies $x - y = k_1 m$, for some integer $k_1$. Likewise, $r - s = k_2 m$, for some integer $k_2$. So,

$$x + r - (y + s) = x - y + r - s = (k_1 + k_2)m.$$

In other words, $[x + r] = [y + s]$, which is what we wanted to show.

The proof that multiplication on $\mathbb{Z}_m$ is also well defined is similar and is left as Exercise 9. □

We now have an 'arithmetic' defined on $\mathbb{Z}_m$. To avoid cumbersome notation, it is common to write the elements of $\mathbb{Z}_m$ as simply $0, 1, \ldots, m - 1$ instead of $[0], [1], \ldots, [m - 1]$. So, in $\mathbb{Z}_5$, $3 + 4 = 2$ and $2 + 3 = 0$. (Thus, $-2 = 3$ and $-3 = 2$.) Bear in mind that the arithmetic is really on residue classes. For the remainder of this chapter we will not omit the brackets, although later we often will.

**Example 3.4**

A first simple example of this arithmetic is in the case where $m = 2$. We then have only two residue classes. In fact, $[0]_2$ is precisely the set of even integers and $[1]_2$ is the set of odd integers. The addition and multiplication tables for $\mathbb{Z}_2$ are given below. The addition table may be simply interpreted as 'The sum of an even and an odd is odd, while the sum of two evens or two odds is even.' The multiplication table may be interpreted as 'The product of two integers is odd only when both integers are odd.'

| + | [0] | [1] |      | · | [0] | [1] |
|---|-----|-----|------|---|-----|-----|
| [0] | [0] | [1] |      | [0] | [0] | [0] |
| [1] | [1] | [0] |      | [1] | [0] | [1] |

addition and multiplication tables for $\mathbb{Z}_2$

---

## 3.3    Properties of Modular Arithmetic

It is illuminating to compare the arithmetic on $\mathbb{Z}_m$ with that on $\mathbb{Z}$. Later in the book (in Chapter 6) we will meet a common abstraction of arithmetic on $\mathbb{Z}$ and on $\mathbb{Z}_m$ that will enable us to pursue this general question in more detail. For now we intend only to suggest a few of the ideas we will meet more formally later.

Arithmetic in $\mathbb{Z}$ depends heavily on the existence of an **additive identity** or **zero**. Zero has the pleasant property in $\mathbb{Z}$ that $0 + n = n$, for all integers $n$. Note that in $\mathbb{Z}_m$ the residue class $[0]$ plays the same role because $[0] + [n] = [0 + n] = [n]$.

Also, each integer $n$ has an **additive inverse** $-n$ in $\mathbb{Z}$, an element which when added to $n$ gives the additive identity 0. This is the formal basis for subtraction, which enables us to solve equations of the form $a + x = b$ in $\mathbb{Z}$ (by simply adding $-a$ to both sides). Notice that additive inverses are available in $\mathbb{Z}_m$ as well. For,

$$[k] + [m - k] = [k + m - k] = [m] = [0],$$

and so $[m - k] = [-k]$ serves as the additive inverse of $[k]$. Consequently, we can always solve equations of the form $[a] + X = [b]$, where here $X$ is an unknown in $\mathbb{Z}_m$.

▷ **Quick Exercise.** Solve the equation $[7]_{12} + X = [4]_{12}$ in $\mathbb{Z}_{12}$, by using the appropriate additive inverse. ◁

We can conveniently summarize the additive arithmetic in $\mathbb{Z}_m$ for a particular $m$ in addition tables. (We have addition tables for $m = 5$ and $m = 6$ below.) Note that these tables reflect the fact that every element of these sets has an additive inverse. (How?)

| + | [0] | [1] | [2] | [3] | [4] |
|---|-----|-----|-----|-----|-----|
| [0] | [0] | [1] | [2] | [3] | [4] |
| [1] | [1] | [2] | [3] | [4] | [0] |
| [2] | [2] | [3] | [4] | [0] | [1] |
| [3] | [3] | [4] | [0] | [1] | [2] |
| [4] | [4] | [0] | [1] | [2] | [3] |

| + | [0] | [1] | [2] | [3] | [4] | [5] |
|---|-----|-----|-----|-----|-----|-----|
| [0] | [0] | [1] | [2] | [3] | [4] | [5] |
| [1] | [1] | [2] | [3] | [4] | [5] | [0] |
| [2] | [2] | [3] | [4] | [5] | [0] | [1] |
| [3] | [3] | [4] | [5] | [0] | [1] | [2] |
| [4] | [4] | [5] | [0] | [1] | [2] | [3] |
| [5] | [5] | [0] | [1] | [2] | [3] | [4] |

addition tables $\mathbb{Z}_5$ and $\mathbb{Z}_6$

What about multiplication? In $\mathbb{Z}$ the integer 1 serves as a **multiplicative identity**, because $1 \cdot n = n$ for all integers $n$, and clearly $[1]$ plays the same role in $\mathbb{Z}_m$.

▷ **Quick Exercise.** Check this. ◁

A multiplicative inverse in $\mathbb{Z}_m$ may be defined analogously to the way we have defined an additive inverse: $[a]$ is the **multiplicative inverse** of $[n]$ if $[a][n] = [1]$. The disadvantage of $\mathbb{Z}$ as opposed to $\mathbb{Q}$ is that no elements have multiplicative inverses (except 1 and $-1$). The consequence is that many equations of the form $ax = b$ are *not* solvable in the integers. But what about in $\mathbb{Z}_m$? Consider the following multiplication tables for our examples $\mathbb{Z}_5$ and $\mathbb{Z}_6$.

| · | [0] | [1] | [2] | [3] | [4] |
|---|-----|-----|-----|-----|-----|
| [0] | [0] | [0] | [0] | [0] | [0] |
| [1] | [0] | [1] | [2] | [3] | [4] |
| [2] | [0] | [2] | [4] | [1] | [3] |
| [3] | [0] | [3] | [1] | [4] | [2] |
| [4] | [0] | [4] | [3] | [2] | [1] |

| · | [0] | [1] | [2] | [3] | [4] | [5] |
|---|-----|-----|-----|-----|-----|-----|
| [0] | [0] | [0] | [0] | [0] | [0] | [0] |
| [1] | [0] | [1] | [2] | [3] | [4] | [5] |
| [2] | [0] | [2] | [4] | [0] | [2] | [4] |
| [3] | [0] | [3] | [0] | [3] | [0] | [3] |
| [4] | [0] | [4] | [2] | [0] | [4] | [2] |
| [5] | [0] | [5] | [4] | [3] | [2] | [1] |

multiplication tables $\mathbb{Z}_5$ and $\mathbb{Z}_6$

Notice the remarkable fact that in $\mathbb{Z}_5$, every element (other than $[0]$) has a multiplicative inverse. For example, the multiplicative inverse of $[3]$ is $[2]$, because $[3][2] = [1]$. Thus, to solve the equation $[3]X = [4]$ in $\mathbb{Z}_5$, we need only multiply both sides of the equation by the multiplicative inverse of $[3]$ (which is $[2]$) to obtain

$$X = [2][3]X = [2][4] = [3].$$

On the other hand, $[3]$ has no multiplicative inverse in $\mathbb{Z}_6$, and there is in fact no solution to the equation $[3]X = [2]$ in $\mathbb{Z}_6$.

▷ **Quick Exercise.** Solve the equation $[4]X = [10]$ in $\mathbb{Z}_{11}$. Then argue that this equation has no solution in $\mathbb{Z}_{12}$. ◁

We have gone far enough here to illustrate the fact that the arithmetic in $\mathbb{Z}_m$ shares similarities with those of $\mathbb{Z}$, but also has some real differences (which depend on the choice of $m$).

# Historical Remarks

The great German mathematician Karl Friedrich Gauss (1777-1855) first introduced the idea of congruence modulo $m$ into the study of integers, in his important book *Disquisitiones Arithmeticae*. Gauss made important contributions to almost all branches of mathematics and did important work in astronomy and physics as well, but number theory (the study of the mathematical properties of the integers) was his first love. The *Disquisitiones* was a landmark work, which systematized and extended the work on number theory done by Gauss's predecessors, Fermat and Euler. Gauss's introduction of the notion of congruence is a good example of the way in which an effective and efficient notation can revolutionize the way a mathematical subject is approached.

---

# Chapter Summary

In this chapter we defined the *residue class* $[a]_m$ of $a$ modulo $m$ (for a positive integer $m$) and characterized the elements of such classes. We then considered the set $\mathbb{Z}_m$ of the $m$ residue classes and defined an *arithmetic* on this set. We proved the following facts about this arithmetic:

- Addition and multiplication are well defined;

- $\mathbb{Z}_m$ has an additive identity $[0]$ and a multiplicative identity $[1]$;

- All elements in $\mathbb{Z}_m$ have additive inverses, but not all elements have multiplicative inverses.

---

# Warm-up Exercises

a. Write out the three residue classes modulo 3 (as we did for $\mathbb{Z}_4$). Write out the addition and multiplication tables for $\mathbb{Z}_3$. Which elements of $\mathbb{Z}_3$ have multiplicative inverses?

b. Does $\{47, 100, -3, 29, -9\}$ contain a representative from every residue class of $\mathbb{Z}_5$? Does $\{-14, -21, -10, -3, -2\}$? Does $\{10, 21, 32, 43, 54\}$?

c. What is the additive inverse of $[13]$ in $\mathbb{Z}_{28}$?

d. What is the relationship between 'clock arithmetic' and modular arithmetic?

e. (a) What time is it 100 hours after 3 o'clock?

   (b) What day of the week is it 100 days after Monday?

f. Solve the following equations, or else argue that they have no solutions:

   (a) $[4] + X = [3]$, in $\mathbb{Z}_6$.
   (b) $[4]X = [3]$, in $\mathbb{Z}_6$.
   (c) $[4] + X = [3]$, in $\mathbb{Z}_9$.
   (d) $[4]X = [3]$, in $\mathbb{Z}_9$.

# Exercises

1. Repeat Warm-up Exercise a for modulo 8.

2. Determine the elements in $\mathbb{Z}_{15}$ that have multiplicative inverses. Give an example of an equation of the form $[a]X = [b]$ $([a] \neq [0])$ that has no solution in $\mathbb{Z}_{15}$.

3. In Exercise c you determined the additive inverse of $[13]$ in $\mathbb{Z}_{28}$. Now determine its multiplicative inverse.

4.  (a) Find an example in $\mathbb{Z}_6$ where $[a][b] = [a][c]$, but $[b] \neq [c]$. How is this example related to the existence of multiplicative inverses in $\mathbb{Z}_6$?

    (b) Repeat this in $\mathbb{Z}_{10}$.

5. If $\gcd(a, m) = 1$, then the GCD identity 2.4 guarantees that there exist integers $u$ and $v$ such that $1 = au + mv$. Show that in this case, $[u]$ is the multiplicative inverse of $[a]$ in $\mathbb{Z}_m$.

6. Now use essentially the reverse of the argument from Exercise 5 to show that if $[a]$ has a multiplicative inverse in $\mathbb{Z}_m$ then $\gcd(a, m) = 1$.

7. According to what you have shown in Exercises 5 and 6, which elements of $\mathbb{Z}_{24}$ have multiplicative inverses? What are the inverses for each of those elements? (The answer is somewhat surprising.)

8. Repeat the previous exercise for $\mathbb{Z}_{10}$. Give the multiplication table for those elements in $\mathbb{Z}_{10}$ that have multiplicative inverses and find an $[n]$ such that all these elements are powers of $[n]$.

9. Prove that the multiplication on $\mathbb{Z}_m$ as defined in the text is well defined, as claimed in Section 3.2.

10. Prove that if all non-zero elements of $\mathbb{Z}_m$ have multiplicative inverses, then multiplicative cancellation holds: that is, if $[a][b] = [a][c]$, then $[b] = [c]$.

11. Consider the following alternate definition of addition of residue classes in $\mathbb{Z}_m$, by defining the set
$$S = \{x + y : x \in [a], y \in [b]\}.$$
Prove that $S = [a] + [b]$ (as defined in Section 3.2); thus, we could have used the definition above to define addition in $\mathbb{Z}_m$.

12. By way of analogy with Exercise 11, one might try to define the multiplication of residue classes in $\mathbb{Z}_m$ by considering the set
$$M = \{xy : x \in [a], y \in [b]\}.$$
Prove that $M \subseteq [a][b]$. Then choose particular $m, a, b$ to show by example that this containment can be proper (that is, $M \subset [a][b]$).

13. In the integers, the equation $x^2 = a$ has a solution only when $a$ is a positive perfect square or zero. For which $[a]$ does the equation $X^2 = [a]$ have a solution in $\mathbb{Z}_7$? What about in $\mathbb{Z}_8$? What about in $\mathbb{Z}_9$?

14. Explain what $a \equiv b \,(\mathrm{mod}\ 1)$ means.

# Chapter 4

## Polynomials with Rational Coefficients

In Chapter 2 we proved that every integer ($\neq 0, \pm 1$) can be written as a product of irreducible integers, and this decomposition is essentially unique. These irreducible integers turn out to be those integers that we call primes. To summarize, in that chapter we proved the following important theorems:

- The Division Theorem for integers (Theorem 2.1),

- Euclid's Algorithm (which yields the gcd of two integers) (Theorem 2.3),

- The GCD identity that $\gcd(a, b) = ax + by$, for some integers $x$ and $y$ (Theorem 2.4),

- Each non-zero integer ($\neq \pm 1$) is either irreducible or a product of irreducibles (Theorem 2.8),

- An integer $p$ is irreducible if and only if $p$ is prime (that is, if $p|ab$, then either $p|a$ or $p|b$) (Theorem 2.7), and

- Each non-zero integer ($\neq \pm 1$) is uniquely (up to order and factors of $-1$) the product of primes (Theorem 2.9).

## 4.1  Polynomials

In this chapter we turn our attention to another algebraic structure with which you are familiar – the polynomials (with unknown $x$) with coefficients from the rational numbers $\mathbb{Q}$. In this chapter and the next we discover that this set of polynomials obeys theorems directly analogous to those we have listed above for the integers.

We denote the set of polynomials with rational coefficients by $\mathbb{Q}[x]$. Let's be careful to define exactly what we mean by a polynomial. A **polynomial** $f \in \mathbb{Q}[x]$ is an expression of the form

$$f = a_0 + a_1 x + a_2 x^2 + \cdots + a_n x^n + \cdots$$

where $a_i \in \mathbb{Q}$, and all but finitely many of the $a_i$'s are 0. We call the $a_i$'s the **coefficients** of the polynomial. When we write down particular polynomials, we will simply omit a term if the coefficient happens to be zero. Thus, such expressions as $2 + x$, $\frac{4}{7} + 2x^2 - \frac{1}{2}x^3$, and 14 are all elements of $\mathbb{Q}[x]$. Henceforth, when we wish to write down a generic polynomial, we will usually be content with an expression of the form $f = a_0 + a_1 x + a_2 x^2 + \cdots + a_n x^n$. This means that we're assuming that $a_m = 0$, for all $m > n$. It may of course be the case that some of the $a_m$ for $m \leq n$ are 0 too.

We say that two polynomials are **equal** if and only if their corresponding coefficients are equal. Thus, $2 + 0x - x^2$, $2 - x^2 + 0x^3$, and $2 - x^2$ are all equal polynomials. The first two polynomials are simply less compact ways of writing the third.

For the most part we deal with polynomials with rational coefficients, but sometimes we wish to restrict our attention to those polynomials whose coefficients are integers; we denote this set by $\mathbb{Z}[x]$. Of course $\mathbb{Z}[x]$ is a proper subset of $\mathbb{Q}[x]$.

Note that $x$ is a formal symbol, not a variable or an indeterminate element of $\mathbb{Q}$. This is probably different from the way you are used to thinking of a polynomial, which is as a function from $\mathbb{Q}$ to $\mathbb{Q}$ (or from $\mathbb{R}$ to $\mathbb{R}$). This is not how we think of them here – we think of a polynomial as a formal expression. In fact, if we consider polynomials with coefficients taken not from $\mathbb{Q}$ but some other number system, two of these new polynomials can be equal as functions but not as polynomials.

▷ **Quick Exercise.**   Consider polynomials with coefficients from $\mathbb{Z}_2$ – denoted by $\mathbb{Z}_2[x]$, naturally. Show that the three different polynomials $x^2 + x + 1$, $x^4 + x^3 + x^2 + x + 1$, and $1$ are indeed the same function from $\mathbb{Z}_2$ to $\mathbb{Z}_2$. (Two functions are equal if they have the same value at all points in their domain.)  ◁

We will nearly always think of polynomials in the formal sense. To emphasize this point of view, when we speak of a particular polynomial we will denote it by a letter like $f$, rather than writing $f(x)$. The one time we wish to consider a polynomial as a function in this chapter will be made explicit, and then we will refer to it as a **polynomial function**.

The **degree** of a polynomial is the largest exponent with corresponding non-zero coefficient. So, a polynomial of degree 0 means the polynomial can be considered an element of $\mathbb{Q}$ (sometimes called a **scalar**). Of course, the zero polynomial has no non-zero coefficients. To cover this special case conveniently, we say that its degree is $-\infty$. We will denote the degree of a polynomial $f$ by $\deg(f)$.

---

## 4.2   The Algebra of Polynomials

We can add, subtract, and multiply polynomials in the ways with which you are already familiar: If $f = a_0 + a_1 x + \cdots + a_n x^n$ and $g = b_0 + b_1 x + \cdots + b_m x^m$ (let's suppose $n > m$), then

$$
\begin{aligned}
f + g \ = \ & (a_0 + b_0) + (a_1 + b_1)x + \cdots \\
& + (a_m + b_m)x^m + a_{m+1}x^{m+1} + \cdots + a_n x^n.
\end{aligned}
$$

The difference, $f - g$, is similarly defined. The definition of product is more difficult to write down abstractly; the following definition actually captures your previous experience in multiplying polynomials:

$$
f \cdot g = a_0 b_0 + (a_0 b_1 + a_1 b_0)x + \cdots + \sum_{i+j=k} (a_i b_j)x^k + \cdots + (a_n b_m)x^{n+m}.
$$

That is, the coefficient of $x^k$ is the sum of all the products of the coefficients of $x^i$ in $f$ with the coefficients of $x^j$ in $g$ where $i$ and $j$ sum to $k$.

### Example 4.1

If $f = 3 + x^4 - 2x^5 + x^6 + 2x^7$ and $g = -1 + 3x + x^2 + 4x^6$, then the coefficient of $x^6$ in $f \cdot g$ is $3 \cdot 4 + 1 \cdot 1 + (-2) \cdot 3 + 1 \cdot (-1) = 6$.

How is degree affected when we add or multiply polynomials? Your previous experience with polynomials suggests the right answer, which is contained in the following theorem.

**Theorem 4.1** *Let $f, g \in \mathbb{Q}[x]$. Then*

a. $\deg(fg) = \deg(f) + \deg(g)$, *where it is understood that $-\infty$ added to anything is $-\infty$.*

b. $\deg(f + g)$ *is less than or equal to the larger of the degrees of $f$ and $g$.*

**Proof:** We prove part (a) first. We consider first the case where one of the polynomials is the zero polynomial. Now, it is evident that $0g = 0$, for any polynomial $g$. Thus,

$$-\infty = \deg(0) = \deg(0g) = -\infty + \deg(g),$$

as required.

We thus may as well assume that neither $f$ nor $g$ is the zero polynomial; suppose that $\deg(f) = n$ and $\deg(g) = m$. Then $f = a_n x^n + f_1$, where $a_n \neq 0$ and $\deg(f_1) < n$. Similarly, $g = b_m x^m + g_1$, where $b_m \neq 0$ and $\deg(g_1) < m$. By the definition of multiplication of polynomials, the coefficient on $x^{n+m}$ is $a_n b_m$, and this is not zero because neither factor is. But all remaining terms in the product have smaller degree than $n + m$, and so

$$\deg(fg) = n + m = \deg(f) + \deg(g),$$

as required.

▷ **Quick Exercise.** You prove part (b). Also show by example that the degree of a sum of polynomials can be strictly smaller than the larger of the degrees. *Hint:* Take two polynomials with the same degree. ◁ □

An important particular case of the first part of this theorem is this: If a product of two polynomials is the zero polynomial, then one of the factors is the zero polynomial.

▷ **Quick Exercise.** Prove this, using the theorem. ◁

---

## 4.3 The Analogy between $\mathbb{Z}$ and $\mathbb{Q}[x]$

We will now begin to prove the theorems analogous to those proved about natural numbers and integers and summarized above. (Actually, in this chapter, *you* will do some of the proving.) You should notice, as you proceed through this chapter and the next, that not only are the theorems similar, but so are the proofs. (You will probably even be able to anticipate some theorems.) This suggests that the integers share properties with $\mathbb{Q}[x]$ that give rise to these theorems – in particular, unique factorization. Later, we will be able to identify these properties and prove unique factorization in a more general setting. This process of generalization is indeed a common theme in mathematics – one sees that A and B both have property C. What is shared by A and B that forces property C on both? For now, we are content to consider the concrete example of $\mathbb{Q}[x]$ and try to build more insight before getting abstract.

Before starting, recall that the proof technique used for most of the important theorems for natural numbers is induction. When considering polynomials, we also frequently use induction, but on the *degree* of the polynomial. Note that since the set of degrees of polynomials is $\{-\infty, 0, 1, 2, \ldots\}$, which is well ordered, induction may be used.

We now start, as with the integers, with the Division Theorem.

**Theorem 4.2    Division Theorem for $\mathbb{Q}[x]$**    *Let $f, g \in \mathbb{Q}[x]$ with $f \neq 0$. Then there are unique polynomials $q$ and $r$, with $\deg(r) < \deg(f)$, such that $g = fq + r$.*

Before proving this theorem, we look at an example.

The actual computation for producing $q$ and $r$, for given polynomials $f$ and $g$, is just long division. For example, let $f = x^2 + 2x - 1$ and $g = x^4 + x^2 - x + 2$.

$$
\begin{array}{r}
x^2 - \phantom{2}2x + \phantom{6}6 \\
x^2 + 2x - 1 \overline{\smash{\big)}\, x^4 + \phantom{2x^3 -} x^2 - \phantom{2}x + 2} \\
\underline{x^4 + 2x^3 - \phantom{2}x^2} \phantom{- x + 2} \\
-2x^3 + 2x^2 - \phantom{2}x + 2 \\
\underline{-2x^3 - 4x^2 + 2x} \phantom{+ 2} \\
6x^2 - \phantom{1}3x + 2 \\
\underline{6x^2 + 12x - 6} \\
- 15x + 8
\end{array}
$$

Hence, $q = x^2 - 2x + 6$ and $r = -15x + 8$. That is,

$$x^4 + x^2 - x + 2 = (x^2 + 2x - 1)(x^2 - 2x + 6) + (-15x + 8).$$

▷ **Quick Exercise.**    Find $q$ and $r$ as guaranteed by the Division Theorem for $g = x^5 + x - 1$ and $f = x^2 + x$. ◁

**Proof of the Division Theorem:**    We first prove the existence of $q$ and $r$, using induction on the degree of $g$. The base case for induction in this proof is $\deg(g) < \deg(f)$. If this is the case, then $g = f \cdot 0 + g$. So, $q = 0$ and $r = g$ satisfy the requirements of the theorem.

We now assume that $f = a_0 + a_1 x + \cdots + a_n x^n$ and $g = b_0 + b_1 x + \cdots + b_m x^m$, and $m = \deg(g) \geq \deg(f) = n$. Our induction hypothesis says that we can find a quotient and remainder whenever the dividend has degree less than $m$.

Let $h = g - (b_m/a_n) x^{m-n} f$. This makes sense because $m \geq n$. Note that the largest non-zero coefficient of $g$ has been eliminated in $h$, so $\deg(h) < \deg(g)$. Hence, by the induction hypothesis, $h = fq' + r$, where $\deg(r) < \deg(f)$. But then,

$$
\begin{aligned}
g &= fq' + r + (b_m/a_n) x^{m-n} f \\
&= f(q' + (b_m/a_n) x^{m-n}) + r.
\end{aligned}
$$

Thus, $q = q' + (b_m/a_n) x^{m-n}$ and $r$ serve as the desired quotient and remainder.

Now we prove the uniqueness of $q$ and $r$ by supposing that $g = fq + r = fq' + r'$, where $\deg(r) < \deg(f)$ and $\deg(r') < \deg(f)$. We will show that $q = q'$ and $r = r'$.

So, because $fq + r = fq' + r'$, we have that $f(q - q') = r' - r$. Because $\deg(f) > \deg(r)$ and $\deg(f) > \deg(r')$, we have $\deg(f) > \deg(r' - r)$. But $\deg(f(q - q')) \geq \deg(f)$, unless $f(q - q') = 0$. Hence, $\deg(f(q - q')) > \deg(r' - r)$ unless both are the zero polynomial. But this must be the case, and so $f(q - q') = 0$, forcing $q - q' = 0$ and $r' - r = 0$, proving uniqueness. □

## 4.4    Factors of a Polynomial

We now make some definitions analogous to those we made for $\mathbb{Z}$. We say a polynomial $f$ **divides** a polynomial $g$ if $g = fq$ for some polynomial $q$. In this case we say that $f$ is a **factor** of $g$, and write $f|g$. In the context of the Division Theorem, $f|g$ means that the remainder obtained is 0. For example, $(x^2+1)|(2x^3-3x^2+2x-3)$, because

$$2x^3 - 3x^2 + 2x - 3 = (x^2+1)(2x-3).$$

Notice that any polynomial divides the zero polynomial because $0 = (f)(0)$.

Suppose now that $a$ is a non-zero constant polynomial, and $f = a_0 + a_1 x + \cdots + a_n x^n$ is any other polynomial. Then $a$ necessarily divides $f$ because

$$f = (a)\left(\frac{a_0}{a} + \frac{a_1}{a}x + \frac{a_2}{a}x^2 + \cdots + \frac{a_n}{a}x^n\right).$$

In Exercise 11 you will prove that the converse of this statement is true.

## 4.5   Linear Factors

In practice, it may be *very* difficult to find all the factors of a given polynomial. However, the following theorem shows how to determine factors of the form $x - a$, where $a \in \mathbb{Q}$.

Note carefully that the next theorem and its corollary are the only times in this chapter where we think of a polynomial as a function. For $f \in \mathbb{Q}[x]$ and $a \in \mathbb{Q}$, we define $f(a)$ to be the result that ensues when we replace $x$ in $f$ by $a$, and then apply the ordinary operations of arithmetic in $\mathbb{Q}$ to simplify the result. Thus, if $f = \frac{1}{3}x^2 - 2x + 1$ and $a = 2$, then

$$f(2) = \frac{1}{3}(2)^2 - 2(2) + 1 = -\frac{5}{3}.$$

This definition obviously gives us a function $f(x)$ which is defined for all rational numbers.

Of particular interest to us is the case when $f(a) = 0$. We say $a \in \mathbb{Q}$ is a **root** of $f \in \mathbb{Q}[x]$ if $f(a) = 0$. Thus, $\frac{2}{3}$ is a root of $g = 3x^3 + 19x^2 - 11x - 2$, because $g\left(\frac{2}{3}\right) = 0$.

**Theorem 4.3   Root Theorem**   *If $f$ is a polynomial in $\mathbb{Q}[x]$ and $a \in \mathbb{Q}$, then $x - a$ divides $f$ if and only if $a$ is a root of $f$.*

**Proof:**   If $x - a$ divides $f$, then $f = (x-a)q$, and so

$$f(a) = (a-a)q(a) = 0.$$

Conversely, suppose $f(a) = 0$. Using the Division Theorem 4.2, we write $f = (x-a)q + r$ where $\deg(r) < \deg(x-a) = 1$. But $\deg(r) < 1$ means $\deg(r) = 0$ or $-\infty$; that is, $r \in \mathbb{Q}$. Thus, when we view $r$ as a function, it is a constant function. We might as well call this constant $r$. Hence, $f(a) = (a-a)q(a) + r$. But, the left-hand side is 0 while the right-hand side is $r$. Hence, $r = 0$, and so $x - a$ divides $f$. □

**Example 4.2**

Consider the polynomial
$$f = x^4 + 2x^3 + x^2 + x - 2.$$

We can conclude that $f$ has a factor of $x + 2$ because $f(-2) = 0$. We need not go through the trouble of long division to verify the fact.

In general, evaluating $f(a)$ is a simpler operation than dividing $f$ by $x - a$, although the latter does have the advantage of giving the factorization if indeed $x - a$ is a factor.

▷ **Quick Exercise.**   Check to see whether $x + 2$ is a factor of the following polynomials: $x^3 - 4x$, $x^2 + 2x - 1$, and $x^{100} - 4x^{98} + x + 2$. ◁

Notice that the Root Theorem 4.3 allows us to determine when any given linear factor, $ax + b$ $(a \neq 0)$, divides a polynomial $f$, because $ax + b$ is a factor if and only if $x + \frac{b}{a}$ is a factor, and the Root Theorem says that this last is a factor if and only if $-\frac{b}{a}$ is a root.

▷ **Quick Exercise.**   Let $f \in \mathbb{Q}[x]$ and $a \neq 0$. Prove that $ax + b$ divides $f$ if and only if $x + \frac{b}{a}$ divides $f$. ◁

**Corollary 4.4** *A non-zero polynomial $f$ of degree $n$ has at most $n$ distinct roots in $\mathbb{Q}$.*

**Proof:**   We proceed by induction on the degree of $f$. If $f$ is non-zero and degree 0, then clearly $f$ has no roots.

Suppose the corollary holds for polynomials with degree smaller than $n$, and $f$ has degree $n > 0$. If $f$ has no roots, the theorem is proved. So, assume $f$ has at least one root, call it $a$. Then by the Root Theorem 4.3, $x - a$ divides $f$. That is, $f = (x - a) \cdot g$. But $\deg(g) = \deg(f) - 1$. And so, by the induction hypothesis, $g$ has at most $n - 1$ roots. Because any root of $f$ other than $a$ must also be a root of $g$, $f$ can have no more than $n$ roots altogether.                                                                       □

---

## 4.6   Greatest Common Divisors

We now turn our attention to finding greatest common divisors of two polynomials, paralleling our development for the integers. Notice that we didn't say *the* greatest common divisor. We will find some differences from the integers here.

If $f$ and $g$ are two polynomials in $\mathbb{Q}[x]$ (not both zero), then a polynomial $d$ in $\mathbb{Q}[x]$ is a **greatest common divisor** (gcd) of $f$ and $g$ if it satisfies the following two criteria:

1. $d$ divides both $f$ and $g$ ($d$ is a **common divisor**), and

2. If a polynomial $e$ divides both $f$ and $g$, then $\deg(e) \leq \deg(d)$; that is, $d$ has largest degree among common divisors.

Notice that according to this definition $x + 1$ is a gcd for $x^2 - 1$ and $x^2 + 4x + 3$, but so are $-\frac{1}{2}x - \frac{1}{2}$, $-x - 1$, and $20x + 20$. This is unlike the integer case, where we have a unique gcd for any pair of integers (not both zero). In fact, for polynomials, we have infinitely many distinct gcds. For if $a$ is any non-zero rational number and $d$ is a gcd of $f$ and $g$, then so is the polynomial $ad$. This follows because $ad$ is also a common divisor of $f$ and $g$, and has the same degree as $d$.

▷ **Quick Exercise.**   Argue that $x + 1$ is indeed a gcd of $x^2 - 1$ and $x^2 + 4x + 3$ as are the other polynomials listed. ◁

As with the integers, Euclid's Algorithm, when applied to polynomials, yields a gcd. (As you might expect, the arithmetic is messier.)

**Example 4.3**

As an example of this, consider the polynomials

$$x^6 - x^5 + 7x^4 + 5x^2 + 9x - 21$$

and

$$2x^4 - x^3 + 5x^2 - 3x - 3.$$

By repeatedly applying the Division Theorem 4.2 we obtain the following:

$$(x^6 - x^5 + 7x^4 + 5x^2 + 9x - 21) =$$
$$(2x^4 - x^3 + 5x^2 - 3x - 3)\left(\frac{x^2}{2} - \frac{x}{4} + \frac{17}{8}\right)$$
$$+ \left(\frac{39}{8}\right)(x^3 - x^2 + 3x - 3)$$

$$(2x^4 - x^3 + 5x^2 - 3x - 3) =$$
$$\left(\left(\frac{39}{8}\right)(x^3 - x^2 + 3x - 3)\right)\left(\left(\frac{8}{39}\right)(2x + 1)\right) + 0.$$

This means that $(39/8)(x^3 - x^2 + 3x - 3)$ is a gcd of the given polynomials; thus, $x^3 - x^2 + 3x - 3$ is one also.

▷ **Quick Exercise.** Apply Euclid's Algorithm to the polynomials

$$x^3 - 3x^2 + 5x - 3 \text{ and } x^4 - 2x^3 + 4x^2 - 2x + 3. \quad ◁$$

We state formally in the next theorem that Euclid's Algorithm gives gcds in the polynomial case. You should notice that the proof of this theorem is nearly identical to the proof of the corresponding theorem (Theorem 2.3) for integers.

**Theorem 4.5  Euclid's Algorithm produces a GCD**
*In Euclid's Algorithm for polynomials $f$ and $g$, the last non-zero remainder is a greatest common divisor for $f$ and $g$.*

**Proof:** The proof of this fact for the integers depended on a lemma, which remains true in this context. Namely, we observe that if $f = gq + r$, then $d$ is a gcd of $f$ and $g$ if and only if $d$ is also a gcd of $g$ and $r$. (As before, this follows because in fact every common divisor of $f$ and $g$ is also a common divisor of $g$ and $r$, and vice versa.)

▷ **Quick Exercise.** Show that $d$ is a gcd of $f$ and $g$ exactly if $d$ is a gcd of $g$ and $r$, following the proof of Lemma 2.2, if necessary. ◁

We proceed by induction on the number of steps required for Euclid's Algorithm to terminate. If Euclid's Algorithm takes one step, $f = gq$ and so $d = g$, which is clearly a gcd of $f$ and $g$ in this case.

So suppose Euclid's Algorithm takes $n \ (> 1)$ steps to obtain $d$ with the first two steps being

$$\begin{aligned} f &= gq_0 + r_0 \\ g &= r_0 q_1 + r_1 \end{aligned}$$

Performing Euclid's Algorithm on $g$ and $r_0$ also yields $d$; the process is exactly the last

$n - 1$ steps of Euclid's Algorithm for $f$ and $g$. But then by the induction hypothesis, $d$ is a gcd of $g$ and $r_0$, and so, by the lemma, $d$ is a gcd of $f$ and $g$. □

We observed above that if $d$ is a gcd of $f$ and $g$, then so is any non-zero scalar multiple of $d$. The converse is also true. That is, if $d$ is any gcd of two polynomials, then all the gcds are simply scalar multiples of $d$. This follows from the next theorem.

**Theorem 4.6** *If $d$ is the gcd of $f$ and $g$ produced by Euclid's Algorithm and $e$ is another common divisor of $f$ and $g$, then $e$ divides $d$.*

**Proof:** The proof of this theorem is nearly identical to the proof that Euclid's Algorithm produces a gcd. The only real addition is that we also check that $e$ divides what it is supposed to. Again, we use induction on $n$, the number of steps in Euclid's Algorithm when obtaining $d$.

If Euclid's Algorithm takes one step, then $f = gq$, and so $d = g$. Clearly, if $e$ divides $f$ and $g (= d)$, then $e$ divides $d$.

Now, suppose Euclid's Algorithm takes $n(> 1)$ steps to obtain $d$, with the first two steps being

$$
\begin{aligned}
f &= gq_0 + r_0 \\
g &= r_0q_1 + r_1.
\end{aligned}
$$

Because $r_0 = f - gq_0$, if $e$ divides $f$ and $g$, $e$ also divides $r_0$. Now $d$ is also the gcd of $g$ and $r_0$ obtained from Euclid's Algorithm. The process is the last $n - 1$ steps of Euclid's Algorithm for $f$ and $g$. So, by the induction hypothesis, because $e$ divides $g$ and $r_0$, it also divides $d$. □

An important consequence of this theorem is that any two gcds of two given polynomials are just scalar multiples of one another.

▷ **Quick Exercise.** Show that if $e$ and $d$ are two gcds of polynomials $f$ and $g$, then $d$ and $e$ are scalar multiples of one another. ◁

Finally, we have the GCD identity for $\mathbb{Q}[x]$.

**Theorem 4.7    GCD identity for polynomials**    *If $d$ is a gcd of polynomials $f$ and $g$, then there exist polynomials $a$ and $b$ so that $d = af + bg$.*

**Proof:** The proof is Exercise 8 (or 9). □

---

# Chapter Summary

In this chapter we introduced the set $\mathbb{Q}[x]$ of all polynomials with rational coefficients and described the arithmetic of this set. In direct analogy with $\mathbb{Z}$, we proved the *Division Theorem*, *Euclid's Algorithm*, and the *GCD identity*. In addition, we described the relationship between the roots of a polynomial and its linear factors; this relationship is called the *Root Theorem*.

# Warm-up Exercises

a. Compute the sum, difference, and product of the polynomials

$$1 - 2x + x^3 - \frac{2}{3}x^4 \quad \text{and} \quad 2 + 2x^2 - \frac{3}{2}x^3$$

in $\mathbb{Q}[x]$.

b. Give the quotient and remainder when the polynomial $2 + 4x - x^3 + 3x^4$ is divided by $2x + 1$.

c. Give two polynomials $f$ and $g$, where the degree of $f + g$ is strictly less than either the degree of $f$ or the degree of $g$.

d. Use the Root Theorem 4.3 to answer the following for polynomials in $\mathbb{Q}[x]$. Does $x - 2$ divide $x^5 - 4x^4 - 4x^3 - x^2 + 4$? Does $x + 1$ divide $x^6 + 2x^5 + x^4 - x^3 + x$? Does $x + 5$ divide $2x^3 + 10x^2 - 2x - 10$? Does $2x - 1$ divide $x^5 + 2x^4 - 3x^2 + 1$?

e. We started this chapter with a list of theorems true about $\mathbb{Z}$. For how many of them have we stated and proved an analogue for $\mathbb{Q}[x]$?

# Exercises

1. Find the gcd of the polynomials $x^6 + x^5 - 2x^4 - x^3 - x^2 + 2x$ and $3x^6 + 4x^5 - 3x^3 - 4x^2$ in $\mathbb{Q}[x]$, and express them as $af + bg$, for some polynomials $a, b$.

2. Divide the polynomial $x^2 - 3x + 2$ by the polynomial $2x + 1$, to obtain a quotient and remainder as guaranteed by the Division Theorem 4.2. Note that although $x^2 - 3x + 2$ and $2x + 1$ are elements of $\mathbb{Z}[x]$, the quotient and remainder are not. Argue that this means that there is no Division Theorem for $\mathbb{Z}[x]$.

3. By Corollary 4.4 we know that a third-degree polynomial in $\mathbb{Q}[x]$ has at most three roots. Give four examples of third-degree polynomials in $\mathbb{Q}[x]$ that have 0, 1, 2, and 3 roots, respectively; justify your assertions. (Recall that here a root must be a rational number!)

4. Your example in the previous exercise of a third-degree polynomial with exactly 2 roots had one **repeated root**; that is, a root $a$ where $(x - a)^2$ is a factor of the polynomial. (Roots may have multiplicity greater than two, of course.) Why can't a third-degree polynomial in $\mathbb{Q}[x]$ have exactly 2 roots where neither is a multiple root?

5. Let $n$ be an odd integer and consider the polynomial

$$\Phi_{n+1} = \frac{x^{n+1} - 1}{x - 1} = x^n + x^{n-1} + \cdots x + 1.$$

Use the Root Theorem 4.3 to argue that $\Phi_{n+1}$ has a linear factor. We call $\Phi_{n+1}$ a **cyclotomic polynomial**; see Exercise 5.17 for more information.

6. Suppose that $f \in \mathbb{Q}[x]$, $q \in \mathbb{Q}$, and $\deg(f) > 0$. Use the Root Theorem 4.3 to prove that the equation $f(x) = q$ has at most finitely many solutions.

7. Let $f \in \mathbb{Q}[x]$. Recall from Exercise 4 above that a rational number $a$ is a repeated root of $f$ if $(x - a)^2$ is a factor of $f$. Given $f = a_0 + a_1 x + \cdots + a_n x^n$, we define the **formal derivative** of $f$ as $f' = na_n x^{n-1} + (n-1)a_{n-1}x^{n-2} + \cdots + a_1$. Prove that $a$ is a repeated root of $f$ if and only if $a$ is a root of both $f$ and $f'$.

8. Prove Theorem 4.7: the GCD identity for $\mathbb{Q}[x]$. Use Euclid's Algorithm 4.5, and the relationship we know between the gcd produced by the algorithm and an arbitrary gcd (Theorem 4.6).

9. One can also prove the GCD identity for $\mathbb{Q}[x]$ with an argument similar to the existential proof of the GCD identity for integers, found in Section 2.3. Try this approach.

10. We say that $p \in \mathbb{Q}[x]$ has a multiplicative inverse if there exists a $q \in \mathbb{Q}[x]$ such that $pq = 1$. Prove that $p \in \mathbb{Q}[x]$ has a multiplicative inverse if and only if $\deg(p) = 0$.

11. Suppose that $g \in \mathbb{Q}[x]$, and $g$ divides all polynomials in $\mathbb{Q}[x]$. Prove that $g$ is a non-zero constant polynomial.

12. Find two different polynomials in $\mathbb{Z}_3[x]$ that are equal as functions from $\mathbb{Z}_3$ to $\mathbb{Z}_3$.

13. Find a non-zero polynomial in $\mathbb{Z}_4[x]$ for which $f(a) = 0$, for all $a \in \mathbb{Z}_4$.

# Chapter 5

## Factorization of Polynomials

We have already seen the Fundamental Theorem of Arithmetic, which says that every integer (other than 0, 1, and $-1$) can be uniquely factored into primes. We wish to come up with a corresponding theorem for the set $\mathbb{Q}[x]$ of polynomials with rational coefficients.

## 5.1 Factoring Polynomials

We note first that uniqueness of factorization cannot be as nice for polynomials as for integers because any factorization in $\mathbb{Q}[x]$ can be adjusted by factoring out scalars. The following example shows what we mean:

$$
\begin{aligned}
x^2 - 4 &= (x+2)(x-2) \\
&= \left(\frac{1}{2}x + 1\right)(2x - 4) \\
&= (2x + 4)\left(\frac{1}{2}x - 1\right),
\end{aligned}
$$

and so on.

But there is a close connection, after all, between the factors $x + 2$ and $\frac{1}{2}x + 1$. Namely, they differ by only a scalar multiple. In fact, we will obtain uniqueness of factorization for polynomials, up to scalar multiples. We now head toward this result. The first order of business is to define irreducible polynomials in a way analogous to our definition of irreducible integers.

A polynomial $p$ is **irreducible** if

a. $p$ is of degree greater than zero, and

b. whenever $p = fg$, then either $f$ or $g$ has degree zero.

In other words, an irreducible is a non-scalar polynomial, whose only factorizations involve scalar factors. We are thus regarding such factorizations as $x + \frac{1}{2} = \frac{1}{2}(2x + 1)$ as trivial, just as we regard factorizations such as $3 = (-1)(-3)$ as trivial in the integer case. So, a *non-trivial factorization* of a polynomial is one that has at least two non-scalar factors. We say that a polynomial is **reducible** if it does have a non-trivial factorization. Thus, $x^4 + 2x^2 + 1$ is reducible.

▷ **Quick Exercise.** Why is $x^4 + 2x^2 + 1$ reducible? ◁

Which polynomials in $\mathbb{Q}[x]$ are irreducible? One immediate consequence of the definition is that all polynomials of degree one are irreducible because if $p$ is of degree 1 and $p = fg$, then exactly one of $f$ and $g$ has degree 0, and is consequently a scalar.

Are there any others? Consider the polynomial $f = x^2 + 2$. If this polynomial had a non-trivial factorization $x^2 + 2 = gh$, then both $g$ and $h$ would be degree one factors. But then the Root Theorem tells us that $x^2 + 2$ would have a rational root. But $f(q) = q^2 + 2 \geq 2$ for all rational numbers $q$, and so $f$ can have no roots.

Consider next the polynomial $x^2 - 2$. By the same reasoning, if this polynomial were reducible, it would have a root, and so there would exist a rational number $q$ so that $q^2 = 2$. There is no such rational number, as you probably know. The fact that $\sqrt{2}$ is an irrational number is a very famous theorem, first proved by the ancient Greeks. You will prove it (and more) in Exercise 13 (see also Exercise 2.14). Thus, $x^2 - 2$ is irreducible.

▷ **Quick Exercise.**  Show that $x^4 + 2$ is irreducible in $\mathbb{Q}[x]$, taking your lead from the discussion of $x^2 + 2$ above. ◁

▷ **Quick Exercise.**  Show that $x^3 - 2$ is irreducible in $\mathbb{Q}[x]$. *Hint:* If $x^3 - 2$ were reducible, then it would have a linear factor. You may assume (see Exercise 13) that there is no rational number $r$ so that $r^3 = 2$. ◁

These examples suggest that there are many irreducible polynomials in $\mathbb{Q}[x]$, and indeed there are, of arbitrarily high degree. (See Exercise 12.) It would be difficult to describe them all, however.

Note that we are interested only in factors which belong to $\mathbb{Q}[x]$ (for the time being, at least). Thus, $x^2 - 2$ is irreducible in $\mathbb{Q}[x]$, even though we *can* factor it if we allow ourselves to use real numbers as coefficients:

$$x^2 - 2 = \left(x - \sqrt{2}\right)\left(x + \sqrt{2}\right).$$

In Chapter 10 we will discuss factorization of polynomials over the real numbers $\mathbb{R}$ and the complex numbers $\mathbb{C}$.

We now claim that every polynomial can be factored into irreducibles:

**Theorem 5.1** *Any polynomial in $\mathbb{Q}[x]$ of degree greater than zero is either irreducible or the product of irreducibles.*

**Proof:**   Prove this in a similar way to the proof for the corresponding theorem for the integers, Theorem 2.8. This proof is Exercise 1.                                           □

---

## 5.2   Unique Factorization

The previous theorem is the first half of the unique factorization theorem that we want for $\mathbb{Q}[x]$. Recall that to obtain the uniqueness of factorization in $\mathbb{Z}$, we required the concept of a *prime* integer. We now make the analogous definition: a polynomial $p$ (with degree bigger than 0) is **prime** if whenever $p$ divides $fg$, then $p$ divides $f$ or $p$ divides $g$.

We claim that $x - 2$ is a prime polynomial. Suppose that $x - 2$ divides the product $fg$. Then by the Root Theorem 4.3, 2 is a root of $fg$, and so $f(2)g(2) = 0$. But $f(2)$ and $g(2)$ are rational numbers, and the only way a product of rational numbers can be zero is if at least one of the factors is zero. That is, $f(2)$ (or $g(2)$) is zero, and so 2 is a root of $f$ (or $g$). Notice that this argument could be modified to show that *any* degree one polynomial is prime.

▷ **Quick Exercise.** Modify the argument in the previous paragraph to prove that all degree one polynomials in $\mathbb{Q}[x]$ are prime. ◁

But we need not pursue any further examples because the concept of prime polynomial turns out to be equivalent to that of irreducible polynomial. This situation is what we discovered for $\mathbb{Z}$, and the proof is the same:

**Theorem 5.2** *A polynomial in $\mathbb{Q}[x]$ is irreducible if and only if it is prime.*

**Proof:** Prove this, again taking your lead from the proof of Theorem 2.7, the analogous result for the integers. This proof is Exercise 2. □

**Corollary 5.3** *If $p$ is an irreducible polynomial that divides*

$$f_1 f_2 \cdots f_n,$$

*then $p$ divides one of the $f_i$.*

**Proof:** Prove this. This proof is Exercise 3. □

So, for both $\mathbb{Z}$ and $\mathbb{Q}[x]$, primeness and irreducibility are equivalent. Be warned that we will eventually examine structures where this is not the case. It is thus important to keep these definitions straight.

Finally, we come to the unique factorization theorem for polynomials. It is similar to the unique factorization theorem for integers, except we must account for the fact that we can always factor out scalars, as noted above. Accordingly, we first make the following convenient definition.

Two polynomials $f$ and $g$ are called **associates** if there is a non-zero scalar $a$ such that $f = ag$. For instance, $x^2 - 3$ and $2x^2 - 6$ are associates. Note that any two non-zero scalars are associates.

▷ **Quick Exercise.** Describe the set of all associates of the polynomial $x^2 - 3$ (there are infinitely many). ◁

**Theorem 5.4    Unique Factorization Theorem for Polynomials**    *If $h$ is a polynomial in $\mathbb{Q}[x]$, and*

$$h = f_1 f_2 \cdots f_n = g_1 g_2 \cdots g_m,$$

*where the $f_i$ and $g_j$ are all irreducible, then $n = m$ and the $g_j$ may be rearranged so that $f_i$ and $g_i$ are associates, for $i = 1, 2, \ldots n$.*

Before we prove this theorem, let's look at an example. Consider the polynomial $h = 2x^3 + 3x^2 + 5x + 2$. Now $h$ can be factored into irreducibles as $(2x + 1)(x^2 + x + 2)$ or as $(x + \frac{1}{2})(2x^2 + 2x + 4)$, or an infinite number of other ways. But, of course, $2x + 1$ and $x + \frac{1}{2}$ are associates, as are $x^2 + x + 2$ and $2x^2 + 2x + 4$.

▷ **Quick Exercise.** Verify that the factors of $h$ given above are really irreducible. ◁

**Proof:** We use induction on $n$, the number of factors $f_i$ of $h$. If $n = 1$, the theorem follows easily. (Right?) So, we assume $n > 1$. By the primeness of the irreducible $f_1$, $f_1$ divides one of the $g_j$ (see the Corollary 5.3). By renumbering the $g_j$, if necessary, we may assume $f_1$ divides $g_1$. So, because $g_1$ is irreducible, $af_1 = g_1$, for some non-zero scalar $a$. That is, $f_1$ and $g_1$ are associates. Therefore, by dividing both sides by $f_1$, we have $f_2 f_3 \cdots f_n = ag_2 g_3 \cdots g_m$. (Because $g_2$ is irreducible, so is $ag_2$, and we consider $ag_2$ as an irreducible factor.) We now have two factorizations into irreducibles and the number of $f_i$ factors is $n - 1$. So, by the induction hypothesis, $n - 1 = m - 1$ and, by renumbering the $g_j$ if necessary, $f_i$ and $g_i$ are associates for $i = 2, 3, \ldots, n$. (Actually, we have that $ag_2$ is an associate of $f_2$. But then $g_2$ is also an associate of $f_2$.) This proves the theorem. □

Notice that the proof of this theorem is nearly identical to the proof of the corresponding Theorem 2.9 for $\mathbb{Z}$. We had only to handle the problem of associates. By now, you should have seen this similarity with *all* the theorems in this chapter and might be wondering the reason for it. We will eventually find a close connection between the integers and $\mathbb{Q}[x]$ that explains this similarity.

---

## 5.3   Polynomials with Integer Coefficients

We close this chapter by discussing the relationship between $\mathbb{Q}[x]$ and $\mathbb{Z}[x]$. You should first note that some of the theorems we have proved about $\mathbb{Q}[x]$ are *false* when we restrict ourselves to polynomials with integer coefficients. In particular, the Division Theorem is false for $\mathbb{Z}[x]$. To see this, merely try to divide $x^2 + 7x$ by $2x + 1$: the Division Theorem 4.2 for $\mathbb{Q}[x]$ provides us with the unique quotient $\frac{1}{2}x + \frac{11}{4}$ and remainder $-\frac{11}{4}$. These are not elements of $\mathbb{Z}[x]$, and so no such quotient/remainder pair can exist in $\mathbb{Z}[x]$. (See also Exercise 4.2.)

Similarly, the GCD identity fails in $\mathbb{Z}[x]$. Consider the polynomials 2 and $x$; a gcd for these polynomials is 1. But we *cannot* write 1 as a linear combination of 2 and $x$ in $\mathbb{Z}[x]$.

▷ **Quick Exercise.**   Show that we cannot write 1 as a linear combination of 2 and $x$. *Hint*: Suppose that $1 = 2f + xg$, where $f, g \in \mathbb{Z}[x]$, and consider the constant term of the right-hand side of the equation.  ◁

It is thus not surprising that we look to $\mathbb{Q}[x]$ (rather than $\mathbb{Z}[x]$) for our analogue to $\mathbb{Z}$. However, the news is not all bad. For it turns out that if a polynomial with integer coefficients can be non-trivially factored into polynomials with rational coefficients, then it can be factored into polynomials with integer coefficients; this result is known as Gauss's Lemma.

For example, consider $2x^2 + 3x - 2$. Because $\frac{1}{2}$ is a root, we have that

$$2x^2 + 3x - 2 = \left( x - \frac{1}{2} \right) (2x + 4).$$

This is a factorization in $\mathbb{Q}[x]$. But by adjusting scalars, we can obtain the factorization

$$2x^2 + 3x - 2 = (2x - 1)(x + 2)$$

in $\mathbb{Z}[x]$.

The idea of the proof below is essentially that of the example we've just given, although it is a bit messy to carry it out in general. Some readers may want to skip this proof on a first reading.

**Theorem 5.5    Gauss's Lemma**    *If $f \in \mathbb{Z}[x]$ and $f$ can be factored into a product of non-scalar polynomials in $\mathbb{Q}[x]$, then $f$ can be factored into a product of non-scalar polynomials in $\mathbb{Z}[x]$; each factor in $\mathbb{Z}[x]$ is an associate of the corresponding factor in $\mathbb{Q}[x]$.*

**Proof:** Suppose $f \in \mathbb{Z}[x]$, and $f = gh$, where $g, h \in \mathbb{Q}[x]$. We can assume that the coefficients of $f$ have no common factors (other than $\pm 1$), or else we factor out the greatest common divisor of those coefficients, apply the following, and then multiply through by the gcd. So, suppose that

$$g = \frac{a_n}{b_n}x^n + \frac{a_{n-1}}{b_{n-1}}x^{n-1} + \cdots + \frac{a_0}{b_0}, \text{ and}$$

$$h = \frac{c_m}{d_m}x^m + \frac{c_{m-1}}{d_{m-1}}x^{m-1} + \cdots + \frac{c_0}{d_0}$$

where the $a_i$'s, $b_i$'s, $c_i$'s, and $d_i$'s are all integers and each fraction is in lowest terms. Multiply the first equation by the product $B$ of the $b_i$'s. We get

$$Bg = A'_n x^n + A'_{n-1} x^{n-1} + \cdots + A'_0.$$

Now divide by the greatest common divisor $A$ of the $A'_i$'s, yielding

$$(B/A)g = A_n x_n + A_{n-1} x^{n-1} + \cdots + A_0,$$

which is a polynomial in $\mathbb{Z}[x]$ whose coefficients have no common divisor (other than $\pm 1$).

Do likewise to the second equation, yielding

$$(D/C)h = C_m x^m + C_{m-1} x^{m-1} + \cdots + C_0,$$

which again is a polynomial whose coefficients have no common divisor (other than $\pm 1$).

Because $f = gh$, we get

$$BDf = (AC)((B/A)g)((D/C)h). \tag{5.1}$$

Because the coefficients of $f$ have no common divisor, we see that $BD$ is the gcd of the coefficients of $BDf$. Because $AC$ is a common divisor of the coefficients of this polynomial (as we see by looking at the right side of equation 5.1) we have that $AC|BD$. Let $E = BD/AC$, which is an integer. We have

$$Ef = (A_n x^n + A_{n-1} x^{n-1} + \cdots + A_0)(C_m x^m + C_{m-1} x^{m-1} + \cdots + C_0). \tag{5.2}$$

We need only show that $E = \pm 1$ to be done, because then $f$ has been written in the required form. If $E \neq \pm 1$, then let $p$ be a prime factor of $E$. Remember that all the $A_i$'s have no common factor and all the $C_i$'s have no common factor. Therefore, there is a smallest $i$ such that $p \nmid A_i$ and a smallest $j$ such that $p \nmid C_j$. We now compute the coefficient of $x^{i+j}$. From the left side of equation 5.2 we see that this coefficient is divisible by $E$, and hence by $p$. From the right side, the coefficient is

$$\sum_{k=0}^{i-1} A_k C_{i+j-k} + \sum_{k=i+1}^{i+j} A_k C_{i+j-k} + A_i C_j.$$

In the first sum, each of $A_k$ is divisible by $p$ because those $k$ are less than $i$. In the second sum, each $C_{i+j-k}$ is divisible by $p$ because those $i + j - k$ are less than $j$. But $A_i C_j$ is not divisible by $p$. Hence, the entire coefficient cannot be divisible by $p$, which is a contradiction. So it must be that $E = \pm 1$ as desired. $\square$

We can rephrase Gauss's Lemma in the following way: To see whether a polynomial in $\mathbb{Z}[x]$ can be factored non-trivially in $\mathbb{Q}[x]$, we need only check to see if it can be factored non-trivially in $\mathbb{Z}[x]$. This latter is presumably an easier task.

For example, consider the case of cubic equations. If a cubic polynomial in $\mathbb{Z}[x]$ can be factored non-trivially in $\mathbb{Z}[x]$, one of the factors must be a degree one polynomial $ax + b$, where $a$ and $b$ are integers. This implies that the polynomial must have a root $-\frac{b}{a}$ in $\mathbb{Q}$. But we can limit the possibilities for $a$ and $b$, which in turn limits which roots are possible. For a specific example of this, consider the polynomial $3x^3 + x + 1$. We wish to factor this into irreducibles in $\mathbb{Q}[x]$ or determine if the polynomial is itself irreducible. If $3x^3 + x + 1$ factors in $\mathbb{Q}[x]$, then, by Gauss's Lemma, it factors in $\mathbb{Z}[x]$ and in fact must have a linear factor in $\mathbb{Z}[x]$. The only possible such factors are of the form $(3x \pm 1)$ or $(x \pm 1)$, in order that the leading and trailing coefficients be correct. But this implies that the polynomial has a root of $\pm 1/3$ or $\pm 1$, which, by inspection, is not the case. We conclude that $3x^3 + x + 1$ is irreducible in $\mathbb{Z}[x]$, and hence in $\mathbb{Q}[x]$.

The argument above can be generalized to obtain a valuable tool for factoring in $\mathbb{Z}[x]$, known as the *Rational Root Theorem*.

**Theorem 5.6    The Rational Root Theorem**    *Suppose that $f = a_0 + a_1 x + \cdots + a_n x^n$ is a polynomial in $\mathbb{Z}[x]$, and $p/q$ is a rational root; that is, $p$ and $q$ are integers, $q \neq 0$, and $f(p/q) = 0$. We may as well assume also that $\gcd(p, q) = 1$. Then $q$ divides the integer $a_n$, and $p$ divides $a_0$.*

**Proof:**    You will prove this in Exercise 6.    □

▷ **Quick Exercise.**    Determine which of the following polynomials has a rational root:

$$3x^4 + 5x^3 + 10, \quad 4x^4 + x^3 - 4x^2 + 7x + 2.$$

◁

▷ **Quick Exercise.**    Is $3x^3 + 2x + 1$ irreducible in $\mathbb{Q}[x]$? Is $x^3 - 3x + 4$? Is $2x^3 + 7x^2 - 2x - 1$?
◁

Another important tool when considering factorization is the following sufficient condition for a polynomial being irreducible in $\mathbb{Z}[x]$.

**Theorem 5.7    Eisenstein's Criterion**    *Suppose that $f \in \mathbb{Z}[x]$, and*

$$f = a_0 + a_1 x + a_2 x^2 + \cdots + a_n x^n.$$

*Let $p$ be a prime integer, and suppose that*

*1. $p$ divides $a_k$, for $0 \leq k < n$,*

*2. $p$ does not divide $a_n$, and*

*3. $p^2$ does not divide $a_0$.*

*Then $f$ is irreducible in $\mathbb{Z}[x]$.*

**Proof:**    Once again, you will prove this theorem in Exercise 16.    □

**Example 5.1**

It is quite evident that the Eisenstein criterion implies that the polynomial $x^5 + 5x + 5$ is irreducible in $\mathbb{Z}[x]$, by using $p = 5$. But note that the criterion does not directly apply to the polynomial $x^5 + 5x + 4$. However, you will discover in Exercise 15 how to apply the criterion to show that this is in fact an irreducible polynomial.

Notice that although Eisenstein's criterion is phrased as a theorem about irreducibility in $\mathbb{Z}[x]$, Gauss's Lemma 5.5 implies that such a polynomial is also irreducible in $\mathbb{Q}[x]$.

**Example 5.2**

Consider the polynomial

$$f = \frac{1}{3}x^4 + \frac{2}{9}x^3 + 4x^2 - \frac{2}{5}x + 2 \in \mathbb{Q}[x].$$

We wish to conclude that $f$ is irreducible in $\mathbb{Q}[x]$. By multiplying through by 45, we obtain the element

$$g = 45f = 15x^4 + 10x^3 + 140x^2 - 18x + 90 \in \mathbb{Z}[x].$$

Note that we could apply the Rational Root Theorem to $g$ to conclude that it has no roots in $\mathbb{Q}$; it would be tedious to carry out the details. However, we can in fact apply Eisenstein's Criterion (with $p = 2$) to reach the strictly stronger conclusion that $g$ is irreducible in $\mathbb{Z}[x]$. But then Gauss's Lemma says that $g$ is in fact irreducible in $\mathbb{Q}[x]$, and so $f$ is irreducible in $\mathbb{Q}[x]$ too.

## Historical Remarks

In the last two chapters we have given a systematic account of the theory of $\mathbb{Q}[x]$, the set of polynomials with rational coefficients, giving due emphasis to its similarity to $\mathbb{Z}$. We have consequently emphasized a formal, algebraic approach to $\mathbb{Q}[x]$, which probably seems foreign to your previous experience with polynomials.

The pieces of this theory were put together over a number of centuries, beginning in earnest with the 16th and 17th-century French algebraists Viète and Descartes and culminating in the work of Gauss. This development was relatively slow, primarily for two reasons. First of all, algebraic notation at the time of Viète was cumbersome and not standardized. Algebra flows much more easily when our notation is clear and efficient. Secondly, in the 17th century people had little agreement about the nature of numbers. Doubts were cast on the 'reality' and utility of irrational numbers, complex numbers, and even negative numbers. We'll discuss this issue more in Chapter 10. The process by which the number system which we use was standardized and widely accepted was long and difficult. The lesson from history is clear: It takes a lot of hard work to become comfortable with the elegant point of view of a modern mathematician!

---

## Chapter Summary

In this chapter we considered factorization in $\mathbb{Q}[x]$ and proved that a polynomial is irreducible if and only if it is prime, and we used this fact to prove the *Unique Factorization Theorem for Polynomials*. We then proved *Gauss's Lemma*, which describes the relationship between factoring in $\mathbb{Z}[x]$ and in $\mathbb{Q}[x]$. We concluded by considering two important tools useful in factoring in $\mathbb{Z}[x]$, called the *Rational Root Theorem* and *Eisenstein's Criterion*.

---

## Warm-up Exercises

  a. Why is a linear polynomial in $\mathbb{Q}[x]$ always irreducible?

  b. Why is a polynomial of the form $x^2 + a \in \mathbb{Q}[x]$, where $a > 0$, always irreducible?

  c. Determine a factorization of $x^4 - 5x^2 + 4$ into irreducibles in $\mathbb{Q}[x]$.

d. Give several distinct factorizations of $x^4 - 5x^2 + 4$ into irreducibles in $\mathbb{Q}[x]$. Why don't these distinct factorizations violate the Unique Factorization Theorem 5.4?

e. We know that 7 is an irreducible integer, but is 7 an irreducible polynomial?

f. Pick your favorite polynomial. What are its associates in $\mathbb{Q}[x]$?

g. Factor $2x^3 + 7x^2 - 2x - 1$ completely into irreducibles in $\mathbb{Q}[x]$, using Gauss's Lemma and the Root Theorem. Adjust your factorization (if necessary) so that all factors belong to $\mathbb{Z}[x]$.

---

# Exercises

1. Prove Theorem 5.1: A polynomial in $\mathbb{Q}[x]$ of degree greater than zero is either irreducible or the product of irreducibles.

2. Prove Theorem 5.2: A polynomial in $\mathbb{Q}[x]$ is irreducible if and only if it is prime.

3. Prove Corollary 5.3: If an irreducible polynomial in $\mathbb{Q}[x]$ divides a product $f_1 f_2 f_3 \cdots f_n$, then it divides one of the $f_i$.

4. Use Gauss's Lemma to determine which of the following are irreducible in $\mathbb{Q}[x]$:
$$4x^3 + x - 2, \quad 3x^3 - 6x^2 + x - 2, \quad x^3 + x^2 + x - 1.$$

5. Show that $x^4 + 2x^2 + 4$ is irreducible in $\mathbb{Q}[x]$.

6. Prove the Rational Root Theorem 5.6.

7. Use the Rational Root Theorem 5.6 to factor
$$2x^3 - 17x^2 - 10x + 9.$$

8. Use the Rational Root Theorem 5.6 to argue that
$$x^3 + x + 7$$
is irreducible over $\mathbb{Q}[x]$. Use elementary calculus to argue that this polynomial does have exactly one *real* root.

9. Use the Rational Root Theorem 5.6 (applied to $x^3 - 2$) to argue that $\sqrt[3]{2}$ is irrational.

10. Suppose that $\alpha$ is a real number (which might not be rational), and suppose that it is a root of a polynomial $p \in \mathbb{Q}[x]$; that is, $p(\alpha) = 0$. Suppose further that $p$ is irreducible in $\mathbb{Q}[x]$. Prove that $p$ has minimal degree in the set
$$\{f \in \mathbb{Q}[x] : f(\alpha) = 0 \text{ and } f \neq 0\}.$$

11. This is a continuation of Exercise 10. Suppose as above that $\alpha$ is a real number, $p \in \mathbb{Q}[x]$ is irreducible, and $p(\alpha) = 0$. Suppose also that $f \in \mathbb{Q}[x]$ with $f(\alpha) = 0$. Prove that $p$ divides $f$.

12. Construct polynomials of arbitrarily large degree, which are irreducible in $\mathbb{Q}[x]$.

13. (a) Prove that the equation $a^2 = 2$ has no rational solutions; that is, prove that $\sqrt{2}$ is irrational. (This part is a repeat of Exercise 2.14.)

    (b) Generalize part (a), by proving that $a^n = 2$ has no rational solutions, for all positive integers $n \geq 2$.

14. Let $f \in \mathbb{Z}[x]$ and $n$ an integer. Let $g$ be the polynomial defined by $g(x) = f(x + n)$. Prove that $f$ is irreducible in $\mathbb{Z}[x]$ if and only if $g$ is irreducible in $\mathbb{Z}[x]$.

15. (a) Apply Eisenstein's criterion 5.7 to check that the following polynomials are irreducible:
    $$5x^3 - 6x^2 + 2x - 14 \text{ and } 4x^5 + 5x^3 - 15x + 20.$$

    (b) Make the substitution $x = y + 1$ to the polynomial $x^5 + 5x + 4$ that appears in Example 5.1. Show that the resulting polynomial is irreducible. Now conclude that the original polynomial is irreducible.

    (c) Use the same technique as in part (b) to find a substitution $x = y + m$ so you can conclude the polynomial
    $$x^4 + 6x^3 + 12x^2 + 10x + 5$$
    is irreducible.

    (d) Show that this technique works in general: Prove that if $f(x) \in \mathbb{Z}[x]$, then $f(x)$ is irreducible if and only if $f(y + m)$ is.

16. Prove Theorem 5.7 (Eisenstein's criterion).

17. Let $p$ be a positive prime integer. Then the polynomial
    $$\Phi_p = \frac{x^p - 1}{x - 1}$$
    is called a **cyclotomic polynomial** (see Exercise 4.5).

    (a) Write out, in the usual form for a polynomial, the cyclotomic polynomials for the first three primes.

    (b) Prove that all cyclotomic polynomials $\Phi_p$ are irreducible over $\mathbb{Z}[x]$, using Eisenstein's criterion 5.7 and Exercise 15d for $m = 1$.

# Section I in a Nutshell

This section examines the integers ($\mathbb{Z}$), the integers modulo $m$ ($\mathbb{Z}_m$), and polynomials with rational coefficients ($\mathbb{Q}[x]$). These structures share many algebraic properties:

- Each has addition defined and the addition is commutative.

- Each has multiplicative defined and the multiplication is commutative.

- Each has an additive identity (0 for $\mathbb{Z}$, [0] for $\mathbb{Z}_m$, and the zero polynomial for $\mathbb{Q}[x]$).

- Each has a multiplicative identity (1 for $\mathbb{Z}$, [1] for $\mathbb{Z}_m$, and the polynomial 1 for $\mathbb{Q}[x]$).

- All elements have additive inverses, but not all elements have multiplicative inverses.

Furthermore, $\mathbb{Z}$ and $\mathbb{Q}[x]$ have some notion of 'size'. The size of an integer is given by its absolute value, while the size of a polynomial is given by its degree.

This notion of size along with their similar algebraic properties, allow us to prove a series of parallel theorems for $\mathbb{Z}$ and $\mathbb{Q}[x]$:

(Theorem 2.1) *Division Theorem for* $\mathbb{Z}$: Let $a, b \in \mathbb{Z}$ with $a \neq 0$. Then there exist unique integers $q$ and $r$ with $0 \leq r < |a|$ such that $b = aq + r$.

(Theorem 4.2) *Division Theorem for* $\mathbb{Q}[x]$: Let $f, g \in \mathbb{Q}[x]$ with $f \neq 0$. Then there are unique polynomials $q$ and $r$ with $\deg(r) < \deg(f)$ such that $g = fq + r$.

These Division Theorems allow us to develop Euclid's Algorithm, which is a method to compute the gcd of two integers (Theorem 2.3) or polynomials (Theorem 4.5).

We then developed the notion of an irreducible integer and irreducible polynomial: An integer (polynomial) $p$ is *irreducible* if $\neq \pm 1$ ($\deg(p) > 0$) and whenever $p = ab$, then either $a$ or $b$ is $\pm 1$ (either $a$ or $b$ has degree 0). There is also a shared notion of primeness: An integer (polynomial) is *prime* if $p$ is not 0, 1, or $-1$ ($\deg(p) > 0$) and whenever $p$ divides $ab$, then either $p$ divides $a$ or $p$ divides $b$. For $\mathbb{Z}$ and $\mathbb{Q}[x]$, the idea of primeness and irreducibility are one and the same: (Theorems 2.7 and 5.2). An integer (polynomial) is irreducible if and only if it is prime. This equivalence enables us to prove the *Fundamental Theorem of Arithmetic*, and its analogue for polynomials: Every integer (other than 0, 1, or $-1$) is irreducible or a product of irreducibles (Theorem 2.8). Similarly, any polynomial in $\mathbb{Q}[x]$ of degree greater than 0 is either irreducible or the product of irreducibles (Theorem 5.1). Both of these factorizations are unique:

(Theorem 2.9) If an integer $x = a_1 a_2 \cdots a_n = b_1 b_2 \cdots b_m$ where the $a_i$ and $b_j$ are all irreducible, then $n = m$ and the $b_j$ may be rearranged so that $a_i = \pm b_i$ for $i = 1, 2, \ldots, m$.

(Theorem 5.4) If the polynomial $p = f_1 f_2 \cdots f_n = g_1 g_2 \cdots g_m$ where the $f_i$ and $g_j$ are all irreducible, then $n = m$ and the $g_j$ may be rearranged so that $f_i$ and $g_j$ are associates (that is, non-zero scalar multiples of one another), for $i = 1, 2, \ldots, n$.

Finally, we examined the polynomials with integer coefficients, $\mathbb{Z}[x]$. Unfortunately, there is no Division Theorem for $\mathbb{Z}[x]$ and no GCD identity. However, $\mathbb{Z}[x]$ does have some interesting factorization properties:

(Theorem 5.5) *Gauss's Lemma* If $f \in \mathbb{Z}[x]$ and $f$ can be factored into a product of non-scalar polynomials in $\mathbb{Q}[x]$, then $f$ can be factored into a product of non-scalar polynomials in $\mathbb{Z}[x]$.

(Theorem 5.6) *Rational Root Theorem* Suppose $f = a_0 + a_1 x + \cdots + a_n x^n \in \mathbb{Z}[x]$ and $p, q \in \mathbb{Z}$ with $q \neq 0$, $\gcd(p, q) = 1$ and $f(p/q) = 0$. Then $q$ divides $a_n$ and $p$ divides $a_n$.

(Theorem 5.7) *Eisenstein's Criterion* Suppose $f = a_0 + a_1 x + \cdots + a_n x^n \in \mathbb{Z}[x]$, and $p$ is a prime where (1) $p$ divides $a_k$ for $0 \leq k < n$, (2) $p$ does not divide $a_n$, and (3) $p^2$ does not divide $a_0$. Then $f$ is irreducible in $\mathbb{Q}[x]$.

# Part II

# Rings, Domains, and Fields

# Chapter 6

## Rings

In the previous chapters we have examined several different algebraic objects: $\mathbb{Z}$, $\mathbb{Q}[x]$, and $\mathbb{Z}_m$ (for integers $m > 1$). You are also probably aquainted with the larger sets of real numbers, $\mathbb{R}$, and complex numbers, $\mathbb{C}$. In each of these cases we have a set of elements, which is equipped with two operations called *addition* and *multiplication*. Each of these is an example of an abstract concept called a *ring*. In this chapter we will give a general definition of this concept and look at some basic properties and examples.

### 6.1    Binary Operations

Before we can do this, we must understand better what we mean by an 'operation' defined on a set. A **binary operation** on a set $S$ is a function $\circ : S \times S \mapsto S$, where

$$S \times S = \{(s, t) : s, t \in S\}$$

is the set of all ordered pairs with entries from $S$. Thus, $\circ$ is a function that takes ordered pairs of elements from $S$ to elements of $S$.

If you think about it, that is exactly what an operation like addition (on $\mathbb{Z}$) does: It takes an ordered pair of elements (such as $(4, 6)$), and assigns to that pair another element ($4 + 6 = 10$).

Because we wish to make this function $\circ$ look like our more familiar operations like addition or multiplication, we write the image of an element $(s, t)$ in $S \times S$, under the function $\circ$ as $s \circ t$.

Thus, addition and multiplication on the set $\mathbb{Z}$ (or for that matter on $\mathbb{Z}_m$, $\mathbb{Q}[x]$, $\mathbb{Q}$, or even the natural numbers $\mathbb{N}$) are binary operations. Notice that although in our discussion above we were using the notation $\circ$ for our generic binary operation, we are perfectly happy to denote addition by $+$ as usual. Subtraction is a binary operation on the first four of these sets but is *not* a binary operation on $\mathbb{N}$ because the function $-$ is not defined on such pairs as $(3, 4)$.

Similarly, division is not a binary operation on any of our sets considered so far because of zero: The function $\div$ has no image defined for such ordered pairs as $(1, 0)$. However, $\div$ is a binary operation on the sets $\mathbb{Q}^+$ of strictly positive rational numbers, and $\mathbb{Q}^*$ of non-zero rational numbers.

For a rather different example of a binary operation, consider any nonempty set $S$ and define $s \circ t = s$. That is, $\circ$ assigns the first entry to any ordered pair it is given.

▷ **Quick Exercise.**   Which of the following are binary operations?

**1.** Matrix multiplication, on the set of all $2 \times 2$ matrices.

**2.** $a \circ b = a + b + ab$, on the set $\mathbb{Z}$.

**3.** Dot product, on the set of all vectors in the plane.

**4.** Cross product, on the set of all vectors in space.

**5.** $A \cap B$, on the set of all subsets of $\{1, 2, 3, 4\}$.

**6.** $a \circ b = \sqrt{ab}$, on the set $\mathbb{R}$.

**7.** $a \circ b = \sqrt{ab}$, on the set $\mathbb{R}^+$, the set of all positive real numbers. ◁

Note the crucial importance of both the set on which the operation is defined, as well as the operation itself. A function can be a binary operation only if it gives a value for all possible ordered pairs in the set. We often say a set is **closed** under an operation if this is the case. Thus, $\mathbb{N}$ is closed under addition and multiplication but is not closed under subtraction or division.

You might already suspect that because the definition of binary operation is so general, we should reserve terms like 'addition' and 'subtraction' for very special binary operations which obey nice rules. This is precisely what we do when we define a ring.

---

## 6.2 Rings

A **ring** $R$ is a set of elements on which two binary operations, addition ($+$) and multiplication ($\cdot$), are defined that satisfy the following properties. (The symbols $a, b$, and $c$ represent any elements from $R$.)

**(Rule 1)** $a + b = b + a$

**(Rule 2)** $(a + b) + c = a + (b + c)$

**(Rule 3)** There exists an element $0$ in $R$ such that $a + 0 = a$

**(Rule 4)** For each element $a$ in $R$, there exists an element $x$ such that $a + x = 0$

**(Rule 5)** $(a \cdot b) \cdot c = a \cdot (b \cdot c)$

**(Rule 6)** $a \cdot (b + c) = a \cdot b + a \cdot c$ , $(b + c) \cdot a = b \cdot a + c \cdot a$

Let's introduce some terminology to describe these rules: Rule 1 says that addition is **commutative**, and Rules 2 and 5 say that addition and multiplication, respectively, are **associative**. Rule 3 says an **additive identity** (or **zero**) exists, and Rule 4 says that each element of the ring has an **additive inverse**. Finally, Rule 6 says that multiplication **distributes** over addition on the right and the left.

We will usually write $ab$ instead of $a \cdot b$.

What are some examples of rings?

### Example 6.1

The integers $\mathbb{Z}$, equipped with the usual addition and multiplication. The properties listed above should all be familiar facts about arithmetic in the integers.

### Example 6.2

The rational numbers $\mathbb{Q}$, with the usual addition and multiplication. Once again, we rely on our previous experience with arithmetic to check that all these properties hold.

## Example 6.3

$\mathbb{Z}_6$, the integers modulo 6. In Chapter 3 we constructed what we mean by this set and defined operations $+$ and $\cdot$ on it. Let's check that addition in $\mathbb{Z}_6$ is commutative: By the definition of addition in $\mathbb{Z}_6$, $[a]_6 + [b]_6 = [a + b]_6$. But because addition in $\mathbb{Z}$ is commutative, $[a + b]_6 = [b + a]_6$. And so (again by the definition of addition), $[a]_6 + [b]_6 = [b + a]_6 = [b]_6 + [a]_6$, as required.

The proof we just performed was admittedly tedious. Note that it could have been paraphrased as follows: Addition in $\mathbb{Z}_6$ is defined in terms of addition in $\mathbb{Z}$, and because addition in $\mathbb{Z}$ is commutative, so is addition in $\mathbb{Z}_6$.

▷ **Quick Exercise.** Check that the other five ring axioms are satisfied by $\mathbb{Z}_6$. Note that we have already discussed the existence of the additive identity and additive inverses for $\mathbb{Z}_6$ in Chapter 3. ◁

## Example 6.4

$\mathbb{Z}_m$, the integers modulo $m$, for any integer $m > 1$, with the addition and multiplication defined in Chapter 3. The proofs in Example 6.3 certainly work for any $m$.

## Example 6.5

The set $\{0\}$, where $0 + 0 = 0 \cdot 0 = 0$. This is the world's most boring ring, called the **zero ring**. Because our set has only a single element, and *any* computation we perform gives us 0, all six axioms must certainly hold.

## Example 6.6

The polynomials with rational coefficients, $\mathbb{Q}[x]$, with the addition and multiplication we defined in Chapter 4. When we add polynomials, we just add corresponding coefficients ($x^2$ terms added together, etc.). But then, because addition of rational numbers is commutative and associative, it follows that addition of polynomials is commutative and associative. The polynomial 0 clearly plays the role of the additive identity in $\mathbb{Q}[x]$, and we can obtain an additive inverse for any polynomial by changing the sign of every term (this amounts to just multiplying by -1).

Thus, to show that $\mathbb{Q}[x]$ is a ring, it remains only to prove that multiplication is associative, and that it distributes over addition. Because the multiplication of polynomials is difficult to describe formally (even though it is very familiar), formal proofs that Rules 5 and 6 are satisfied by $\mathbb{Q}[x]$ are exceedingly tedious. We consequently omit them here, and refer you to Exercises 21 and 22.

## Example 6.7

The polynomials with integer coefficients, $\mathbb{Z}[x]$, with the addition and multiplication as in Chapter 5. We first note that whenever we add or multiply two polynomials with integer coefficients, we get another one. That is, this set is closed under addition and multiplication. Because the rational polynomials satisfy Rules 1, 2, 5, and 6, it is quite evident that $\mathbb{Z}[x]$ does too because $\mathbb{Z}[x] \subseteq \mathbb{Q}[x]$. Rule 3 holds because the polynomial $0 \in \mathbb{Z}[x]$, and Rule 4 holds because multiplying a polynomial with integer coefficients by -1 gives more integer coefficients.

**Example 6.8**

The set of all even integers, which we abbreviate as $2\mathbb{Z}$, together with ordinary addition and multiplication.

▷ **Quick Exercise.**   Use an argument modelled on that we used for Example 6.7 to show that $2\mathbb{Z}$ is a ring. ◁

**Example 6.9**

Let $\mathbb{Z} \times \mathbb{Z}$ be the set of ordered pairs with integer entries. That is,

$$\mathbb{Z} \times \mathbb{Z} = \{(a, b) : a, b \in \mathbb{Z}\}.$$

Define addition and multiplication *point-wise*; that is,

$$
\begin{aligned}
(n, m) + (r, s) &= (n + r, m + s) \\
(n, m) \cdot (r, s) &= (nr, ms).
\end{aligned}
$$

Then $\mathbb{Z} \times \mathbb{Z}$ is a commutative ring. You will verify the details in Exercise 13.

**Example 6.10**

Let $R$ and $S$ be arbitrary rings, and let $R \times S$ be the set of ordered pairs with first entry from $R$, and second entry from $S$. Then if we define addition and multipication point-wise (as in Example 6.9), we have created a new ring, called the **direct product** of the rings $R$ and $S$. You will check the details, and further generalize this, in Exercise 15.

▷ **Quick Exercise.**   Write out the addition and multiplication tables for the direct product ring $\mathbb{Z}_2 \times \mathbb{Z}_3$. ◁

It might also be worthwhile to provide a few examples of sets equipped with two operations, which are not rings:

**Example 6.11**

The set $\mathbb{N}$ of natural numbers, equipped with the usual addition and multiplication. This structure satisfies Rules 1, 2, 5, and 6, but Rules 3 and 4 are false.

▷ **Quick Exercise.**   Check that these assertions are true. ◁

**Example 6.12**

> The set $\mathbb{Z}$, with the usual addition, and the operation $\circ$ defined by $a \circ b = a$. This structure satisfies the first five rules. Furthermore, $\circ$ distributes over addition from the right, because
>
> $$(b + c) \circ a = b + c = b \circ a + c \circ a.$$
>
> However, $\circ$ does not distribute over addition from the left. To see this, we need only provide an example:
>
> $$2 \circ (3 + 4) = 2 \neq 2 + 2 = 2 \circ 3 + 2 \circ 4.$$
>
> Thus, $\mathbb{Z}$ equipped with these operations is not a ring.

Let us now look a little more carefully at the rules determining a ring. Rules 1, 2, 5, and 6 specify that addition and multiplication are to satisfy certain nice properties (thus distinguishing them from general binary relations). These properties are universal statements applying to all elements of the ring. Thus, addition is required to be commutative and associative, multiplication is to be associative, and multiplication should distribute over addition. Note that we do not require that multiplication be commutative. If that is the case, we say that we have a **commutative ring**. The rings in Examples 6.1–6.9 are all commutative. Example 6.13 below is a non-commutative ring.

Rule 3 is quite different. It asserts that an element of a particular kind (the **additive identity**) exists in the ring. Without Rule 3, the empty set would qualify as a ring. Rule 4 specifies that additive inverses exist for each element we find in the ring, but it certainly does not require that a ring have any other elements than 0, because $0 + 0 = 0$ means that 0 is its own additive inverse. We already saw in Example 6.5 that a ring can consist of the additive identity only.

**Example 6.13**

> Let $M_2(\mathbb{Z})$ be the set of $2 \times 2$ matrices with integer entries. We equip this set with the usual addition and multiplication of matrices. (In case you have not seen these operations before, or don't remember them well, we have relegated a discussion of them to Exercises 6 and 7.) Note that when we add or multiply two matrices with integer entries, we obtain another one, and so this set is closed under the operations. We claim that $M_2(\mathbb{Z})$ is a ring:
>
> Rules 1 and 2 follow easily because they hold in $\mathbb{Z}$. The zero of $M_2(\mathbb{Z})$ is $\begin{pmatrix} 0 & 0 \\ 0 & 0 \end{pmatrix}$, which can easily be verified. The additive inverse of $\begin{pmatrix} a & b \\ c & d \end{pmatrix}$ is $\begin{pmatrix} -a & -b \\ -c & -d \end{pmatrix}$. That multiplication is associative and that the distributive laws hold are left as Exercise 7.
>
> But note that $M_2(\mathbb{Z})$ is not commutative. For example,
>
> $$\begin{pmatrix} 1 & 2 \\ 3 & 4 \end{pmatrix} \begin{pmatrix} 0 & 1 \\ 1 & 0 \end{pmatrix} = \begin{pmatrix} 2 & 1 \\ 4 & 3 \end{pmatrix} \neq$$
>
> $$\begin{pmatrix} 3 & 4 \\ 1 & 2 \end{pmatrix} = \begin{pmatrix} 0 & 1 \\ 1 & 0 \end{pmatrix} \begin{pmatrix} 1 & 2 \\ 3 & 4 \end{pmatrix}.$$

The previous example shows the relevance of the ring concept to students of linear algebra. Our final example of the chapter connects algebra to the study of calculus and analysis:

**Example 6.14**

> Let $C[0,1]$ be the set of real-valued functions defined on the closed unit interval $[0,1] = \{x \in \mathbb{R} : 0 \leq x \leq 1\}$, which are continuous. Define the sum and product of two functions point-wise: $(f+g)(x) = f(x) + g(x)$ and $(fg)(x) = f(x)g(x)$. You can use theorems from calculus to show that this set is a commutative ring. (See Exercise 9.)

---

## 6.3   Arithmetic in a Ring

We can now begin to talk about the arithmetic in an arbitrary ring. The following theorem shows some of the simple arithmetic operations we're used to doing in $\mathbb{Z}$, which we can now perform in an arbitrary ring:

**Theorem 6.1** *Suppose $R$ is a ring, and $a, b, c \in R$.*

   a. *(Additive cancellation) If $a + b = a + c$, then $b = c$.*

   b. *(Solution of equations) The equation $a + x = b$ always has a unique solution in $R$.*

   c. *(Uniqueness of additive inverse) Every element of $R$ has exactly one additive inverse.*

   d. *(Uniqueness of additive identity) There is only one element of $R$ which satisfies the equations $z + a = a$, for all $a$; namely, the element 0.*

**Proof:**   We will proceed very carefully from the rules defining a ring for the first of these proofs, and then argue less formally. The reader is invited to fill in the careful details.

▷ **Quick Exercise.**   Fill in the details for the proofs below of parts (b), (c), and (d) of the theorem. ◁

   (a): Suppose that $R$ is a ring, $a, b, c \in R$, and $a + b = a + c$. By Rule 4, we know there exists an element $x \in R$ for which $a + x = 0$. By Rule 1 we know that $0 = a + x = x + a$. We can now add $x$ to both sides of our given equation, to obtain $x + (a + b) = x + (a + c)$. By two applications of Rule 2, we then have that $(x + a) + b = (x + a) + c$. But then we have $0 + b = 0 + c$, and so by Rules 1 and 3, we conclude that $b = c$, as required.

   (b): We will now proceed less formally than in the proof for (a). Note that $x = d + b$ will do the job, where $d$ is an additive inverse of $a$. Suppose that $e$ is some other solution to the equation. Then $a + e = a + (d + b)$, and so, by the Additive Cancellation property, $e = d + b$. Thus, our solution is unique.

   (c): Suppose $a$ has two additive inverses: say $x$ and $y$. Then $a + x = 0 = a + y$, and so, by the Additive Cancellation property, $x = y$.

   (d): This is left as Exercise 2.                                                                      □

Note well that although additive cancellation holds in any ring (by part (a) of Theorem 6.1), multiplicative cancellation can certainly fail. For instance, we have already seen that in $\mathbb{Z}_6$, $3 \cdot 2 = 3 \cdot 4 \, (= 0)$, but $2 \neq 4$. Because the axioms in the definition of a ring require less of multiplication, we should expect fewer nice properties for multiplication than for addition.

There are other familiar properties from arithmetic that hold in an arbitrary ring. For example, multiplication by 0 always gives 0, and minus times plus is minus. We leave these for you to prove (see Exercises 1, 3, and 4).

A very careful reader may still be concerned about the rigor of the proof we offered above for part (a). How can we justify adding $x$ to both sides of the given equation? First of all, note that elements $x + (a + b)$ and $x + (a + c)$ exist in the ring, because the ring is closed under addition. The fact that they are equal follows, not from one of our rules for rings, but rather from a property of equality, known as the *Substitution Rule for Equality*. This says that if $a = b$ (that is, if $a$ and $b$ are identical elements of the ring), then we can safely replace any appearances of $a$ in an expression by $b$ and get a new expression equal to the old. This is actually a rule of logic, rather than of ring theory, and in this book we have made no attempt to carefully axiomatize the rules of logic, since this would take us too far afield from algebra. In proofs you write you should feel free to use the ordinary properties of equality with which you have been long familiar, including the *Substitution Rule*. In particular, note that

1. $a = a$, for all $a$ (*reflexivity*),

2. if $a = b$ then $b = a$ (*symmetry*), and

3. if $a = b$ and $b = c$, then $a = c$ (*transitivity*).

(A relation satisfying these three properties is called an **equivalence relation**.)

---

## 6.4 Notational Conventions

Because additive inverses are unique, we can thus denote *the* additive inverse of element $a$ by the unambiguous notation $-a$.

You should exercise some care in using this notation, however. We are *not* interpreting $-a$ as meaning the product of $-1$ and $a$. In an arbitrary ring we have no guarantee that there exist such elements as $-1$ or $1$ (see Example 6.8).

We can now make sense of *subtraction* in an arbitrary ring, simply by interpreting $a - b$ as $a + (-b)$. In Exercise 5, you will show that subtraction in an arbitrary ring obeys rules like those found in $\mathbb{Z}$.

Here are some further handy notational conventions: For $n \in \mathbb{N}$ and ring element $a$, $na = a + a + \ldots + a$ and $a^n = a \cdot a \cdot \ldots \cdot a$, where $a$ appears $n$ times on the right-hand side of the equations. Thus, $3a$ is shorthand for $a + a + a$, and $a^4$ is shorthand for $a \cdot a \cdot a \cdot a$.

Care must be exercised in the use of these conventions. For example, in $2\mathbb{Z}$, we can write

$$2 + 2 + 2 = 3(2),$$

and interpret this as a calculation inside $2\mathbb{Z}$, even though there is no element 3 belonging to the ring.

▷ **Quick Exercise.** In the ring $M_2(\mathbb{Z})$, what are the elements

$$3 \begin{pmatrix} 1 & 2 \\ 3 & 4 \end{pmatrix} \quad \text{and} \quad \begin{pmatrix} 1 & 2 \\ 3 & 4 \end{pmatrix}^3 ? \ \triangleleft$$

## 6.5  The Set of Integers Is a Ring

In Chapters 1 and 2 we relied on your previous experience with the integers when dealing with arithmetic and admitted that this was a flaw if we were attempting to make a careful axiomatic development of the properties of the integers. We now have the language necessary to rectify this logical flaw by stating carefully what we've been assuming all along.

**Axiom of Arithmetic** *The integers $\mathbb{Z}$ under ordinary addition and multiplication is a ring.*

This axiom (together with the Well-ordering Principle) is all we need to prove what we have about the integers. In the future we will be showing that many other rings have 'integer-like' properties.

### Historical Remarks

In this chapter we defined the abstract concept of ring, as a way of generalizing the arithmetic properties possessed by our particular examples $\mathbb{Z}$ and $\mathbb{Q}[x]$. Historically, this definition was a long time in coming, but, in broad strokes, we are being accurate to the historical development in basing our definition on $\mathbb{Z}$ and $\mathbb{Q}[x]$. This is true because in the 19th century, the formal definition of ring (and its ensuing popularity in mathematical circles) grew out of two subjects, related respectively to $\mathbb{Z}$ and $\mathbb{Q}[x]$. The first subject is *number theory*. Such mathematicians as the Germans Ernst Kummer and Richard Dedekind discovered that number-theoretic questions about $\mathbb{Z}$ are related to such rings as $\mathbb{Z}[i]$ (described in Exercise 12). We will follow this topic further in ensuing chapters. The second subject is the geometry related to polynomial equations (especially in more than one variable). $\mathbb{Q}[x]$ is a starting point for this subject (called *algebraic geometry*), which we won't pursue in this book. The crucial historical figure here is another German, named David Hilbert. Hilbert was a great believer in the power of axiomization and abstraction, and his success profoundly influenced the course of 20th-century mathematics. If you are having some difficulty understanding the utility of such an abstract concept as ring, you should take solace in the fact that many of Hilbert's late 19th-century colleagues were also reluctant to follow him down the road of abstraction! But this road has many wonders, just around the corner.

### Chapter Summary

In this chapter we defined what we mean by a *ring*: a set equipped with two operations called addition and multiplication, which satisfy certain natural axioms. We examined numerous examples of rings (including $\mathbb{Z}$ and $\mathbb{Q}[x]$) and began the study of the arithmetic of an arbitrary ring.

### Warm-up Exercises

a. Explain why our definition of a binary operation guarantees that the set is closed under the operation.

b. Are the following binary operations?

(a) $a * b = 1$, on the set $\mathbb{Z}$.

(b) $a * b = a/b$, on the set $\mathbb{Q}$.

(c) $a * b = a + bi$, on the set $\mathbb{R}$.

c. Give examples of binary operations satisfying the following:

    (a) A non-commutative binary operation.

    (b) A non-associative binary operation.

d. Give examples of rings satisfying the following:

    (a) A ring with finitely many elements.

    (b) A non-commutative ring.

e. Are the following rings?

    (a) $3\mathbb{Z}$, the set of all integers divisible by 3, together with ordinary addition and multiplication.

    (b) The set of all irreducible integers, together with ordinary addition and multiplication.

    (c) $\mathbb{R}$, with the operations of addition and division.

    (d) The set $\mathbb{R}^*$ of non-zero real numbers, with the operations of multiplication, and the operation $a \circ b = 1$. *Note*: We are trying to use ordinary multiplication as the 'addition' in this set!

    (e) The set of polynomials in $\mathbb{Q}[x]$ where the constant term is an integer, with the usual addition and multiplication of polynomials.

    (f) The set of all matrices in $M_2(\mathbb{Z})$ whose lower left-hand entry is zero, with the usual matrix addition and multiplication.

f. Compute $4a$ and $a^4$ for the following elements $a$ of the following rings:

    (a) $1/2 \in \mathbb{Q}$.

    (b) $2 \in \mathbb{Z}_8$.

    (c) $2 \in \mathbb{Z}_{16}$.

    (d) $2 \in \mathbb{Z}_3$.

    (e) $1 + 3x^2 \in \mathbb{Q}[x]$.

    (f) $\begin{pmatrix} 1 & 2 \\ -1 & 3 \end{pmatrix} \in M_2(\mathbb{Z})$.

# Exercises

1. Show that in a ring, $0a = a0 = 0$.

2. Prove part (d) of Theorem 6.1: Show that in a ring the additive identity is unique, by supposing that both $0$ and $0'$ satisfy Rule 3 and proving that $0 = 0'$.

3. Show that in a ring, $(-a)b = a(-b) = -(ab)$.

4. Show that in a ring, $(-a)(-b) = ab$.

5. Prove the following facts about subtraction in a ring $R$, where $a, b, c \in R$:

   (a) $a - a = 0$.

   (b) $a(b - c) = ab - ac$.

   (c) $(b - c)a = ba - ca$.

6. Given two matrices in $M_2(\mathbb{Z})$,

$$A = \begin{pmatrix} a_{11} & a_{12} \\ a_{21} & a_{22} \end{pmatrix} \quad \text{and} \quad B = \begin{pmatrix} b_{11} & b_{12} \\ b_{21} & b_{22} \end{pmatrix},$$

   define their **matrix sum** to be the matrix

$$A + B = \begin{pmatrix} a_{11} + b_{11} & a_{12} + b_{12} \\ a_{21} + b_{21} & a_{22} + b_{22} \end{pmatrix}.$$

   Verify that this is a binary operation, which is associative, commutative, has an additive identity, and has additive inverses.

7. Given two matrices in $M_2(\mathbb{Z})$,

$$A = \begin{pmatrix} a_{11} & a_{12} \\ a_{21} & a_{22} \end{pmatrix} \quad \text{and} \quad B = \begin{pmatrix} b_{11} & b_{12} \\ b_{21} & b_{22} \end{pmatrix},$$

   define their **matrix product** to be the matrix

$$AB = \begin{pmatrix} a_{11}b_{11} + a_{12}b_{21} & a_{11}b_{12} + a_{12}b_{22} \\ a_{21}b_{11} + a_{22}b_{21} & a_{21}b_{12} + a_{22}b_{22} \end{pmatrix}.$$

   Verify that this is a binary operation, which is associative and distributes over matrix addition, but is not commutative.

8. We generalize Exercises 6 and 7: Let $R$ be any commutative ring (other than the zero ring). Define $M_2(R)$ as the set of $2 \times 2$ matrices with entries from $R$. Show that $M_2(R)$ is a ring which is not commutative. (Note that for the most part the proofs in Exercises 6 and 7 lift over without change.)

9. Check that Example 6.14 is indeed a ring; that is, let $C[0, 1]$ be the set of functions defined from the closed unit interval $[0, 1]$ to the real numbers that are continuous. Define the sum and product of two functions point-wise: $(f + g)(x) = f(x) + g(x)$ and $(fg)(x) = f(x)g(x)$. Show that $C[0, 1]$ is a commutative ring. (You may use theorems from calculus.)

10. Let $\mathcal{D}$ be the set of functions defined from the real numbers to the real numbers that are differentiable. Define addition and multiplication of functions point-wise, as in the previous exercise. Show that $\mathcal{D}$ is a commutative ring. (You may use theorems from calculus.)

11. Let $\mathbb{C}$ be the **complex numbers**. That is,

$$\mathbb{C} = \{a + bi : a, b \in \mathbb{R}\},$$

where $i$ is the square root of $-1$ (that is, $i \cdot i = -1$). Here,

$$(a + bi) + (c + di) = (a + c) + (b + d)i$$

and

$$(a + bi)(c + di) = (ac - bd) + (ad + bc)i.$$

Show that $\mathbb{C}$ is a commutative ring.

12. Let

$$\mathbb{Z}[i] = \{a + bi \in \mathbb{C} : a, b \in \mathbb{Z}\}.$$

Show that $\mathbb{Z}[i]$ is a commutative ring (see Exercise 11). This is called the ring of **Gaussian integers**.

13. Verify that Example 6.9 is a ring. Recall that $\mathbb{Z} \times \mathbb{Z}$ is the set of ordered pairs with integer entries. That is,

$$\mathbb{Z} \times \mathbb{Z} = \{(a, b) : a, b \in \mathbb{Z}\}.$$

Define addition and multiplication coordinate-wise; that is,

$$
\begin{aligned}
(n, m) + (r, s) &= (n + r, m + s) \\
(n, m) \cdot (r, s) &= (nr, ms).
\end{aligned}
$$

Show that $\mathbb{Z} \times \mathbb{Z}$ is a commutative ring.

14. We generalize Exercise 13: Let $\mathbb{Z}^n$ be the set of ordered $n$-tuples with integer entries. Define addition and multiplication on $\mathbb{Z}^n$ coordinate-wise and show $\mathbb{Z}^n$ is a commutative ring. Similarly, define $R^n$ for any ring $R$. Show $R^n$ is a ring and is commutative if $R$ is.

15. Verify that Example 6.10 is a ring. Namely, let $R$ and $S$ be arbitrary rings. Define addition and multiplication appropriately to make $R \times S$ a ring, where $R \times S$ is the set of ordered pairs with first entry from $R$ and second entry from $S$. Now generalize this to the set $R_1 \times R_2 \times \cdots \times R_n$ of $n$-tuples with entries from the rings $R_i$. This new ring is called the **direct product** of the rings $R_i$.

16. Find an example in $M_2(\mathbb{Z})$ to show that $(a+b)^2$ is *not* necessarily equal to $a^2 + 2ab + b^2$. (Recall that $2ab = ab + ab$.) What is the correct expansion of $(a + b)^2$ for an arbitrary ring? What can you say if the ring is commutative?

17. (This exercise extends the discussion of Exercise 16.) Let $R$ be a commutative ring and $a, b \in R$. Then prove the *Binomial Theorem* for $R$, by induction on $n$: Namely, show that

$$(a + b)^n = \sum_{k=0}^{n} \binom{n}{k} a^{n-k} b^k.$$

(This is essentially a repeat of Exercise 1.14c, in a more general context.)

18. Suppose that $a \cdot a = a$ for every element $a$ in a ring $R$. (Elements $a$ in a ring where $a^2 = a$ are called **idempotent**.)

    (a) Show that $a = -a$.

    (b) Now show that $R$ is commutative.

19. Let $S = \{(x_1, x_2, x_3, \ldots) : x_i \in \mathbb{R}\}$, the real-valued sequences. Define addition and multiplication on $S$ coordinate-wise (see Exercises 13 and 14). Show that $S$ is a commutative ring.

20. Let $X$ be some arbitrary set, and $P(X)$ the set of all subsets of $X$. In Example 1.1 we proved that if $X$ has $n$ elements, then $P(X)$ has $2^n$ elements; we are here allowing the possibility that $X$ (and hence $P(X)$) has *infinitely* many elements. Define operations on $P(X)$ as follows, where $a, b \in P(X)$:

$$a + b = (a \cup b) \backslash (a \cap b) \quad \text{and} \quad ab = a \cap b.$$

(Addition here is often called the **symmetric difference** of the two sets $a, b$.) Prove that $P(X)$ is a commutative ring. ($P(X)$ is called the **power set** for the set $X$.)

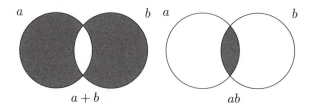

21. Prove that multiplication is associative in $\mathbb{Q}[x]$, as claimed in Example 6.6. Suppose that

$$f = a_0 + a_1 x + \cdots + a_n x^n,$$
$$g = b_0 + b_1 x + \cdots + b_n x^m,$$

and

$$h = c_0 + c_1 x + \cdots + c_n x^r.$$

Now prove that $f(gh) = (fg)h$.

22. Prove that multiplication distributives over addition in $\mathbb{Q}[x]$, as claimed in Example 6.6.

23. Let $R$ be any commutative ring. Let $R[x]$ be the collection of polynomials with coefficients from $R$. Show that $R[x]$ is a ring.

24. Let $\mathcal{B}$ be the set of functions defined from the real numbers to the real numbers that are continuous and **bounded**; given a continuous function $f$, it is bounded if there exists a positive number $M$ for which $f(x) \leq M$, for all real numbers $x$. With the usual addition and multiplication of functions, prove that $\mathcal{B}$ is a ring.

25. In a manner entirely parallel to Example 6.8 you can show $3\mathbb{Z}$ is a ring. Is $2\mathbb{Z} \cup 3\mathbb{Z}$ a ring?

26. Show that $2\mathbb{Z} \cap 3\mathbb{Z}$ is a ring. What is a more concise way to denote this ring?

27. Suppose that $R$ is a ring and that $a^3 = a$ for all $a \in R$. Prove that $R$ is commutative. (Compare this to Exercise 18.)

# Chapter 7

## Subrings and Unity

Consider Example 6.7: We discovered that it was relatively easy to show that $\mathbb{Z}[x]$ is a ring, because it is a subset of $\mathbb{Q}[x]$, which we had already shown is a ring. Because the operations in $\mathbb{Z}[x]$ are the same as in $\mathbb{Q}[x]$, we didn't have to check again the associative laws, the distributive laws, or that addition was commutative. Because the addition and multiplication of $\mathbb{Q}[x]$ have these properties, the addition and multiplication of $\mathbb{Z}[x]$ inherit them automatically. What did need to be checked was that addition and multiplication were closed in $\mathbb{Z}[x]$, that the additive identity of $\mathbb{Q}[x]$ was also in $\mathbb{Z}[x]$, and that the additive inverses of elements of $\mathbb{Z}[x]$ were also in $\mathbb{Z}[x]$. Similarly, in Example 6.8 you showed that $2\mathbb{Z}$ is a ring, taking advantage of the fact that $2\mathbb{Z} \subseteq \mathbb{Z}$.

## 7.1 Subrings

We generalize this situation: A subset $S$ of a ring $R$ is said to be a **subring** of $R$ if $S$ is itself a ring under the operations induced from $R$.

### Example 7.1

$\mathbb{Z}[x]$ is a subring of $\mathbb{Q}[x]$.

### Example 7.2

$2\mathbb{Z}$ is a subring of $\mathbb{Z}$.

### Example 7.3

$\mathbb{Z}$ is a subring of $\mathbb{Q}$, which is in turn a subring of $\mathbb{R}$.

### Example 7.4

The Gaussian integers $\mathbb{Z}[i]$ are a subring of $\mathbb{C}$. (See Exercises 6.11 and 6.12.)

### Example 7.5

Let $R$ be any ring. Then $\{0\}$ and $R$ are always subrings of $R$. We call $\{0\}$ the **trivial** subring, and $R$ the **improper** subring. All subrings other than $R$ we call **proper** subrings.

In the general case, what exactly do we need to check to see that a subset of a ring is a subring? We need not go through all the work we did in showing that $\mathbb{Z}[x]$ is a subring of $\mathbb{Q}[x]$, as listed above. The following theorem provides a simpler answer. We will often use this theorem to check whether something is a ring (by considering it as a subset of a larger well-known ring):

**Theorem 7.1     The Subring Theorem**     *A non-empty subset of a ring is a subring under the same operations if and only if it is closed under multiplication and subtraction.*

**Proof**:     It is obvious that a subring is closed under multiplication and subtraction. For the converse, suppose that $R$ is a ring and $S$ a non-empty subset, which is closed under multiplication and subtraction. We wish to show that $S$ is a ring. Now, because $S$ is non-empty, we can then choose an element of it, which we call $s$. First note that because $S$ is closed under subtraction, then $s - s = 0 \in S$. That is, the additive identity belongs to $S$ (Rule 3). Next, suppose that $a \in S$. Because $S$ is closed under subtraction, $-a = 0 - a \in S$. This means that $S$ is closed under taking additive inverses (Rule 4). Now suppose that $a, b \in S$. Then we've just seen that $-b \in S$. But then

$$a + b = a - (-b) \in S,$$

and so $S$ is closed under addition as well.

To show that $S$ is a ring, it remains to show that addition is commutative, that addition and multiplication are associative, and that multiplication distributes over addition. But all these properties hold in $R$, and so are automatically inherited for $S$.     □

The most important use of this theorem is to check that some set is in fact a ring, by viewing it as a subring of some larger previously known ring. We will see this principle illustrated repeatedly in the examples below.

Because commutativity is automatically inherited by a subset, it follows that a subring of a commutative ring is also commutative. Note that a subring of a non-commutative ring may be commutative, because the zero ring is a commutative subring of *any* ring. Example 7.9 is a more interesting example of this.

**Example 7.6**

Let $m\mathbb{Z} = \{mn : n \in \mathbb{Z}\}$, where $m$ is an integer greater than 1. That is, $m\mathbb{Z}$ is the set of integer multiples of $m$. We have already seen this example in case $m = 2$ as Example 7.2. We claim that $m\mathbb{Z}$ is a subring of $\mathbb{Z}$. For if $ma, mb \in m\mathbb{Z}$, then

$$ma - mb = m(a - b) \in m\mathbb{Z}$$

and so $m\mathbb{Z}$ is closed under subtraction. Similarly,

$$(ma)(mb) = m(mab) \in m\mathbb{Z},$$

and so $m\mathbb{Z}$ is closed under multiplication.

**Example 7.7**

$6\mathbb{Z}$ is a subring of both $3\mathbb{Z}$ and $2\mathbb{Z}$.

▷ **Quick Exercise.**   Check that $6\mathbb{Z}$ is a subring of $3\mathbb{Z}$ and $2\mathbb{Z}$, modelling your proof on that of Example 7.6. ◁

We generalize this example in Exercise 5.

## Example 7.8

{0, 2, 4} is a subring of $\mathbb{Z}_6$. It is easy to check directly that this set is closed under subtraction and multiplication.

## Example 7.9

Let $D_2(\mathbb{Z})$ be the set of all 2-by-2 matrices with entries from $\mathbb{Z}$ and with all entries off the main diagonal being zero. We call these the **diagonal matrices**. We claim that this is a subring of $M_2(\mathbb{Z})$.

▷ **Quick Exercise.** Check that the set of diagonal matrices is closed under subtraction and multiplication. ◁

Note that $D_2(\mathbb{Z})$ is in fact a commutative ring:

$$\begin{pmatrix} a & 0 \\ 0 & b \end{pmatrix} \begin{pmatrix} c & 0 \\ 0 & d \end{pmatrix} = \begin{pmatrix} ac & 0 \\ 0 & bd \end{pmatrix} = \begin{pmatrix} c & 0 \\ 0 & d \end{pmatrix} \begin{pmatrix} a & 0 \\ 0 & b \end{pmatrix}.$$

Thus, $D_2(\mathbb{Z})$ is a commutative subring (bigger than the zero ring) of a noncommutative ring.

## Example 7.10

The direct product $\mathbb{Z} \times \mathbb{Q}$ is a subring of the ring $\mathbb{R} \times \mathbb{R}$. (See Example 6.10.)

Here are a few examples of subsets of rings which are *not* subrings. Note how many different ways a subset can fail to be a subring.

## Example 7.11

Consider the set $\mathbb{Q}^+$ of all strictly positive elements from $\mathbb{Q}$. This is clearly not a subring, because the additive identity 0 does not belong to $\mathbb{Q}^+$.

## Example 7.12

Consider the set $\mathbb{Q}^+ \cup \{0\} \subseteq \mathbb{Q}$. This set does include 0 and is in fact closed under multiplication. But it is *not* closed under subtraction and so is not a subring.

## Example 7.13

Consider the set

$$\pi\mathbb{Z} = \{\pi n : n \in \mathbb{Z}\},$$

that is, the set of all integer multiples of the real number $\pi$. This is a subset of the ring $\mathbb{R}$ of all real numbers. It is closed under subtraction.

▷ **Quick Exercise.** Check that $\pi\mathbb{Z}$ is closed under subtraction. ◁

However, it is not closed under multiplication, for if it were, then $\pi^2$ would be an integer multiple of $\pi$. But then $\pi$ itself would be an integer, which is false.

**Example 7.14**

Consider $\mathbb{Z}_4 = \{0, 1, 2, 3\}$. Is it a subring of $\mathbb{Z}$? Clearly not, because although our notation makes it *look* as if $\mathbb{Z}_4$ is a subset of $\mathbb{Z}$, this is not actually the case. Recall from Section 3.2 that when working with the rings $\mathbb{Z}_n$ we often drop the square brackets around the integers, if it is clear from context what we mean. In $\mathbb{Z}_4$, 3 means a residue class, and so is an infinite set of integers. This infinite set of integers is certainly not the same as the individual integer 3.

## 7.2   The Multiplicative Identity

We now raise another topic, which further illustrates the different roles that addition and multiplication play in a ring. It is an important part of the definition of a ring that it have an additive identity or 0. We make no such assumption about a multiplicative identity; many rings do possess a multiplicative identity, however. We call an element $u$ of a ring $R$ a **unity** or **multiplicative identity** if $ua = au = a$ for all elements $a \in R$.

**Example 7.15**

Obviously, the integer 1 is the unity of $\mathbb{Z}$. Similarly, 1 is the unity for the rings $\mathbb{Q}$, $\mathbb{R}$, and $\mathbb{C}$. In $\mathbb{Q}[x]$, the constant polynomial 1 is the unity. In the ring $M_2(\mathbb{Z})$, the unity is $\begin{pmatrix} 1 & 0 \\ 0 & 1 \end{pmatrix}$. In Chapter 3, we were careful to point out that the residue class [1] plays the role of multiplicative identity in $\mathbb{Z}_m$. On the other hand, the ring $2\mathbb{Z}$ has no unity, because in the integers $2a = 2$ holds exactly when $a = 1$, an element which $2\mathbb{Z}$ lacks. More generally, $m\mathbb{Z}$ lacks unity for all $m > 1$.

Note that for the particular examples we've discussed, we have spoken as if the unity element of a ring (if it exists) is unique. This is in fact the case and can be proved by a proof very similar to the one we used to show that the zero of a ring is unique. We leave this proof as Exercise 17. Because unity is unique, for an arbitrary ring we can thus denote it unambiguously by 1.

There is a surprising difficulty with this, however. Consider the zero ring $\{0\}$. In this ring 0 is both the additive identity *and* the multiplicative identity, and so in this case $0 = 1$. In order to avoid this trivial case, we will henceforth reserve the term 'unity' for rings other than the zero ring. In Exercise 16, you will prove that in a ring with more than one element, the additive identity and multiplicative identity cannot be the same.

## 7.3   Surjective, Injective, and Bijective Functions

Algebraists place more emphasis on structure and properties than on details of notation. When discussing examples of an abstract structure like a ring, they often assert two rings are essentially the same, except for the details of how they are described. This is made precise by the idea of a *ring isomorphism*.

In order to understand this, we need to review some basic concepts and terminology regarding functions. For this purpose, suppose that $X$ and $Y$ are arbitrary sets, and $\varphi : X \to Y$ is a function. Recall that we call the set $X$ the **domain** of the function, and $Y$ the **range**.

It may well be the case that $Y$ contains elements that do not occur as $\varphi(x)$ for any $x \in X$. Writing $\{\varphi(x) : x \in X\}$ as $\varphi(X)$, we are saying that it may be the case that $\varphi(X) \subset Y$. If $\varphi(X) = Y$, we say that $\varphi$ is a **surjective**. Alternatively, we say that $\varphi$ is an **onto** function.

For each element $y$ of $\varphi(X)$, there exists an element $x$ of $X$ such that $\varphi(x) = y$ (this is just the definition of $\varphi(X)$). If this element is unique for each such $y \in \varphi(X)$, we say that $\varphi$ is an **injective** function. Alternatively, we call $\varphi$ a **one-to-one** function. Another way of saying this is to assert that $\varphi$ never takes two distinct elements of $X$ to the same element of $Y$.

If the function $\varphi$ is both surjective and injective, we say that it is a **bijective** function. Alternatively, we say that $\varphi$ is a **one-to-one correspondence**. Note that if $X$ is a finite set and $\varphi$ is a bijection, then $X$ and $Y$ have the same number of elements. Establishing a one-to-one correspondence is exactly the way we count sets!

In the case that $\varphi : X \to Y$ is a bijection, then we can define a function $\psi : Y \to X$ by letting $\psi(y)$ be the unique element $x$ of $X$ for which $\varphi(x) = y$. Because $\varphi$ is surjective, there exists at least one such $x$; since $\varphi$ is injective, there is exactly one such, and so $\psi$ is well-defined as a function. We usually call $\psi$ the **inverse function** of $\varphi$ and even sometimes write this as $\psi = \varphi^{-1}$.

---

## 7.4 Ring Isomorphisms

We now wish to make precise the idea that two rings $R$ and $S$ are essentially the same; we do this by using a function between them. We say that two rings $R$ and $S$ are **isomorphic** if there is a function $\varphi$ with $\varphi : R \to S$ which preserves algebra ($\varphi(a+b) = \varphi(a) + \varphi(b)$ and $\varphi(ab) = \varphi(a)\varphi(b)$ for all $a, b \in R$), and which preserves set theory ($\varphi$ is bijective). These properties together mean that there is a one-to-one correspondence between the rings as sets, and that furthermore all algebra in $R$ corresponds to algebra in $S$. We call the function $\varphi$ a **ring isomorphism**.

Here is a simple example that illustrates why this idea is useful. Consider the ring $\mathbb{Q}$ of rational numbers, and the subring $C$ of $\mathbb{Q}[x]$ consisting of the constant polynomials, that is polynomials $f = a_0 + a_1 x + a_2 x^2 + \cdots$ where $a_1 = a_2 = \cdots = 0$. It seems intuitively clear that these two rings are essentially the same, even though the first ring is a set of numbers and the second is a set of polynomials. We make this precise by defining a bijection $\varphi$ defined on $\mathbb{Q}$ where $\varphi(q) = q + 0x + 0x^2 + \cdots$. It is clear that this function preserves both addition and multiplication, and so we can say that the rings $\mathbb{Q}$ and $C$ are isomorphic. This first example seems quite obvious, but does show how the concept of ring isomorphism gives us the language to express such ideas clearly and precisely.

### Example 7.16

Here is a more interesting example. Consider the subring $S$ of $M_2(\mathbb{Z})$ consisting of those matrices which have zero entries everywhere except in the upper lefthand corner.

▷ **Quick Exercise.** Check using the subring theorem that $S$ is in fact a subring of $M_2(\mathbb{Z})$. ◁

As you compute in the ring $S$ it quickly becomes clear that all the action occurs in the upper lefthand corner, and that for all intents and purposes we are just doing arithmetic in $\mathbb{Z}$. We make this more precise by defining the function $\varphi : \mathbb{Z} \to S$ by setting

$$\varphi(n) = \begin{pmatrix} n & 0 \\ 0 & 0 \end{pmatrix}.$$

This function is clearly bijective. Note that $\varphi$ also preserves both addition and multiplication:

$$\varphi(n+m) = \begin{pmatrix} n+m & 0 \\ 0 & 0 \end{pmatrix} = \begin{pmatrix} n & 0 \\ 0 & 0 \end{pmatrix} + \begin{pmatrix} m & 0 \\ 0 & 0 \end{pmatrix} = \varphi(n) + \varphi(m).$$

A similar calculation shows $\varphi(nm) = \varphi(n)\varphi(m)$.

▷ **Quick Exercise.** Show this. ◁

We thus say that $\mathbb{Z}$ and $S$ are isomorphic rings, and mean by that that they are algebraically 'essentially the same.'

## Example 7.17

Consider the ring $D_2(\mathbb{Z})$ from Example 7.9. By defining the obvious ring isomorphism, we can show that this ring is isomorphic to the ring $\mathbb{Z} \times \mathbb{Z}$, from Example 6.9.

▷ **Quick Exercise.** Carry this out. ◁

The definition of ring isomorphism is strong enough to imply that *all* significant algebraic structure is preserved. For example, suppose we have a ring isomorphism $\varphi : R \to S$ for two rings $R$ and $S$. If we denote the additive identities of these rings by $0_R$ and $0_S$, we claim that $\varphi(0_R) = 0_S$. To prove this, pick any $s \in S$. Because $\varphi$ is onto, there exists $r \in R$ with $\varphi(r) = s$. But then

$$\varphi(0_R) + s = \varphi(0_R) + \varphi(r) = \varphi(0_R + r) = \varphi(r) = s,$$

and so $\varphi(0_R) = 0_S$, because the additive identity in a ring is unique. In Exercise 26 you will be asked to use a similar proof to show that if $R$ has unity $1_R$, then $\varphi(1_R)$ is unity for the ring $S$.

Our definition of ring isomorphism appears to depend on the order of the rings, because the ring isomorphism above is defined as a function from $R$ to $S$. But this is actually symmetric, because we can always define a ring isomorphism $\psi : S \to R$, where $\psi$ is the inverse function we have discussed in Section 7.3 above. We know that $\psi$ makes sense as a bijection, and it is straightforward to prove that $\psi$ preserves addition and multiplication as well; you will do this in Exercise 27.

---

## Chapter Summary

In this chapter we defined the notion of *subring* and proved that the subrings of a ring are exactly those non-empty subsets closed under subtraction and multiplication. We provided many examples of subrings.

We also defined the notion of *unity* (or multiplicative identity) and observed that many rings have unity, and many don't.

Finally, we introduced the notion of *ring isomorphism*, to provide us with precise language for saying that two rings are 'essentially the same.'

# Warm-up Exercises

a. Give examples of the following:

    (a) A non-empty subset of a ring, closed under subtraction, but not multiplication.

    (b) A non-empty subset of a ring, closed under multiplication, but not subtraction.

b. Are the following subsets subrings?

    (a) $\mathbb{Z} \subseteq \mathbb{Q}$.

    (b) $\mathbb{Q}^* \subseteq \mathbb{Q}$; recall that $\mathbb{Q}^*$ is the set of non-zero rational numbers.

    (c) $\mathbb{Q}^+ \subseteq \mathbb{Q}$; recall that $\mathbb{Q}^+$ is the set of strictly positive rational numbers.

    (d) The set of irrational numbers, a subset of $\mathbb{R}$.

    (e) $\{0, 1, 2, 3\} \subseteq \mathbb{Z}_8$.

    (f) The linear polynomials, a subset of $\mathbb{Q}[x]$.

c. Find all the subrings of these rings: $\mathbb{Z}_5$, $\mathbb{Z}_6$, $\mathbb{Z}_7$, $\mathbb{Z}_{12}$.

d. Give examples of the following (or explain why they don't exist):

    (a) A commutative subring of a non-commutative ring.

    (b) A non-commutative subring of a commutative ring.

    (c) A subring without unity, of a ring with unity. (See Exercise 22 of this chapter for the converse possibility.)

    (d) A ring (with more than one element) whose only subrings are itself, and the zero subring. *Hint:* Look at an earlier Warm-up Exercise.

e. What is the unity of the power set ring $P(X)$ considered in Exercise 6.20?

f. What is the unity of the ring $\mathbb{Z} \times \mathbb{Z}$? (See Example 6.9.) What about of $R \times S$, where $R$ and $S$ are rings with unity? (See Example 6.10.)

g. Consider the one-to-one correspondence $\varphi : \mathbb{Z} \to 2\mathbb{Z}$ given by $\varphi(n) = 2n$. Explain why this is not a ring isomorphism.

h. Consider the subring of $\mathbb{Z} \times \mathbb{Z}$ consisting of all elements of the form $(n, 0)$. Argue that this ring is isomorphic to $\mathbb{Z}$.

i. Consider the functions $f, g, h$, and $k$ with range and domain $\mathbb{R}$, defined by

$$f(x) = x^2, \quad g(x) = x^3, \quad h(x) = e^x \quad k(x) = x^3 - x.$$

Which of these functions is injective? Which of these functions is surjective? (*Note:* We are *not* claiming that these functions are ring homomorphisms.)

j. Suppose that $f : X \to Y$ is an injective function, and $X$ is a finite set with $n$ elements. What can you say about the number of elements in $Y$?

k. Suppose that $f : X \to Y$ is a surjective function, and $X$ is a finite set with $n$ elements. What can you say about the number of elements in $Y$?

# Exercises

1. Let $\mathbb{Z}[\sqrt{2}] = \{a + b\sqrt{2} : a, b \in \mathbb{Z}\}$. Show that $\mathbb{Z}[\sqrt{2}]$ is a commutative ring by showing it is a subring of $\mathbb{R}$.

2. We generalize Exercise 1: Let $\mathbb{Z}[\sqrt{n}] = \{a + b\sqrt{n} : a, b \in \mathbb{Z}\}$, where $n$ is some fixed integer (positive or negative). Show $\mathbb{Z}[\sqrt{n}]$ is a commutative ring by showing it is a subring of $\mathbb{C}$. (See Exercise 6.12 for the case $n = -1$.)

3. Let
$$\alpha = \sqrt[3]{5}$$
and
$$\mathbb{Z}[\alpha] = \{a + b\alpha + c\alpha^2 : a, b, c \in \mathbb{Z}\} \subseteq \mathbb{R}.$$
Prove that $\mathbb{Z}[\alpha]$ is a subring of $\mathbb{R}$.

4. Show that $m\mathbb{Z}$ is a subring of $n\mathbb{Z}$ if and only if $n$ divides $m$. (See Example 7.7.)

5. (a) Show that
$$4\mathbb{Z} \cap 6\mathbb{Z} = 12\mathbb{Z}.$$

   (b) Let $m$ and $n$ be two positive integers. Show that
$$m\mathbb{Z} \cap n\mathbb{Z} = l\mathbb{Z},$$
   where $l$ is the least common multiple of $m$ and $n$. (See Exercise 2.11.)

6. Let $S$ be the set of all polynomials in $\mathbb{Q}[x]$ which have 0 as constant term (that is, polynomials of the form $a_1x + a_2x^2 + \cdots a_nx^n$). Show that $S$ is a subring of $\mathbb{Q}[x]$.

7. Let $f$ be some polynomial with rational coefficients, with $\deg(f) > 0$, and let $S$ be the set of all polynomials $g$ in $\mathbb{Q}[x]$ for which $f$ divides $g$. Show that $S$ is a subring of $\mathbb{Q}[x]$. How is this exercise related to the previous exercise?

8. (a) Show that the set
$$\{(a, a) : a \in \mathbb{Z}\}$$
   is a subring of $\mathbb{Z} \times \mathbb{Z}$.

   (b) Show that the subring in part (a) is isomorphic to the ring $\mathbb{Z}$.

   (c) Now consider the set
$$\{(a, -a) : a \in \mathbb{Z}\}.$$
   Show that this set is closed under subtraction, but not closed under multiplication, and so is *not* a subring of $\mathbb{Z} \times \mathbb{Z}$.

9. Show that the intersection of any two subrings of a ring is a subring.

10. Show by example that the union of any two subrings of a ring need *not* be a subring. *Hint:* You can certainly find such an example by working in $\mathbb{Z}$.

11. Let
$$\mathbb{Z}_{\langle 2 \rangle} = \left\{q \in \mathbb{Q} : q = \frac{a}{b}, \quad a, b \in \mathbb{Z}, \ b \text{ is odd}\right\}.$$

    (a) Why is $\mathbb{Z}_{\langle 2 \rangle}$ a proper subset of $\mathbb{Q}$?

    (b) Show that $\mathbb{Z}_{(2)}$ is a subring of $\mathbb{Q}$.

    (c) Show that

$$\left\{q \in \mathbb{Q} : q = \frac{a}{b}, \quad a, b \in \mathbb{Z}, \ a \text{ is odd}, \ b \neq 0\right\}$$

    is *not* a subring of $\mathbb{Q}$.

    (d) Show that

$$\left\{q \in \mathbb{Q} : q = \frac{a}{2^n}, \quad a \in \mathbb{Z}, \ n = 0, 1, 2, \cdots\right\}$$

    is a subring of $\mathbb{Q}$.

12. Let $R$ be an arbitrary ring, and define

$$Z(R) = \{r \in R : rx = xr, \text{for all } x \in R\};$$

    this subset is called the **center** of the ring $R$. Show that $Z(R)$ is a subring. What is $Z(R)$ if $R$ is a commutative ring?

13. Find $Z(M_2(\mathbb{Z}))$, the center of $M_2(\mathbb{Z})$. (See the previous exercise.)

14. Let $R$ be a ring, and $s$ a particular fixed element of $R$. Let

$$Z_s(R) = \{r \in R : rs = sr\}.$$

    (a) Prove that $Z_s(R)$ is a subring of $R$.

    (b) Recall the definition of $Z(R)$ from Exercise 12. Prove that

$$Z(R) = \cap\{Z_s(R) : s \in R\}.$$

15.   (a) An element $a$ of a ring is **nilpotent** if $a^n = 0$ for some positive integer $n$. Given a ring $R$, denote by $N(R)$ the set of all nilpotent elements of $R$. If $R$ is any commutative ring, show that $N(R)$ is a subring. (This subring is called the **nilradical** of the ring.)

    (b) Determine $N(\mathbb{Z}_{10})$, the nilradical of $\mathbb{Z}_{10}$.

    (c) Determine $N(\mathbb{Z}_8)$, the nilradical of $\mathbb{Z}_8$.

16. Suppose that $R$ is a ring with unity, and $R$ has at least two elements. Prove that the additive identity of $R$ is not equal to the multiplicative identity.

17. Show that if a ring has unity, it is unique.

18.   (a) Let $R$ be a ring, and consider the set $R \times \mathbb{Z}$ of all ordered pairs with entries from $R$ and $\mathbb{Z}$. Equip this set with operations

$$(r, n) + (s, m) = (r + s, n + m)$$

    and

$$(r, n)(s, m) = (rs + mr + ns, nm).$$

    Prove that these operations make $R \times \mathbb{Z}$ a ring. (Note that this is *not* the same ring discussed in Example 6.10.)

    (b) Show that $R \times \mathbb{Z}$ under these operations has unity, even if $R$ does not.

    (c) Show that $R \times \{0\}$ is a subring of the ring $R \times \mathbb{Z}$. Argue that this subring is isomorphic to $R$. This means that any ring without unity can essentially be found as a subring of a ring which has unity.

19. Consider the set

$$\left\{ \begin{pmatrix} a & b \\ 0 & d \end{pmatrix} : a, b, c \in \mathbb{Z} \text{ and } a \text{ is even} \right\}.$$

Prove that this set is a subring of $M_2(\mathbb{Z})$. Does it have unity?

20. Some students wonder why we require that the addition in a ring be commutative; this exercise shows why. Suppose that $R$ is a set with two operations $+$ and $\circ$, which satisfy the rules defining a ring, except for Rule 1; that is, we do not assume that the addition is commutative. Suppose that $R$ also has a multiplicative identity 1. Then prove that the addition in $R$ must in fact be commutative, and so $R$ under the given operations is a ring.

21. Consider the ring $S$ of all real-valued sequences, as discussed in Exercise 6.19. Let $\Sigma$ be the set of all sequences $(x_1, x_2, x_3, \cdots)$ where at most finitely many of the entries $x_i$ are non-zero. Prove that $\Sigma$ is a subring of $S$. Does $\Sigma$ have unity?

22. Let $\Sigma$ be the ring of finitely non-zero real-valued sequences, considered in the previous exercise. Now let $I_1$ consist of all sequences $(x_1, x_2, x_3 \cdots)$ where $0 = x_2 = x_3 = \cdots$. Show that $I_1$ is a subring that has unity, even though the larger ring $\Sigma$ does not.

23. Let $R$ be a ring, and let

$$S = \{r \in R : r + r = 0\}.$$

Prove that $S$ is a subring of $R$.

24. Generalize Exercise 23. That is, let $R$ be a ring, and let $n$ be a fixed positive integer. Let

$$S = \{r \in R : nr = 0\}.$$

Prove that $S$ is a subring of $R$. (Recall that $nr$ means to add $r$ to itself $n$ times; the integer $n$ need not belong to $R$.)

25. Let $R$ be a commutative ring with unity. An element $e \in R$ is **idempotent** if $e^2 = e$. Note that the elements 0 and 1 are idempotent. Throughout this problem assume that $e$ is idempotent in $R$.

    (a) Find a commutative ring with unity with at least one idempotent element other than 0 and 1.

    (b) Let $f = 1 - e$. Prove that $f$ is idempotent, too.

    (c) Let $Re = \{re : r \in R\}$. Prove that $Re$ is a subring of $R$, and that $e$ is unity for this subring.

    (d) Prove that $Re \cap Rf = \{0\}$ (where $f$ is the idempotent from part (b)).

    (e) Prove that for all $r \in R$, $r = a + b$, where $a \in Re$ and $b \in Rf$.

26. Suppose that $R$ is a ring with unity $1_R$, and $\varphi : R \to S$ is a ring isomorphism. Show that $\varphi(1_R)$ is unity for the ring $S$.

27. Suppose that $\varphi : R \to S$ is a ring isomorphism, and $\psi : S \to R$ is the inverse function as defined in section 7.3 of the text. Show that if $s, t \in S$, then $\psi(s + t) = \psi(s) + \psi(t)$ and $\psi(st) = \psi(s)\psi(t)$. That is, $\psi$ is a ring isomorphism too.

28. Suppose that $n$ is a positive integer with prime factorization

$$n = p_1^{k_1} p_2^{k_2} \dots p_j^{k_j}.$$

Consider $\mathbb{Z}_n = \{0, 1, 2, \dots, n-1\}$. Prove that $m \in N(\mathbb{Z}_n)$, the nilradical of $\mathbb{Z}_n$ (see Exercise 15) if and only if $p_k$ divides $m$, for all $k = 1, 2, \dots m$. Then determine how many elements there are in $N(\mathbb{Z}_n)$.

# Chapter 8

## Integral Domains and Fields

Something surprising happens in the arithmetic in $\mathbb{Z}_6$: Two non-zero elements 2 and 3 give 0 when multiplied together. This is something that never happens in $\mathbb{Z}$ (or $\mathbb{Q}$ or $\mathbb{R}$). In fact, this never happens in $\mathbb{Z}_5$.

▷ **Quick Exercise.** Use the multiplication table for $\mathbb{Z}_5$ in Chapter 3 to check that the product of two non-zero elements in $\mathbb{Z}_5$ is always non-zero. ◁

The fact that $2 \cdot 3 = 0$ in $\mathbb{Z}_6$ has undesirable consequences. For example, it means that $2 \cdot 3 = 2 \cdot 0$, and so we *cannot* cancel the 2 from each side of this equation.

---

## 8.1   Zero Divisors

We make a definition to explore this situation: Let $R$ be a commutative ring. An element $a \neq 0$ is a **zero divisor** if there exists an element $b \in R$ such that $b \neq 0$ and $ab = 0$. Of course, then $b$ is a zero divisor also.

### Example 8.1

Thus, the elements 2 and 3 in $\mathbb{Z}_6$ are zero divisors because $2 \cdot 3 = 0$. For another example, consider the elements $(1, 0)$ and $(0, 1)$ in $\mathbb{Z} \times \mathbb{Z}$. They are zero divisors because $(1, 0)(0, 1) = (0, 0)$. On the other hand, 2 is *not* a zero divisor in $\mathbb{Z}$ because the equation $2x = 0$ has only $x = 0$ as a solution.

▷ **Quick Exercise.** Determine the set of all zero divisors for the rings $\mathbb{Z}_6$, $\mathbb{Z}_5$, $\mathbb{Z} \times \mathbb{Z}$, and $\mathbb{Z}$. ◁

You should have just concluded that the rings $\mathbb{Z}$ and $\mathbb{Z}_5$ have no zero divisors. This desirable property we highlight by a definition: A commutative ring with unity that has no zero divisors is called an **integral domain**, or simply a **domain**.

### Example 8.2

Thus, $\mathbb{Z}$ and $\mathbb{Z}_5$ are integral domains, as are rings $\mathbb{Q}$, $\mathbb{R}$, and $\mathbb{C}$.

### Example 8.3

What about the ring $\mathbb{Q}[x]$? We observed in Chapter 4 that the product of two non-zero polynomials is non-zero, because the degree of the product is the sum of the degrees (Theorem 4.1). Thus, $\mathbb{Q}[x]$ (and similarly $\mathbb{Z}[x]$) is a domain, too.

**Example 8.4**

On the other hand, we have seen that the rings $\mathbb{Z} \times \mathbb{Z}$ and $\mathbb{Z}_6$ are *not* domains, simply because we have already exhibited the existence of zero divisors in them. In fact, if $n$ is not prime, then $\mathbb{Z}_n$ is not a domain.

▷ **Quick Exercise.** Show that $\mathbb{Z}_n$ is not a domain when $n$ is not prime by exhibiting zero divisors in each such ring. ◁

Notice that in the definition of domain we require both that it be commutative and have unity. These restrictions are standard in the subject, and we will consequently adhere to them. But note that this means that $2\mathbb{Z}$ is not a domain, even though it is commutative and has no zero divisors.

The arithmetic in integral domains is much simpler than in arbitrary commutative rings. An important example of this is the content of the following theorem: We can cancel multiplicatively in a domain. Of course, we already saw above that multiplicative cancellation fails in $\mathbb{Z}_6$.

**Theorem 8.1    Multiplicative Cancellation**     *Suppose $R$ is an integral domain and $a, b, c$ are elements of $R$, with $a \neq 0$. If $ab = ac$, then $b = c$.*

**Proof:**    Suppose that $R$ is a domain, $a \neq 0$, and $ab = ac$. Then $ab - ac = 0$. But then $a(b - c) = 0$, and because $R$ is a domain with $a \neq 0$, we must have $b - c = 0$, or $b = c$, as required.                                                                                                   □

If $a, b, c$ had been rational numbers in the previous proof, we would have been inclined to multiply both sides of the equation $ab = ac$ by $1/a$, the multiplicative inverse of $a$. However, the proof of the theorem holds even in the absence of multiplicative inverses (as is the case for most elements in $\mathbb{Z}$).

---

## 8.2    Units

Of course, when multiplicative inverses *do* exist, life is much simpler. Let us introduce some terminology to deal with this case. Suppose $R$ is a ring with unity 1. Let $a$ be any non-zero element of $R$. We say $a$ is a **unit** if there is an element $b$ of $R$ such that $ab = ba = 1$. In this case, $b$ is a **(multiplicative) inverse** of $a$. (Of course, $b$ is also a unit with inverse $a$.)

First note that the unity 1 is always a unit, because $1 \cdot 1 = 1$. What other elements are units? In $\mathbb{Z}$, the units are just 1 and -1, because the only integer solutions of $ab = 1$ are $\pm 1$. In $\mathbb{Q}$ and $\mathbb{R}$, *all* non-zero elements are units. In $\mathbb{Z}_6$ we have $5 \cdot 5 = 1$, and so 5 is a unit, as well as 1. Furthermore, there are no other elements $a$ and $b$ for which $ab = 1$ is true (see the multiplication table for $\mathbb{Z}_6$ in Chapter 3).

Now consider the ring $\mathbb{Q}[x]$; here unity is the constant polynomial 1. The units consist of exactly the non-zero constant polynomials (that is, polynomials of degree zero). This follows easily because the degree of the product of two polynomials is the sum of the degrees (Theorem 4.1).

▷ **Quick Exercise.** Find the units of these rings: $\mathbb{Z}_5$, $\mathbb{Z}_{12}$, $\mathbb{Z} \times \mathbb{Z}$, $\mathbb{R} \times \mathbb{R}$, $\mathbb{Z}[x]$. ◁

Note that the concept of multiplicative inverse makes perfectly good sense in non-commutative rings. For example, in the (non-commutative) ring of matrices $M_2(\mathbb{R})$, the elements

$$\begin{pmatrix} 1 & 2 \\ 3 & 4 \end{pmatrix} \quad \text{and} \quad \begin{pmatrix} -2 & 1 \\ \frac{3}{2} & -\frac{1}{2} \end{pmatrix}$$

are units, because their product (in either order) is the multiplicative identity. In Exercise 2 you will obtain all the units in this ring.

We now claim that multiplicative inverses, if they exist, are unique. To show this, suppose that $a$ has multiplicative inverses $b$ and $c$; that is, $1 = ba = ca$. But now multiply through these equations on the right by $b$: We then have $b = bab = cab = c$ (where the last equation holds because $ab = 1$ also). But then $b = c$, as required. Consequently, we will denote the (unique) inverse of an element $a$ (if it exists) by $a^{-1}$. This is of course consistent with the ordinary notation for multiplicative inverse which we use in $\mathbb{R}$ (and for matrix inverses in linear algebra too).

We denote the set of units of a ring $R$ by $U(R)$. Thus, $U(\mathbb{Z}) = \{1, -1\}$, $U(\mathbb{Q}) = \mathbb{Q}\backslash\{0\}$, and $U(\mathbb{Z}_6) = \{1, 5\}$.

The set $U(R)$ has some very nice properties, only some of which we can fully exploit right now. What we wish to observe immediately is that $U(R)$ is closed under multiplication.

▷ **Quick Exercise.** Check that this is true for $\mathbb{Z}$, $\mathbb{Q}$, and $\mathbb{Z}_6$. ◁

To show that $U(R)$ is closed under multiplication, suppose that $a, b \in U(R)$. Then we claim $ab$ is also a unit. But this is easy to see, because its inverse is just $b^{-1}a^{-1}$:

$$(ab)(b^{-1}a^{-1}) = a(bb^{-1})a^{-1} = aa^{-1} = 1,$$

and similarly for the product in the other order.

You might be surprised to see that taking the multiplicative inverse in a non-commutative ring reverses the order of multiplication. But interpret $a$ as putting on socks, and $b$ as putting on shoes. To reverse the operation $ab$ of putting on both socks and shoes, you must reverse the order: You take off shoes first, and so the inverse operation is $b^{-1}a^{-1}$. You will explore the importance of the order of the multiplication of the elements $a^{-1}$ and $b^{-1}$ in non-commutative rings in Exercise 3.

Students often confuse the unity of a ring with the concept of units, and you should take care to understand the definition. The unity of a ring is unique if it exists, and it is of course a unit (because it is its own multiplicative inverse). And a ring with units must necessarily have unity (because otherwise the concept of multiplicative inverse makes no sense). However, most rings have many units other than unity, as our examples above make clear.

---

## 8.3 Associates

In our study of factorization in the rings $\mathbb{Z}$ and $\mathbb{Q}[x]$, we disregarded as insignificant the factors $\pm 1$ in $\mathbb{Z}$ and the non-zero scalars in $\mathbb{Q}[x]$ – that is, exactly the factors that are units. We can generalize this to any commutative ring with unity, by saying that two elements $a$ and $b$ of a commutative ring with unity are **associates** if there exists a unit $u$ such that $a = ub$. Thus, if we speak intuitively, two elements are associates if they differ by only a 'trivial' factor. Note that if $a = ub$ where $u$ is a unit, then $b = u^{-1}a$, where $u^{-1}$ is a unit. That is, the relation of being associates is symmetric, as we might expect.

**Example 8.5**

The set of units of $\mathbb{Z}$ consists of exactly $\{1, -1\}$, and so two elements are associates if they differ only by a factor of $\pm 1$.

**Example 8.6**

The set of units of $\mathbb{Q}[x]$ consists of the non-zero constant polynomials (which we can identify with the non-zero rational numbers) and two polynomials are associates if they differ only by a (non-zero) rational multiple. Of course, this is exactly the definition we made of 'associate' in Section 5.2, and so our general definition is consistent.

**Example 8.7**

In $\mathbb{Z}_{12}$, the units are $\{1, 5, 7, 11\}$; 4 and 8, for example, are associates, because $5 \cdot 4 = 8$.

▷ **Quick Exercise.**   What are the other associates of 4 in $\mathbb{Z}_{12}$? What are the associates of 4 in $\mathbb{Z}_{20}$? ◁

**Example 8.8**

In $\mathbb{Q}$, *all* non-zero elements are units, and so in this case all (non-zero) elements are associates of one another.

---

## 8.4   Fields

Of course, rings like $\mathbb{Q}$ and $\mathbb{R}$ where *all* non-zero elements have multiplicative inverses seem particularly attractive, and we emphasize them by making a definition: A commutative ring with unity in which every non-zero element is a unit is called a **field**. Another way to look at the definition of a field is this: It is a commutative ring in which one can always solve equations of the form $ax = b$ (when $a \neq 0$). The solution is, of course, $x = a^{-1}b$. In other words: In all rings we can add, subtract, and multiply, but in fields we can also *divide*. Or to say this yet another way, a field is a commutative ring with unity for which all non-zero elements are associates of one another.

**Example 8.9**

Thus, $\mathbb{Q}$ and $\mathbb{R}$ are fields. By the first Quick Exercise in Section 8.2, $\mathbb{Z}_5$ is also a field.

Every field is a domain. This follows immediately from the next theorem.

**Theorem 8.2** *A field has no zero divisors.*

**Proof:**   Suppose $F$ is a field and $a \in F$, with $a \neq 0$. Now suppose that $ab = 0$. Then $0 = a^{-1} \cdot 0 = a^{-1}ab = b$. Thus, $a$ is not a zero divisor.  □

Note that the above proof actually shows that a unit of any ring cannot be a zero divisor. For example, in $\mathbb{Z}_6$, the units are 1 and 5, while the zero divisors are 2, 3, and 4.

## 8.5 The Field of Complex Numbers

Another important example of a field is the commutative ring $\mathbb{C}$ of complex numbers. (See Exercise 6.11.) To see that this ring is actually a field, we must compute the multiplicative inverse of the arbitrary non-zero complex number $\alpha = a + bi$, where $a$ and $b$ are real numbers (not both zero). We do this first by means of a bit of algebraic trickery.

The complex number $a - bi$ we call the **complex conjugate** of $a + bi$. We write it as $\bar{\alpha}$.

▷ **Quick Exercise.** Determine the complex conjugates of the following complex numbers:
$$1 + i, \ 4 - \frac{1}{2}i, \ 6i, \ 7 \ \triangleleft$$

It is easy to see that the product of a complex number with its conjugate is a real number:
$$\alpha\bar{\alpha} = (a + bi)(a - bi) = a^2 - (bi)^2 = a^2 + b^2.$$
But this means that
$$1 = (a + bi)(a - bi)\left(\frac{1}{a^2 + b^2}\right) = (a + bi)\left(\frac{a}{a^2 + b^2} - \frac{b}{a^2 + b^2}i\right),$$

and so any non-zero complex number has a multiplicative inverse. Thus, $\mathbb{C}$ is a field. Note also that $\mathbb{R}$ is a subring of this field.

▷ **Quick Exercise.** Determine the multiplicative inverses of the following complex numbers:
$$1 + i, \ 4 - \frac{1}{2}i, \ 6i, \ 7 \ \triangleleft$$

There is a geometric approach to understanding these computations. Because a complex number $a + bi$ is determined by an ordered pair $(a, b)$ of real numbers, it is only natural to associate with each complex number a point in the plane. The first coordinate gives us the **real part** of $\alpha$, and the second coordinate gives us the **imaginary part** of $\alpha$. We call the plane interpreted in this way the **complex plane**.

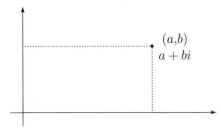

We can now talk about the **length** (or **modulus**) of a complex number. By the Pythagorean Theorem, this is evidently just
$$\sqrt{a^2 + b^2} = \sqrt{\alpha\bar{\alpha}}.$$

We abbreviate this by $|\alpha|$. This generalizes the notion of *absolute value* for real numbers.

▷ **Quick Exercise.** Compute the moduli of the following complex numbers:
$$1 + i, \ 4 - \frac{1}{2}i, \ 6i, \ 7 \ \triangleleft$$

It is certainly true that the absolute value of a product of real numbers is the product of the absolute values. That is,

$$|ab| = |a||b|.$$

This generalizes to complex numbers:

**Theorem 8.3** *For any complex numbers $\alpha$ and $\beta$,*

$$|\alpha\beta| = |\alpha||\beta|.$$

**Proof:** Suppose that $\alpha = a + bi$, and $\beta = c + di$ are complex numbers. Then

$$\begin{aligned} |\alpha\beta|^2 &= |(ac - bd) + (ad + bc)i|^2 = (ac - bd)^2 + (ad + bc)^2 \\ &= a^2c^2 + b^2d^2 + a^2d^2 + b^2c^2 = (a^2 + b^2)(c^2 + d^2) \\ &= |\alpha|^2|\beta|^2. \end{aligned}$$

By taking the (positive) square root of both sides, we obtain what we wish. $\square$

But a geometric understanding of complex multiplication extends even further than this. Note that the radial line from the origin to the point $(a, b)$ makes an angle with the positive real axis. Let's call this angle $\theta$, and always choose it in the interval $-\pi < \theta \leq \pi$. (We will leave this angle undefined for the complex number 0.) We call this angle the **argument** of $\alpha$ and write it as $\arg(\alpha)$.

$\triangleright$ **Quick Exercise.** Compute the arguments of the following complex numbers (you may have to use a calculator to approximate the angle):

$$1 + i, \ 4 - \frac{1}{2}i, \ 6i, \ 7 \ \triangleleft$$

As the following diagram suggests, this means that we can now write any non-zero complex number as a product of its modulus (a positive real number), and a complex number of length 1, which can be written trigonometrically:

$$\begin{aligned} \alpha &= (\sqrt{a^2 + b^2}) \left( \frac{a}{\sqrt{a^2 + b^2}} + \frac{b}{\sqrt{a^2 + b^2}}i \right) \\ &= |\alpha|(\cos(\arg(\alpha)) + i\sin(\arg(\alpha))). \end{aligned}$$

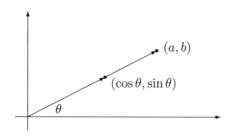

$\triangleright$ **Quick Exercise.** Put together your results from the last two Quick Exercises to write $1 + i, \ 4 - \frac{1}{2}i, \ 6i, \ 7$ in this trigonometric form. $\triangleleft$

We can now interpret multiplication in the field $\mathbb{C}$ as a geometric operation. Suppose that $\alpha$ and $\beta$ are non-zero complex numbers, and we have factored them as

$$\alpha = |\alpha|(\cos\theta + i\sin\theta)$$

and
$$\beta = |\beta|(\cos \varphi + i \sin \varphi),$$

where $\theta$ and $\varphi$ are, respectively, the arguments of $\alpha$ and $\beta$. To multiply these numbers, we first multiply the moduli to obtain the modulus of the product. Now let's deal with the trigonometric part. A surprising thing happens.

$$(\cos \theta + i \sin \theta)(\cos \varphi + i \sin \varphi) =$$

$$(\cos \theta \cos \varphi - \sin \varphi \sin \theta) + i(\cos \theta \sin \varphi + \cos \varphi \sin \theta) =$$

$$\cos(\theta + \varphi) + i \sin(\theta + \varphi).$$

(We have used two familiar trigonometric identities.) This means that the argument of a product is the *sum* of the arguments (except that we may have to adjust the angle by $2\pi$ if $\theta + \varphi$ does not fall between $-\pi$ and $\pi$).

▷ **Quick Exercise.** Verify this statement about the argument of a product by computing the arguments of the products $(1 + i)(6i)$ and $(-1 + i)(1 + \sqrt{3}i)$ in two ways. ◁

We record this statement precisely for future reference:

**Theorem 8.4    DeMoivre's Theorem**    *Let $\theta$ and $\varphi$ be two angles. Then*

$$(\cos \theta + i \sin \theta)(\cos \varphi + i \sin \varphi) =$$

$$\cos(\theta + \varphi) + i \sin(\theta + \varphi).$$

This theorem is often written in a very compact way by making use of exponential notation. If we define $e^{i\theta} = \cos \theta + i \sin \theta$, then DeMoivre's Theorem takes the form $e^{i\theta} e^{i\varphi} = e^{i(\theta + \varphi)}$. Remarkably enough, this expression actually makes sense analytically, where $e = 2.71828 \cdots$ is the base of the natural logarithm. In this book we will use this only as a formal shorthand for the more explicit expression involving the sine and cosine function. In Exercise 21 you will explore a derivation of the exponential form of DeMoivre's Theorem that depends on power series from calculus.

We can now interpret the computations of inverses in the field $\mathbb{C}$ geometrically. If we wish to compute the multiplicative inverse of the non-zero complex number

$$\alpha = |\alpha|(\cos \theta + i \sin \theta) = |\alpha| e^{i\theta},$$

we need a complex number whose modulus is $1/|\alpha|$ (because the modulus of 1 is 1), and whose argument is $-\theta$ (because the argument of 1 is 0). What this means geometrically is this: Flip through the $x$-axis (which obtains the complex conjugate), and then adjust the length.

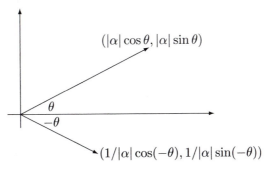

Notice how simple this appears if we use exponential notation. For if $\alpha = |\alpha|e^{i\theta}$, then

$$\alpha^{-1} = |\alpha|^{-1}\left(e^{i\theta}\right)^{-1} = |\alpha|^{-1}e^{-i\theta}.$$

For us, this is not really a rigorous calculation though, because we are assuming that the complex exponentiation obeys ordinary rules of exponentiation. This is actually true, and the calculation remains illuminating and suggestive.

In future chapters we will return often to the field of complex numbers, as a very important example, both practically and historically.

We have based our discussion of the field of complex numbers on your previous experience with the real numbers: The arithmetic of $\mathbb{C}$ comes entirely out of our understanding of the field properties of $\mathbb{R}$. Actually, it is possible to carefully prove that $\mathbb{Q}$ and $\mathbb{R}$ are fields, by basing their arithmetic on the arithmetic of the ring $\mathbb{Z}$. In this book we will not enquire into these details, and rely instead on your previous experience with these sets of numbers. We will continue to use $\mathbb{Q}$ and $\mathbb{R}$ (and now $\mathbb{C}$ too) as some of our most important examples of fields.

## 8.6    Finite Fields

To provide further examples of fields, we now turn our attention to the important collection of commutative rings, $\mathbb{Z}_n$. We will determine which of these are fields. From our examples above, we know that $\mathbb{Z}_5$ is a field while $\mathbb{Z}_6$ is not. These examples suggest the following theorem:

**Theorem 8.5** *$\mathbb{Z}_n$ is a field if and only if $n$ is prime.*

**Proof:**    If $n$ is not prime, then $n = xy$ for some $0 < x, y < n$. But then in $\mathbb{Z}_n$, $[x][y] = [0]$. So, $\mathbb{Z}_n$ could not be a field by Theorem 8.2.

Conversely, suppose $n$ is prime and let $0 < x < n$. We need to establish the existence of $[x]^{-1}$. That is, we must find a $y$ with $[x][y] = [1]$. Because $n$ is prime, we know that $\gcd(n, x) = 1$. Then by the GCD identity 2.4, there exist $r, s$ such that $rn + sx = 1$. But then $[s][x] = [1] - [rn] = [1]$, and so $[x]^{-1} = [s]$.                □

The proof above has an interesting application: We can use it to compute multiplicative inverses in $\mathbb{Z}_p$, where $p$ is prime.

**Example 8.10**

Let's compute $[23]^{-1}$ in $\mathbb{Z}_{119}$. First apply Euclid's Algorithm to obtain the equation $1 = (6)(119) + (-31)(23)$. But then $[23]^{-1} = [-31] = [88]$.

▷ **Quick Exercise.**   Check directly that $[23][88] = [1]$ in $\mathbb{Z}_{119}$. ◁

Thus, the general recipe for computing $[x]^{-1}$ in $\mathbb{Z}_p$ is this: Given $x$ and $p$, apply Euclid's Algorithm to show that $\gcd(p, x) = 1$, and then work backward through the resulting equations to obtain $r$ and $s$. Reducing $s$ modulo $p$ then gives the appropriate residue class. Notice that this method applies even if $p$ is not prime, as long as $\gcd(p, x) = 1$. We can then conclude the following:

**Theorem 8.6** *Let $0 < x < m$. Then $[x]$ is a unit in the $\mathbb{Z}_m$ if and only if $\gcd(x, m) = 1$.*

**Proof:** To show that $[x]$ has an inverse, merely repeat the argument above.

Conversely, if $\gcd(x, m) = d$ and $d \neq 1$, then $m = rd$ and $x = sd$, where $r$ and $s$ are integers with $m > r, s > 1$. But then $[x][r] = [sdr] = [sm] = [0]$. That is, $[x]$ is a zero divisor, and so cannot be a unit.

(Notice that if you have done Exercises 3.5 and 3.6, then you have already completed this proof!) □

For an alternative approach to computing $[x]^{-1}$ in $\mathbb{Z}_p$, we consider a beautiful and important theorem due to the 17th-century French mathematician Fermat.

**Theorem 8.7    Fermat's Little Theorem**    *If $p$ is prime and $0 < x < p$, then* $x^{p-1} \equiv 1 (mod\ p)$. *Hence, in $\mathbb{Z}_p$, $[x]^{-1} = [x]^{p-2}$.*

**Example 8.11**

In $\mathbb{Z}_5$, this theorem asserts that $[3]^4 = [1]$. But then $[3][3]^3 = [1]$, and so $[3]^{-1} = [3]^3 = [27] = [2]$.

**Proof:**    Suppose that $p$ is prime and $0 < x < p$. Then $[x]$ is a non-zero element of the field $\mathbb{Z}_p$. Consider the set $S$ of non-zero multiples of $[x]$ in $\mathbb{Z}_p$:

$$S = \{[x \cdot 1], [x \cdot 2], \ldots, [x \cdot (p-1)]\}.$$

Because a field has no zero divisors, each element of $S$ is non-zero. Because a field satisfies multiplicative cancellation, no two of these elements are the same.

▷ **Quick Exercise.**    Why? ◁

Thus, the set $S$ consists of $p - 1$ distinct non-zero elements and so must consist of the set

$$\{[1], [2], \cdots, [p-1]\}$$

of all non-zero elements in $\mathbb{Z}_p$, in some order. (It might be helpful for you at this point in the proof to verify this fact in a particular case; see Exercise 11.)

We now multiply all the elements of $S$ together; this is the same as multiplying all the non-zero elements of $\mathbb{Z}_p$ together. We thus obtain the following equation in $\mathbb{Z}_p$:

$$[1][2] \cdots [p-1][x]^{p-1} = [1][2] \cdots [p-1].$$

But by multiplicative cancellation in the domain $\mathbb{Z}_p$, we may cancel the non-zero elements $[1], [2], \cdots, [p-1]$ from each side of this equation, leaving $[x]^{p-1} = [1]$. Or in other words, $x^{p-1} \equiv 1 (mod\ p)$, as claimed. □

**Example 8.12**

As another example of this theorem, consider the element $[3]$ in $\mathbb{Z}_7$. To compute $[3]^{-1}$ we do the following:

$$3^5 \equiv ((3^2)^2)(3) \equiv (9^2)(3) \equiv (2^2)(3) \equiv (4)(3) \equiv 12 \equiv 5.$$

Thus, $[3]^{-1} = [5]$; you can check directly that $[3][5] = [1]$.

We have now exhibited infinitely many distinct finite fields ($\mathbb{Z}_p$ for any prime $p$). It is natural to ask whether this is a complete list. We shall discover later that there are fields with finitely many elements which are not of the form $\mathbb{Z}_p$, but we will need to know a lot more about field theory. (See Chapter 45 and also Exercise 15.)

Note that we have *not* exhibited a finite domain that is not a field. There is good reason for this: The next theorem asserts that all finite domains are actually fields. Of course, the finiteness here is important; $\mathbb{Z}$ is an example of an infinite domain that is not a field.

**Theorem 8.8** *All finite domains are fields.*

**Proof:** Suppose $D$ is a finite domain with $n$ elements. Then we can list the elements of $D$ as $0, d_2, \cdots, d_n$. (Somewhere in this list must occur the multiplicative identity 1.) Suppose now that $a$ is a *non-zero* element of $D$. It is then one of the $d_i$, where $i \geq 2$. Consider now the list of $n-1$ elements $ad_2, ad_3, \cdots, ad_n$. Because $D$ is a domain, none of these elements are 0. Because multiplicative cancellation holds in a domain, no two of these elements are the same (because if $ad_i = ad_j$, then $d_i = d_j$ by cancellation). There are thus $n-1$ distinct elements in this list. This means the list consists of *all* the non-zero elements of $D$. But then $ad_i = 1$ for some $i$, because 1 is certainly one of the non-zero elements of $D$. This means that $a$ is a unit. Hence, all non-zero elements of $D$ are units, and so $D$ is a field. $\square$

You should think carefully about why the proof of the above theorem does not work when the domain in question has infinitely many elements. (See Exercise 13.)

## Historical Remarks

Pierre Fermat was one of the most important mathematicians of the 17th century, even though he was an amateur. A lawyer by trade, he did most of his mathematical work in the evenings, as an intellectually stimulating recreation. Consequently, almost no mathematics under his name was published until after his death. He did engage actively in the mathematical life of his day, by corresponding with most of the great names in physics and mathematics of the era; such men as Huygens and Descartes were his correspondents. In this way, many of Fermat's results became well known to the mathematical community of the time. In the present day, mathematicians (and other scientists) communicate their results in widely available scholarly journals. In this way, scientific results can be easily used (and checked) by others. This sort of wide access to scientific knowledge was not yet available in the 17th century. The evolution toward that system began only with the founding of such scholarly societies as the Royal Society in Britain, near the end of the seventeenth century.

Fermat's Little Theorem is so called to distinguish it from his Great or Last Theorem. This asserts that there are no solutions in non-zero integers to the equation

$$x^n + y^n = z^n,$$

whenever $n > 2$. (Of course, there are many such solutions when $n = 2$; for example, $3^2 + 4^2 = 5^2$.) This assertion was found, without proof, in a margin of a book of Fermat's after his death. He noted that the margin was insufficiently large to contain his 'truly marvelous demonstration.' Because no mathematician for 350 years was able to prove this assertion, it seems highly unlikely that Fermat could prove it. Attempts to prove what might more properly be called Fermat's Last Conjecture have had an important impact on the development of abstract algebra. But note that if there had been a system of mathematical journals at that time, we might now have an answer to this famous problem, or at least know whether Fermat's proof was fallacious! In 1993 the British mathematician Andrew

Wiles thrilled the mathematical world by announcing that he had proved Fermat's Last Theorem, basing his results on a large body of modern mathematical work. A gap in his proof was discovered as his work underwent the present-day scrutiny of peer review, but Wiles and his colleague Richard Taylor were able to fill in the gap, and complete the proof of a 350 year old conjecture.

## Chapter Summary

In this chapter, we defined *integral domains* (commutative rings with unity without zero divisors) and *fields* (commutative rings in which all elements have multiplicative inverses).

We proved the following theorems about fields and domains:

- Multiplicative cancellation holds in integral domains.

- Every field is an integral domain.

- Every finite integral domain is a field.

- The ring $\mathbb{Z}_n$ is a field exactly if $n$ is prime.

- Fermat's Little Theorem.

This last theorem gives a method for computing multiplicative inverses in $\mathbb{Z}_p$, for $p$ prime.

## Warm-up Exercises

a. Determine the units and the zero divisors in the following rings:

$$\mathbb{Z} \times \mathbb{Z}, \quad \mathbb{Z}_{20}, \quad \mathbb{Z}_4 \times \mathbb{Z}_2, \quad \mathbb{Z}_{11}, \quad \mathbb{Z}[x].$$

b. Suppose that $a$ is a unit in a ring. Is $-a$ a unit? Why or why not?

c. Determine whether the following pairs of elements are associates:

    (a) $x$ and $3x$ in $\mathbb{Q}[x]$.

    (b) $x$ and $3x$ in $\mathbb{Z}[x]$.

    (c) 4 and 2 in $\mathbb{Z}_{10}$.

    (d) 4 and 2 in $\mathbb{Z}_8$.

    (e) 4 and 2 in $\mathbb{Z}_7$.

    (f) $(2, -1)$ and $(4, 1)$ in $\mathbb{Z} \times \mathbb{Z}$.

    (g) $(2, -1)$ and $(4, 1)$ in $\mathbb{Q} \times \mathbb{Z}$.

d. Discover which elements of $\mathbb{Z}_{15}$ are associates of which.

e. What are the units in $\mathbb{Q} \times \mathbb{Q}$? What are the associates in this ring of $(1, 2)$? of $(1, 0)$?

f. Find two non-zero matrices $A$ and $B$ in $M_2(\mathbb{Z})$ so that $AB = 0$; that is, find some zero divisors.

g. Find a non-zero matrix $A$ in $M_2(\mathbb{Z})$ so that $A^2 = 0$. Then $A$ is a zero divisor. (A ring element $a$ so that $a^n = 0$, for some positive integer $n$ is called *nilpotent*. See Exercise 7.15.)

h. Suppose that $D$ is a domain. Show that the direct product $D \times D$ is not a domain.

i. Compute the argument, modulus, and multiplicative inverse of the complex number $3 - 4i$.

j. What is the argument of a (non-zero) real number?

k. A complex number with modulus 1 is said to lie on the unit circle. Why?

l. Choose two complex numbers on the unit circle (see Exercise k). Why is their product also on the unit circle?

m. Consider the complex numbers $1 - \sqrt{3}i$ and $2 + 2i$.

    (a) Compute their product twice: First do the arithmetic directly; then determine the arguments and moduli of these numbers, and compute the product using Theorems 8.3 and 8.4.

    (b) Compute their multiplicative inverses twice: First do the arithmetic directly, and then use Theorems 8.3 and 8.4 instead.

n. Write out the following complex numbers in the form $a + bi$, giving exact values for $a$ and $b$ if possible, or by using a calculator if necessary:

$$e^{\pi i}, \quad e^{\frac{\pi i}{4}}, \quad e^i, \quad 2e^{\frac{2\pi i}{3}}$$

o. Give examples of the following, or explain why they don't exist:

    (a) A finite field.

    (b) A finite field that isn't a domain.

    (c) A finite domain that isn't a field.

    (d) An infinite field.

    (e) An infinite domain that isn't a field.

p. Does there exist an integer $m$ for which $\mathbb{Z}_m$ is a domain, but not a field? Explain.

q. Use Euclid's Algorithm to compute the multiplicative inverse of $[2]$ in $\mathbb{Z}_9$.

r. Use Fermat's Little Theorem 8.7 to compute the multiplicative inverse of $[2]$ in $\mathbb{Z}_5$.

---

# Exercises

1. Prove that if $R$ is a commutative ring and $a \in R$ is a zero divisor, then $ax$ is also a zero divisor or 0, for all $x \in R$.

2. Consider the set $M_2(\mathbb{R})$ of all $2 \times 2$ matrices with entries from the real numbers $\mathbb{R}$; by Exercise 6.8, we know this is a ring. Prove that the units of $M_2(\mathbb{R})$ are precisely those matrices

$$\begin{pmatrix} a & b \\ c & d \end{pmatrix}$$

such that $ad - bc \neq 0$. In this case, you can find a formula for the multiplicative inverse of the matrix. We call $ad - bc$ the **determinant** of the matrix. And so the units of $M_2(\mathbb{R})$ are those matrices with non-zero determinant; we will explore the notion of determinant further in Example 22.5.

3. Find two non-commuting units $A, B$ in $M_2(\mathbb{R})$, and check that $(AB)^{-1} = B^{-1}A^{-1}$ and $(AB)^{-1} \neq A^{-1}B^{-1}$.

4. Generalizing Example 7.9, consider the ring $D_2(\mathbb{R})$ of diagonal matrices with real entries (a subring of $M_2(\mathbb{R})$). What are the units of $D_2(\mathbb{R})$?

5. Determine $U(\mathbb{Z}[i])$, the units of the Gaussian integers (see Exercise 6.12).

6. Show that $\alpha = 1 + \sqrt{2} \in U(\mathbb{Z}[\sqrt{2}])$ (see Exercise 7.2). Then consider the elements $\alpha, \alpha^2, \alpha^3, \ldots$ and argue that $U(\mathbb{Z}[\sqrt{2}])$ is an infinite set. Then find infinitely many more units in this ring.

7. Suppose that $R$ and $S$ are commutative rings with unity, and they are isomorphic via the ring isomorphism $\varphi : R \to S$. Let $r \in R$. Prove that:

   (a) $r$ is a zero divisor if and only if $\varphi(r)$ is a zero divisor.

   (b) $r$ is a unit if and only if $\varphi(r)$ is a unit.

   (c) $R$ is an integral domain if and only if $S$ is an integral domain.

   (d) $R$ is a field if and only if $S$ is a field.

8. Let $\alpha = e^i = \cos 1 + i \sin 1$. (Note that the argument here is 1 *radian*.) What is the modulus of $\alpha^n$, where $n$ is a positive integer? Prove that $\alpha^n \neq \alpha^m$, whenever $n$ and $m$ are positive integers and $n \neq m$.

9. Use Fermat's Little Theorem 8.7 to find $[6]^{-1}$ in $\mathbb{Z}_{19}$.

10. Use Euclid's Algorithm to find $[36]^{-1}$ in $\mathbb{Z}_{101}$.

11. Verify explicitly the key idea in the proof of Fermat's Little Theorem 8.7 for $x = 3$ and $p = 17$; that is, check that the set $S$ consists of the 16 non-zero elements of $\mathbb{Z}_{17}$.

12. Suppose that $b \in R$, a non-commutative ring with unity. Suppose that $ab = bc = 1$; that is, $b$ has a **right inverse** $c$ and a **left inverse** $a$. Prove that $a = c$ and that $b$ is a unit.

13. Try to apply the proof that every finite integral domain is a field to the integral domain $\mathbb{Z}$. Where does it go awry?

14. Let $R$ be a commutative ring with unity. Suppose that $n$ is the least positive integer for which we get 0 when we add 1 to itself $n$ times; we then say $R$ has **characteristic n**. If there exists no such $n$, we say that $R$ has **characteristic 0**. For example, the characteristic of $\mathbb{Z}_5$ is 5 because $1 + 1 + 1 + 1 + 1 = 0$, whereas $1 + 1 + 1 + 1 \neq 0$. (Note that here we have suppressed '[' and ']'.)

(a) Show that, if the characteristic of a commutative ring with unity $R$ is $n$ and $a$ is *any* element of $R$, then $na = 0$. (Recall that $na = \underbrace{a + a + \cdots + a}_{n \text{ times}}$.)

(b) What are the characteristics of $\mathbb{Q}$, $\mathbb{R}$, $\mathbb{Z}_{17}$?

(c) Prove that if a field $F$ has characteristic $n$, where $n > 0$, then $n$ is a prime integer.

(d) Suppose that $R$ and $S$ are commutative rings with unity that are isomorphic. Prove that $R$ and $S$ have the same characteristic.

15. Consider the commutative ring

$$F = \{0, 1, \alpha, 1 + \alpha\},$$

where $0$ is the additive identity, $1$ is the multiplicative identity, $x + x = 0$, for all $x \in F$, and $\alpha^2 = \alpha + 1$.

(a) Write out explicitly the addition and multiplication tables for $F$.

(b) Prove that $F$ is a field.

(c) Because $F$ has four elements, you might expect that $F$ would be isomorphic to the other ring we know with 4 elements, $\mathbb{Z}_4$. Show this is false, by computing the characteristics of $F$ and $\mathbb{Z}_4$ (see the previous exercise). Alternatively, use Exercise 7.

16. Suppose that $R$ is a commutative ring, and $a$ is a non-zero nilpotent element. (See Exercise 7.15; this means that $a^n = 0$ for some positive integer $n$.) Prove that $1 - a$ is a unit. *Hint:* You can actually obtain a formula for the inverse.

17. Prove that $\mathbb{Z}_m$ is the union of three mutually disjoint subsets: its zero divisors, its units, and $\{0\}$. Show by example that this is false for an arbitrary commutative ring.

18. Prove that if $p$ is prime, then $[(p - 1)!] = [-1]$ in $\mathbb{Z}_p$. *Note:* In number theory, this result is known as Wilson's Theorem.

19. Suppose that

$$A = \begin{pmatrix} a & b \\ c & d \end{pmatrix} \in M_2(\mathbb{R})$$

is a non-zero element, which is not a unit. Show that $A$ is actually a zero divisor. (You might want to look at Exercise 2.) What is the relationship of this result to Exercise 17?

20. Consider the ring $C[0, 1]$. (See Example 6.14.)

(a) What are the units in this ring? (You need a theorem from calculus to prove your answer is correct.)

(b) Given $f \in C[0, 1]$, define $Z(f) = \{x \in [0, 1] : f(x) = 0\}$. Prove that if $f, g$ are associates, then $Z(f) = Z(g)$.

21. Recall the Taylor series expansions centered at 0 for the three functions $\sin x$, $\cos x$, $e^x$. (These are also called the MacLaurin series for these functions.) In calculus we discover that these series converge to their functions absolutely for all real numbers $x$. Let's assume that these series also make sense for imaginary numbers $ix$ . By replacing $x$ in $e^x$ by $ix$, and rearranging terms of the series, verify that $e^{ix} = \cos x + i \sin x$, for all real numbers $x$. (This verification can be made rigorous, using analytic techniques we will avoid here.)

22. Suppose $F_1$ and $F_2$ are fields. Is $F_1 \times F_2$ a field? Why or why not?

# Chapter 9

## Ideals

In this chapter we will introduce a particularly nice kind of subring, called a *principal ideal*, and then generalize yet further to define *ideals* in an arbitrary commutative ring. It turns out that we will be able to better understand the arithmetic in a given ring by changing our perspective from the properties of individual elements, to considering ideals instead. This was a great insight brought to the subject by such nineteenth century mathematicians as Ernst Kummer and Richard Dedekind: Difficulties in some rings can be better understood by thinking of some 'ideal' elements for our ring, in addition to the ordinary elements we have. We will thus begin by translating elementwise properties into statements about corresponding principal ideals, and then discover that non-principal ideals really do fill in certain gaps in our understanding.

## 9.1 Principal Ideals

We discovered in Examples 7.6 and 7.7 (and in Exercises 7.4 and 7.5) that subrings of $\mathbb{Z}$ of the form $m\mathbb{Z}$ carry the divisibility information for the particular integer $m$. We generalize this: Let $R$ be any commutative ring, and $a \in R$. We then let

$$\langle a \rangle = \{b \in R : b = ax, \text{for some } x \in R\}.$$

We then call $\langle a \rangle$ a **principal ideal**, and call $a$ a **generator** for $\langle a \rangle$.

We claim that $\langle a \rangle$ is a subring. We first show that it is closed under subtraction. For if $x, y \in \langle a \rangle$, then $x = ar$ and $y = as$, for some $r, s \in R$. But then $x - y = ar - as = a(r - s)$, which is clearly a multiple of $a$, and so an element of $\langle a \rangle$. To show that $\langle a \rangle$ is closed under multiplication, suppose that $x, y \in \langle a \rangle$. Then $x = ar, y = as$, and so $xy = aras = a(ras)$. This latter element is a multiple of $a$, and so belongs to $\langle a \rangle$.

### Example 9.1

Let's discuss the principal ideals in $\mathbb{Z}$. Here is a complete list:

$$\langle 0 \rangle = \{0\}, \quad \langle 1 \rangle = \langle -1 \rangle = \mathbb{Z},$$
$$\langle 2 \rangle = \langle -2 \rangle = \{x \in \mathbb{Z} : x \text{ is even}\}, \quad \langle 3 \rangle = \langle -3 \rangle, \quad \langle 4 \rangle = \langle -4 \rangle, \cdots.$$

Note that we had earlier referred to these latter principal ideals as $2\mathbb{Z}, 3\mathbb{Z}, 4\mathbb{Z}$, etc.

### Example 9.2

Clearly, the principal ideal $\langle 3 \rangle$ in $\mathbb{Z}_{12}$ consists of the elements $\{0, 3, 6, 9\}$.

**Example 9.3**

Consider the principal ideal $\langle \sqrt{5} \rangle$ in $\mathbb{Z}[\sqrt{5}]$ (if necessary, see Exercise 7.2 where we introduce such rings). What does a typical element of this ideal look like? All such elements are multiples of $\sqrt{5}$, and this naturally leads one to think of such elements as $\sqrt{5}, -\sqrt{5}, 2\sqrt{5}$, and so forth. However, these are merely integer multiples of $\sqrt{5}$, and we must allow multiples of $\sqrt{5}$ by any element from the ring in question, namely $\mathbb{Z}[\sqrt{5}]$. Thus, a typical element of this principal ideal is an element of the form $(a+b\sqrt{5})(\sqrt{5}) = 5b + a\sqrt{5}$. This means that

$$\langle \sqrt{5} \rangle = \{5r + s\sqrt{5} : r, s \in \mathbb{Z}\}.$$

That is, $\langle \sqrt{5} \rangle$ consists of all elements of $\mathbb{Z}[\sqrt{5}]$ whose "rational part" is divisible by 5.

**Example 9.4**

Let's now look at the principal ideal $\langle x^2 \rangle$ in $\mathbb{Q}[x]$. This consists of all multiples of $x^2$; that is, those polynomials with $x^2$ as a factor. Equivalently, $\langle x^2 \rangle$ is the set of all polynomials whose constant and degree 1 coefficients are zero. More generally, $\langle f \rangle$ in $\mathbb{Q}[x]$ consists of all polynomials with $f$ as a factor.

**Example 9.5**

Consider the principal ideal $\langle 2 \rangle$ in the ring $2\mathbb{Z}$ of even integers. In this example, when we form the multiples of 2, we are restricted only to even numbers. Thus,

$$\langle 2 \rangle = \{\cdots, 0, 4, 8, 12, \cdots\}.$$

Notice that in this case $2 \notin \langle 2 \rangle$. This situation arises because $2\mathbb{Z}$ lacks unity.

Note that if our ring does have unity (as will usually be the case), we then have that $a = a1 \in \langle a \rangle$, thus avoiding the perversity of the previous example.

We also observe that for any commutative ring with unity the trivial and improper subrings are always principal ideals, because $\langle 0 \rangle = \{0\}$ and $\langle 1 \rangle = R$ (because any element of $R$ is a multiple of 1). (We saw that this was true in $\mathbb{Z}$ in the example above.)

Now, one of our goals in this chapter is change our focus from individual elements to principal ideals (and ideals in general, when we define them below). Our first step in that direction is the observation that divisibility in the ring can be rephrased in terms of containment of principal ideals. That is, if $r, s, t \in R$, a commutative ring, then $r = st$ if and only if $\langle r \rangle \subseteq \langle s \rangle$.

▷ **Quick Exercise.** Prove this. ◁

Note that 'smaller' elements in terms of factorization lead to bigger ideals, in the sense of set containment. We record this as a simple theorem:

**Theorem 9.1** *Let $R$ be a commutative ring with unity, with $r \in R$. Then $\langle r \rangle \subseteq \langle s \rangle$ if and only if $r = st$, for some $t \in R$.*

Our goal now is to better understand when two principal ideals are equal. First, let's consider the case when $\langle r \rangle = R$, where $R$ is a commutative ring with unity, and $r \in R$. This means that 1 is a multiple of $r$; but if $rs = 1$, then $r$ is a unit. And conversely, if $r$ is a unit, then $1 = rt$, for some $t \in R$. But then $x = r(tx)$, for any $x \in R$. Thus, all elements of $R$ are multiples of $r$. We have thus shown that in a commutative ring with unity, $\langle r \rangle = R$ if and only if $r$ is a unit. We record this as a theorem:

**Theorem 9.2** *Let $R$ be a commutative ring with unity, with $r \in R$. Then $\langle r \rangle = R$ if and only if $r$ is a unit.*

▷ **Quick Exercise.** Consider the unit 5 in $\mathbb{Z}_{12}$. Verify explicitly that $\langle 5 \rangle = \mathbb{Z}_{12}$. (That is, show by computation that every element of the ring is a multiple of 5.) ◁

To illuminate our discussion, let's return to the principal ideals in $\mathbb{Q}[x]$.

**Example 9.6**

Consider the ideal $\langle x \rangle$ in $\mathbb{Q}[x]$. Clearly, this consists of those polynomials with no constant term. But what about the ideal $\langle 3x \rangle$? We claim that $\langle x \rangle = \langle 3x \rangle$. It is clear that $\langle 3x \rangle \subseteq \langle x \rangle$, but the reverse inclusion holds too, because $x = \frac{1}{3}(3x)$, and so $x \in \langle 3x \rangle$.

This example (and also our discussion of the principal ideals in $\mathbb{Z}$) suggests an answer to the question of when two elements of a commutative ring generate the same principal ideal. The answer given in the next theorem for domains does not hold for all commutative rings.

**Theorem 9.3** *Let $R$ be a domain, and $r, s \in R$. Then $\langle r \rangle = \langle s \rangle$ if and only if $r$ and $s$ are associates.*

**Proof:** Suppose first that $r$ and $s$ are associates. Then $r = su$, where $u$ is a unit. Thus, $r$ is a multiple of $s$, and so $r \in \langle s \rangle$. Hence, $\langle r \rangle \subseteq \langle s \rangle$. But $s = ru^{-1}$, and so similarly $\langle s \rangle \subseteq \langle r \rangle$.

For the converse, suppose that $\langle r \rangle = \langle s \rangle$. Then $r$ is a multiple of $s$ and vice versa; that is, $r = sx$ and $s = ry$, for some $x$ and $y$. But then $r = ryx$, and so by the multiplicative cancellation property for domains, $1 = yx$. That is, $y$ and $x$ are units, and so $r$ and $s$ are associates. □

**Example 9.7**

In the Gaussian integers, we know that $1 + i$ is a non-unit, and so $\langle 1 + i \rangle \subset \mathbb{Z}[i]$ but $\langle 1 + i \rangle = \langle 1 - i \rangle$.

▷ **Quick Exercise.** Verify these assertions. ◁

**Example 9.8**

Consider the ring $\mathbb{Z}[\sqrt{2}]$. (See Exercise 7.1.) In Exercise 8.6 we discovered that $(1 + \sqrt{2})^2 = 3 + 2\sqrt{2}$ is a unit in this ring. Now $\sqrt{2}$ is not a unit, and we have that $\sqrt{2}(3 + 2\sqrt{2}) = 4 + 3\sqrt{2}$. Thus $\sqrt{2}$ and $4 + 3\sqrt{2}$ are associates, and so $\langle \sqrt{2} \rangle = \langle 4 + 3\sqrt{2} \rangle$. This is not so obvious at first glance!

---

## 9.2 Ideals

The concept of a principal ideal is really quite concrete: It is just the set of all multiples of a given element, which we call the generator. We've then seen that this is always a subring. But to move further we must make an abstract definition of an ideal in general:

Let $R$ be a commutative ring with unity. An **ideal** of $R$ is a non-empty subset $I$ of $R$ satisfying the following criteria:

1. $I$ is closed under subtraction;

2. If $a \in I$ and $r \in R$, then $ar \in I$.

Thus, an ideal is a subring that satisfies the stronger multiplicative closure property (2). This property says that $I$ 'absorbs' multiplication by *any* ring element. We remind ourselves of this by calling (2) the **multiplicative absorption property** of ideals.

We have made this definition only for commutative rings, because we deal primarily with such rings in this book. However, the definition above can be modified to account for the notion of ideal in the non-commutative case: Simply strengthen (2) to require that both $ar$ and $ra$ belong to $I$ if $a \in I$.

We first make the expected observation: A principal ideal is in fact an ideal. That is, a principal ideal satisfies the multiplicative absorption property. But this is quite clear:

▷ **Quick Exercise.** Check this. ◁

With our new definition of ideal, we can now think of a principal ideal $\langle a \rangle \subseteq R$ more conceptually: It is the smallest ideal of $R$ containing $a$. This merely means that $\langle a \rangle$ is a subset of *any* ideal of $R$ containing $a$. But this is obvious because if $a \in I$, where $I$ is an ideal, then by the multiplicative absorption property for $I$, all multiples of $a$ are elements of $I$.

### Example 9.9

> Consider now the ring $\mathbb{Q}$. We know that $\mathbb{Z}$ is a subring of $\mathbb{Q}$. Is it an ideal of $\mathbb{Q}$? Suppose it is; we now apply the multiplicative absorption property to the element 1 of $\mathbb{Z}$. Choose any rational number $q$. Then by multiplicative absorption, $(q)(1) = q$ would be an element of $\mathbb{Z}$, which is certainly not always true. Thus, $\mathbb{Z}$ is a subring of $\mathbb{Q}$ which is *not* an ideal.

This example can clearly be generalized: If an ideal contains any unit, it is the whole ring. Now, because all non-zero elements in a field are units, a field has only the two ideals that all commutative rings possess: namely, the zero ideal and the improper ideal. The converse is also true:

**Corollary 9.4** *A commutative ring with unity is a field if and only if its only ideals are the trivial and improper ideals.*

**Proof:** We have already proved half of this corollary. For the converse, suppose that $R$ is a commutative ring with unity whose only ideals are the trivial and improper ideals. Choose any non-zero element $r \in R$. Because $r$ is non-zero, then $\langle r \rangle \neq \{0\}$, and so $\langle r \rangle = R$. But then $1 \in \langle r \rangle$, and so $1 = rs$, for some element $s$. But then $r$ is a unit. Thus, all non-zero elements of $R$ are units, and so $R$ is a field. □

## 9.3    Ideals That Are Not Principal

Before proceeding, we should first convince you that there do exist ideals that are *not* principal. It turns out that we can find such ideals in the domain $\mathbb{Z}[x]$.

To describe such an example we first introduce a little more general notation. Suppose that $R$ is a commutative ring with unity and $a, b \in R$. We shall define $\langle a, b \rangle$ as

$$\{ ax + by : x, y \in R \}.$$

We leave it as Exercise 11 below for you to prove that this is an ideal and is in fact the smallest ideal of $R$ that contains both $a$ and $b$. (In fact, this idea can be generalized to ideals generated by any finite number of elements; see Exercise 12.) Note that $\langle a, b \rangle$ is really just the set of all linear combinations of $a$ and $b$ (where the coefficients on $a$ and $b$ are allowed to be any ring elements).

### Example 9.10

In the ring $\mathbb{Z}$, the ideal $\langle 12, 9 \rangle$ consists of the set of all linear combinations of 12 and 9. We know from our work about the integers that this is the set of all multiples of $\gcd(12, 9) = 3$ (by the GCD identity), and so

$$\langle 12, 9 \rangle = \langle 3 \rangle.$$

This is a principal ideal. In $\mathbb{Z}$ we can always find a single generator for an ideal of the form $\langle a, b \rangle$ using what we know about the arithmetic in that ring.

### Example 9.11

The ring $\mathbb{Q}[x]$ also has a GCD identity. We can then infer that

$$\langle x^2 + 3x, x^3 - x^2 + 3x - 3 \rangle$$

is a principal ideal, by computing its single generator.

▷ **Quick Exercise.** Do this. ◁

### Example 9.12

For a first example of a non-principal ideal, let's work in the ring $\mathbb{Z}[x]$. We claim that the ideal $\langle 2, x \rangle$ is not principal. To see this, we will first look more closely at arbitrary elements of this ideal. Any element of it is of the form $2f + xg$, where $f$ and $g$ are arbitrary polynomials from $\mathbb{Z}[x]$. Let's consider the constant term of this polynomial. The polynomial $xg$ has no constant term, and so the constant term for $2f + xg$ is equal to the constant term of $2f$. This constant must be even. Thus, every element in $\langle 2, x \rangle$ has even constant term. But conversely, consider any polynomial in $\mathbb{Z}[x]$ with even constant term. We can write such a polynomial as $2n + xh \in \langle 2, x \rangle$, where $n$ is an integer, and $h$ is a polynomial. Thus, $\langle 2, x \rangle$ consists of all polynomials with even constant term. (Note of course that zero is an even integer.)

We now claim that $\langle 2, x \rangle$ is not a principal ideal. If it were, with generator $f$, then all polynomials in the ideal would be multiples of $f$. In particular, 2 and $x$ would be multiples of $f$. Because 2 is a multiple of $f$, the degree of $f$ must be zero. But the only zero-degree polynomials that have $x$ as a multiple are $\pm 1$. Neither of these could be $f$, because their constant terms are odd. Thus, $\langle 2, x \rangle$ is not a principal ideal. (You might want to compare this discussion to the first Quick Exercise in Section 5.3.)

The contrast between these two examples is important: In $\mathbb{Z}$, two non-zero elements always have a greatest common divisor, and that element serves as the single generator of the ideal $\langle 12, 9 \rangle$. In contrast, in $\mathbb{Z}[x]$ we cannot always make sense of a greatest common divisor, and so an ideal such as $\langle 2, x \rangle$ has no single generator.

## 9.4 All Ideals in $\mathbb{Z}$ Are Principal

We can now express in abstract terms the underlying reason why the arithmetic in $\mathbb{Z}$ is so nice: *All ideals are actually principal.*

**Theorem 9.5** *All ideals of $\mathbb{Z}$ are principal.*

**Proof**: Suppose that $I$ is an ideal of $\mathbb{Z}$. We wish to show that $I$ is principal. If $I$ is the zero ideal this is already obvious; so suppose $I$ has more elements than just 0. What element of $I$ might serve as a generator for $I$? We answer this question by using the Well-ordering Principle: Choose the smallest positive element $m$ of $I$.

▷ **Quick Exercise.** Why need $I$ have *any* positive elements? ◁

We claim that $\langle m \rangle = I$. Because $I$ is an ideal (and so satisfies the multiplicative absorption property), it is clear that $\langle m \rangle \subseteq I$. Suppose now that $b \in I$. We claim $b \in \langle m \rangle$; that is, we claim that $b$ is a multiple of $m$. To check this, we will use the Division Theorem 2.1 to obtain a quotient $q$ and a remainder $r$ where $b = qm + r$. Obviously, we hope that the remainder is zero. But what do we know about $r$? Because $r = b - qm$, $r \in I$, using both of the defining properties of ideal. But we also know that $0 \leq r < m$, and because $m$ is the *smallest* positive element of $I$, $r = 0$, as required. Thus for $\mathbb{Z}$, all ideals are principal. □

We are leaving it to you in Exercise 2 below to show that the same fact holds true for $\mathbb{Q}[x]$. By this time you should not be surprised by this analogy between $\mathbb{Z}$ and $\mathbb{Q}[x]$. The proof you will construct is similar, except you will choose $m$ to be a non-zero polynomial with smallest *degree* in the ideal.

We close with a definition: An integral domain is a **principal ideal domain** (or **PID**) if all its ideals are principal. Thus we can say that $\mathbb{Z}$ and $\mathbb{Q}[x]$ are PIDs while $\mathbb{Z}[x]$ is not.

## Chapter Summary

In this chapter we define a *principal ideal* in a commutative ring, which is just the set of multiples of a given particular element. Divisibility properties in the ring can be rephrased in terms of principal ideals. The idea of principal ideal leads to a more general definition of an *ideal*, which is a subring that also satisfies the *multiplicative absorption property*. In some integral domains, all ideals are principal (these are called *principal ideal domains*), but this is not always the case.

## Warm-up Exercises

a. Are the following ideals?

    (a) $\mathbb{Q}$ in $\mathbb{R}$.

    (b) $\{0, 2, 4, 6\}$ in $\mathbb{Z}_8$.

    (c) $\mathbb{Z}[x]$ in $\mathbb{Q}[x]$.

    (d) $\mathbb{Z}$ in $\mathbb{Z}[i]$.

    (e) $\{ni : n \in \mathbb{Z}\}$ in $\mathbb{Z}[i]$.

   (f) $\mathbb{Z} \times \{0\}$ in $\mathbb{Z} \times \mathbb{Z}$.

   (g) $\{5n + 5m\sqrt{2} : n, m \in \mathbb{Z}\}$ in $\mathbb{Z}[\sqrt{2}]$.

   (h) The set of all polynomials with even degree (together with the zero polynomial), in $\mathbb{Q}[x]$.

b. Are the following equalities true or false?

   (a) $\langle x \rangle = \langle 3x \rangle$, in $\mathbb{Z}[x]$.

   (b) $\langle x \rangle = \langle 3x \rangle$, in $\mathbb{Q}[x]$.

   (c) $\langle i \rangle = \mathbb{Z}[i]$.

   (d) $\langle 2 + i \rangle = \mathbb{Z}[i]$.

   (e) $\langle 2 \rangle = \langle 10 \rangle$, in $\mathbb{Z}_{14}$.

c. Suppose that $R$ is a commutative ring with unity, and $I$ is an ideal of $R$. If $1 \in I$, what can you say about $I$?

d. Give a nice description of the elements of the ideal $\langle \sqrt{7} \rangle$ in the ring $\mathbb{Z}[\sqrt{7}]$.

e. Give examples of the following (or explain why they don't exist):

   (a) A non-zero proper ideal of a finite ring.

   (b) A non-zero proper ideal of $\mathbb{C}$.

   (c) A non-zero proper ideal of $\mathbb{Z}[i]$.

   (d) A non-zero proper ideal of $\mathbb{Z}[x]$.

f. Write the ideal $\langle 35, 15 \rangle$ in $\mathbb{Z}$ as a principal ideal. Same for $\langle 12, 20, 15 \rangle$.

g. Describe the ideal $\langle 4, x \rangle$ in $\mathbb{Z}[x]$. Same for $\langle 4, x^2 \rangle$.

h. Write the ideal $\langle x^2 - 3x + 2, x^2 - 2x + 1 \rangle$ in $\mathbb{Q}[x]$ as a principal ideal.

i. What is the ideal $\langle (1, 0), (0, 1) \rangle$ in $\mathbb{Z} \times \mathbb{Z}$? What about $\langle (1, 1) \rangle$?

j. Why is a field always a PID, practically by default?

k. Give examples of the following (or explain why they don't exist):

   (a) A domain that is not a PID.

   (b) Elements $f, g \in \mathbb{Z}[x]$, so that $f \neq g$, but $\langle f \rangle = \langle g \rangle$.

   (c) Elements $f, g \in \mathbb{Q}[x]$, so that $\langle f, g \rangle = \langle f \rangle$ but $\langle f, g \rangle \neq \langle g \rangle$.

## Exercises

1. In Examples 7.1–7.10, determine in each case whether the subrings described there are ideals.

2. Prove that all ideals in $\mathbb{Q}[x]$ are principal, using a similar proof to that we used for $\mathbb{Z}$ (Theorem 9.5).

3. Consider the set
$$I = \{f \in \mathbb{Q}[x] : f(i) = 0\}.$$
(Here, $i$ is the usual complex number.)

   (a) Prove that $I$ is an ideal.

   (b) We know that $I$ is a principal ideal (why?). Find a generator for $I$, and prove that it works. *Hint:* The generator should be an element in $I$ with the smallest degree greater than 0.

4. Consider the set
$$I = \{f \in \mathbb{C}[x] : f(i) = 0\}.$$
Repeat Exercise 3 for this set.

5. Consider the set
$$I = \{f \in \mathbb{Q}[x] : f(3) = 0 \text{ and } f\left(\sqrt{3}\right) = 0\}.$$
Repeat Exercise 3 for this set.

6. Determine all the principal ideals of $\mathbb{Z}_{12}$, and draw a diagram describing their containment relations. Prove that $\mathbb{Z}_{12}$ has no other ideals.

7. Consider the ring $\mathbb{Z}[\alpha]$, where $\alpha = \sqrt[3]{5}$, as described in Exercise 7.3. Describe the elements of the principal ideals $\langle \alpha \rangle$ and $\langle 2 \rangle$.

8. Consider
$$I = \left\{ \begin{pmatrix} a & 0 \\ b & 0 \end{pmatrix} \in M_2(\mathbb{Z}) : a, b \in \mathbb{Z} \right\}.$$

   (a) Show that $I$ is a subring of $M_2(\mathbb{Z})$.

   (b) Show that $I$ is *not* an ideal of $M_2(\mathbb{Z})$ (using the stronger definition of ideal used for non-commutative rings).

9. Consider the principal ideal $\langle 3 + \sqrt{5} \rangle$ in $\mathbb{Z}[\sqrt{5}]$ (see Exercise 7.2). Prove that
$$\langle 3 + \sqrt{5} \rangle = \{c + d\sqrt{5} : 4 | (c + d)\}.$$

10. Suppose that $p, q$ are distinct prime integers in $\mathbb{Z}$. Prove that
$$\langle p \rangle \cap \langle q \rangle = \langle pq \rangle.$$

11. Let $R$ be a commutative ring with unity and $a, b \in R$. Prove that
$$\langle a, b \rangle = \{ax + by : x, y \in R\}$$
is an ideal; furthermore, show that it is the smallest ideal of $R$ that contains $a$ and $b$.

12. Let $R$ be a commutative ring with unity and $a_1, a_2, \cdots, a_n \in R$. Prove that

$$\langle a_1, a_2, \cdots, a_n \rangle =$$

$$\{a_1 x_1 + a_2 x_2 + \cdots + a_n x_n : x_i \in R, i = 1, 2, \cdots, n\}$$

is an ideal; furthermore, show that it is the smallest ideal of $R$ that contains all the $a_i$'s. We call the $a_i$'s the **generators** of the ideal; we say that the ideal is **finitely generated**. Note that each element of the ideal can be expressed as a *linear combination* of its generators.

13. Consider the domain $\mathbb{Z}[\sqrt{3}]$. Prove that in this ring,

$$\langle 1 + \sqrt{3} \rangle = \{x + y\sqrt{3} : x + y \text{ is an even integer}\}.$$

14. Consider the ring $\mathbb{Z}[\sqrt{2}]$. Prove that

$$\langle 3 + 8\sqrt{2}, 7 \rangle = \langle 3 + \sqrt{2} \rangle.$$

15. Suppose that $a, b, c \in \mathbb{Z}$ and $c = \gcd(a, b)$. Show that $\langle a, b \rangle = \langle c \rangle$. (See Example 9.10 for an illustration of this.)

16. Let $R$ be a commutative ring, and suppose that $I$ and $J$ are ideals. Prove that $I \cap J$ is an ideal. (Note the many examples of intersections of ideals provided in Exercise 10. Also compare this exercise to Exercise 7.9.) Now describe the ideal in Exercise 5 as an intersection of two proper ideals.

17. Suppose that $R$ is a commutative ring, $I$ and $J$ are ideals of $R$, and $I \subseteq J$. Prove that $I$ is an ideal of the ring $J$.

18. Let $R$ be a commutative ring, with ideals $I$ and $J$. Let

$$I + J = \{a + b : a \in I, \ b \in J\}.$$

Prove that $I + J$ is an ideal of $R$. For obvious reasons, we call the ideal $I + J$ the **sum** of the ideals $I$ and $J$.

19. Suppose that $R$ is a commutative ring with unity, and $I$ and $J$ are ideals. Define

$$I \cdot J = \left\{ \sum_{k=1}^{n} a_k b_k : \ a_k \in I, \ b_k \in J, \ n \in \mathbb{N} \right\}.$$

(That is, $I \cdot J$ consists of all possible finite sums of products of elements from $I$ and $J$.)

(a) Prove that $I \cdot J$ is an ideal. We call the ideal $I \cdot J$ the **product** of the ideals $I$ and $J$.

(b) Prove that $I \cdot J \subseteq I \cap J$.

(c) Prove that if $a, b \in R$, we have that $\langle a \rangle \cdot \langle b \rangle = \langle ab \rangle$.

(d) Show by example in $R = \mathbb{Z}$ that we can have $I \cdot J \subset I \cap J$.

20. Suppose that $I, J, K$ are ideals in $R$, a commutative ring with unity. Prove that

$$I \cdot (J + K) = I \cdot J + I \cdot K.$$

21. Let $X$ be a set; consider the power set ring $P(X)$. See Exercise 6.20, where we made $P(X)$ a ring, by equipping it with an addition (symmetric difference) and a multiplication (intersection). Pick a fixed subset $a$ of $X$, and let

$$I = \{b \in P(X) : b \subseteq a\}.$$

    (a) Prove that $I$ is an ideal of $P(X)$.

    (b) Prove that $I$ is a principal ideal (find the generator!).

22. Let $X$ be an arbitrary set, and consider the power set ring $P(X)$. (In the previous exercise we discussed a principal ideal of $P(X)$; that exercise is relevant to the present problem but not strictly necessary.)

    (a) Let $a \in P(X)$. Describe the elements of the principal ideal $\langle a \rangle$.

    (b) Suppose that $X$ has more than one element. Show that $P(X)$ is not a domain.

    (c) Suppose that $X$ has infinitely many elements. Let

    $$I = \{a \in P(X) : a \text{ has finitely many elements}\}.$$

    Prove that $I$ is an ideal of $P(X)$. Show that $I$ is not a principal ideal.

23. Prove that the nilradical of a commutative ring with unity is an ideal. (See Exercise 7.15, where you proved it is a subring.)

24. Suppose that $R$ is a commutative ring with unity, and $r \in R$. Let $A(r) = \{s \in R : rs = 0\}$. (This set is called the **annihilator** of $r$.) Prove that $A(r)$ is an ideal.

25. Generalize Exercise 24: Suppose that $R$ is a commutative ring with unity, and $I$ is an ideal. Let
    $$A(I) = \{s \in R : rs = 0, \text{ for all } r \in I\};$$
    we call $A(I)$ the **annihilator** of $I$. Prove that $A(I)$ is an ideal.

26. Suppose that $R$ and $S$ are domains. Determine all possible ideals of the direct product $R \times S$, which occur as annihilators $A((r, s))$. *Hints:* See Exercise 24 for the definition of annihilator. In this problem there are only four cases.

27. Suppose that $R$ is a commutative ring with unity, and $e$ is an idempotent. In Exercise 7.25, we defined what this means and also considered the subring $Re$; we now know that this subring is the principal ideal $\langle e \rangle$. Prove that the annihilator $A(e)$ is a principal ideal. (See Exercise 24 for the definition of annihilator.)

28. Generalize Theorem 9.3: Suppose that $R$ is a commutative ring with unity and $r, s \in R$, with $r$ not a zero divisor. Prove that $\langle r \rangle = \langle s \rangle$ if and only if $r, s$ are associates.

29. Consider the ring $S$ of real-valued sequences discussed in Exercise 6.19.

    (a) Let $n$ be a fixed positive integer, and let

    $$I_n = \{(s_1, s_2, \ldots) \in S : s_m = 0, \text{ for all } m > n\}.$$

    Prove that $I_n$ is an ideal of $S$. (You proved that $I_1$ is a subring of $S$ in Exercise 7.22.)

(b) Let
$$\Sigma = \{(s_1, s_2, \ldots) \in S : \text{at most finitely many } s_i \neq 0\};$$

prove that $\Sigma$ is an ideal of $S$. (You proved that $\Sigma$ is a subring in Exercise 7.21.) What is the relationship between $\Sigma$ and the $I_i$'s?

(c) Prove that $\Sigma$ is not finitely generated. Recall from Exercise 12 that by this we mean that
$$\Sigma \neq \langle \vec{s}_1, \vec{s}_2, \cdots, \vec{s}_n \rangle,$$

for any finite set of sequences $\vec{s}_i \in \Sigma$.

30. Consider again the ring $S$ of real-valued sequences. Let
$$B = \{(s_1, s_2, \ldots) \in S : \text{there exists } M \in \mathbb{R} \text{ with } |s_n| \leq M, \text{ for all } n\}.$$

These are the **bounded** sequences. Note that $M$ is not fixed in the definition of $B$; that is, different sequences may require different bounds. Prove that $B$ is a subring, but is nonetheless *not* an ideal of $S$.

31. Consider $\mathbb{Q}[x, y]$, the set of all polynomials with coefficients from $\mathbb{Q}$, in the two indeterminates $x$ and $y$. In Exercise 6.23, you showed that the ring of polynomials $R[x]$ with coefficients in any given commutative ring makes sense. Applying this construction with coefficients from $\mathbb{Q}[x]$, where we then have to use a different symbol $y$ for the new indeterminate, gives us the ring $\mathbb{Q}[x, y]$. It turns out (though we won't verify all details here) that addition and multiplication in this ring behave just as you would expect. Formally then, an element of $\mathbb{Q}[x, y]$ can be viewed as an element of the form
$$a_{0,0} + a_{1,0}x + a_{0,1}y + a_{2,0}x^2 + a_{1,1}xy + a_{0,2}y^2 + \cdots,$$

where the $a_{i,j}$'s are rational numbers, and only finitely many of them are not zero.

(a) Provide a nice description of the elements in the ideal $\langle x, y \rangle$.

(b) Show that the ideal $\langle x, y \rangle$ is not principal, thus showing that $\mathbb{Q}[x, y]$ is not a PID.

# Chapter 10

## Polynomials over a Field

In this chapter we generalize what we've learned about $\mathbb{Q}[x]$, the ring of polynomials with rational coefficients, by replacing the rational numbers by entries from an arbitrary field. We will discover that the resulting rings behave very similarly to $\mathbb{Q}[x]$, and hence to the ring $\mathbb{Z}$ as well.

## 10.1 Polynomials with Coefficients from an Arbitrary Field

Suppose we consider polynomials with coefficients not from $\mathbb{Q}$, but from some arbitrary field $F$. We denote this set of polynomials by $F[x]$. Addition and multiplication are defined as in $\mathbb{Q}[x]$, but when coefficients are added or multiplied, it is done in $F$.

**Example 10.1**

Consider $\mathbb{Z}_2[x]$, the set of polynomials with coefficients from $\mathbb{Z}_2$. (So coefficients are either 0 or 1.) Consider the sum and product of $x^2 + x + 1$ and $x^2 + 1$:

$$(x^2 + x + 1) + (x^2 + 1) = (x^2 + x^2) + x + (1 + 1) = x;$$

and

$$\begin{aligned}(x^2 + x + 1)(x^2 + 1) &= (x^4 + x^3 + x^2) + (x^2 + x + 1) \\ &= x^4 + x^3 + (x^2 + x^2) + x + 1 \\ &= x^4 + x^3 + x + 1.\end{aligned}$$

Notice how various of the above terms disappear, because $1 + 1 = 0$ in the ring $\mathbb{Z}_2$ of coefficients. Similar care must be taken with other finite fields.

▷ **Quick Exercise.** Compute the sum, difference, and product of $x^2 + 2x + 1$ and $x^2 + x + 2$ in $\mathbb{Z}_3[x]$ (here, the coefficients consist of 0, 1, or 2). List all the units in $\mathbb{Z}_3[x]$. Then write down all the associates of $x^2 + 2x + 2$ in $\mathbb{Z}_3[x]$. ◁

It is clear that $F[x]$ (like $\mathbb{Q}[x]$) is a commutative ring with unity; in fact, it is an integral domain. But these two rings have much more in common than that.

▷ **Quick Exercise.** Convince yourself that $F[x]$ is an integral domain. ◁

If you return to Chapters 4 and 5 and look at the theorems (and their proofs), you'll see that every theorem about $\mathbb{Q}[x]$ is valid if $\mathbb{Q}$ is replaced by a field $F$ of your choice. The properties of $\mathbb{Q}$ that are important in those theorems are the field properties of $\mathbb{Q}$. As before, the theorem that is the driving force is the Division Theorem.

▷ **Quick Exercise.** Prove the Division Theorem for $F[x]$, by carefully re-reading the proof of this theorem for $\mathbb{Q}[x]$. You will see that multiplicative inverses for coefficients are required in this proof, but nothing else special about $\mathbb{Q}$.   ◁

So, the Division Theorem, the Root Theorem, Euclid's Algorithm, the GCD identity, and the Unique Factorization Theorem hold for $\mathbb{R}[x]$, $\mathbb{C}[x]$, and $\mathbb{Z}_p[x]$ (for prime $p$), as well as $F[x]$ where $F$ is any other field. Notice that to make sense of these theorems in this new and more general context, we must be careful also to define the terms as **irreducible** and **prime**.

▷ **Quick Exercise.** Check that our definitions for these terms for $\mathbb{Q}[x]$ still make sense for $F[x]$.   ◁

▷ **Quick Exercise.** State the Unique Factorization Theorem for $F[x]$, where $F$ is an arbitrary field.   ◁

It is perhaps just as well to pause here a moment, and make certain that you are aware of what we have just done. We have just asserted (and you have checked!) that the statement and proofs of the Division Theorem, the Root Theorem, Euclid's Algorithm, the GCD identity, and the Unique Factorization Theorem generalize when $\mathbb{Q}$ is replaced by any field whatsoever. This is a striking example of the power of generalization and abstraction in algebra. We will devote the rest of this chapter to examining many examples of these theorems in action, using various fields (other than $\mathbb{Q}$) for the coefficients on our polynomials.

**Example 10.2**

Let's find a gcd of $x^3 + x + 1$ and $x^2 + 1$ in $\mathbb{Z}_2[x]$, using Euclid's Algorithm:

$$\begin{aligned} x^3 + x + 1 &= (x^2 + 1)x + 1 \\ x^2 + 1 &= 1(x^2 + 1) + 0. \end{aligned}$$

Therefore, 1 is a gcd of $x^3 + x + 1$ and $x^2 + 1$. Now any other gcd of these two polynomials would be an associate of 1. That is, it would be a non-zero scalar multiple of 1. But in $\mathbb{Z}_2[x]$, the only non-zero scalar is 1 itself. Thus, in $\mathbb{Z}_2[x]$, 1 is the *unique* gcd of $x^3 + x + 1$ and $x^2 + 1$.

▷ **Quick Exercise.** Write 1 as a linear combination of $x^3 + x + 1$ and $x^2 + 1$ in $\mathbb{Z}_2[x]$.   ◁

**Example 10.3**

Let's now compute a gcd of $x^5 + x^2 + 1$ and $x^2 + 2$ in $\mathbb{Z}_3[x]$:

$$\begin{aligned} x^5 + x^2 + 1 &= (x^2 + 2)(x^3 + x + 1) + (x + 2) \\ x^2 + 2 &= (x + 2)(x + 1) + 0. \end{aligned}$$

Therefore, $x + 2$ is a gcd of $x^5 + x^2 + 1$ and $x^2 + 2$ in $\mathbb{Z}_3[x]$.

▷ **Quick Exercise.** Divide $x^5 + x^2 + 1$ by $x + 2$ in $\mathbb{Z}_3[x]$.   ◁

▷ **Quick Exercise.** List *all* the gcds of $x^5 + x^2 + 1$ and $x^2 + 2$ in $\mathbb{Z}_3[x]$.   ◁

## 10.2 Polynomials with Complex Coefficients

Given a quadratic polynomial with real coefficients, you probably recall the **quadratic formula** that provides the roots for the polynomial; you may recall that the formula is proved by means of an algebraic technique called *completing the square*.

To demonstate this, suppose that $f = ax^2 + bx + c \in \mathbb{R}[x]$, with $a \neq 0$. To solve the equation $ax^2 + bx + c = 0$ we can do a little algebra to obtain $x^2 + \frac{b}{a}x = -\frac{c}{a}$. To make the left side of the equation a perfect square, we add the term $\left(\frac{b}{2a}\right)^2$ to both sides of the equation. With a little further algebraic simplification (see Exercise 1), we obtain the usual form for the quadratic formula:

$$x = \frac{-b \pm \sqrt{b^2 - 4ac}}{2a}.$$

The quantity $D = b^2 - 4ac$ is called the **discriminant** of $f$. Clearly the two roots are equal if $D = 0$, and we get two distinct real roots if $D > 0$.

But if $D < 0$, the quadratic formula gives two distinct complex roots, which are conjugates of one another. Now by the Root Theorem these two roots also give a factorization of $f$ (in $\mathbb{C}[x]$) into linear polynomials:

$$f = ax^2 + bx + c = a\left(x - \frac{-b + \sqrt{b^2 - 4ac}}{2a}\right)\left(x - \frac{-b - \sqrt{b^2 - 4ac}}{2a}\right).$$

But in this case it is now clear that $f$ *cannot* be non-trivially factored in $\mathbb{R}[x]$. For if it did so factor, it would then also have real roots, making more than two roots for $f$ in $\mathbb{C}$, which is impossible.

We have thus shown that if $f \in \mathbb{R}[x]$ is a quadratic polynomial with negative discriminant, then it is an irreducible element of $\mathbb{R}[x]$. But any such quadratic polynomial *can* be factored further in $\mathbb{C}[x]$.

**Example 10.4**

Let's factor the polynomial
$$f = x^3 - 7x^2 + 17x - 15 \in \mathbb{R}[x],$$
into irreducibles, in both $\mathbb{R}[x]$ and $\mathbb{C}[x]$. By the Root Theorem, $x - 3$ is a factor of $f$ because $f(3) = 0$. We can then factor $x - 3$ out of $f$:
$$f = (x - 3)(x^2 - 4x + 5).$$
Because $x - 3$ is linear, it is irreducible. Now consider $x^2 - 4x + 5$. Its discriminant is $D = -4 < 0$, and so $x^2 - 4x + 5$ is the other irreducible factor of $f$ in $\mathbb{R}[x]$. But if we apply the quadratic formula we obtain two linear irreducible factors in $\mathbb{C}[x]$:
$$f = (x - 3)(x - 2 + i)(x - 2 - i).$$

The example we've just examined, showing that more factorization is possible in $\mathbb{C}[x]$ than in $\mathbb{R}[x]$, is a particular example of an important general fact about $\mathbb{C}[x]$: *Every non-constant polynomial in $\mathbb{C}[x]$ can be factored into linear factors.* This fact is known as the *Fundamental Theorem of Algebra*, a very important theorem first proved rigorously by Gauss. We won't prove this theorem here. The easiest proof of the Fundamental Theorem of Algebra uses complex analysis and is accessible only to those who have taken an introductory course in that subject. We will state the Fundamental Theorem in an (apparently weaker) form, and then prove the statement we've made above as a corollary.

**Theorem 10.1    Fundamental Theorem of Algebra**        *Every non-constant polynomial in $\mathbb{C}[x]$ has a root in $\mathbb{C}$.*

**Proof**:    We omit this proof, and refer the reader to any introductory text in complex analysis.                                                                                    $\square$

**Corollary 10.2** *Every non-constant polynomial with degree $n$ in $\mathbb{C}[x]$ is linear or can be factored as a product of $n$ linear factors. Thus, the irreducibles in $\mathbb{C}[x]$ consist exactly of the linear polynomials.*

**Proof**:    Let $f$ be a polynomial in $\mathbb{C}[x]$. The essence of this proof is to apply the Fundamental Theorem of Algebra to $f$ and its factors, over and over again. We make this precise by means of an induction argument on $n$, the degree of $f$. If $n = 1$, then the polynomial is itself a linear polynomial. Now suppose $n > 1$. By the Fundamental Theorem of Algebra, $f$ has a root $\alpha$. By the Root Theorem, $x - \alpha$ is a factor of $f$, and so $f = (x - \alpha)g$, where $\deg(g) = n - 1$. By the induction hypothesis, $g$ is either linear or a product of linear factors, and thus $f$ itself is a product of linear factors.

We've already observed that linear polynomials are *always* irreducible. Any polynomial in $\mathbb{C}[x]$ of higher degree can be factored and so is not irreducible.                        $\square$

One important observation needs to be made about the Fundamental Theorem of Algebra: It merely asserts the *existence* of the linear factors. Neither in the statement of the theorem nor in its (omitted) proof is an effective method provided for actually finding them. We will return to this subject in the final section of this book, on what is called *Galois Theory*.

---

## 10.3    Irreducibles in $\mathbb{R}[x]$

We now use the Fundamental Theorem of Algebra to find all irreducible polynomials in $\mathbb{R}[x]$. We have already proved half of the following theorem:

**Theorem 10.3** *The irreducible polynomials in $\mathbb{R}[x]$ are precisely the linear polynomials, and those quadratic polynomials with negative discriminant.*

**Proof**:    We know already that linear polynomials are irreducible. And in our discussion of the quadratic formula in Section 10.2 above we have argued already that if a quadratic polynomial has a negative discriminant, then it is irreducible in $\mathbb{R}[x]$. We now will argue that linear polynomials and quadratics with negative discriminant are the *only* irreducibles in $\mathbb{R}[x]$.

Suppose that $f$ is a non-linear irreducible in $\mathbb{R}[x]$. Then $f$ can have no real roots (else it would have a linear factor, by the Root Theorem). But $f$ has complex roots, by the Fundamental Theorem of Algebra. Let $\alpha$ be a complex root of $f$. Suppose that $\alpha = s + ti$, where $s$ and $t$ are real and $t \neq 0$. We now use $\bar{\alpha} = s - ti$, the complex conjugate of $\alpha$. We define

$$g = (x - \alpha)(x - \bar{\alpha}) \;=\; (x - (s + ti))(x - (s - ti))$$
$$=\; x^2 - 2sx + (s^2 + t^2),$$

which is a polynomial in $\mathbb{R}[x]$. We now apply the Division Theorem for $\mathbb{R}[x]$. By this theorem, $f = gq + r$, where $f$, $g$, $q$, and $r$ all have real coefficients. Now think of these polynomials as being in $\mathbb{C}[x]$. So,

$$0 = f(\alpha) = g(\alpha)q(\alpha) + r(\alpha).$$

Because $g(\alpha) = 0$, we have $r(\alpha) = 0$. But $\deg(r) < 2$ and so $r = cx + d$ with $c, d \in \mathbb{R}$. But $r(\alpha) = 0$; that is, $c\alpha + d = 0$. But $\alpha$ is not real, and therefore we must have that $c = d = 0$. Hence, $f = gq$, and so $f$ is not irreducible unless $q$ is a scalar. But then $f$ is of degree two with negative discriminant, which proves what we wanted. $\qquad\square$

---

## 10.4  Extraction of Square Roots in $\mathbb{C}$

We now know that we can use the quadratic formula to factor *any* quadratic polynomial in $\mathbb{R}[x]$, over the complex numbers. But what about quadratic polynomials in $\mathbb{C}[x]$? We know by the Fundamental Theorem of Algebra that such polynomials factor and thus must factor into two linear factors. Surely the quadratic formula should still work. But we must extract a square root when using the quadratic formula, and if the original quadratic polynomial belongs to $\mathbb{C}[x]$, this means extracting the square root of an arbitrary complex number.

**Example 10.5**

Consider the quadratic equation

$$x^2 - ix - (1 + i) = 0.$$

If we proceed with the quadratic formula, we obtain the following:

$$x = \frac{i \pm \sqrt{3 + 4i}}{2}.$$

▷ **Quick Exercise.**  Check the above computation. ◁

Unfortunately, we have a complex number under the radical sign. If we are lucky enough to observe that $(2 + i)^2 = 3 + 4i$, we can then obtain the two roots $x = -1$ and $x = 1 + i$.

▷ **Quick Exercise.**  Check our arithmetic, and verify that the given complex numbers are roots of the original equation. ◁

In the previous example we were lucky. But how can we extract square roots of arbitrary complex numbers? To do this, we make use of the geometric representation for complex numbers we discussed in Chapter 8.

Suppose that $\alpha = a + bi \in \mathbb{C}$. We wish to find roots to the polynomial $x^2 - \alpha$. From Chapter 8, we can factor $\alpha$ as

$$\alpha = |\alpha|(\cos\theta + i\sin\theta) = |\alpha|e^{i\theta},$$

where $\theta$ is the argument of $\alpha$. We wish to find a complex number whose square is this. But because the modulus of a product is the product of the moduli (Theorem 8.3), the modulus of a square root of $\alpha$ must be $\sqrt{|\alpha|}$. And one argument that will surely work is $\theta/2$. Thus,

$$\beta = \sqrt[4]{a^2 + b^2}(\cos(\theta/2) + i\sin(\theta/2))$$

is a square root of the complex number $\alpha$, and of course $-\beta$ is another. Because the polynomial $x^2 - \alpha$ can have at most two roots, these must be the *only* roots it has; these roots are distinct unless $\alpha = 0$. Thus, every non-zero complex number has exactly two square roots.

**Example 10.6**

Let's compute the square roots of $i$. Its modulus is 1, and so its square roots will also have modulus 1. Its argument is $\pi/2$, and so one square root must be

$$\cos(\pi/4) + i\sin(\pi/4) = \frac{1}{\sqrt{2}} + \frac{1}{\sqrt{2}}i$$

and the negative of this is the other.

$\triangleright$ **Quick Exercise.** Check this explicitly by squaring both these complex numbers. $\triangleleft$

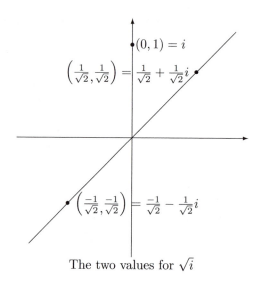

The two values for $\sqrt{i}$

$\triangleright$ **Quick Exercise.** Apply this method to compute the square roots of $1 + \sqrt{3}i$. $\triangleleft$

In Exercise 23, you will extend this method to compute the $n$th roots of an arbitrary complex number.

You might suppose that we could apply the quadratic formula in any field, but we would then have to be able to take the square root in this more general context, and there is nothing in the field axioms which guarantees we can do this. However, such generalizations are possible, and you will explore this in the context of $\mathbb{Z}_7$ in Exercise 29.

## Historical Remarks

Naturally, given a particular polynomial with real coefficients it may be *very* difficult to come up with the factorization into irreducibles, particularly if the degree of the polynomial is large. Of course, the quadratic formula allows us to find such a factorization for all polynomials of degree two, and there exist progressively more complicated cubic and quartic formulas, which allow us to factor all polynomials of degree 3 and degree 4. (See Exercise 12 for the degree 3 case, and Exercise 20 for degree 4.) However, a difficult theorem due to the 19th-century Norwegian mathematician Abel states that there are polynomials of degree 5 or higher for which *no* explicit determination of the roots is possible, using the ordinary operations of addition, subtraction, multiplication, division, and extraction of roots. We will encounter this theorem in the last chapter in this book.

The Fundamental Theorem of Algebra was one of Karl Gauss's favorites; he returned to it several times over the course of his career, proving it by different means. He had his first proof in hand before his twentieth year. The theorem had been conjectured (and believed) by such predecessors of Gauss as Lambert and Legendre. One of the most important obstacles preventing a proof before the time of Gauss was that there still remained doubts in the minds of most mathematicians as to the status of complex numbers. You may yourself when first encountering complex numbers have doubted they were as 'real' as real numbers; it is precisely this attitude which is reflected in the terms 'real' and 'imaginary'. Gauss was among the first mathematicians to use complex numbers with confidence and rigor. He made full use of the geometric interpretation we have discussed. In part, this reduction of complex arithmetic to geometry gave some reassurance to those who doubted the possibility of such calculations. Such doubts (which had been the rule in mathematical circles since the 16th century) were forgotten in the generation of mathematicians after Gauss.

## Chapter Summary

In this chapter we discussed how the ring $F[x]$ of polynomials with coefficients from a field $F$ has properties analogous to those of $\mathbb{Q}[x]$. In particular, this means that the *Division Theorem*, *Euclid's Algorithm*, the *GCD Identity*, the *Unique Factorization Theorem*, and the *Root Theorem* hold for $F[x]$.

We also considered the important examples $\mathbb{C}[x]$ and $\mathbb{R}[x]$. We stated the *Fundamental Theorem of Algebra* and inferred from it that the irreducible polynomials in $\mathbb{R}[x]$ are exactly the linear polynomials, and the quadratic polynomials with negative discriminant.

## Warm-up Exercises

a. Calculate the quotient and remainder for the following, in various rings $F[x]$:

   (a) $x^3 + 4x + 1$ by $x + 2$, in $\mathbb{Z}_5[x]$.
   (b) $x^3 + 4x + 1$ by $2x + 1$, in $\mathbb{Z}_5[x]$.
   (c) $x^3 - (2 + i)x^2 + 5$ by $ix - 1$, in $\mathbb{C}[x]$.
   (d) $x^4 - 2x^3 + \frac{1}{3}$ by $\pi x + 1$, in $\mathbb{R}[x]$.

b. Use the Root Theorem to check for roots of $x^4 + 4$ in $\mathbb{Z}_5[x]$. Use your result to completely factor this polynomial.

c. Why are $x + i$ and $(1 + i)x + (-1 + i)$ associates in $\mathbb{C}[x]$?

d. List all associates of $2x^2 + 3x + 3$ in $\mathbb{Z}_5[x]$. Is it clearer how to factor this polynomial, if you consider one of its associates?

e. Why is $x^2 + 2$ irreducible in $\mathbb{Z}_5[x]$?

f. Factor $x^3 - 2$ into irreducibles in

$$\mathbb{Q}[x], \ \mathbb{R}[x], \ \mathbb{C}[x].$$

Repeat this problem for $x^2 + \pi$.

g. Extract the square roots of $-3 + \sqrt{3}i$ in $\mathbb{C}$.

h. Let $F$ be a field. Could the ring $F[x]$ be a field? Why or why not?

---

# Exercises

1. Use the method of completing the square to complete the proof of the quadratic formula for finding roots of polynomials of degree two in $\mathbb{R}[x]$, as begun at the beginning of Section 10.2.

2. (a) Why does every non-zero complex number have exactly two square roots?

   (b) Given part (a), check that the proof of the quadratic formula obtained in Exercise 1 still holds in $\mathbb{C}[x]$.

   (c) Use the quadratic formula to compute the roots of the polynomials
   $$x^2 - (3 + 2i)x + (1 + 3i) \text{ and } x^2 - (1 + 3i)x + (-2 + 2i).$$

3. Give examples of two different polynomials in $\mathbb{Z}_5[x]$ that are identical as functions over $\mathbb{Z}_5$. This shows that equality of polynomials in $F[x]$ cannot be thought of as equality of the corresponding polynomial *functions*. (See the Quick Exercise in Section 4.1 for the $F = \mathbb{Z}_2$ case, and Exercise 4.12 for the case $F = \mathbb{Z}_3$.)

4. Consider the polynomial $f = x^3 + 3x^2 + 2x \in \mathbb{Z}_6[x]$. Show that this polynomial has more than three roots in $\mathbb{Z}_6$. Why doesn't this contradict the Root Theorem?

5. Find a gcd of $x^3 + 4x^2 + 4x + 9$ and $x^2 + x - 2$ in $\mathbb{Q}[x]$; in $\mathbb{R}[x]$; in $\mathbb{C}[x]$.

6. Find a gcd of $x^3 + x^2 + x$ and $x^2 + x + 1$ in $\mathbb{Z}_3[x]$; in $\mathbb{Z}_5[x]$; in $\mathbb{Z}_{11}[x]$.

7. Write the gcd you found in Exercise 5 as a linear combination of the two polynomials involved.

8. Show that if $f$ is a polynomial with real coefficients and $\alpha = s + ti$ is a root of $f$ in $\mathbb{C}$, then so is $\bar{\alpha} = s - ti$.

9. Use Exercise 8 to construct a polynomial in $\mathbb{R}[x]$ with roots $\frac{1}{2}$, $i$, $2 - i$.

10. Factor $x^4 + x^3 + 2x^2 + 1$ into irreducibles in $\mathbb{Z}_3[x]$. Be sure to prove that your factors are in fact irreducible.

11. Determine all the irreducible elements of $\mathbb{Z}_2[x]$, with degree less than or equal to 4.

12. In this exercise, we describe the cubic formula for factoring an arbitrary polynomial of degree 3 in $\mathbb{R}[x]$. This version of the formula is called the *Cardano-Tartaglia* formula, after two 16th-century Italian mathematicians involved in its discovery. Consider the polynomial $f = x^3 + ax^2 + bx + c \in \mathbb{R}[x]$ (by dividing by the lead coefficient if necessary, we have assumed without loss of generality that it is 1).

    (a) Show that the change of variables $x = y - \frac{1}{3}a$ changes $f$ into a cubic polynomial that lacks a square term; that is, a polynomial of the form $g = f(y - \frac{1}{3}a) = y^3 + py + q = 0$. *Note:* This process is called *depressing the conic*. Clearly we can solve $f = 0$ for $x$ if and only if we can solve $g = 0$ for $y$.

(b) Find explicit solutions $u, v$ to the pair of simultaneous equations

$$v^3 - u^3 = q$$
$$uv = \frac{1}{3}p.$$

*Hint:* These equations reduce to a *quadratic* equation in $u^3$ or $v^3$.

(c) Prove the identity

$$(u - v)^3 + 3uv(u - v) + (v^3 - u^3) = 0$$

and use it to show that $y = u - v$ is a solution to the cubic equation $y^3 + py + q = 0$.

(d) Let $D = q^2 + \frac{4p^3}{27}$. (This is called the *discriminant* of the conic.) Conclude that

$$y = \sqrt[3]{\frac{-q + \sqrt{D}}{2}} - \sqrt[3]{\frac{q + \sqrt{D}}{2}}$$

is a root for $g = 0$. (This is just $u - v$.)

13. In Exercise 12, there is an apparent ambiguity arising from the plus or minus when extracting the square root of $D$ to obtain values for $u^3$ and $v^3$. However, show that we obtain the same value for the root $u - v$, regardless of which choice is made.

14. Apply the Cardano-Tartaglia formula to find a root of the cubic equation

$$x^3 = 6x + 9.$$

Then factor $x^3 - 6x - 9$, and use the quadratic formula to obtain the remaining two roots.

15. Suppose as in Exercise 12 that $g = y^3 + py + q$ is a cubic polynomial with real coefficients, and $y = u - v$ is the root given by the Cardano-Tartaglia formula. Suppose that $D > 0$. (Thus $u$ and $v$ are real numbers.) Let $\zeta = e^{\frac{2\pi}{3}}$ be a cube root of unity (called the primitive cube root of unity in Exercise 25 below). Argue that the other two distinct roots of $g = 0$ are the complex conjugates $u\zeta - v\zeta^2$ and $u\zeta^2 - v\zeta$. *Note:* Be sure and check both that these are roots and that they are necessarily distinct.

16. Apply Exercise 15 to obtain all three of the roots for the cubic $y^3 + 3y + 1$.

17. An interesting and surprising conclusion one can draw from Exercise 15 is that if the discriminant $D > 0$, then the cubic polynomial $y^3 + py + q \in \mathbb{R}[x]$ necessarily has exactly one real root, and a conjugate pair of complex roots. In this exercise you will use elementary calculus to verify this fact again:

(a) Consider the function $g(y) = y^3 + py + q$. Suppose that $p > 0$. Compute the derivative $g'(y)$, and use it to argue that $g$ has exactly one real root, and consequently two complex roots.

(b) Suppose now that $p = 0$. Then conclude that $q \neq 0$. In this simple case, what are the roots of $g$?

(c) Now suppose that $p < 0$. Compute the two roots of $g'(y) = 0$. Argue that the values of $g$ at these two roots are both positive (using the assumption that $D > 0$). Why does this mean that $g$ has exactly one real root?

18. (a) Find a cubic polynomial whose roots are $1 + \sqrt{3}, 1 - \sqrt{3}, -3$. *Hint:* Use the Root Theorem.

    (b) Apply the Cardano-Tartaglia cubic formula derived in Exercise 12 to solve the cubic obtained in part (a).

    (c) The answer you have obtained should be one of the roots you started with, but it does not *appear* to be. Can you explain this?

    (d) Obtain the other two roots for this polynomial, by using the same strategy as in Exercise 15. Note that we must use complex arithmetic to obtain these three real numbers!

19. Exercise 18 is a particular example of what is called the *irreducible case* for a real cubic. Show that in case $D < 0$, we obtain a real root for the polynomial $g = y^3 + py + q$ by an appropriate choice for $u$ and $v$.

20. In this problem we will explore Ferrari's approach to solving the general quartic equation. Consider the arbitrary quartic

$$f = x^4 + a_3 x^3 + a_2 x^2 + a_1 x + a_0 \in \mathbb{R}[x].$$

    (a) Find a linear change of variables $y = x + m$ so as to *depress* the quartic – that is, to eliminate the cubic term, as we eliminated the quadratic term in Exercise 12.

    (b) We may by part (a) assume that our quartic equation is of the form $x^4 = px^2 + qx + r$, where $p, q, r \in \mathbb{R}$. Add the term $2bx^2 + b^2$ to both sides of this equation. Clearly this makes the left-hand side of the equation a perfect square. We would like to choose $b$ so that the right-hand side is also a perfect square. Obtain an equation for $b$ (in terms of $p, q, r$) that makes this true. The equation you obtain should be a cubic equation in $b$. Explain why a (real) solution to this cubic will always lead to a solution to the quartic equation. How do you then get all four solutions?

21. Carry out Ferrari's method for solving quartic equations, for the equation $x^4 = -x^2 - 4x + 3$.

22. Use a calculator and our algorithm to compute (approximations of) the square roots of $5 + 11.6i$ in $\mathbb{C}$.

23. Suppose that the complex number $\alpha = a + bi$ has been factored as

$$\alpha = \sqrt{a^2 + b^2}(\cos\theta + i\sin\theta) = |\alpha|e^{i\theta},$$

as we did in the text when computing the square roots of $\alpha$.

    (a) Show that

$$\beta_k = \sqrt[2n]{a^2 + b^2}\left(\cos\left(\frac{\theta + 2\pi k}{n}\right) + i\sin\left(\frac{\theta + 2\pi k}{n}\right)\right),$$

    for $k = 0, 1, 2, \cdots n - 1$, are all $n$th roots of $\alpha$.

    (b) Show that these are all distinct roots.

    (c) Why is this the *complete* list of $n$th roots of $\alpha$?

24. What are the five complex fifth roots of 1? What are the five complex fifth roots of $1 + i$?

25. We generalize the previous exercise. Let $p$ be a positive prime integer. In this problem we will describe the $p$th roots of 1, which are customarily called the $p$th **roots of unity**. If we set

$$\zeta = \cos(2\pi/p) + i\sin(2\pi/p) = e^{\frac{2\pi i}{p}},$$

show that each of the following are $p$th roots of unity:

$$1, \ \zeta, \ \zeta^2, \ \cdots \zeta^{p-1}.$$

Be sure to argue that these numbers are all distinct. (We call $\zeta$ the **primitive** $p$th root of unity. Note that these numbers (other than 1) are the roots of the cyclotomic polynomial

$$\Phi_p = \frac{x^p - 1}{x - 1} = x^{p-1} + x^{p-2} + \cdots + x + 1$$

first encountered in Exercises 4.5 and 5.17.)

26. A field $F$ is said to be **algebraically closed** if every polynomial $f \in F[x]$ with $\deg(f) \geq 1$ has a root in $F$; we can rephrase this definition roughly by saying that a field is algebraically closed if it satisfies the Fundamental Theorem of Algebra. Thus, $\mathbb{C}$ is algebraically closed, while $\mathbb{R}$ and $\mathbb{Q}$ are not. Show that for every prime $p$, the field $\mathbb{Z}_p$ is not algebraically closed.

27. Show that the field in Exercise 8.15 is not algebraically closed. (See the previous exercise for a definition.)

28. Show that, if $F$ is a field with infinitely many elements, then $f(x) = g(x)$ for all $x \in F$ implies that $f = g$ as polynomials. (We have already seen that this is not the case if $F$ is a finite field. For example, consider $x^2 + x + 1$ and 1 in $\mathbb{Z}_2[x]$.)

29. In this problem we explore the quadratic formula, in the context of the field $\mathbb{Z}_7$.

    (a) By brute force, square all the elements of $\mathbb{Z}_7$, and thus discover exactly which elements in this field have a square root. How many square roots does each such non-zero field element have?

    (b) Let's now suppose that we have a quadratic equation $x^2 + bx + c = 0$ in $\mathbb{Z}_7$. (We will assume without lost of generality that the $x^2$ coefficient is 1.) We can rewrite this as $x^2 + bx = -c$. What element should we add to both sides to make the left side a perfect square?

    (c) Choose values for $b, c$ so that the constant on the right side has a square root in $\mathbb{Z}_7$, thus obtaining two solutions to your quadratic equation.

    (d) Choose values for $b, c$ so that the constant on the right side has no square root in $\mathbb{Z}_7$. What does this mean about the original equation?

    (e) Choose values for $b, c$ where your quadratic equation has a repeated root.

# Section II in a Nutshell

This section defines three important algebraic structures: rings, integral domains, and fields.

Well-known objects ($\mathbb{Z}$, $\mathbb{Q}[x]$, $\mathbb{Z}_m$, $\mathbb{Q}$, $\mathbb{R}$, and $\mathbb{C}$) share many algebraic properties. These properties define an abstract object called a *ring*:

A *ring* $R$ is a set of elements on which two binary operations, addition ($+$) and multiplication ($\cdot$), are defined that satisfy the following properties for all $a, b, c \in R$:

1. (*Addition is commutative*) $a + b = b + a$

2. (*Addition is associative*) $(a + b) + c = (a + (b + c)$

3. (*Additive identity exists*) There exists an element $0$ in $R$ such that $a + 0 = a$.

4. (*Additive inverses exist*) For each element $a$ in $R$, there exists an element $x$ such that $a + x = 0$.

5. (*Multiplication is associative*) $(a \cdot b) \cdot c = a \cdot (b \cdot c)$

6. (*Multiplication distributes over addition*) $a \cdot (b+c) = a \cdot b + a \cdot c$, $(b+c) \cdot a = b \cdot a + c \cdot a$

Note that the multiplication in a ring need not be commutative. A ring where multiplication is commutative is called a *commutative ring*, naturally. All the examples of rings we've listed above are commutative rings. An example of a non-commutative ring is $M_2(\mathbb{Z})$, the collection of $2 \times 2$ matrices with integer entries.

Addition in rings has some useful properties:

(Theorem 6.1) Suppose $R$ is a ring and $a, b \in R$.

1. (*Additive Cancellation*) If $a + b = a + c$, then $b = c$.

2. (*Solution of equations*) The equation $a + x = b$ always has a unique solution in $R$.

3. (*Uniqueness of additive inverses*) Every element of $R$ has exactly one additive inverse.

4. (*Uniqueness of additive identity*) There is only one element of $R$ that satisfies the equations $z + a = a$, for all $a$; namely, the element $0$.

A *subring* $S$ of ring $R$ is subset of $R$ that is itself a ring under the operations induced from $R$. To determine whether a subset of a ring is a subring, the *Subring Theorem* asserts that it is not necessary to verify all the rules that define a ring:

(Theorem 7.1) A non-empty subset of a ring is a subring under the same operations if and only if it is closed under multiplication and subtraction.

Some rings $R$ have a *unity*, or *multiplicative identity*; that is, an element $u \in R$ where $au = ua = a$ for all $a \in R$. The number $1$ is the unity in $\mathbb{Z}$, $\mathbb{Q}$, $\mathbb{R}$, and $\mathbb{C}$. The scalar polynomial $1$ is the unity in $\mathbb{Q}[x]$ and $\mathbb{Z}[x]$. And the matrix $\begin{pmatrix} 1 & 0 \\ 0 & 1 \end{pmatrix}$ is the unity in $M_2(\mathbb{Z})$. The residue class $[1]$ is the unity in $\mathbb{Z}_m$. But $2\mathbb{Z}$ has no unity. If the unity exists, it is unique.

An element $a \neq 0$ is a *zero divisor* if there is an element $b \neq 0$ with $ab = 0$. A commutative ring with unity is an *integral domain* if it has no zero divisors. $\mathbb{Z}$, $\mathbb{Q}$, $\mathbb{R}$, $\mathbb{C}$, $\mathbb{Q}[x]$, and $\mathbb{Z}[x]$ are all integral domains. $\mathbb{Z}_n$ is an integral domain if and only if $n$ is prime. $\mathbb{Z} \times \mathbb{Z}$ is a commutative ring with unity which is not an integral domain.

Integral domains have the nice property of multiplicative cancellation:

(Theorem 8.1) If $R$ is an integral domain and $a, b, c \in R$ with $a \neq 0$, then $ab = ac$ implies that $b = c$.

If $R$ is a ring with unity 1, then an element $a \in R$ is a *unit* if there exists $b \in R$ such that $ab = 1$. In this case, $b$ is said to be the *multiplicative inverse* of $a$. If all the non-zero elements of a commutative ring with unity are units, then we say the ring is a *field*. The rings $\mathbb{Q}$, $\mathbb{R}$, and $\mathbb{C}$ are all fields but $\mathbb{Z}$, $\mathbb{Q}[x]$, and $\mathbb{Z}[x]$ are not fields. All fields are integral domains (Theorem 8.2). $\mathbb{Z}_p$ is a field, for prime $p$ (Theorem 8.5). Indeed, all finite integral domains are fields (Theorem 8.8).

The field $\mathbb{Z}_p$ is the setting for the important Fermat's Little Theorem:

(Theorem 8.7) If $p$ is prime and $0 < x < p$, then $x^{p-1} \equiv 1 \pmod{p}$. Thus, $x^{p-2} \equiv x^{-1}$ in $\mathbb{Z}_p$.

A map $\varphi : X \to Y$ between sets $X$ and $Y$ is said to be *surjective* if it is onto, *injective* if it is one-to-one, and *bijective* if it is both one-to-one and onto. Furthermore if $R$ and $S$ are both rings, the map $\varphi : R \to S$ is a *ring isomorphism* if it is a bijection where $\varphi(r + s) = \varphi(r) + \varphi(s)$ and $\varphi(rs) = \varphi(r)\varphi(s)$ for all $r, s \in R$. Here we say that $\varphi$ preserves the ring operations. The existence of such a map tells us that as rings, $R$ and $S$ are essentially the same.

For a commutative ring $R$ and $a \in R$ the *principal ideal* generated by $a$ is $\langle a \rangle = \{b \in R : b = ax, \text{ for some } x \in R\}$.

(Theorem 9.1) Let $R$ be a commutative ring with unity, with $r \in R$. Then $\langle r \rangle = R$ if and only if $r$ is a unit.

To generalize, if $R$ is a commutative ring with unity, an *ideal* of $R$ is a non-empty set $I$ of $R$ where (1) $I$ is closed under subtraction, and (2) if $a \in I$ and $r \in R$, then $ar \in I$. This last property says that $I$ absorbs multiplication by any element in $R$. $I$ is a subring of $R$. $\{0\}$ is the *trivial* ideal and $R$ is the *improper ideal* of $R$. Every non-zero ring has these two ideals.

Fields have a paucity of ideals:

(Corollary 9.4) A commutative ring with unity is a field if and only if its only ideals are the trivial and improper ideals.

A principal ideal is an ideal and $\langle a \rangle$ is the smallest ideal of $R$ containing $a$. Not all ideals are principal. But if all ideals of an integral domain are indeed principal, we call the domain a *principal ideal domain (PID)*. Important PIDs are $\mathbb{Z}$ and $\mathbb{Q}[x]$. $\mathbb{Z}[x]$ is not a PID.

Finally, consider $F[x]$, the polynomials over an arbitrary field $F$. $F[x]$ is an integral domain. The Division Theorem, the Root Theorem, Euclid's Algorithm, the GCD identity and Unique Factorization all hold for $F[x]$. A particularly important example is $F = \mathbb{C}$, the field of complex numbers. $\mathbb{C}[x]$ satisfies the Fundamental Theorem of Algebra 10.1: every non-constant polynomial in $\mathbb{C}$ has a root.

# Part III

# Ring Homomorphisms and Ideals

# Chapter 11

## Ring Homomorphisms

Up to now we have mostly examined the relationship between rings by looking at properties they might have in common. But some pairs of rings can actually be placed into a rather closer relationship by means of a function between them. The most important example of this idea is the relationship between $\mathbb{Z}$ and $\mathbb{Z}_n$, as given by the residue function $\varphi : \mathbb{Z} \to \mathbb{Z}_n$ defined by $\varphi(m) = [m]_n$. Another example is the evaluation function $\psi : \mathbb{Q}[x] \to \mathbb{Q}$ defined by $\psi(f) = f(a)$ (where $a$ is some fixed rational number). A third more general example is the idea of ring isomorphism we introduced in Chapter 7: A bijection between two rings that preserves addition and multiplication. Exploring the general context of these examples will then give us a new tool with which to study rings.

## 11.1 Homomorphisms

Now consider the functions $\varphi$ and $\psi$ defined above. They certainly have rings as their ranges and domains; however, they also carry significant parts of the structure of the domain ring over to the range ring. We make this precise in the following definition: Let $R$ and $S$ be rings and $\varphi : R \to S$ a function such that

$$\varphi(a + b) = \varphi(a) + \varphi(b)$$

and

$$\varphi(ab) = \varphi(a)\varphi(b),$$

for all $a, b \in R$. We call this a **ring homomorphism**. Speaking more colloquially, we say that $\varphi$ *preserves addition* and $\varphi$ *preserves multiplication*.

Later in the book we will consider homomorphisms between other algebraic structures than rings and will then have to be careful to refer to *ring* homomorphisms, as opposed to others we might discuss; but for now the term homomorphism will suffice. The word homomorphism comes from Greek roots meaning 'same shape'. We shall see as we explore this concept how appropriate this term is.

Notice that the examples of ring isomorphisms we encountered in Chapters 7 and 8 are of course ring homomorphisms too. Any bijective ring homomorphism is called a **ring isomorphism**. The prefix 'iso' means equal, and thus indicates a closer relationship between the two rings.

It is time now to look at some examples of ring homomorphisms. We will begin with specific examples of the functions with which we started the chapter.

**Example 11.1**

Consider the residue function $\varphi : \mathbb{Z} \to \mathbb{Z}_4$ defined by $\varphi(m) = [m]_4$. Thus, for example,

$\varphi(7) = [3]$. Is this function a homomorphism? That is, does it preserve addition and multiplication? Let's check addition:

$$\varphi(a + b) = [a + b] = [a] + [b] = \varphi(a) + \varphi(b).$$

The crucial step here is the second equal sign; it holds because of the way we defined addition in $\mathbb{Z}_4$, back in Chapter 3. The proof for multiplication works just the same. We can view the homomorphism $\varphi$ as a somewhat more abstract and precise way of conveying the fact that we already know: The operations on $\mathbb{Z}_4$ are related to those on $\mathbb{Z}$. It is obvious that the function takes on all values in the range $\mathbb{Z}_4$: It is onto. But different integers are taken to the same element of $\mathbb{Z}_4$: It is not one-to-one. To rephrase: The residue function is a surjection but not an injection. Notice of course that there's nothing special about 4: The residue function is a surjective homomorphism, for any modulus.

## Example 11.2

Consider the evaluation function $\varphi : \mathbb{Q}[x] \to \mathbb{Q}$ defined by $\varphi(f) = f(2)$: evaluation of polynomials at 2. For example, if $f = x^2 - 3x + 1$, then $\varphi(f) = 4 - 6 + 1 = -1$. This is also a homomorphism:

$$\varphi(f + g) = (f + g)(2) = f(2) + g(2) = \varphi(f) + \varphi(g).$$

Once again, the crucial step is the second equal sign. It holds because of the way we defined the addition of polynomials in Chapter 4. The proof for multiplication is just the same. It is easy to see that this homomorphism is surjective: To get any rational number $q$, just evaluate the constant polynomial with value $q$ at 2. Notice of course that there's nothing special about 2: The evaluation function is a surjective homomorphism, for any element of $\mathbb{Q}$. Even more generally, we could define an evaluation homomorphism for any ring of polynomials $R[x]$, where $R$ is a commutative ring, and we evaluate at any fixed $r \in R$.

## Example 11.3

Consider the function $\pi : \mathbb{Z} \times \mathbb{Z} \to \mathbb{Z}$ defined by $\pi(a, b) = a$. It is easy to see that this is a homomorphism. For example, it preserves multiplication because

$$\pi((a, b) \cdot (c, d)) = \pi(ac, bd) = ac = \pi(a, b) \cdot \pi(c, d).$$

This function is called a **projection** homomorphism (we are projecting on the first component of the ordered pair); it is clearly surjective. See the diagram below.

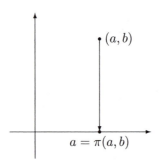

**Example 11.4**

Consider instead the function $\iota : \mathbb{Z} \to \mathbb{Z} \times \mathbb{Z}$ defined by $\iota(a) = (a, a)$. It is easy to see that this is a homomorphism. For example, it preserves addition because

$$\iota(a + b) = (a + b, a + b) = (a, a) + (b, b) = \iota(a) + \iota(b).$$

This function is injective but not surjective. Note however that we can restrict the range of this function to the subring $D = \{(a, a) \in \mathbb{Z} \times \mathbb{Z} : a \in \mathbb{Z}\}$, and then $\iota : \mathbb{Z} \to D$ is a bijection, and an example of a ring isomorphism. This example is essentially the content of Exercise 7.8b. We will discuss the restriction of the range of a ring homomorphism in general in Section 11.4 below.

**Example 11.5**

Consider the function $\varphi : \mathbb{C} \to \mathbb{C}$ defined by $\varphi(\alpha) = \bar{\alpha}$, which takes a complex number to its complex conjugate. For example, $\varphi(3 - i) = 3 + i$. This preserves addition:

$$\begin{aligned}
\varphi((a + bi) + (c + di)) &= \varphi((a + c) + (b + d)i) = \\
(a + c) - (b + d)i &= (a - bi) + (c - di) = \\
&\varphi(a + bi) + \varphi(c + di).
\end{aligned}$$

It also preserves multiplication:

$$\begin{aligned}
\varphi((a + bi)(c + di)) &= \varphi((ac - bd) + (ad + bc)i) = \\
(ac - bd) - (ad + bc)i &= (a - bi)(c - di) = \\
&\varphi(a + bi)\varphi(c + di).
\end{aligned}$$

This function is both one-to-one and onto; it is a bijection.

▷ **Quick Exercise.** Verify these assertions. ◁

The conjugation function is thus a ring isomorphism, from the ring $\mathbb{C}$ onto itself.

Of course, we can define many functions whose ranges and domains are rings that are not homomorphisms; here is one example:

**Example 11.6**

Consider the function $\rho : \mathbb{Z} \to \mathbb{Z}$ defined by $\rho(n) = 3n$; this function merely multiplies by 3. Note that $\rho(nm) = 3nm$, while $\rho(n)\rho(m) = 9nm$, and so this function certainly does not preserve multiplication; it is consequently not a ring homomorphism.

▷ **Quick Exercise.** Does this function preserve addition? Is it surjective? Is it injective? ◁

## 11.2   Properties Preserved by Homomorphisms

Although we have in our definition required only that a homomorphism preserve the ring operations addition and multiplication, a homomorphism actually preserves much more of the ring structure. We catalog such properties of a homomorphism in the following theorem; at the end of Chapter 7 (and Exercise 8.7) we foreshadowed this theorem, in the case of ring isomorphisms.

**Theorem 11.1** *Let $\varphi : R \to S$ be a homomorphism between the rings $R$ and $S$. Let $0_R$ and $0_S$ be the additive identities of $R$ and $S$, respectively.*

   *a. $\varphi(0_R) = 0_S$.*

   *b. $\varphi(-a) = -\varphi(a)$, and so $\varphi(a - b) = \varphi(a) - \varphi(b)$.*

   *c. If $R$ has unity $1_R$ and $\varphi$ is onto $S$, and $S$ is not the zero ring, then $\varphi(1_R)$ is unity for $S$.*

   *d. If $a$ is a unit of $R$ and $\varphi$ is onto $S$, and $S$ is not the zero ring, then $\varphi(a)$ is a unit of $S$. In this case, $\varphi(a)^{-1} = \varphi(a^{-1})$.*

    In other words, a homomorphism always preserves the zero of the ring and the additive inverses. Furthermore, a homomorphism preserves unity and units *if the homomorphism is onto*. This theorem thus asserts that the assumption that a function preserve addition and multiplication is enough to conclude that it preserve other additive and multiplicative structures. Note that a ring always has an additive identity and additive inverses, and so (a) and (b) are phrased to reflect that fact. On the other hand, a ring need not have unity or units, and so (c) and (d) are phrased rather differently.

**Proof:**   (a): Now $\varphi(0_R) + \varphi(0_R) = \varphi(0_R + 0_R) = \varphi(0_R)$. By subtracting $\varphi(0_R)$ from both sides, we obtain $\varphi(0_R) = 0_S$.

    (b): We have $\varphi(-a) + \varphi(a) = \varphi(-a + a) = \varphi(0_R) = 0_S$; thus by the definition and uniqueness of $-\varphi(a)$, we have that $\varphi(-a) = -\varphi(a)$. But then

$$\varphi(a - b) = \varphi(a + (-b)) = \varphi(a) + \varphi(-b) = \varphi(a) - \varphi(b),$$

as claimed.

    (c): If $S$ is the zero ring, we have previously decreed that we won't call its unique element $0$ unity; we've excluded this case in our hypothesis. Let $s$ be any element of $S$; then there exists $r \in R$ with $\varphi(r) = s$, because $\varphi$ is onto. Then

$$s\varphi(1_R) = \varphi(r)\varphi(1_R) = \varphi(r \cdot 1_R) = \varphi(r) = s,$$

and so $\varphi(1_R)$ must be the unity of $S$.

    (d): Suppose that $aa^{-1} = 1_R$. Then

$$\varphi(a)\varphi(a^{-1}) = \varphi(aa^{-1}) = \varphi(1_R) = 1_S,$$

and so $\varphi(a)$ is a unit of $S$ with inverse $\varphi(a^{-1})$.       □

---

## 11.3   More Examples

We now look at some further examples of homomorphisms, which will illustrate applications and limitations of Theorem 11.1. In each case you should check that the example given does preserve multiplication and addition. The first three examples seem rather trivial but are important.

## Example 11.7

Let $R$ be a ring with unity 1, and $S$ another ring. Define $\zeta : R \to S$ by $\zeta(r) = 0$, for all $r \in R$ (this is called the **zero homomorphism**). If $S$ has unity, note that $\zeta(1) = 0 \neq 1$, and in this case the function is not onto, and so (c) of Theorem 11.1 does not apply. Notice that this homomorphism is (notoriously) not injective.

## Example 11.8

Let $R$ be a ring and define $\iota : R \to R$ by $\iota(r) = r$ (this is called the **identity homomorphism**). It of course preserves *all* structure of the ring $R$; it is a ring isomorphism.

## Example 11.9

Let $R$ be a subring of the ring $S$. Consider the function $\iota : R \to S$ defined by $\iota(r) = r$ (this is called the **inclusion homomorphism**). It is always injective, but not onto if $R \subset S$.

## Example 11.10

Consider the rings $\mathbb{Z}_4$ and $\mathbb{Z}_2 \times \mathbb{Z}_2$; they both have 4 elements, and so there exists a bijection between them, just as sets. Is there any such function that preserves addition and multiplication? We claim something stronger: There isn't even an injection that preserves addition. If there were, 0 would have to go to $(0,0)$. (See Theorem 11.1a.) There are then three possibilities for where 1 might be mapped: $(1,0),(0,1),(1,1)$. But if the correspondence preserves addition, then the images of

$$2 = 1+1 \text{ and } 3 = 1+1+1$$

under the correspondence are forced by choice of the image of 1. It is easy to see that none of the three choices listed above allow a one-to-one function.

▷ **Quick Exercise.** Check that each of the three choices above yields a function that is not one-to-one. ◁

To summarize: $\mathbb{Z}_4$ and $\mathbb{Z}_2 \times \mathbb{Z}_2$ are two rings with 4 elements, but they are not isomorphic as rings. Furthermore, if we consider Exercise 8.15 we have yet another ring with 4 elements which is not isomorphic to the other two. This is clear, because the ring in that example is a field.

## Example 11.11

Reconsider the residue map $\varphi : \mathbb{Z} \to \mathbb{Z}_4$. Notice that although 2 is not a zero divisor in $\mathbb{Z}$, $\varphi(2)$ *is* a zero divisor in $\mathbb{Z}_4$ because $[2][2] = [4] = [0]$. Another thing to notice about this function is that although the domain is infinite, the range is finite; although surjective, it is not injective.

## Example 11.12

Consider the rings $\mathbb{Q}$ and $\mathbb{Q} \times \mathbb{Q}$ (recall from Example 6.10 how the direct product $\mathbb{Q} \times \mathbb{Q}$ is made into a ring). Define the function $\varphi : \mathbb{Q} \to \mathbb{Q} \times \mathbb{Q}$ by setting $\varphi(r) = (r,0)$. This homomorphism is one-to-one, but not onto. Now 3 is a unit in $\mathbb{Q}$ because $3 \cdot \frac{1}{3} = 1$. But the unity of $\mathbb{Q} \times \mathbb{Q}$ is $(1,1)$, and so $\varphi(3) = (3,0)$ cannot be a unit in $\mathbb{Q} \times \mathbb{Q}$, because there is no $(z,w)$ such that $(3,0)(z,w) = (1,1)$. In fact, $\varphi(3)$ (and $\varphi(1)$) are zero divisors.

**Example 11.13**

Reconsider the evaluation map $\varphi : \mathbb{Q}[x] \to \mathbb{Q}$ given by $\varphi(f) = f(2)$, which we discussed in Example 11.2. This homomorphism is surjective but not injective. Notice that the polynomial $x$ in $\mathbb{Q}[x]$ is not a unit, but $\varphi(x) = 2$ is a unit in $\mathbb{Q}$.

**Example 11.14**

Let $C[0,1]$ be the ring of continuous real-valued functions with domains $[0,1]$ (we discussed this ring in Example 6.14). Consider the evaluation map $\psi : C[0,1] \to \mathbb{R}$ defined by $\psi(f) = f(\frac{1}{4})$. Define the elements $f, g$ of $C[0,1]$ like this:

$$f(x) = \begin{cases} \frac{1}{2} - x, & \text{if } 0 \le x \le \frac{1}{2} \\ 0, & \text{if } \frac{1}{2} \le x \le 1 \end{cases}$$

and

$$g(x) = \begin{cases} 0, & \text{if } 0 \le x \le \frac{1}{2} \\ x - \frac{1}{2}, & \text{if } \frac{1}{2} \le x \le 1. \end{cases}$$

(The graphs of $f$ and $g$ are shown below.) Then $fg = 0$, and so $f$ is a zero divisor. However, $\psi(f) = \frac{1}{4}$, which is a unit of $\mathbb{R}$.

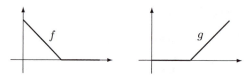

▷ **Quick Exercise.** Verify the claims made in the above examples. ◁

## 11.4    Making a Homomorphism Surjective

Sometimes we wish to modify a homomorphism so that it becomes onto. Given homomorphism $\varphi : R \to S$, if we restrict the range to the set $\varphi(R)$, the new function $\varphi : R \to \varphi(R)$ (which we still call $\varphi$) is obviously surjective. But is it still a homomorphism? Because we have not affected preservation of addition and multiplication by restricting the range, this new function must still be a homomorphism, *if $\varphi(R)$ is a ring*. But this is always the case, as we prove next:

**Theorem 11.2** *Let $\varphi : R \to S$ be a homomorphism between rings $R$ and $S$. Then $\varphi(R)$ is a subring of $S$.*

**Proof:**    We need only check that the set $\varphi(R)$ is closed under subtraction and multiplication; then we'll know by the Subring Theorem 7.1 that it is a subring of $S$. For this purpose, choose $x, y \in \varphi(R)$; there exist $a, b \in R$ such that $\varphi(a) = x$ and $\varphi(b) = y$. Then

$$x - y = \varphi(a) - \varphi(b) = \varphi(a - b) \in \varphi(R),$$

and similarly, $xy = \varphi(ab) \in \varphi(R)$. □

## Historical Remarks

The functional point of view is a very powerful one in all branches of mathematics. Analysts and topologists generally restrict themselves to *continuous* functions, because these are the functions that preserve analytic or topological structure. Similarly, in algebra we generally restrict ourselves to *homomorphisms*, because they are the functions that preserve algebraic structure.

The great 19th-century German mathematician Felix Klein was the first to use the word homomorphism in the context of a function preserving algebraic structure. He was actually talking about groups rather than rings; a group is a set endowed with algebraic structure that we will meet in Chapter 19.

## Chapter Summary

In this chapter we introduced the idea of *ring homomorphism*, a function between two rings which preserves algebraic structure. We looked at numerous examples of such functions.

## Warm-up Exercises

a. Give an example of a ring homomorphism $\varphi : R \to S$ satisfying each of the following (or explain why they cannot exist):

   (a) $\varphi$ is surjective but not injective.

   (b) $\varphi$ is injective but not surjective.

   (c) $\varphi$ is both injective and surjective.

   (d) $\varphi$ is neither injective nor surjective.

   (e) $R$ has unity 1, but $\varphi(1)$ is not unity for $S$.

   (f) $R$ has additive identity 0, but $\varphi(0)$ is not the additive identity for $S$.

   (g) $a$ is a zero divisor for $R$, but $\varphi(a)$ is not a zero divisor for $S$.

   (h) $a$ is a unit for $R$, but $\varphi(a)$ is not a unit for $S$.

   (i) $a$ is not a zero divisor for $R$, but $\varphi(a)$ is a zero divisor for $S$.

   (j) $a$ is not a unit for $R$, but $\varphi(a)$ is a unit for $S$.

b. We often write $\mathbb{Z}_4$ as $\{0, 1, 2, 3\}$. Using this notation, define the function $\varphi : \mathbb{Z}_4 \to \mathbb{Z}$ by $\varphi(n) = n$. Is this a ring homomorphism?

c. The function $\alpha : \mathbb{Z} \to \mathbb{Z}$ defined by $\alpha(n) = |n|$ is not a ring homomorphism. Why not?

d. Suppose that $R$ and $S$ are rings, and $\varphi : R \to S$ is a surjective ring homomorphism.

   (a) If $R$ is a domain, is $S$ a domain?

   (b) If $R$ is commutative, is $S$ commutative?

## Exercises

In Exercises 1–24, functions are defined whose domains and ranges are rings. In each case determine whether the function is a ring homomorphism, and whether it is injective, surjective, or bijective. Justify your answers:

1. Define $\varphi : \mathbb{Z} \to m\mathbb{Z}$ by $\varphi(n) = mn$.

2. Define $\varphi : \mathbb{Q}[x] \to \mathbb{Q}[x]$ by $\varphi(f) = f'$. ($f'$ is the formal derivative of $f$, discussed in Exercise 4.7.)

3. Define $\varphi : M_2(\mathbb{Z}) \to \mathbb{Z}$ by mapping a matrix to its determinant. (The determinant function is discussed in Exercise 8.2.)

4. Let $R$ and $S$ be rings and define $\pi_1 : R \times S \to R$ by $\pi_1(r, s) = r$, the **projection homomorphism** from $R \times S$ to $R$. Similarly, we can define $\pi_2 : R \times S \to S$. (This generalizes Example 11.3.)

5. Let $R$ be a ring, and define $\varphi : R \to R \times R$ by $\varphi(r) = (r, r)$. (This generalizes Example 11.4.)

6. Let $R$ be a ring with unity and define $\varphi(r) = -r$. *Hint*: Consider Exercises 6.3 and 6.4.

   Case 1: There is at least one element $a \in R$ with $a + a \neq 0$.

   Case 2: For all $r \in R$, $r + r = 0$.

7. Recall from Example 7.9 that $D_2(\mathbb{Z})$ is the ring of 2-by-2 matrices with entries from $\mathbb{Z}$, with all entries off the main diagonal being zero. Define

$$\varphi : D_2(\mathbb{Z}) \to \mathbb{Z} \times \mathbb{Z}$$

by $\varphi \begin{pmatrix} a & 0 \\ 0 & b \end{pmatrix} = (a, b)$.

8. Define $\varphi : \mathbb{Z}[\sqrt{2}] \to \mathbb{Z}$ by $\varphi(a + b\sqrt{2}) = a$. (See Exercise 7.1 for $\mathbb{Z}[\sqrt{2}]$.)

9. Define $\varphi : \mathbb{Z}[\sqrt{2}] \to \mathbb{Z}[\sqrt{2}]$ by $\varphi(a + b\sqrt{2}) = a - b\sqrt{2}$.

10. Let $X$ be an arbitrary set; recall the power set ring $P(X)$ of subsets of $X$ (described in Exercise 6.20). Choose some fixed $x \in X$. Then define $\varphi : P(X) \to \mathbb{Z}_2$ by setting

$$\varphi(a) = \begin{cases} [1], & x \in a \\ [0], & x \notin a. \end{cases}$$

11. Define $\varphi : \mathbb{Z} \times \mathbb{Z} \to \mathbb{Z}$ by $\varphi(a, b) = a + b$.

12. Define $\varphi : \mathbb{Z} \times \mathbb{Z} \to \mathbb{Z}$ by $\varphi(a, b) = ab$.

13. Recall the ring $S$ of all real-valued sequences. (See Exercise 6.19.) Define $\varphi : S \to S$ by

$$\varphi((s_1, s_2, \cdots)) = (s_2, s_3, \cdots).$$

(That is, $\varphi$ obtains the new sequence merely by dropping the first term of the sequence.)

14. Recall the ring $\mathbb{Z}[\alpha]$, where $\alpha = \sqrt[3]{5}$, as described in Exercise 7.3. Define $\varphi : \mathbb{Z}[\alpha] \to \mathbb{Z}_4$ by $\varphi(a + b\alpha + c\alpha^2) = [a + b + c]_4$.

15. Define $\varphi : \mathbb{Z} \times \mathbb{Z} \to \mathbb{Z}_6$ by $\varphi(a, b) = [a - b]_6$.

16. Define $\varphi : \mathbb{Z} \times \mathbb{Z} \to \mathbb{Z}_6$ by $\varphi(a, b) = [a + b]_6$.

17. Define $\varphi : \mathbb{Z}_2 \times \mathbb{Z}_3 \to \mathbb{Z}_6$ by $\varphi([a]_2, [b]_3) = [3a + 2b]_6$.

18. Define $\varphi : \mathbb{Z}_2 \times \mathbb{Z}_3 \to \mathbb{Z}_6$ by $\varphi([a]_2, [b]_3) = [3a + 4b]_6$.

19. Define $\varphi : \mathbb{Z} \times \mathbb{Z} \to \mathbb{Z}_6$ by $\varphi(a, b) = [3a + 4b]_6$.

20. Define $\varphi : \mathbb{Z}_6 \to \mathbb{Z}_4$ by $\varphi([a]_6) = [a]_4$.

21. Define $\varphi : \mathbb{C} \to \mathbb{R}$ by $\varphi(a + bi) = a$.

22. Define $\varphi : C[0, 1] \to \mathbb{R} \times \mathbb{R}$ by $\varphi(f) = (f(0), f(1))$.

23. Define $\varphi : \mathbb{Q}[x] \to \mathbb{Q}$ by letting $\varphi(f)$ be the $x$-coefficient in $f$.

24. Define $\varphi : \mathbb{Q}[x] \to M_2(\mathbb{R})$ by

$$\varphi(f) = \begin{pmatrix} f(0) & f'(0) \\ 0 & f(0) \end{pmatrix}.$$

25. Let $R$ be a commutative ring, and suppose that $\varphi : R \to R$ is a ring homomorphism. Consider

$$S = \{r \in R : \varphi(r) = r\}.$$

   (a) Show that $S$ is a subring of $R$.

   (b) Show that $S$ might not be an ideal.

26. Show that the rings

$$\mathbb{Z}_8, \ \mathbb{Z}_4 \times \mathbb{Z}_2, \text{ and } \mathbb{Z}_2 \times \mathbb{Z}_2 \times \mathbb{Z}_2$$

   are all non-isomorphic, even though each of these rings has the same number of elements.

27. Suppose that $R$ and $S$ are arbitrary rings, and $\varphi : R \to S$ is a surjective ring homomorphism. Prove that $\varphi(Z(R)) \subseteq Z(S)$. (Recall from Exercise 7.12 that $Z(R)$ is called the center of the ring $R$.)

28. Suppose that $R$ and $S$ are commutative rings, and $\varphi : R \to S$ is a ring homomorphism. Let $N(R)$ and $N(S)$ be the nilradicals of $R$ and $S$, respectively. (See Exercise 7.15.) Prove that $\varphi(N(R)) \subseteq N(S)$.

29. Let $R$ and $S$ be rings, with $\varphi : R \to S$ a ring homomorphism. Suppose that $T$ is a subring of $S$. Let

$$\varphi^{-1}(T) = \{r \in R : \varphi(r) \in T\}.$$

   Prove that $\varphi^{-1}(T)$ is a subring of $R$.

30. Suppose that $R$ is a (not necessarily commutative) ring with unity, and $a$ is a unit in $R$. Define the function $\varphi : R \to R$ by $\varphi(r) = ara^{-1}$. Prove that $\varphi$ is a ring isomorphism from $R$ to $R$.

31. Consider the ring $S$ of real-valued sequences. (See Exercise 6.19.) Let

$$C = \{(x_1, x_2, x_3, \ldots) : \lim_{n \to \infty} x_n \text{ exists}\}.$$

This is the set of **convergent** sequences.

(a) Prove that $C$ is a subring of $S$.

(b) Define $\varphi : C \to \mathbb{R}$ by setting

$$\varphi((x_1, x_2, x_3, \ldots)) = \lim_{n \to \infty} x_n.$$

Prove that this is a ring homomorphism.

32. In this exercise we extend Theorem 11.2 to ideals.

(a) Suppose that $R$ and $S$ are commutative rings, and $\varphi : R \to S$ is a surjective ring homomorphism. Let $I$ be an ideal of $R$. Prove that $\varphi(I)$ is an ideal of $S$.

(b) Show by example that part (a) is false, if we do not require that $\varphi$ is onto.

# Chapter 12

## The Kernel

Let's consider the residue homomorphism $\varphi : \mathbb{Z} \to \mathbb{Z}_4$. There are exactly four residue classes modulo 4; namely,

$$\{\cdots, -4, 0, 4, 8, \cdots\},$$
$$\{\cdots, -3, 1, 5, 9, \cdots\},$$
$$\{\cdots, -2, 2, 6, 10, \cdots\}, \text{ and}$$
$$\{\cdots, -1, 3, 7, 11, \cdots\}.$$

How do they relate to the function $\varphi$? The answer to this question is reasonably obvious: They consist of the four **pre-images** of the elements of $\mathbb{Z}_4$. Namely, the residue classes consist of the four sets

$$\varphi^{-1}([0]) = \{n \in \mathbb{Z} : \varphi(n) = [0]\},$$
$$\varphi^{-1}([1]), \ \varphi^{-1}([2]), \text{ and } \ \varphi^{-1}([3]).$$

Furthermore, these sets are rather nicely related: $\varphi^{-1}([0]) = \langle 4 \rangle$, the set of multiples of 4 (and principal ideal generated by 4), and the other three can be obtained from $\langle 4 \rangle$ by adding a fixed element to every element of $\langle 4 \rangle$;

$$\begin{aligned}
\varphi^{-1}([1]) &= \{a + 1 : a \in \langle 4 \rangle\}, \\
\varphi^{-1}([2]) &= \{a + 2 : a \in \langle 4 \rangle\}, \text{ and} \\
\varphi^{-1}([3]) &= \{a + 3 : a \in \langle 4 \rangle\}.
\end{aligned}$$

We say that we have obtained $\varphi^{-1}([1])$, $\varphi^{-1}([2])$, and $\varphi^{-1}([3])$ by **additively translating** $\langle 4 \rangle$ by 1, 2, and 3, respectively.

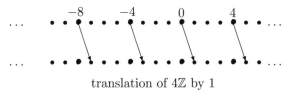

translation of $4\mathbb{Z}$ by 1

## 12.1 The Kernel

We'd like to generalize this situation as far as possible to arbitrary rings. This leads us to the following definition: Let $\varphi : R \to S$ be a homomorphism between the rings $R$ and $S$. The **kernel** of $\varphi$ is

$$\varphi^{-1}(0) = \{r \in R : \varphi(r) = 0\};$$

we will denote this set by $\ker(\varphi)$. In other words, the kernel is the pre-image of 0.

Thus, the kernel of the residue homomorphism above is exactly $\langle 4 \rangle$. Notice that the kernel always contains the additive identity of $R$ (because homomorphisms take zero to zero). But as the example above shows, the kernel can include a great many other elements as well.

Let's go back to some of the examples from the previous chapter and compute their kernels.

## Example 12.1

(From Example 11.2) This is the homomorphism that evaluates each rational polynomial at 2, and so its kernel is the set of all polynomials $f$ such that $f(2) = 0$. By the Root Theorem, this is the set of all polynomials of the form $(x - 2)g$, where $g$ is some arbitrary element of $\mathbb{Q}[x]$. Note that this set is precisely $\langle x - 2 \rangle$.

## Example 12.2

(From Example 11.3.) This is the projection homomorphism onto the first component, from $\mathbb{Z} \times \mathbb{Z}$ to $\mathbb{Z}$. When is $\pi(a, b) = 0$? The answer is precisely when $a = 0$; thus,

$$\ker(\pi) = \{(0, b) : b \in \mathbb{Z}\}.$$

Note that this is $\langle (0, 1) \rangle$.

## Example 12.3

(From Example 11.5.) This is the homomorphism $\varphi$ taking each complex number to its conjugate. Thus, $a + bi = \alpha \in \ker(\varphi)$ exactly if $a - bi = \bar{\alpha} = 0$. But this means that both $a$ and $b$ are zero, and so $\alpha = 0$. Thus,

$$\ker(\varphi) = \{0\} = \langle 0 \rangle.$$

## Example 12.4

(From Example 11.7.) This is the zero homomorphism $\zeta : R \to S$, and its kernel is quite evidently the entire ring $R$.

## Example 12.5

(From Example 11.14.) This is the homomorphism $\psi : C[0, 1] \to \mathbb{R}$ that sends a continuous function to its value at $1/4$. Here,

$$\ker(\psi) = \{f \in C[0, 1] : f(1/4) = 0\}.$$

There is no nice elementwise description of this set, which includes infinitely many functions, including for example $\sin\left(x - \frac{1}{4}\right)$, $e^{4x} - e$, $\left(x - \frac{1}{4}\right)^2$, etc.

▷ **Quick Exercise.**   Compute the kernels of the other examples of ring homomorphisms from the previous chapter. ◁

## 12.2 The Kernel Is an Ideal

Note that every kernel from the above examples is an ideal: A subring that has the multiplicative absorption property. In the case of Example 12.4, this is evident because the kernel is the entire ring. In the case of Examples 12.1, 12.2, and 12.3, the kernel is actually a principal ideal.

For Example 12.5, let's check explicitly that

$$\ker(\psi) = \{f \in C[0,1] : f(1/4) = 0\}$$

is an ideal. We first see that it is a non-empty set, by noting that the zero function is in $\ker(\psi)$. It is easy to check that $\ker(\psi)$ is closed under subtraction.

▷ **Quick Exercise.** Show that the kernel from Example 12.5 is closed under subtraction. ◁

This kernel also has the absorption property: If $f$ is in $\ker(\psi)$ and $g$ is any function in $C[0,1]$, then

$$(fg)(1/4) = f(1/4) \cdot g(1/4) = 0 \cdot g(1/4) = 0.$$

In other words, $fg$ is also in $\ker(\psi)$.

With the examples above before us, you may be ready to believe the following theorem:

**Theorem 12.1** *Let* $\varphi : R \to S$ *be a homomorphism between the commutative rings $R$ and $S$. Then $\ker(\varphi)$ is an ideal of $R$.*

**Proof:** To show that $\ker(\varphi)$ is an ideal, we must first check that it is a non-empty set. But evidently, the zero element of $R$ belongs to the kernel. We next check that the kernel is closed under subtraction. So suppose $a$ and $b$ are elements of the kernel. We need to check that $a - b \in \ker(\varphi)$. But $\varphi(a - b) = \varphi(a) - \varphi(b) = 0 - 0 = 0$. That is, $a - b \in \ker(\varphi)$, as required. Similarly, if $r \in R$, then $\varphi(ar) = \varphi(a)\varphi(r) = 0\varphi(r) = 0$, and so $ar \in \ker(\varphi)$. □

## 12.3 All Pre-images Can Be Obtained from the Kernel

The residue homomorphism example suggests that we can capture *all* pre-images by additively translating the kernel. Let's look at the pre-images of Example 12.1, the evaluation map

$$\varphi : \mathbb{Q}[x] \to \mathbb{Q}$$

defined by $\varphi(f) = f(2)$. This is more complicated than the residue homomorphism example because the range $\mathbb{Q}$ is infinite, and so there are infinitely many pre-images. Recall that the kernel of $\varphi$ is $\langle x - 2 \rangle$. We wish to show that it is also the case in this example that each pre-image can be obtained by additively translating the kernel. That is, we claim that each pre-image can be written as

$$\langle x - 2 \rangle + g = \{f \in \mathbb{Q}[x] : f = h + g \text{ where } h \in \langle x - 2 \rangle\},$$

for some choice of $g$. To show this, choose $a \in \mathbb{Q}$ and consider $g \in \varphi^{-1}(a)$. This means that $\varphi(g) = g(2) = a$. We claim that every $f$ in $\varphi^{-1}(a)$ can be written in the form $f = h + g$

where $h \in \langle x - 2 \rangle$. This would show that $\varphi^{-1}(a) = \langle x - 2 \rangle + g$. Now because $f(2) = a$, and $g(2) = a$, we have that $(f - g)(2) = 0$. That is, $f - g \in \ker(\varphi)$. But $\ker(\varphi) = \langle x - 2 \rangle$, so $f - g = h$, where $h$ is some multiple of $x - 2$. But then $f = h + g$, as we wish.

Note that the $g$ we picked to represent $\varphi^{-1}(a)$ was chosen arbitrarily from all the elements of $\varphi^{-1}(a)$. Another element in $\varphi^{-1}(a)$ would have served just as well. The form of the translation would be different but would give the same set. In other words, if $g_1$ and $g_2$ are both in $\varphi^{-1}(a)$, then

$$\varphi^{-1}(a) = \langle x - 2 \rangle + g_1 = \langle x - 2 \rangle + g_2.$$

This is analogous to our freedom of choice in representing residue classes in $\mathbb{Z}_m$: for instance, $[4]_6 = [10]_6 = [16]_6$; or, to express this using the notation of ideals,

$$\langle 6 \rangle + 4 = \langle 6 \rangle + 10 = \langle 6 \rangle + 16.$$

It turns out that we can always capture all the pre-images of a ring homomorphism by additively translating the kernel. To state this formally, we need a definition. Let $I$ be an ideal of the ring $R$. Given $r \in R$, the **coset** of I determined by $r$ consists of the set $\{a + r : a \in I\}$, which we write as $I + r$. Thus, $\langle 4 \rangle + 3$ is a coset of the ideal $\langle 4 \rangle$ in $\mathbb{Z}$.

▷ **Quick Exercise.**   Consider the function $\varphi$ in Example 12.1; describe the coset $\ker(\varphi) + 5$. What values do the polynomials in this set take on at 2? ◁

**Theorem 12.2** *Let $\varphi : R \to S$ be a homomorphism between the rings $R$ and $S$, and $s \in \varphi(R)$. Then $\varphi^{-1}(s)$ equals the coset $\ker(\varphi) + r$, where $r$ is any given element of $\varphi^{-1}(s)$.*

**Proof:**   Let $s \in \varphi(R)$, and choose any $r \in \varphi^{-1}(s)$ (which is non-empty by assumption). We must show that the sets $\ker(\varphi) + r$ and $\varphi^{-1}(s)$ are equal. Choose an arbitrary element $a + r \in \ker(\varphi) + r$, where $a \in \ker(\varphi)$. Then $\varphi(a + r) = \varphi(a) + \varphi(r) = 0 + \varphi(r) = s$. Thus, $a + r \in \varphi^{-1}(s)$, as claimed.

Conversely, choose $t \in \varphi^{-1}(s)$. Then consider $t - r$;

$$\varphi(t - r) = \varphi(t) - \varphi(r) = s - s = 0,$$

and so $t - r \in \ker(\varphi)$. But then $t = (t - r) + r \in \ker(\varphi) + r$, as required.   □

Now because pre-images of distinct elements are clearly disjoint from one another, we have that the set of cosets of the kernel decomposes the domain ring into a set of pairwise disjoint subsets. That is, the set of cosets of the kernel partitions the ring. The unique one of these sets containing 0 is an ideal (because an ideal *has* to contain 0, it is clear that only one of the cosets can be an ideal). Notice that we can (and will) think of the ideal itself as a coset, namely, as $\ker(\varphi) + 0$.

### Example 12.6

Let's explicitly compute this decomposition for the function $\varphi : \mathbb{Z}_{12} \to \mathbb{Z}_4$ defined by $\varphi([a]_{12}) = [a]_4$.

▷ **Quick Exercise.**   Check that this function is a ring homomorphism. ◁

The kernel of $\varphi$ is

$$\varphi^{-1}(0) = \{0, 4, 8\} = \langle 4 \rangle = \langle 4 \rangle + 0 = \langle 4 \rangle + 4 = \langle 4 \rangle + 8,$$

(where we have omitted the square brackets for simplicity's sake). The other distinct cosets of $\langle 4 \rangle$ are

$$\varphi^{-1}(1) \;=\; \{1, 5, 9\} = \langle 4 \rangle + 1 = \langle 4 \rangle + 5 = \langle 4 \rangle + 9,$$
$$\varphi^{-1}(2) \;=\; \{2, 6, 10\} = \langle 4 \rangle + 2 = \langle 4 \rangle + 6 = \langle 4 \rangle + 10, \text{ and}$$
$$\varphi^{-1}(3) \;=\; \{3, 7, 11\} = \langle 4 \rangle + 3 = \langle 4 \rangle + 7 = \langle 4 \rangle + 11.$$

Notice that in Example 12.6, each of the cosets has exactly the same number of elements. This is true in general because there is a bijection between any pair of cosets of an ideal: Given two cosets $I + a$ and $I + b$, the function $\alpha : I + a \to I + b$ defined by $\alpha(x) = x - a + b$ is a bijection.

▷ **Quick Exercise.** Check this; that is, show that $\alpha$ is both one-to-one and onto. ◁

It is important to make clear that the function $\alpha$ is *not* a homomorphism; it does *not* preserve the operations, and the domain and range (except in the special case of the coset $I$) are not even rings. Two sets have the same number of elements exactly if there is a one-to-one correspondence between them; this is the situation in our example above. (Although we will not inquire into this here, this bijection is even useful if the ideal (and hence its cosets) is an infinite set, because it turns out that not all infinite sets can be put into one-to-one correspondence with one another; some are 'bigger' than others.)

We record our result formally:

**Theorem 12.3** *Let $I$ be an ideal of the commutative ring $R$, and let $I + a$ and $I + b$ be any two cosets of $I$. Then there is a bijection between the elements of $I + a$ and $I + b$. In particular, if these sets are finite, they have the same number of elements.*

## 12.4 When Is the Kernel Trivial?

An important special case to consider of this bijection is when the kernel consists of only a single element; namely, when the ideal is $\{0\}$ (because an ideal is a subring this is the only possible one element ideal). Thus, *every* coset consists of a single element. Because the cosets consist of the pre-images of elements from the range ring, this means that a homomorphism with kernel $\{0\}$ is necessarily one-to-one. To rephrase: If a homomorphism takes *only* zero to zero, then there is only a single element that it takes to *any* of its values – the homomorphism is injective. We state this formally as a corollary:

**Corollary 12.4** *A ring homomorphism is injective if and only if its kernel is $\{0\}$.*

## 12.5 A Summary and Example

We have thus concluded that each homomorphism gives rise to an ideal, and this ideal in essence determines the pre-images of elements from the range of the function. We thus know (from just knowing the kernel) which elements in the ring are sent by the homomorphism to the same elements.

For example, consider the ring $\mathbb{Z} \times \mathbb{Z}$ and the ideal $I = \{(x, 0) : x \in \mathbb{Z}\}$.

▷ **Quick Exercise.** Check that this is an ideal. ◁

Suppose we are told that this is the kernel of some homomorphism. Then we know that all elements in the coset

$$I + (3, 4) = I + (-5, 4) = \{(x, 4) : x \in \mathbb{Z}\}$$

*must* be sent by the homomorphism to the same element in the range ring (whatever that might be). Can you in fact think of a homomorphism with domain $\mathbb{Z} \times \mathbb{Z}$ of which $I$ is the kernel? (If not, rest assured that in Example 14.3 we will return to this example.)

This all suggests that knowing the kernel of a homomorphism in essence gives us the homomorphism. This is precisely the content of the next chapter, where we will prove that *every* ideal of a ring is the kernel of some homomorphism. That is, we will prove the converse of Theorem 12.1, which asserts that the kernel of a homomorphism is always an ideal.

---

## Chapter Summary

In this chapter we defined the notion of the *kernel* of a ring homomorphism and proved that the kernel is always an *ideal.* Furthermore, the pre-images of a ring homomorphism are exactly the *cosets* of the kernel.

---

## Warm-up Exercises

a. What is the kernel of

$$\varphi : \mathbb{Z}_6 \to \mathbb{Z}_2$$

given by $\varphi([a]_6) = [a]_2$?

b. Write down the pre-images from the previous exercise, and check that they are cosets.

c. Consider the function $f : \mathbb{R} \to \mathbb{R}$ defined by $f(x) = x^2$. How many real numbers are there for which $f(x) = 0$? How about for $f(x) = 4$? Why does this tell us that $f$ is *not* a homomorphism?

d. Give examples of ring homomorphisms $\varphi$ satisfying the following (or say why you can't):

   (a) $\varphi$ is surjective, and its kernel is $\{0\}$.

   (b) $\varphi$ is surjective, and its kernel is not $\{0\}$.

   (c) $\varphi$ is injective, and its kernel is $\{0\}$.

   (d) $\varphi$ is injective, and its kernel is not $\{0\}$.

e. Give an example of a subring that is not the kernel of any ring homomorphism (if you can).

f. Give an example of a kernel of a ring homomorphism that is not a subring (if you can).

---

## Exercises

1. Consider the homomorphism given in Exercise 11.5. What is its kernel? What does this mean about the homomorphism?

2. Consider the homomorphism given by Exercise 11.7. What is its kernel? What does this mean about the homomorphism?

3. Consider the homomorphism given by Exercise 11.10. What is its kernel? Can you describe this kernel as a principal ideal $\langle b \rangle$, for some subset $b \in P(X)$?

4. What is the kernel of the homomorphism given by Exercise 11.13?

5. Consider the homomorphism $\varphi$ given by Exercise 11.14. Prove that $\ker(\varphi) = \langle 3+\alpha, 4 \rangle$.

6. What is the kernel of the homomorphism given by Exercise 11.18?

7. What is the kernel of the homomorphism given by Exercise 11.22?

8. Consider again the ring homomorphism of Example 12.1, namely, the function $\varphi :$ $\mathbb{Q}[x] \to \mathbb{Q}$ defined by $\varphi(f) = f(2)$. Consider the translates of $\langle x - 2 \rangle$ by each of the constant polynomials (one for each rational). Show that each of these gives rise to a different pre-image of $\varphi$.

9. (Continuation of Exercise 8.) Now show that all the pre-images of $\varphi$ can be obtained in this manner.

10. If $F$ is a field and $\varphi$ is a ring homomorphism, is $\varphi(F)$ also a field? If yes, prove it. If no, give a counterexample.

11. Consider the projection homomorphism $\pi_1 : R \times S \to R$ given in Exercise 11.4. What is the kernel of $\pi_1$? When do two elements of $R \times S$ get mapped to the same element of $R$? The set of pre-images of $\pi_1$ is naturally in one-to-one correspondence with what ring?

12. If $I$ is an ideal of a commutative ring $R$, then show that two cosets of $I$ (say, $I + a$ and $I + b$) are either equal or disjoint. (That is, the set of all translates of $I$ *partition* $R$.) If you've previously encountered the idea of *equivalence relation*, rephrase this result in terms of that concept.

13. Consider the homomorphism given by Exercise 11.24. What is the kernel of this homomorphism? Can you describe this kernel as a principal ideal?

14. Let $R$ be a commutative ring with unity, and suppose that $e$ is an idempotent element. (That is, $e^2 = e$.) See Exercise 7.25 for more about idempotents. Define $\varphi : R \to R$ by setting $\varphi(r) = er$. Prove that $\varphi$ is a ring homomorphism. Describe its kernel (using $f = 1 - e$).

15. This is a converse for Exercise 14. Suppose that $R$ is a commutative ring with unity, $a \in R$, and $\varphi(r) = ar$ defines a ring homomorphism. Prove that $a$ is idempotent.

16. Let $R$ be a finite commutative ring, with $n$ elements. Suppose that $I$ is an ideal of $R$ with $m$ elements. Prove that $m$ divides $n$. (This result will be put in a more general context in Theorem 24.2, which is called Lagrange's Theorem.)

17. Suppose that $F$ is a finite field with characteristic 2. (See Exercise 8.14 for a definition of characteristic.)

   (a) Prove that $\varphi : F \to F$, defined by $\varphi(r) = r^2$ is a ring isomorphism.

   (b) One example of a field with characteristic 2 is $\mathbb{Z}_2$. Describe the isomorphism $\varphi$ explicitly in this case.

(c) Another example of a field with characteristic 2 is the field described in Exercise 8.15, which consists of the elements

$$\{0, 1, \alpha, 1 + \alpha\}.$$

Describe the isomorphism $\varphi$ explicitly in this case.

18. Generalize Exercise 17. That is, suppose that $F$ is a finite field with characteristic $p$.

(a) Prove that $\varphi : F \to F$ defined by $\varphi(r) = r^p$ is a ring isomorphism. This function is called the **Frobenius isomorphism**.

(b) Suppose now that the finite field $F$ is the field $\mathbb{Z}_p$. Explain why Fermat's Little Theorem 8.7 implies that in this case the Frobenius isomorphism is actually the identity map. (Note that in Chapter 45 we will encounter many finite fields that are not of the form $\mathbb{Z}_p$.)

# Chapter 13

## Rings of Cosets

In practice when we think of the ring $\mathbb{Z}_4$ we think of the four elements $[0], [1], [2], [3]$ (or even $0, 1, 2, 3$) together with the appropriate operations. But technically, when we first considered $\mathbb{Z}_4$ and its operations in Chapter 3, we defined those elements as infinite sets of integers; that is,

$$[a] = \{a + 4m : m \in \mathbb{Z}\} = \{n \in \mathbb{Z} : 4|(a - n)\}.$$

We then defined addition and multiplication by setting

$$[a] + [b] = [a + b] \quad \text{and} \quad [a][b] = [ab].$$

While these definitions look innocuous enough, what we first had to do was to check that they make sense (or are *well defined*): If $[c] = [a]$ and $[d] = [b]$, do $[c + d] = [a + b]$ and $[cd] = [ab]$? We now wish to follow the same line of thought for the cosets of an arbitrary ideal of a commutative ring.

---

## 13.1   The Ring of Cosets

We first need some notation. Let $R$ be a commutative ring with ideal $I$. Then we denote the set $\{I + r : r \in R\}$ of all cosets of $I$ in $R$ by $R/I$; we read this as '$R$ modulo $I$'. Note that by defining $R/I$ in this way, it looks as if $R/I$ has as many elements as $R$ does. But this is certainly not the case, for different elements of $R$ may well give rise to the same coset.

For example, we saw in Section 12.3 that we can write

$$\mathbb{Z}/\langle 4 \rangle = \{\langle 4 \rangle + n : n \in \mathbb{Z}\} = \{\langle 4 \rangle + 0, \langle 4 \rangle + 1, \langle 4 \rangle + 2, \langle 4 \rangle + 3\}.$$

Another example we discussed in the previous chapter is this:

$$\mathbb{Q}[x]/\langle x - 2 \rangle = \{\langle x - 2 \rangle + f : f \in \mathbb{Q}[x]\} = \{\langle x - 2 \rangle + q : q \in \mathbb{Q}\}.$$

What we are going to do now is define an addition and multiplication on $R/I$ in such a way as to make it a commutative ring. We do this so that $\mathbb{Z}/\langle 4 \rangle$ becomes a ring essentially the same as $\mathbb{Z}_4$, and $\mathbb{Q}[x]/\langle x - 2 \rangle$ becomes a ring essentially the same as $\mathbb{Q}$.

Given a commutative ring $R$ with ideal $I$, we define

$$(I + a) + (I + r) = I + (a + r) \quad \text{and} \quad (I + a)(I + r) = I + ar.$$

We must now check that these definitions make sense. As when checking that addition and multiplication are well defined on $\mathbb{Z}_m$, we must show these definitions for addition and multiplication on cosets of $I$ are well defined by showing that different representations of the cosets yield the same sum (and product). That is, suppose that $I + a = I + c$ and $I + r = I + s$: Is

$$(I + a) + (I + r) = (I + b) + (I + s)?$$

Is

$$(I + a)(I + r) = (I + b)(I + s)?$$

What does such an assumption as $I + a = I + b$ mean? In other words: When do two elements determine the same coset of $I$? This is important enough to characterize in the following theorem. But before stating the theorem, think again about the example $\mathbb{Z}/\langle 4 \rangle$. When do two integers $a$ and $b$ determine the same coset (or in this case, residue class modulo 4)? The answer is exactly if their difference $a - b$ belongs to the ideal $\langle 4 \rangle$. This is the answer in general; we shall prove this now, along with some other important observations about cosets, in the Coset Theorem.

**Theorem 13.1     The Coset Theorem**     *Let $I$ be an ideal of the commutative ring $R$ with $a, b \in R$.*

a.  *If $I + a \subseteq I + b$, then $I + a = I + b$.*

b.  *If $I + a \cap I + b \neq \emptyset$, then $I + a = I + b$.*

c.  *$I + a = I + b$ if and only if $a - b \in I$.*

d.  *There exists a bijection between any two cosets $I + a$ and $I + b$. Thus, if $I$ has finitely many elements, every coset has that same number of elements.*

Notice that if you have done Exercise 12.12, you have already checked that parts (a) and (b) are true.

**Proof:**     (a) Suppose that $I$ is an ideal of the commutative ring $R$, and $a$ and $b$ are elements of the ring for which $I + a \subseteq I + b$. Then

$$a = 0 + a \in I + a \subseteq I + b,$$

and so there exists $x \in I$ such that $a = x + b$. But then $b = -x + a \in I + a$. Now, if $k \in I$, $k + b = (k - x) + a \in I + a$, and so $I + b \subseteq I + a$. That is, $I + a = I + b$.

(b): Suppose that $I + a \cap I + b \neq \emptyset$. Choose $c$ in this intersection. Then $c \in I + a$, and so $I + c \subseteq I + a$. But then by part (a), $I + c = I + a$. But similarly, $I + c = I + b$, and so $I + a = I + b$.

(c): If $I + a = I + b$, then $a = 0 + a \in I + a = I + b$, and so there exists $k \in I$ such that $a = k + b$. But then $a - b = k \in I$, as required. Conversely, suppose if $a - b \in I$, then $a = (a - b) + b \in I + b$. But then $a \in I + a \cap I + b$, and so by part (b), $I + a = I + b$.

(d): In the Quick Exercise following Example 12.6, you argued that the function $\alpha : I + a \to I + b$ defined by $\alpha(x) = x - a + b$ is bijective.     □

We now use the Coset Theorem 13.1, as promised, to show that the addition and multiplication we have defined above on $R/I$ are well defined:

**Proof that Operations are Well Defined:**     We can now return to the task of checking that the above definitions of addition and multiplication for $R/I$ make sense. Suppose that $I + a = I + b$ and $I + r = I + s$; we claim first that $I + (a + r) = I + (b + s)$. But this amounts to claiming that $(a + r) - (b + s) \in I$, and because $(a + r) - (b + s) = (a - b) + (r - s)$, this is clear. We claim next that $I + ar = I + bs$, or in other words, that $ar - bs \in I$. But

$$ar - bs = ar - br + br - bs = (a - b)r + b(r - s).$$

Because $I$ has the multiplicative absorption property, $(a - b)r \in I$ and $b(r - s) \in I$, and so therefore $ar - bs \in I$.     □

Notice that in order to show that multiplication makes sense for cosets, we needed the full strength of the definition of ideal. You will see by example in Exercise 9, that multiplication of cosets does *not* make sense if $I$ is merely a subring.

Now that we have the appropriate operations defined on $R/I$, the rest of the following theorem is easy:

**Theorem 13.2** *Let $I$ be an ideal of the commutative ring $R$. Then the set $R/I$ of cosets of $I$ in $R$, under the operations defined above, is a commutative ring.*

**Proof:** We know from above that the addition and multiplication defined on $R/I$ are in fact binary operations. They are associative and commutative because the corresponding operations on $R$ are:

$$
\begin{aligned}
(I+a)((I+b)(I+c)) &= (I+a)(I+bc) = I+a(bc) \\
&= I+(ab)c = ((I+a)(I+b))(I+c),
\end{aligned}
$$

and similarly for addition. In a similar fashion the distributive law for $R/I$ carries over from the distributive law for $R$. You should check that the additive identity for $R/I$ is $I+0$, and the additive inverse of $I+a$ is $I+(-a)$.

▷ **Quick Exercise.** Perform these verifications. ◁

□

The ring $R/I$ is called the **ring of cosets**, or the **quotient ring of $R$ modulo $I$**.

**Example 13.1**

Let's look at an example of this construction in the commutative ring $\mathbb{Z}_{12}$. Consider the ideal
$$\langle 4 \rangle = \{0, 4, 8\}.$$
This ideal has 4 distinct cosets:
$$\mathbb{Z}_{12}/\langle 4 \rangle = \{\langle 4 \rangle, \langle 4 \rangle + 1, \langle 4 \rangle + 2, \langle 4 \rangle + 3\}.$$

▷ **Quick Exercise.** Write down the multiplication and addition tables for the ring $\mathbb{Z}_{12}/\langle 4 \rangle$, and thus check quite explicitly that this is a ring. ◁

---

## 13.2 The Natural Homomorphism

Suppose that $R$ is a commutative ring with ideal $I$. Consider the function $\nu : R \to R/I$ defined by
$$\nu(a) = I + a.$$
Because of the definition of the operations on $R/I$, it is obvious that this function preserves them both. Thus, $\nu$ is a homomorphism from $R$ *onto* $R/I$. We call it the **natural homomorphism from $R$ onto $R/I$**.

Our experience in the previous chapter suggests the following question: What is the kernel of the natural homomorphism? If $\nu(a) = I + 0$ (the additive identity of $R/I$), then $I + a = I + 0$. But by the Coset Theorem 13.1, this is true exactly if $a = a - 0 \in I$. Thus, the kernel of the natural homomorphism from $R$ to $R/I$ is exactly $I$. This is very easy but is important enough to record as a theorem:

**Theorem 13.3** *Let $R$ be a commutative ring with ideal $I$, and $\nu : R \to R/I$ the natural homomorphism. Then $\ker(\nu) = I$.*

In the last chapter we saw that every kernel of a homomorphism is an ideal. In this chapter we have seen that every ideal is the kernel of some homomorphism, namely, the natural homomorphism from $R$ to $R/I$. We have thus obtained a completely different way of thinking about ideals: *Ideals are kernels of homomorphisms.* This is the sort of surprise that mathematicians particularly enjoy: Two apparently quite different ideas (in this case, homomorphisms and ideals) turn out to be inextricably linked! But to really say that the ideas of homomorphisms and ideals amount to the same thing, we need a bit more. To see this, let's look again at an example.

Consider again the evaluation homomorphism $\psi : \mathbb{Q}[x] \to \mathbb{Q}$ where $\psi(f) = f(2)$. We know that $\psi(\mathbb{Q}[x]) = \mathbb{Q}$. Now $\psi$ has kernel $\langle x-2 \rangle$, which is an ideal. In this chapter we have constructed a ring and a homomorphism of which the ideal $\langle x-2 \rangle$ is the kernel; namely, we have the ring of cosets $\mathbb{Q}[x]/\langle x-2 \rangle$ and the natural homomorphism $\nu : \mathbb{Q}[x] \to \mathbb{Q}[x]/\langle x-2 \rangle$. The two functions $\psi$ and $\nu$ are certainly different; the range of one is the set of rational numbers, while the range of the other is a certain set of subsets of $\mathbb{Q}[x]$. However, in structure these two ranges (and the functions $\psi$ and $\nu$ which connect them to the domain $\mathbb{Q}[x]$) *seem essentially the same.* The ring of cosets $\mathbb{Q}[x]/\langle x-2 \rangle$ appears to be just our old friend the rationals, in disguised garb. Our goal in the next chapter is to show that this is no accident.

---

# Chapter Summary

In this chapter we saw how to make a ring out of the set of cosets of an ideal in a commutative ring, generalizing the construction of $\mathbb{Z}_n$ in Chapter 3.

---

# Warm-up Exercises

a. How many elements does the ring
$$\mathbb{Z}_{16}/\langle 4 \rangle$$
have? Which one is its additive identity? Does this ring have unity? Does it have any zero divisors?

b. Compute
$$(\langle 4 \rangle + 3)(\langle 4 \rangle + 2)$$
in $\mathbb{Z}_{12}/\langle 4 \rangle$ twice, using different representatives from these two cosets.

c. What is the additive inverse of
$$\langle x - 2 \rangle + (x + 2)$$
in $\mathbb{Q}[x]/\langle x - 2 \rangle$?

d. What is the multiplicative inverse of
$$\langle x - 2 \rangle + (x + 2)$$
in $\mathbb{Q}[x]/\langle x - 2 \rangle$? Hint: $x + 2 = (x - 2) + 4$.

e. Let $R$ be a commutative ring and $I$ one of its ideals. What is the nature of the elements belonging to the set $R/I$?

f. What is the kernel of the natural homomorphism $\nu : R \to R/I$?

g. Is the natural homomorphism always onto? Why or why not?

h. Is the natural homomorphism always one-to-one? Why or why not?

i. Is every ideal of a commutative ring a kernel of some homomorphism? Why or why not?

---

# Exercises

1. Compute the multiplicative inverse of the element

$$\langle x + 1 \rangle + (x^3 + 3x^2 - 7x + 1)$$

   in the ring $\mathbb{Q}[x]/\langle x + 1 \rangle$.

2. Consider the ideal $\langle x^2 \rangle$ in $\mathbb{Q}[x]$.

   (a) Prove that for each $f \in \mathbb{Q}[x]$, there exists $a + bx \in \mathbb{Q}[x]$, so that

   $$\langle x^2 \rangle + f = \langle x^2 \rangle + (a + bx).$$

   (b) Show that $\mathbb{Q}[x]/\langle x^2 \rangle$ is not a domain.

3. Consider the ring $\mathbb{Z}/I$, for some ideal $I$. For what ideals $I$ does this ring have infinitely many elements? For what ideals $I$ does this ring have finitely many elements?

4. Consider the ideal $\langle 1 + i \rangle$ in $\mathbb{Z}[i]$.

   (a) Show that
   $$\langle 1 + i \rangle = \{a + bi : a + b \text{ is even}\}.$$
   Hint: $2 = (1 + i)(1 - i)$.

   (b) Use part (a) to show that if $a + bi \in \mathbb{Z}[i]$, there exists $c \in \mathbb{Z}$, such that

   $$\langle 1 + i \rangle + (a + bi) = \langle 1 + i \rangle + c.$$

   (c) Use part (b) to prove that $\mathbb{Z}[i]/\langle 1 + i \rangle$ has only two elements.

5. Consider the ideal
   $$I = \{f \in C[0, 1] : f(1/4) = 0\}$$
   in the commutative ring $C[0, 1]$; we considered this ideal in Example 12.5.

   (a) Prove that
   $$I + f = I + g \text{ if and only if } f(1/4) = g(1/4).$$

   (b) Prove that $C[0, 1]/I$ is a field.

6. Consider the ideal $I = \langle (3, 4) \rangle$ of the ring $\mathbb{Z} \times \mathbb{Z}$. Prove that $(\mathbb{Z} \times \mathbb{Z})/I$ is not a domain.

7. Prove that $J = \{(x, 0) : x \in \mathbb{Z}\}$ is an ideal in $\mathbb{Z} \times \mathbb{Z}$. Prove that $(\mathbb{Z} \times \mathbb{Z})/J$ is a domain (even though $\mathbb{Z} \times \mathbb{Z}$ is *not* a domain).

8. Consider the ring $\mathbb{Z}[\alpha]$ described in Exercise 7.3.

   (a) Prove that any element

   $$\langle 1 + \alpha \rangle + (a + b\alpha + c\alpha^2) \in \mathbb{Z}[\alpha]/\langle 1 + \alpha \rangle$$

   can be written in the form $\langle 1 + \alpha \rangle + m$, where $m$ is an integer.

   (b) Show that $1 + \alpha$ divides 6.

   (c) Use (a) and (b) to show that

   $$\mathbb{Z}[\alpha]/\langle 1 + \alpha \rangle \;\; = \;\; \{\langle 1 + \alpha \rangle + 0, \langle 1 + \alpha \rangle + 1, \langle 1 + \alpha \rangle + 2,$$
   $$\cdots, \langle 1 + \alpha \rangle + 5\}.$$

   (d) Use (c) to show that $\mathbb{Z}[\alpha]/\langle 1 + \alpha \rangle$ is not a domain.

9. The ring $\mathbb{Z}$ is a subring of $\mathbb{Q}$ but is not an ideal. Therefore, it makes no sense to speak of the ring of cosets $\mathbb{Q}/\mathbb{Z}$. Show by explicit example that multiplication makes no sense for $\mathbb{Q}/\mathbb{Z}$. (We will see in Chapter 25 that addition *does* make sense. This means that $\mathbb{Q}/\mathbb{Z}$ is an additive group, but not a ring.)

10. Consider the ring $S$ of real-valued sequences, and $\Sigma$ the ideal of $S$, considered in Exercise 9.29.

    (a) Give a nice description of the elements of the coset $\Sigma + (1, 1, 1, \ldots)$.

    (b) Show by explicit computation that $S/\Sigma$ is not a domain.

    (c) Show that the ring $S/\Sigma$ has infinitely many distinct idempotents. (Recall from Exercise 7.25 that an idempotent is an element $e$ for which $e^2 = e$.)

11. Let $R$ be a commutative ring; recall from Exercise 7.15 that the nilradical $N(R)$ of $R$ is a subring.

    (a) Prove that $N(R)$ is an ideal of $R$.

    (b) Prove that the ring $R/N(R)$ has no non-zero nilpotent elements.

    (c) Check explicitly that part (b) is true for the ring $R = \mathbb{Z}_8$. To what ring is $\mathbb{Z}_8/N(\mathbb{Z}_8)$ isomorphic?

12. Consider the following alternate definition of addition of the cosets of an ideal $I$ of a commutative ring $R$.

    $$(I + a) + (I + b) = \{x + y : x \in I + a, y \in I + b\}.$$

    Prove that this definition is equivalent to the definition of addition on $R/I$ given in the text. (Compare this exercise to Exercise 3.11.)

# Chapter 14

## The Isomorphism Theorem for Rings

In this chapter we bring together the ideas of the last two chapters, by showing that the range of a surjective ring homomorphism is always isomorphic to the ring of cosets formed from the kernel of the homomorphism. This says informally that homomorphisms and ideals convey essentially the same information.

## 14.1   An Illustrative Example

We return to Example 11.2, the homomorphism from $\mathbb{Q}[x]$ to $\mathbb{Q}$ which evaluates any polynomial at 2. This is a surjection: The result of applying the function to all polynomials is the set of all rational numbers. But we claim that if we form the ring $\mathbb{Q}[x]/\ker(\varphi) = \mathbb{Q}[x]/\langle x-2\rangle$, we also get the rational numbers. To show this, consider the function

$$\lambda : \mathbb{Q} \to \mathbb{Q}[x]/\langle x - 2\rangle$$

defined by $\lambda(r) = \langle x - 2\rangle + r$. We claim this is a ring isomorphism. Because of the way addition and multiplication are defined on the ring of cosets, this function clearly preserves them.

▷ **Quick Exercise.**   Verify that the function $\lambda$ preserves addition and multiplication.  ◁

We next claim that $\lambda$ is injective. To do this, we show that its kernel is $\{0\}$. For that purpose, suppose that $r \in \mathbb{Q}$ and $\lambda(r) = 0$; we must show that $r = 0$. But $\langle x - 2\rangle + r = \langle x - 2\rangle + 0$ means that $r \in \langle x - 2\rangle$; that is, the rational number $r$ is a multiple of the polynomial $x - 2$. But $\deg(r) < \deg(x - 2)$ forces us to conclude that $r = 0$.

Finally, we claim that this function is onto. Suppose that

$$\langle x - 2\rangle + f \in \mathbb{Q}[x]/\langle x - 2\rangle.$$

We must find an element of $\mathbb{Q}$ whose value is $\langle x-2\rangle + f$. By applying the Division Theorem 4.2 we have that $f = (x - 2)q + r$, where the degree of the remainder $r$ is less than the degree of $x - 2$; this means that $r$ is a constant. But then

$$\lambda(r) = \langle x - 2\rangle + r = \langle x - 2\rangle + (x - 2)q + r = \langle x - 2\rangle + f.$$

This example is a concrete illustration of the general theorem that explains this; this theorem is called the *Fundamental Isomorphism Theorem for Commutative Rings*.

## 14.2    The Fundamental Isomorphism Theorem

We're now ready to show the true equivalence of the notions of homomorphism and ideal by proving the theorem we've been leading up to for some time. Given a surjective homomorphism $\varphi : R \to S$ between commutative rings $R$ and $S$, we know that $\ker(\varphi)$ is an ideal, and so we have another homomorphism $\nu : R \to R/\ker(\varphi)$. We claim that the ranges of these homomorphisms are 'essentially the same', and what's more, the functions themselves are 'essentially the same'. We use the language of isomorphism to state this formally.

**Theorem 14.1    The Fundamental Isomorphism Theorem for Commutative Rings**    *Let $\varphi : R \to S$ be a surjective homomorphism between rings, and let $\nu : R \to R/\ker(\varphi)$ be the usual natural homomorphism. Then there exists an isomorphism $\mu : R/\ker(\varphi) \to S$ such that $\mu \circ \nu = \varphi$.*

We exhibit this situation in the following diagram:

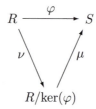

For simplicity's sake, in this book we are restricting our attention in this theorem to *commutative* rings, because we've really only looked at the concept of ideal in this case. However, at the expense of only slightly more work, a more general such theorem actually remains true in the case of arbitrary rings; we won't pursue this matter here.

**Proof**:    Suppose that $R$ and $S$ are commutative rings, $\varphi : R \to S$ is a surjective homomorphism, and $\nu : R \to R/\ker(\varphi)$ is the natural homomorphism. Clearly what we need to do is define a function $\mu$, and prove that it has the desired properties. Once we've obtained the appropriate definition, the rest of the proof will flow smoothly (if a bit lengthily, because there are a lot of properties for $\mu$ we have to verify).

Choose an arbitrary element of $R/\ker(\varphi)$; it is a coset of the form $\ker(\varphi) + r$, where $r \in R$. What element of $S$ should it correspond to? If the composition of functions required in the theorem is to work, we must have that $\mu(\ker(\varphi) + r) = \varphi(r)$; this is our *definition* of $\mu$.

There is an immediate problem we must solve: This definition apparently *depends on the particular coset representative $r$*. That is, our function does not appear to be unambiguously defined. But suppose that $\ker(\varphi) + r = \ker(\varphi) + s$. Then $r - s \in \ker(\varphi)$, and so $\varphi(r) = \varphi(s)$. Thus, it really didn't matter what coset representative we chose, and so our function is well defined. Furthermore, it has been defined precisely so that $\mu \circ \nu(r) = \mu(\ker(\varphi) + r) = \varphi(r)$.

We must now check that $\mu$ is an isomorphism; we take each of the required properties in turn:

$\mu$ *is a homomorphism*: But

$$\begin{aligned}
\mu((\ker(\varphi) + r)(\ker(\varphi) + s)) &= \mu(\ker(\varphi) + rs) \\
&= \varphi(rs) = \varphi(r)\varphi(s) \\
&= \mu(\ker(\varphi) + r)\mu(\ker(\varphi) + s),
\end{aligned}$$

and similarly for addition.

$\mu$ *is surjective*: But $\varphi$ is onto, and so for any $s \in S$, there exists $r \in R$ such that $\varphi(r) = s$. Then $\mu(\ker(\varphi) + r) = \varphi(r) = s$.

$\mu$ is *injective*: Suppose that $\mu(\ker(\varphi) + r) = \mu(\ker(\varphi) + s)$; then $\varphi(r) = \varphi(s)$, and so $r - s \in \ker(\varphi)$. But then $\ker(\varphi) + r = \ker(\varphi) + s$, as required. (You should note that this argument is just the reverse of the argument proving that $\mu$ is well defined.)

▷ **Quick Exercise.** Think about why the parenthetic remark is true.

We thus have the isomorphism required by the theorem. □

## 14.3 Examples

### Example 14.1

Let's look yet again at Example 11.2 where $\varphi : \mathbb{Q}[x] \to \mathbb{Q}$ is the function that evaluates a polynomial at 2. Because this homomorphism is onto and its kernel is $\langle x-2 \rangle$, Theorem 14.1 asserts that the rings $\mathbb{Q}[x]/\langle x - 2 \rangle$ and $\mathbb{Q}$ are isomorphic, via the map $\mu$ which is defined by $\mu(\langle x - 2 \rangle + f) = f(2)$. In Section 14.1 we proved directly that these two rings are isomorphic, via the map $\lambda : \mathbb{Q} \to \mathbb{Q}[x]/\langle x - 2 \rangle$ defined by $\lambda(q) = \langle x - 2 \rangle + q$. What is the relationship between the functions $\lambda$ and $\mu$? They are inverse functions. For

$$\mu(\lambda(q)) = \mu(\langle x - 2 \rangle + q) = \mu \circ \nu(q) = \varphi(q) = q(2) = q.$$

(Notice this last equality holds because we are thinking of $q$ as a constant polynomial.)

### Example 14.2

If we apply the theorem to the residue homomorphism $\varphi : \mathbb{Z} \to \mathbb{Z}_4$, we obtain what we already suspected: $\mathbb{Z}/\langle 4 \rangle$ and $\mathbb{Z}_4$ are isomorphic rings.

### Example 14.3

Consider the function $\varphi : \mathbb{Z} \times \mathbb{Z} \to \mathbb{Z}$ defined by $\varphi(x, y) = y$; you can easily check that this is a surjective homomorphism. (Indeed, this was verified in a more general context as Exercise 11.4.) The kernel of this homomorphism is clearly $\{(x, 0) : x \in \mathbb{Z}\}$, which we might write as $\mathbb{Z} \times \{0\}$. (Note that in Section 12.5 we asked you for a homomorphism with this ideal as kernel.) Our theorem thus asserts that $(\mathbb{Z} \times \mathbb{Z})/(\mathbb{Z} \times \{0\})$ is isomorphic to $\mathbb{Z}$. Notice also that in Exercise 13.7 you proved that this quotient ring is a domain; because we've now shown that it is isomorphic to $\mathbb{Z}$, this result should not be surprising.

### Example 14.4

Let's now consider two general but trivial examples of the theorem. Given any commutative ring $R$, it always possesses two ideals; namely, the trivial ideal $\{0\}$ and the improper ideal $R$ itself. These are certainly the kernels of the identity isomorphism $\iota : R \to R$ and of the zero homomorphism $\zeta : R \to \{0\}$, respectively. Theorem 14.1 then asserts in the one case that $R$ is isomorphic to $R/\{0\}$, and in the other that $R/R$ is isomorphic to $\{0\}$. Speaking informally, this says that if we mod zero out of a ring, we have left it unaffected, while if we mod out the entire ring, we are left with the zero ring.

**Example 14.5**

As another example of the theorem, consider the function $\varphi : \mathbb{R}[x] \to \mathbb{C}$ defined by $\varphi(f) = f(i)$. Notice that evaluating a polynomial with real coefficients at the complex number $i$ will certainly give a complex number. As usual, an evaluation map of this sort is a homomorphism. What is the kernel of this function? It consists of $\{f \in \mathbb{R}[x] : f(i) = 0\}$, but these are precisely those polynomials which have $x^2 + 1$ as a factor. That is, the kernel here is again a principal ideal, this time $\langle x^2 + 1 \rangle$. Thus, we have that the field of complex numbers is isomorphic to $\mathbb{R}[x]/\langle x^2 + 1 \rangle$.

The previous example is actually an algebraically elegant way of describing how we obtain the complex numbers from the real numbers. The goal in obtaining the complex numbers is after all to be able to solve more equations (such as $x^2 + 1 = 0$). We will inquire more carefully into this example soon, but we will first have to characterize those ideals that lead to fields as their rings of cosets. This characterization is one of the goals of the next chapter.

**Example 14.6**

For another more sophisticated example of a homomorphism onto a field, consider the function

$$\varphi : \mathbb{Z}[\sqrt{-5}] \to \mathbb{Z}_3$$

defined as $\varphi(a + b\sqrt{-5}) = [a - b]_3$. (Recall Exercise 7.2 for a description of rings of the form $\mathbb{Z}[\sqrt{n}]$.) Let's first check that this is a homomorphism. We see that $\varphi$ preserves addition:

$$\varphi((a + b\sqrt{-5}) + (c + d\sqrt{-5})) = [(a + c) - (b + d)] =$$
$$[a - b] + [c - d] = \varphi(a + b\sqrt{-5}) + \varphi(c + d\sqrt{-5}).$$

Similarly, $\varphi$ preserves multiplication:

$$\begin{aligned}
\varphi((a + b\sqrt{-5})(c + d\sqrt{-5})) &= \varphi((ac - 5bd) + (ad + bc)\sqrt{-5}) \\
&= [ac - 5bd - ad - bc] \\
&= [ac - ad + bd - bc] \\
&= [(a - b)(c - d)] \\
&= \varphi(a + b\sqrt{-5})\varphi(c + d\sqrt{-5})
\end{aligned}$$

(where we've used the fact that $[-5] = [1]$). The kernel of this function is

$$\{a + b\sqrt{-5} : 3 | (a - b)\}.$$

Since this is evidently the kernel, we know that this is an ideal, although that is not entirely obvious (see Exercise 9). Thus, we have that

$$\mathbb{Z}[\sqrt{-5}]/\{a + b\sqrt{-5} : 3 | (a - b)\}$$

is isomorphic to the field $\mathbb{Z}_3$.

This would be a good deal more satisfying if we had a more concrete description of this kernel. We claim that it is actually the smallest ideal of $\mathbb{Z}[\sqrt{-5}]$ that contains the elements 3 and $1 + \sqrt{-5}$; that is, it is $\langle 3, 1 + \sqrt{-5} \rangle$. To show this, we note first that clearly

$$3, \ 1 + \sqrt{-5} \in \{a + b\sqrt{-5} : 3 | (a - b)\},$$

and so therefore the smallest ideal containing these two elements is a subset of the kernel. For the reverse inclusion, suppose that $a - b$ is divisible by 3. Then $a - b = 3k$, and so

$$\begin{aligned}
a + b\sqrt{-5} &= a(1 + \sqrt{-5}) + (b - a)\sqrt{-5} \\
&= a(1 + \sqrt{-5}) - k(\sqrt{-5})(3) \in \langle 3, 1 + \sqrt{-5} \rangle.
\end{aligned}$$

That is,

$$\langle 3, 1 + \sqrt{-5} \rangle = \{x + y\sqrt{-5} : 3 \text{ divides } (x - y)\},$$

and so $\mathbb{Z}[\sqrt{-5}]/\langle 3, 1 + \sqrt{-5} \rangle$ is isomorphic to $\mathbb{Z}_3$.

Now what is even more interesting about this kernel is that this description is as good as we can do – it is not a principal ideal! You will verify this in Exercise 10.

## Historical Remarks

The equivalence between homomorphisms and kernels, as expressed in the Fundamental Isomorphism Theorem for Commutative Rings, is a crucial idea in abstract algebra; there are corresponding theorems for many other algebraic structures. Indeed, we will encounter the corresponding theorem for groups in Chapter 26. The crisp, abstract formulation of this theorem contained in this chapter is very much an artifact of the 20th century axiomatic approach to algebra. The ideas behind this theory were encountered in many specific situations in the 19th century, but the sort of formulation we give here was not possible until the now accepted axiomatics for such algebraic structures as rings, groups, and fields were firmly in place. A lot of work and thought by many mathematicians went into these definitions. The French mathematician Camille Jordan was probably the first to consider clearly the notion of a quotient structure such as our ring of cosets $R/I$; he was working in the specific context of permutation groups. (See Chapters 18 and 27.) It is important for you to keep in mind that the abstract and efficient structure of modern algebra was not born overnight, but rather was the result of painstaking study of examples, from all over mathematics. The lesson of this history is clear: We should appreciate the generality of our theorems but must always return to specific examples to understand their purpose and application.

---

## Chapter Summary

In this chapter we proved the *Fundamental Isomorphism Theorem for Commutative Rings*, which asserts that every onto homomorphism can be viewed as a natural homomorphism onto the ring modulo the kernel.

---

## Warm-up Exercises

a. For a commutative ring $R$, what is $R/R$ isomorphic to? How about $R/\{0\}$?

b. Is every commutative ring isomorphic to a ring of cosets?

c. Consider the evaluation homomorphism $\varphi : \mathbb{Z}[x] \to \mathbb{Z}[\sqrt{2}]$ defined by $\varphi(f) = f(\sqrt{2})$. What is the kernel of this homomorphism? What two rings are isomorphic, according to the fundamental theorem?

d. Let $F$ be a field. How many ideals does it have? So (up to isomorphism) what are the only possible rings $\varphi(F)$, where $\varphi$ is a ring homomorphism?

e. What is the complete list of ideals for $\mathbb{Z}$? (See Theorem 9.5.) So (up to isomorphism) what are the only possible rings $\varphi(\mathbb{Z})$, where $\varphi$ is a ring homomorphism?

f. Let's work in the ring $\mathbb{Z}[x]/\langle 2, x \rangle$. Pick any $f = a_0 + a_1 x + a_2 x^2 + \cdots a_n x^n \in \mathbb{Z}[x]$. Justify the following string of equalities:

$$\langle 2, x \rangle + f = \langle 2, x \rangle + a_0 = \begin{cases} \langle 2, x \rangle + 0 & a_0 \text{ is even} \\ \langle 2, x \rangle + 1 & a_0 \text{ is odd.} \end{cases}$$

So to what ring is $\mathbb{Z}[x]/\langle 2, x \rangle$ isomorphic? Show this using the Fundamental Theorem.

g. Suppose that $\varphi : R \to S$ is a ring homomorphism between commutative rings $R$ and $S$. Is $R/\ker(\varphi)$ isomorphic to $S$? Be careful! What *is* true in this situation?

h. In what sense are the ideas of 'ideal' and 'homomorphism' equivalent?

---

# Exercises

1. Prove that $\mathbb{Z}[i]/\langle 1 + i \rangle$ is isomorphic to $\mathbb{Z}_2$, by defining a homomorphism from $\mathbb{Z}[i]$ onto $\mathbb{Z}_2$ whose kernel is $\langle 1 + i \rangle$, and using the Fundamental Theorem 14.1. (Compare this to Exercise 13.4.)

2. Let $I = \{f \in C[0,1] : f(1/4) = 0\}$. Prove that $C[0,1]/I$ is isomorphic to $\mathbb{R}$ by defining the appropriate homomorphism from $C[0,1]$ onto $\mathbb{R}$ and using the Fundamental Theorem 14.1. (Compare this to Exercise 13.5.)

3. Let $I = \langle (3,4) \rangle \subseteq \mathbb{Z} \times \mathbb{Z}$. Prove that $(\mathbb{Z} \times \mathbb{Z})/I$ is isomorphic to $\mathbb{Z}_{12}$, by using a homomorphism of the form $\varphi(a, b) = [ax + by]_{12}$ (for some fixed $x, y$). (Compare this to Exercise 13.6.)

4. Use the Fundamental Theorem to prove that

$$(\mathbb{Q}[x] \times \mathbb{Z})/\langle (x, 2) \rangle$$

is isomorphic to $\mathbb{Q} \times \mathbb{Z}_2$.

5. Let $R$ be the set $\mathbb{Q} \times \mathbb{Q}$. We will equip this set with operations, other than the usual ones for the direct product, as described in Example 6.10. Namely, define the operations

$$(a, b) + (c, d) = (a + c, b + d)$$

(this is the usual addition), and

$$(a, b)(c, d) = (ac, ad + bc).$$

(a) Prove from first principles that $R$ is a ring.

(b) Use the Fundamental Theorem to prove that $R$ is isomorphic to $\mathbb{Q}[x]/\langle x^2 \rangle$.

6. Prove that $\mathbb{Z}_2 \times \mathbb{Z}_3$ is isomorphic to $\mathbb{Z}_6$ by showing that the homomorphism $\varphi([a]_2, [b]_3) = [3a + 4b]_6$ is onto and has zero kernel. (Or, simply that it is onto and $\mathbb{Z}_2 \times \mathbb{Z}_3$ and $\mathbb{Z}_6$ both have 6 elements; this is an application of the *pigeonhole principle* discussed in Section 16.1.) What gets mapped to $[1]_6$? To $[2]_6$? $[3]_6$? $[4]_6$? $[5]_6$?

7. Let $X$ be a set with $n$ elements; consider the power set ring $P(X)$ of subsets of $X$ (described in Exercise 6.20). Consider also the ring

$$\mathbb{Z}_2 \times \mathbb{Z}_2 \times \cdots \times \mathbb{Z}_2$$

of $n$-tuples whose entries are taken from $\mathbb{Z}_2$. Prove that these two rings are isomorphic.

8. Consider the ring $\mathbb{Z}[\alpha]$, described in Exercise 7.3. Prove that

$$\mathbb{Z}[\alpha]/\langle 1 + \alpha \rangle$$

   is isomorphic to the ring $\mathbb{Z}_6$, by defining a homomorphism $\varphi : \mathbb{Z}[\alpha] \to \mathbb{Z}_6$ with the appropriate kernel. (Compare this to Exercise 13.8.)

9. In Example 14.6, we concluded that $\{a + b\sqrt{-5} : 3|(a - b)\}$ is an ideal in $\mathbb{Z}[\sqrt{-5}]$ by observing that it is the kernel of a ring homomorphism. In this exercise, you should show directly that this is an ideal, by verifying that this set is closed under subtraction and multiplicative absorption.

10. In Example 14.6, we show that the kernel described in the previous exercise can be more simply described as $\langle 3, 1 + \sqrt{-5} \rangle$. In this exercise you will show that this ideal is not principal. Suppose by way of contradiction that it is principal with generator $\alpha = a + b\sqrt{-5}$. Then both 3 and $1 + \sqrt{-5}$ must be multiples of $\alpha$. Use the complex absolute value and Theorem 8.3 to conclude that $|\alpha|^2$ must divide both 9 and 6. Obtain a contradiction from this.

11. Let $C$ be the ring of convergent real-valued sequences. Consider the homomorphism $\varphi : C \to \mathbb{R}$ given in Exercise 11.31. What two rings are isomorphic, according to the Fundamental Theorem?

12. Use the Fundamental Theorem to prove that $\mathbb{Q}[x]/\langle x^2 \rangle$ and

$$\left\{ \begin{pmatrix} a & b \\ 0 & a \end{pmatrix} \in M_2(\mathbb{Q}) : a,\ b \in \mathbb{Q} \right\}$$

   are isomorphic. (Compare this to Exercise 5 in this chapter.)

13. In this problem you will prove that the fields $\mathbb{R}$ and $\mathbb{C}$ are *not* isomorphic. Suppose by way of contradiction that there exists a ring isomorphism $\varphi : \mathbb{C} \to \mathbb{R}$. Now answer the following questions (justifying your answers, of course): What is $\varphi(1)$? What is $\varphi(-1)$? What does this tell you about $\varphi(i)$?

14. In this problem you will prove that the fields $\mathbb{R}$ and $\mathbb{Q}$ are *not* isomorphic. Suppose by way of contradiction that there exists a ring isomorphism $\varphi : \mathbb{R} \to \mathbb{Q}$. Now answer the following questions (justifying your answers, of course): What is $\varphi(1)$? What is $\varphi(2)$? What does this tell you about $\varphi(\sqrt{2})$?

15. Use the ideas of Exercise 13 (or Exercise 14) to prove that $\mathbb{Q}$ and $\mathbb{C}$ are not isomorphic.

16. If you have encountered the ideas of *countably infinite* and *uncountably infinite* sets, another proof of Exercise 14 (and 15) is possible. What is it?

17. Suppose that rings $R$ and $S$ are isomorphic, via the isomorphism $\varphi : R \to S$.

    (a) Show that the rings $R[x]$ and $S[x]$ are isomorphic, by defining a ring isomorphism $\bar{\varphi} : R[x] \to S[x]$ which extends $\varphi$; by this we mean that if $r \in R$, then $\varphi(r) = \bar{\varphi}(r)$.

    (b) Suppose now that the rings $R$ and $S$ are fields. Prove that $r \in R$ is a root of $f \in R[x]$ if and only if $\varphi(r)$ is a root of $\bar{\varphi}(f)$.

    (c) Suppose again that $R$ and $S$ are fields. Prove that $f \in R[x]$ is irreducible if and only if $\bar{\varphi}(f) \in S[x]$ is irreducible.

18. Suppose that the commutative rings $R$ and $S$ are isomorphic, via the isomorphism $\varphi : R \to S$. Let $r \in R$. Use the Fundamental Isomorphism Theorem to prove that the rings $R/\langle r \rangle$ and $S/\langle \varphi(r) \rangle$ are isomorphic.

19. Extend Exercise 18. Suppose that the commutative rings $R$ and $S$ are isomorphic via the isomorphism $\varphi : R \to S$. Let $I$ be an ideal of $R$; by Exercise 11.32 we know that $\varphi(I)$ is an ideal of $S$. Prove that the rings $R/I$ and $S/\varphi(I)$ are isomorphic.

20. Suppose that the commutative rings $R$ and $S$ are isomorphic via the isomorphism $\varphi : R \to S$. Combine Exercises 17 and 19 to conclude that if $f \in R[x]$, then the rings $R[x]/\langle f \rangle$ and $S[x]/\langle \bar{\varphi}(f) \rangle$ are isomorphic.

21. Suppose that $R$ is a commutative ring, with ideals $A$ and $I$, where $I \subseteq A$.

    (a) Prove that $A/I$ is an ideal in the ring $R/I$.

    (b) Prove the converse of part (a). Namely, show that any ideal of $R/I$ is of the form $A/I$, where $A$ is an ideal of $R$ containing $I$.

    (c) Prove that $(R/I)/(A/I)$ is isomorphic to $R/A$.

22. Let $R = C[0,1]$, $I = \{f \in C[0,1] : \ f(0) = f(1) = 0\}$ and $A = \{f \in C[0,1] : \ f(0) = 0\}$. Check that $R, A, I$ satisfy the hypotheses of Exercise 21. Then exhibit explicitly the isomorphism given by that exercise.

23. Let $R$ be a commutative ring with ideals $I$ and $J$. Then $I \cap J$ and $I + J$ are also ideals of $R$. (See Exercises 9.16 and 9.18.) Furthermore, $I \cap J$ is an ideal of $I$, and $J$ is an ideal of $I + J$. (See Exercise 9.17.) Prove that the rings $I/(I \cap J)$ and $(I + J)/J$ are isomorphic.

24. Let $R$ be $\mathbb{Z}$, $I = \langle 12 \rangle$ and $J = \langle 8 \rangle$. Check that $R, I, J$ satisfy the hypotheses of Exercise 23. Then exhibit explicitly the isomorphism given by that exercise.

25. The essence of the Fundamental Isomorphism Theorem is that if $\varphi : R \to S$ is a surjective homomorphism, then the rings $S$ and $R/\ker(\varphi)$ are isomorphic, and our proof defines an isomorphism $\mu : R/\ker(\varphi) \to S$. It is possible instead to prove this theorem by defining an isomorphism $\lambda : S \to R/\ker(\varphi)$. Carry this out, making sure your function $\lambda$ is well-defined, preserves addition and multiplication, and is bijective.

# Chapter 15

## Maximal and Prime Ideals

Let's return to Example 14.5. We saw that the ring homomorphism (the evaluation homomorphism) $\varphi : \mathbb{R}[x] \to \mathbb{C}$ given by $\varphi(f) = f(i)$ has kernel $\langle x^2 + 1 \rangle$. This homomorphism is surjective, and so $\mathbb{R}[x]/\langle x^2 + 1 \rangle$ is isomorphic to the *field* $\mathbb{C}$.

Now compare this to the residue homomorphism $\varphi : \mathbb{Z} \to \mathbb{Z}_7$. This homomorphism is surjective, and so $\mathbb{Z}/\langle 7 \rangle$ is isomorphic to the *field* $\mathbb{Z}_7$.

We would like to find out what sort of ideal leads to a ring of cosets that is a field.

---

### 15.1   Irreducibles

Notice that $x^2 + 1$ is an *irreducible* element of $\mathbb{R}[x]$ (that is, it is a polynomial that cannot be further factored in $\mathbb{R}[x]$), and 7 is an *irreducible* element of $\mathbb{Z}$ (that is, it is an integer that cannot be further factored in $\mathbb{Z}$). This suggests we should make a more general definition.

A non-zero element $p$ of a commutative ring with unity $R$ is **irreducible** if

**(1)** it is a non-unit, and

**(2)** if whenever $p = ab$, then (exactly) one of $a$ and $b$ is a unit.

Let's explore some examples of this concept.

#### Example 15.1

The irreducibles in $\mathbb{Z}$ are exactly the prime numbers (and their negatives).

#### Example 15.2

The irreducibles in $\mathbb{R}[x]$ are exactly the linear polynomials, together with the quadratics with negative discriminant. (See Theorem 10.3.)

#### Example 15.3

The irreducibles in $\mathbb{Q}[x]$ include all linear polynomials, in addition to such polynomials as $x^2 + 1$, $x^3 - 2$, and others.

#### Example 15.4

Consider the case of a field. Here, *all* non-zero elements are units, and so in a field there are no irreducible elements.

## Example 15.5

What are irreducible elements in $\mathbb{Z}[x]$? Note that although $3x+6$ is irreducible in $\mathbb{Q}[x]$, it is not irreducible in $\mathbb{Z}[x]$. This is because 3 is a non-unit in $\mathbb{Z}[x]$, and so the factorization $3x+6 = 3(x+2)$ is non-trivial. Thus, the coefficients of an irreducible polynomial in $\mathbb{Z}[x]$ can have no common integer factor (other than $\pm 1$). Such polynomials in $\mathbb{Z}[x]$ are called **primitive**. Gauss's Lemma 5.5 asserts that primitive polynomials (with degree greater than zero) of $\mathbb{Z}[x]$ are irreducible if and only if they are irreducible in $\mathbb{Q}[x]$. And what about polynomials of degree zero in $\mathbb{Z}[x]$ (that is, the integers)? Any factorization of such a polynomial would also be factorization in $\mathbb{Z}$. Thus, 3 (and any other prime integer, or its negative) will be an irreducible element of $\mathbb{Z}[x]$.

## Example 15.6

What are the irreducible elements of $\mathbb{Z}_4$? In this ring, 1 and 3 are units, and so the only possible irreducible is 2. But the only factorizations of 2 in $\mathbb{Z}_4$ are $2 = 1 \cdot 2$ and $2 = 3 \cdot 2$, which are both trivial; thus, 2 is irreducible. Note in this case that 2 has no associates other than itself.

For an integral domain $R$, we are now going to translate the statement that $p$ is an irreducible element in terms of the principal ideal $\langle p \rangle$. We do this in the spirit of our discussion of principal ideals in chapter 9. Since $p$ is a non-unit, this means that $\langle p \rangle$ is a proper ideal. But if $\langle p \rangle \subseteq \langle a \rangle$, then $p = ab$. Since $p$ is irreducible, we may as well assume that $b$ is a unit. But then $p$ and $a$ are associates, and $\langle p \rangle = \langle a \rangle$. The upshot of this is that there is no proper principal ideal strictly larger than $\langle p \rangle$. Conversely, suppose that the principal ideal $\langle p \rangle$ is *maximal among all proper principal ideals*. To show that $p$ is irreducible, suppose that $p = ab$. Since $p$ is not a unit, at least one of $a, b$ is also a non-unit; we may as well assume that $a$ is a non-unit. Now clearly $\langle p \rangle \subseteq \langle a \rangle$, and by maximality these must be equal, because $\langle a \rangle \neq R$, since $a$ is not a unit. But then $p$ and $a$ are associates by Theorem 9.3, and so $b$ is a unit.

We record as a theorem the conclusion we have reached:

**Theorem 15.1** *Let $R$ be an integral domain. Then the element $p \in R$ is an irreducible element if and only if there is no strictly larger principal ideal in $R$.*

Let's now look at the whole ideal structure of $\mathbb{Z}$ in light of the above discussion. We know that the principal ideals which admit no larger proper principal ideals are given exactly by the ideals of the form $\langle p \rangle$, where $p$ is an irreducible (or prime) integer, and containments reflect divisibility. But note from Theorem 9.5 that *all* ideals of $\mathbb{Z}$ are principal, and so these ideals are maximal among *all* proper ideals. A proper ideal of any commutative ring with unity for which there is no larger proper ideal is called a **maximal ideal**. We will explore this important concept more in the following section. Remember that an integral domain for which all ideals is principal is called a *principal ideal domain*, or *PID* for short (see Section 9.4). So in a PID an ideal $\langle p \rangle$ is maximal if and only if $p$ is an irreducible element.

Here then is a picture of the ideals in $\mathbb{Z}$:

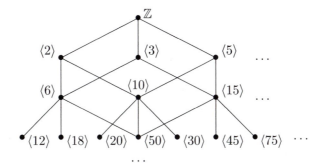

In contrast, consider a domain like $\mathbb{Z}[x]$; see our discussion above in Example 15.5. Because the element $x$ is irreducible, we have that $\langle x \rangle$ is certainly maximal among principal ideals, but this is not a maximal ideal, because the non-principal ideal $\langle 2, x \rangle$ is strictly larger – see Example 9.12. In fact, we claim that the ideal $\langle 2, x \rangle$ is a maximal ideal. For if we choose any element $f$ outside $\langle 2, x \rangle$, then $f$ will have an odd constant term, and so $f = 2n + 1 + xg$, where $n$ is an integer and $g \in \mathbb{Z}[x]$. But then $1 = f - 2n - xg$ and so any ideal containing $f, 2, x$ must include unity 1, and so is the whole ring.

Now what is the meaning in a computational sense of the maximality of a principal ideal in a domain? If $\langle p \rangle$ is maximal and $a \notin \langle p \rangle$, then the ideal $\langle p, a \rangle$ must be the entire domain, because it is certainly a strictly larger ideal than $\langle p \rangle$. But $\langle p, a \rangle$ is the entire domain exactly if it contains (any) unit, and in particular if $1 \in \langle p, a \rangle$. That is, $1 = px + ay$, for some $x, y$ in the domain.

## Example 15.7

Consider the maximal ideal $\langle 7 \rangle$ in $\mathbb{Z}$. Now, $12 \notin \langle 7 \rangle$. But

$$3 \cdot 12 + (-5) \cdot 7 = 1,$$

and so $\langle 7, 12 \rangle = \mathbb{Z}$. Of course, the fact that there exist $x$ and $y$ such that $12x + 7y = 1$ is just an example of the GCD identity for $\mathbb{Z}$. This reveals a more general principle operating in the ideal structure of $\mathbb{Z}$: $\langle a, b \rangle = \langle c \rangle$, where $c = \gcd(a, b)$. (See Exercise 9.15.)

## Example 15.8

Because $x$ is an irreducible element in $\mathbb{Q}[x]$ and this ring is a PID (Exercise 9.2), we know that $\langle x \rangle$ is a maximal ideal; notice the contrast to our discussion about $\mathbb{Z}[x]$ above. To see this computationally, choose $f \notin \langle x \rangle$. Because $\langle x \rangle$ consists of the polynomials in $\mathbb{Q}[x]$ with zero constant term, $f$ must have a non-zero constant term $c$, and so $f$ can be written as $f = c + xg$, where $g$ is some polynomial in $\mathbb{Q}[x]$. But direct computation shows that

$$1 = c^{-1}f + (-c^{-1}g)x.$$

This last expression is a linear combination of $f$ and $x$, and so $1 \in \langle x, f \rangle$. Thus, $\langle x, f \rangle = \mathbb{Q}[x]$, and so $\langle x \rangle$ is a maximal ideal, as claimed.

## 15.2  Maximal Ideals

We now would like to focus our attention on maximal ideals in any commutative ring with unity. We observed in the introduction to this chapter that both $\mathbb{R}[x]/\langle x^2 + 1 \rangle$ and $\mathbb{Z}/\langle 7 \rangle$ are fields. This suggests the following theorem:

**Theorem 15.2** *Let $R$ be a commutative ring with unity. Then $M$ is a maximal ideal of $R$ if and only if $R/M$ is a field.*

**Proof**:    Let $R$ be a commutative ring with unity, and suppose first that $M$ is a maximal ideal. To show that $R/M$ is a field, we consider any $M + a \in R/M$, with $M + a \neq M + 0$; we must show that $M + a$ has a multiplicative inverse. But because $M + a \neq M$, we know that $a \notin M$. Consider

$$\langle M, a \rangle = \{m + ab : m \in M, b \in R\};$$

we are using the notation $\langle M, a \rangle$ to suggest that this is in fact the smallest ideal of $R$ that contains both $M$ and $a$. In Exercise 1, you prove exactly this.

   Furthermore, $\langle M, a \rangle$ properly contains $M$, because $a \in \langle M, a \rangle$ but $a \notin M$. Thus, because $M$ is maximal, $\langle M, a \rangle = R$. So there exist $b \in R$ and $m \in M$ such that $m + ab = 1$. But then

$$M + 1 = M + (m + ab) = M + ab = (M + a)(M + b),$$

and so $M + b$ is the required multiplicative inverse.

   Conversely, suppose that $R/M$ is a field. Let $I$ be an ideal with $R \supseteq I \supset M$; we must show that $I = R$. Choose $r$, an element in $I$ but not in $M$. Because $R/M$ is a field, there exists $s \in R$ such that $M + rs = M + 1$. That is, $M = M + 1 - rs$ or $1 - rs = m$ for some $m$ in $M$. But $1 = rs + m$ implies $1 \in I$, because $r \in I$ and $m \in I$. Now $1 \in I$ implies that $I = R$. Hence, $M$ is maximal.                    □

### Example 15.9

> An important if relatively trivial example of this theorem occurs in case $R$ itself is a field. Because all non-zero elements are units, we've observed before that the only proper ideal a field has is $\{0\}$; hence, $\{0\}$ is (practically by default) a *maximal* ideal, and so the theorem applies. That is, the ring of cosets $R/\{0\}$ is a field; but of course this ring is just (isomorphic to) $R$ itself. But we can also apply the converse: If $R$ is a commutative ring with unity without ideals (other than $\{0\}$ and $R$), then $R$ is necessarily a field. (Of course, we've seen this result before as Corollary 9.4.)

### Example 15.10

> In Example 14.5 we proved that $\mathbb{R}[x]/\langle x^2 + 1 \rangle$ is isomorphic to the field $\mathbb{C}$. This means that $\langle x^2 + 1 \rangle$ is a maximal ideal in $\mathbb{R}[x]$.

### Example 15.11

> In Example 14.6 we proved that $\mathbb{Z}[\sqrt{-5}]/\langle 3, 1 + \sqrt{-5} \rangle$ is isomorphic to the field $\mathbb{Z}_3$. This means that $\langle 3, 1 + \sqrt{-5} \rangle$ is a maximal ideal of $\mathbb{Z}[\sqrt{-5}]$.

**Example 15.12**

Consider the irreducible polynomial $x^2 - 2$ in $\mathbb{Q}[x]$ (note that this is *not* irreducible in $\mathbb{R}[x]$). Now $\mathbb{Q}[x]$ is a principal ideal domain, and so like the examples at the start of the chapter, we have that $\langle x^2 - 2 \rangle$ is a maximal ideal in $\mathbb{Q}[x]$. Thus, by Theorem 15.2, we know that $\mathbb{Q}[x]/\langle x^2 - 2 \rangle$ is a field. But what is this field? Consider the evaluation homomorphism $\varphi : \mathbb{Q}[x] \to \mathbb{R}$, given by $\varphi(f) = f(\sqrt{2})$. Quite evidently, the kernel of this homomorphism is exactly $\langle x^2 - 2 \rangle$. But it seems clear that $\varphi$ is not onto $\mathbb{R}$. For example, $\sqrt{3} \notin \varphi(\mathbb{Q}[x])$: We cannot obtain $\sqrt{3}$ by plugging $\sqrt{2}$ into a rational polynomial (this statement is probably plausible, but we won't prove it here – see Example 37.2). But if $\varphi$ is not onto, we cannot directly apply the Fundamental Isomorphism Theorem 14.1. What the Fundamental Isomorphism Theorem does tell us is that $\mathbb{Q}[x]/\langle x^2 - 2 \rangle$ is isomorphic to $\varphi(\mathbb{Q}[x])$, a proper subfield of the field $\mathbb{R}$.

Furthermore, this field $\mathbb{Q}[x]/\langle x^2 - 2 \rangle$ contains a subring isomorphic to $\mathbb{Q}$: Consider the function

$$\iota : \mathbb{Q} \to \mathbb{Q}[x]/\langle x^2 - 2 \rangle$$

defined by $\iota(q) = \langle x^2 - 2 \rangle + q$.

▷ **Quick Exercise.** Check that this is an injective homomorphism (which is *not* onto). ◁

Thus, $\iota(\mathbb{Q})$ is the isomorphic copy of $\mathbb{Q}$ contained in the field $\mathbb{Q}[x]/\langle x^2 - 2 \rangle$. This suggests that we can use the previous theorem to build bigger fields. In this case we seem to have constructed a field strictly between $\mathbb{Q}$ and $\mathbb{R}$. We'll have a lot to say about this in Chapters 41 and 42.

An alternative proof for Theorem 15.2 is possible, using ideas from Exercise 14.21. There we discover that the ideals of a ring of cosets $R/I$ are exactly of the form $A/I$, where $A \supseteq I$ is an ideal of $R$. But furthermore, we know that a commutative ring with unity is a field if and only if it has no proper ideals except the zero ideal (Corollary 9.4 or Example 15.9). But then $R/I$ has no proper ideals except the zero ideal $I/I$, if and only if there are no ideals $A$ with $R \supset A \supset I$, that is, if and only if $I$ is a maximal ideal of $R$.

---

## 15.3 Prime Ideals

We now consider a class of ideals closely related to the class of maximal ideals. A proper ideal $I$ of a commutative ring with unity is a **prime ideal** if whenever $ab \in I$, then either $a \in I$ or $b \in I$. Let's look at an example of a prime ideal.

**Example 15.13**

Consider the ideal $\langle x \rangle$ in $\mathbb{Z}[x]$. We claim that it is prime: For if $pq \in \langle x \rangle$, then $pq$ is a polynomial with no constant term. But the constant term of a product of polynomials is the product of their constant terms, and so this means that at least one of $p$ and $q$ must also lack a constant term; that is, at least one of $p$ and $q$ belongs to $\langle x \rangle$. This is what we wished to show.

Thus, $\langle x \rangle$ is a prime ideal in $\mathbb{Z}[x]$. Now consider another evaluation homomorphism, this time at zero: If $\psi(f) = f(0)$, then $\ker(\psi) = \langle x \rangle$. Of course, $\psi$ is here a homomorphism onto $\mathbb{Z}$, and so $\mathbb{Z}[x]/\langle x \rangle$ is isomorphic to $\mathbb{Z}$. Thus, we conclude that $\langle x \rangle$ is *not* a maximal ideal (because $\mathbb{Z}$ is *not* a field). More directly, we can conclude that $\langle x \rangle$ is not a maximal ideal, by noting that

$$\langle x \rangle \subset \langle 2, x \rangle \subset \mathbb{Z}[x].$$

However, the homomorphic image $\mathbb{Z}$ is at least a *domain*; this observation suggests the following theorem:

**Theorem 15.3** *Let $R$ be a commutative ring with unity. Then $P$ is a prime ideal if and only if $R/P$ is an integral domain.*

**Proof:** Let $R$ be a commutative ring with unity, and suppose first that $P$ is a prime ideal. To show that $R/P$ is a domain (because it obviously has unity) we need only show that it has no zero divisors. So suppose that $(P + a)(P + b) = P + 0$. But then $ab \in P$, and so because $P$ is prime $a$ or $b$ belongs to $P$, and so $P + a = P + 0$ or $P + b = P + 0$, as required. In the Quick Exercise below you will do the converse argument, which is essentially the reverse of what we've just done. □

▷ **Quick Exercise.** Prove the converse of the above theorem. ◁

Note in particular the special case when $R$ itself is a domain. We conclude that a commutative ring with unity is a domain if and only if $\{0\}$ is a prime ideal.

Because all fields are domains, we can conclude that maximal ideals are necessarily prime.

For principal ideals, there is an element-wise characterization of when such an ideal is prime; you will prove this characterization in Exercise 2. These ideas will be important when we study factorization theory in domains, in Chapter 32.

**Example 15.14**

As an example of Theorem 15.3, consider (from Example 11.3) the ring $\mathbb{Z} \times \mathbb{Z}$ and the homomorphism $\pi(x, y) = y$, which is onto $\mathbb{Z}$. Because $\mathbb{Z}$ is a domain, the kernel $\mathbb{Z} \times \{0\}$ is a prime ideal of $\mathbb{Z} \times \mathbb{Z}$ (that is also not maximal).

## 15.4 An Extended Example

We will now look at all of these ideas in the context of the ideal $\langle 3x + 3 \rangle$, both in $\mathbb{Z}[x]$ and in $\mathbb{Q}[x]$. This will provide some further insight into all of the theorems of this chapter.

Now in $\mathbb{Q}[x]$ the polynomial $3x + 3$ is irreducible (since any linear polynomial is – see Section 5.1). The extra factor 3 is a unit in $\mathbb{Q}[x]$, and so $3x + 3$ and $x + 1$ are associates and $\langle x + 1 \rangle = \langle 3x + 3 \rangle$. Since $x + 1$ is irreducible, we know know that this principal ideal is maximal among all other principal ideals. But $\mathbb{Q}[x]$ is a PID (see Exercise 9.2), and so $\langle x+1 \rangle$ is a maximal ideal. Of course, this also follows because it is the kernel of the evaluation homomorphism $\varphi_1 : \mathbb{Q}[x] \to \mathbb{Q}$ defined by $\varphi_1(f) = f(-1)$, which is onto the field $\mathbb{Q}$.

The situation is rather different in $\mathbb{Z}[x]$. Here both 3 and $x + 1$ are non-units, and so $3x+3 = 3(x+1)$ is not irreducible. This means that the ideal $\langle 3x+3 \rangle$ is not maximal among all principal ideals, and it is indeed quite clear that $\langle 3x+3 \rangle \subset \langle 3 \rangle$ and also $\langle 3x+3 \rangle \subset \langle x+1 \rangle$.

▷ **Quick Exercise.** Why are these set containments proper? ◁

Now $\langle 3 \rangle$ is quite evidently the set of all polynomials in $\mathbb{Z}[x]$ for which every non-zero coefficient is divisible by 3. What kind of ideal is this? It is actually the kernel of the ring homomorphism

$$\varphi_2 : \mathbb{Z}[x] \to \mathbb{Z}_3[x]$$

defined by

$$\varphi_2(a_0 + a_1 x + \cdots + a_n x^n) = [a_0] + [a_1]x + \cdots + [a_n]x^n,$$

where the square brackets indicate that we should reduce each coefficient modulo 3. (You will verify the details of this homomorphism, including the fact that its kernel is $\langle 3 \rangle$ in Exercise 23.) Now $\mathbb{Z}_3$ is a field, and so $\mathbb{Z}_3[x]$ is an integral domain that is not a field. This means that the ideal $\langle 3 \rangle$ is a prime ideal that is not maximal.

What about the ideal $\langle x + 1 \rangle$? This is the kernel of the same evaluation homomorphism $\varphi_1$ (except restricted to $\mathbb{Z}[x]$). This is not entirely obvious, because we cannot simply prove this by dividing an arbitrary polynomial by $x+1$ using the Division Theorem: Since $\mathbb{Z}$ is not a field, this ring does not have a division theorem. However, Gauss's Lemma (Theorem 5.5) allows us to conclude this (you will verify these details in Exercise 24). But the evaluation homomorphism restricted to $\mathbb{Z}[x]$ is onto $\mathbb{Z}$, which is a domain and not a field. This means that $\langle x + 1 \rangle$ is also a prime ideal that is not maximal.

But we can see that both of these principal ideals are proper subsets of the ideal $\langle 3, x+1 \rangle$. Note that this latter ideal is not principal – a proof very similar to the proof we used to show that $\langle 2, x \rangle$ is not principal is possible here (see Example 9.12).

▷ **Quick Exercise.** Carry out this modified proof. ◁

We now claim furthermore that $\langle 3, x + 1 \rangle$ is a maximal ideal. We can show this by claiming that it is the kernel of the homomorphism

$$\varphi_3 : \mathbb{Z}[x] \to \mathbb{Z}_3.$$

Here is how we define this homomorphism. Given any $f \in \mathbb{Z}[x]$, we let $\varphi_3(f) = [f(-1)]_3$.

▷ **Quick Exercise.** Check that this is a surjective homomorphism. ◁

It is clear that $\langle 3, x + 1 \rangle$ is a subset of the kernel, because each of the generators is sent to 0 by this homomorphism. But conversely, by Exercise 24 any element of the kernel must be of the form $3k + (x + 1)h$, where $k \in \mathbb{Z}$ and $h \in \mathbb{Z}[x]$, and this is obviously an element of $\langle 3, x + 1 \rangle$.

Consequently, we can now conclude that $\langle 3, x+1 \rangle$ is a maximal, non-principal ideal that contains the ideals $\langle 3 \rangle$ and $\langle x + 1 \rangle$.

But what about the original ideal $\langle 3x + 3 \rangle$? What kind of ideal is this? While clearly not maximal, could it be prime? Consider the following easy computation in $\mathbb{Z}[x]/\langle 3x + 3 \rangle$:

$$(\langle 3x + 3 \rangle + 3)(\langle 3x + 3 \rangle + (x + 1)) = \langle 3x + 3 \rangle + (3x + 3) = \langle 3x + 3 \rangle + 0.$$

We have thus shown that this ring has zero divisors and is consequently not an integral domain. Consequently, $\langle 3x + 3 \rangle$ is not a prime ideal either.

## 15.5 Finite Products of Domains

Let's generalize Example 15.14. Suppose

$$R_1, R_2, \cdots, R_n$$

are rings and $\prod R_i = R_1 \times R_2 \times \cdots \times R_n$ is their $n$-fold direct product. Recall that

$$\prod R_i = \{(r_1, r_2, \cdots, r_n) : r_i \in R_i, \text{for } i = 1, 2, \cdots, n\}.$$

(See Exercise 6.15.) Now for each $j$ consider the function

$$\pi_j : \prod R_i \to R_j$$

defined by $\pi_j(r_1, r_2, \cdots, r_n) = r_j$. This map is called the *$j$th projection*. Because of the definition of the ring operations on $\prod R_i$ (which are component-wise) it is obvious that $\pi_j$ is an onto homomorphism; furthermore, its kernel is

$$R_1 \times R_2 \times \cdots \times R_{j-1} \times \{0\} \times R_{j+1} \times \cdots \times R_n.$$

Note that if $R_j$ is a domain, then this kernel is a prime ideal. Similarly, if $R_j$ is a field, then this kernel is a maximal ideal.

▷ **Quick Exercise.** Why? ◁

Even if all the $R_i$ are domains, then $\prod R_i$ certainly is not: For

$$(a_1, a_2, \cdots, a_n)(b_1, b_2, \cdots, b_n) = (0, 0, \cdots, 0)$$

if and only if $a_i b_i = 0$ for each $i$. But because the $R_i$ are domains, either $a_i$ or $b_i$ is 0. We thus have the product in $\prod R_i$ being zero if and only if the set $\{i : a_i \neq 0\}$ is disjoint from the set $\{j : b_j \neq 0\}$. For example, $(3, 1, 0, 0)(0, 0, 0, 5) = (0, 0, 0, 0)$. Thus, although $\prod R_i$ is not a domain, determination of which elements are zero divisors is certainly an easy matter. This means that representing a given ring as (isomorphic) to a product of rings is often a real help toward understanding the arithmetic of the ring. We will see a dramatic and famous example of this in the next chapter.

---

## Chapter Summary

In this chapter we discussed the concept of *irreducible* elements of a commutative ring with unity. We showed that in a PID, an element is irreducible if and only if its principal ideal is a *maximal ideal*. We then showed that an ideal $I$ of a commutative ring $R$ is maximal if and only if $R/I$ is a field. Similarly, we discussed *prime ideals* and showed that an ideal $I$ is prime if and only if $R/I$ is a domain.

---

## Warm-up Exercises

a. Determine which of the following elements are irreducible:

    (a) $2x^2 + 4$ in $\mathbb{Z}[x]$.

    (b) $2x^2 + 4$ in $\mathbb{Q}[x]$.

    (c) $x^2 + 4$ in $\mathbb{Z}_7[x]$.

    (d) $(1, 3)$ in $\mathbb{Z} \times \mathbb{Z}$.

    (e) $(0, 3)$ in $\mathbb{Z} \times \mathbb{Z}$.

    (f) $(1, -1)$ in $\mathbb{Z} \times \mathbb{Z}$.

(g) $(1, 3)$ in $\mathbb{Z} \times \mathbb{Q}$.

b. Give examples of the following (that is, specify both a domain and an ideal of it):

    (a) A principal ideal that is maximal.

    (b) A principal ideal that is not maximal.

    (c) A maximal ideal that is not principal.

    (d) An ideal that is neither maximal nor principal.

c. Why is the following statement silly? Every commutative ring has exactly one maximal ideal because the whole ring is an ideal, which is obviously as large as possible.

d. Give examples of a commutative ring with unity and an ideal satisfying the following (or argue that no such example exists):

    (a) The ideal is prime but not maximal.

    (b) The ideal is maximal but not prime.

    (c) The ideal is prime and the ring has zero divisors.

e. Explain why $\langle x^2 - 2 \rangle$ is a maximal ideal of $\mathbb{Q}[x]$, but not of $\mathbb{R}[x]$.

f. Explain why $\langle x - 2 \rangle$ is a prime ideal of $\mathbb{Z}[x]$ but is not maximal.

g. Give the quickest possible proof that a maximal ideal is prime.

h. What are the zero divisors of $\mathbb{Z} \times \mathbb{Z}$? What about of $\mathbb{Z}_4 \times \mathbb{Z}$?

i. Let $R$ be a commutative ring with unity. Under what circumstances is $\{0\}$ a prime ideal? A maximal ideal?

j. Let $R$ be a commutative ring with unity. Why isn't the ideal $R$ maximal or prime?

---

## Exercises

1. Let $R$ be a commutative ring with unity, $I$ an ideal of $R$, and $a \in R$. Define

$$\langle I, a \rangle = \{k + ab : \ k \in I, \ b \in R\}.$$

Prove that $\langle I, a \rangle$ is an ideal. Then show that $I \subseteq \langle I, a \rangle$ and $a \in \langle I, a \rangle$. Furthermore, show that $\langle I, a \rangle$ is the smallest ideal of $R$ that contains both $I$ and $a$. (This result is needed in the proof of Theorem 15.2.)

2. Let $R$ be a commutative ring with unity. Let $0 \neq p \in R$, and suppose $p$ is a non-unit. We say that $p$ is **prime** if whenever $p$ divides a product $ab$, then $p$ divides $a$ or $b$. Prove that $\langle p \rangle$ is a prime ideal if and only if $p$ is a prime element.

3. Show that there are no irreducible elements in $\mathbb{Z}_6$.

4. Show that 2 is irreducible in $\mathbb{Z}_8$, and that every non-unit in $\mathbb{Z}_8$ is irreducible or a product of irreducibles.

5. Determine all irreducible elements of $\mathbb{Z} \times \mathbb{Z}$.

6. Suppose that $R$ is a commutative ring with unity, and consider the ring $R[x]$ of polynomials with coefficients from $R$. (See Exercise 6.23.) Let $r$ be an irreducible element of $R$. Prove that $r$ is an irreducible element of $R[x]$.

7. Suppose that $a, b, c \in \mathbb{Z}$, and $c = \gcd(a, b)$. Show that $\langle a, b \rangle = \langle c \rangle$. (See Example 15.7 for an illustration of this.)

8. Let $p$ be a prime integer in $\mathbb{Z}$. Prove directly that $\langle p \rangle$ is a maximal ideal.

9. Find all prime ideals of $\mathbb{Z}$. Find all maximal ideals of $\mathbb{Z}$.

10. (a) Why is $x^2 + 1$ irreducible in $\mathbb{Q}[x]$?

    (b) By part (a) we now know that $\langle x^2 + 1 \rangle$ is a maximal ideal in $\mathbb{Q}[x]$. Provide a direct argument for this, similar to the corresponding argument for $\langle x \rangle$ in Example 15.8.

11. Find all prime ideals of $\mathbb{Z}_{20}$. Find all maximal ideals of $\mathbb{Z}_{20}$.

12. Find all prime ideals of $\mathbb{Z} \times \mathbb{Z}$. Find all maximal ideals of $\mathbb{Z} \times \mathbb{Z}$.

13. Find all prime ideals of $\mathbb{Z}_2 \times \mathbb{Z}_3$; find all maximal ideals of $\mathbb{Z}_2 \times \mathbb{Z}_3$. Now do the same for $\mathbb{Z}_2 \times \mathbb{Z}_4$.

14. Find all prime ideals of $\mathbb{Q}$ and $\mathbb{Q} \times \mathbb{Q}$; do likewise for all maximal ideals.

15. Describe the elements of the ideal $\langle 3, x \rangle$ in $\mathbb{Z}[x]$. Show that $\langle 3, x \rangle$ is not a principal ideal.

16. Show that $\langle 3, x \rangle$ is a maximal ideal in $\mathbb{Z}[x]$.

17. Show that $\langle 9, x \rangle$ is an ideal in $\mathbb{Z}[x]$, which is neither principal nor maximal.

18. Consider $\mathbb{Q}[x, y]$, the set of all polynomials in indeterminates $x, y$ with coefficients from $\mathbb{Q}$ (see the Exercise 9.31).

    (a) Consider the function
    $$\psi : \mathbb{Q}[x, y] \to \mathbb{Q}$$
    defined by $\psi(f) = f(0, 0)$. Why is this a ring homomorphism? Why is the kernel of $\psi$ a maximal ideal? Determine explicitly a nice description of the kernel.

    (b) Consider the function
    $$\rho : \mathbb{Q}[x, y] \to \mathbb{Q}[x]$$
    defined by $\rho(f) = f(0, x)$. Why is this a ring homomorphism? Why is the kernel of $\rho$ a prime ideal? Determine explicitly a nice description of the kernel. Find a larger proper ideal, thus showing concretely that this ideal is not maximal.

    (c) Consider the function
    $$\varphi : \mathbb{Q}[x, y] \to \mathbb{Q}[x]$$
    defined by $\varphi(f) = f(x, x)$. Prove that this is a ring homomorphism. Why is the kernel of $\varphi$ prime but not maximal? Determine explicitly a nice description of the kernel. Find a larger proper ideal, thus showing concretely that this ideal is not maximal.

19. Consider the element $x + y$ in $\mathbb{Q}[x, y]$. Argue that $x + y$ is irreducible, and so $\langle x + y \rangle$ is maximal among all principal ideals in $\mathbb{Q}[x, y]$. Show that $\langle x + y \rangle$ is not maximal among all ideals of $\mathbb{Q}[x, y]$.

20. Why is

$$\langle 2x + 1 \rangle$$

maximal among all principal ideals in $\mathbb{Z}[x]$? Now show that this ideal is *not* maximal among all ideals, by considering the ideal

$$\langle 2x + 1, 3 \rangle.$$

21. Consider the function

$$\varphi : \mathbb{Z} \to \mathbb{Z} \times \mathbb{Z}$$

defined by $\varphi(n) = (n, n)$. Check that this is a homomorphism. Now, $\ker(\varphi) = \{0\}$. Argue directly from the definition that $\{0\}$ is a prime ideal of $\mathbb{Z}$. Now $\mathbb{Z} \times \mathbb{Z}$ is obviously not a domain; why does this not contradict the theorem in the text characterizing prime ideals?

22. Consider the function

$$\varphi : \mathbb{Q}[x] \to \mathbb{C}$$

defined by $\varphi(f) = f(\sqrt{2}i)$. Check that this is a homomorphism.

   (a) Describe the elements of the ring $\varphi(\mathbb{Q}[x])$.
   (b) This ring is clearly a domain (because it is a subring of $\mathbb{C}$). Is it a field?
   (c) What is the kernel of this homomorphism? Is it prime or maximal?
   (d) The kernel of this homomorphism is a principal ideal. How could we use Theorem 15.1 to answer the question in part (c)?

23. Consider the function

$$\varphi_2 : \mathbb{Z}[x] \to \mathbb{Z}_3[x]$$

defined by

$$\varphi_2(a_0 + a_1 x + \cdots + a_n x^n) = [a_0] + [a_1]x + \cdots + [a_n]x^n,$$

where the square brackets indicate that we should reduce each coefficient modulo 3.

   (a) Prove that this is an onto ring homomorphism.
   (b) Why is the kernel of $\varphi_2$ a prime ideal? Give an explicit description of its kernel (it is a principal ideal).

24. In this problem you will check that if $\varphi_1 : \mathbb{Z}[x] \to \mathbb{Z}$ is the evaluation homomorphism $\varphi_1(f) = f(-1)$, then $\ker(\varphi_1) = \langle x + 1 \rangle$.

   (a) Show that $\ker(\varphi_1) \supseteq \langle x + 1 \rangle$.
   (b) Suppose now that $f \in \ker(\varphi_1)$. Argue that $x + 1$ divides $f$, in $\mathbb{Q}[x]$.
   (c) Now use Gauss's Lemma (Theorem 5.5) to argue that $x + 1$ divides $f$, in $\mathbb{Z}[x]$; this shows the reverse inclusion holds, and so we have completed our proof.

25. Consider the ideal $\langle x^2 + 1 \rangle$ in $\mathbb{R}[x]$, discussed in Example 15.10. Prove directly that this is a maximal ideal, without using either Theorem 15.1 or Theorem 15.2.

26. Consider the function $\varphi : \mathbb{Z}[x] \to \mathbb{Q}$ defined by $\varphi(f) = f(-1/2)$.

   (a) What is the kernel of $\varphi$?

(b) We know that $\mathbb{Q}$ is a field. Does this mean that $\ker(\varphi)$ is a maximal ideal? Be careful!

(c) Give a nice description of the elements $\varphi(\mathbb{Z}[x])$.

27. Consider the homomorphism discussed in Exercise 11.31 and Exercise 14.11. What ideal do we now know is maximal? Prove that this ideal is maximal directly, from the definition.

28. Consider the homomorphism discussed in Exercise 11.10. What ideal do we now know is maximal? Prove that this ideal is maximal directly, from the definition.

29. Let $R$ be a commutative ring with unity, and $I$ a prime ideal of $R$. Note that $I[x]$ is a subring of $R[x]$. Prove that $I[x]$ is a prime ideal of $R[x]$. Now suppose that $I$ is a maximal ideal; show by example that $I[x]$ need not be a maximal ideal.

30. Let $R$ and $S$ be commutative rings. Prove that if the direct product $R \times S$ is a domain, then exactly one of $R$ and $S$ is the zero ring.

31. It turns out that in a commutative ring with unity, every proper ideal is a subset of a maximal ideal. (The proof of this theorem requires more sophisticated set theory than we wish to enquire into here.) We call this the **Maximal Ideal Theorem**. Use this theorem to establish the following:

    (a) Prove that every commutative ring with unity has a field as a homomorphic image.

    (b) If the commutative ring $R$ with unity has a unique maximal ideal $M$, then $M$ consists of exactly the set of non-units.

32. Let $P$ be an ideal of a commutative ring with unity. Prove that $P$ is a prime ideal if and only if, whenever $P \supseteq I \cdot J$, where $I$ and $J$ are ideals, then $P \supseteq I$ or $P \supseteq J$. (For the definition of the product of two ideals, see Exercise 9.19.)

33. Prove that in a PID, a non-zero proper ideal is prime if and only if it is maximal.

34. Prove that in a finite commutative ring with unity, a proper ideal is prime if and only if it is maximal.

35. Let $R$ be a commutative ring with unity, for which every element is an idempotent; such a ring is called **Boolean**. (See Exercise 7.25 for more about idempotents.) Prove that in $R$, a proper ideal is prime if and only if it is maximal.

36. Let $R$ be a commutative ring with unity, and define the **Jacobson Radical** $J(R)$ to be

$$J(R) = \bigcap \{M : M \text{ is a maximal ideal of } R\}.$$

    (a) Prove that $J(R)$ is an ideal.

    (b) What are $J(\mathbb{Z})$, $J(\mathbb{Z}_2 \times \mathbb{Z}_4)$, $J(\mathbb{Z}_8)$?

    (c) Prove that the nilradical of a commutative ring with unity is always a subset of the Jacobson radical (see Exercises 7.15 and 9.23).

    (d) Prove that

    $$J(R) = \{s \in R : 1 - rs \text{ is a unit, for all } r \in R\}.$$

    To show that $J(R)$ is a subset of the given set, you will need to use the Maximal Ideal Theorem (see Exercise 31).

    (e) For any commutative ring with unity, show that $J(R/J(R))$ is the zero ring.

# Chapter 16

## The Chinese Remainder Theorem

We can cast considerable light on the arithmetic of the rings $\mathbb{Z}_m$, by making use of the theory of the previous chapters. We will in the process encounter a famous and ancient result known as the Chinese Remainder Theorem.

## 16.1 Some Examples

Let's begin by looking at an example.

**Example 16.1**

Consider the ring $\mathbb{Z}_6$. The function $\varphi : \mathbb{Z}_6 \to \mathbb{Z}_3$ defined by $\varphi([a]_6) = [a]_3$ obviously preserves addition and multiplication and hence is a homomorphism, if it is well defined. But if $[a]_6 = [b]_6$, then $6|(a-b)$, and so $3|(a-b)$; thus,

$$\varphi([a]_6) = [a]_3 = [b]_3 = \varphi([b]_6)$$

and so $\varphi$ is well defined. Note also that $\varphi$ is onto. Because $\mathbb{Z}_3$ is a field, we have that $\ker(\varphi) = \langle [3] \rangle$ is a maximal ideal.

Now the corresponding function $\psi : \mathbb{Z}_6 \to \mathbb{Z}_2$ is also a homomorphism, and its kernel $\langle [2] \rangle$ is also a maximal ideal.

Let's put these two homomorphisms together, using the idea of direct product. Namely, define $\mu : \mathbb{Z}_6 \to \mathbb{Z}_3 \times \mathbb{Z}_2$ by setting

$$\mu([a]_6) = (\varphi([a]_6), \psi([a]_6)) = ([a]_3, [a]_2).$$

It is easy to see that this is a homomorphism, because $\varphi$ and $\psi$ are.

▷ **Quick Exercise.** Check this. ◁

Because the zero element of the direct product $\mathbb{Z}_3 \times \mathbb{Z}_2$ is the element $([0], [0])$, an element of $\mathbb{Z}_6$ belongs to the kernel of $\mu$ only if it belongs to the kernels of *both* $\varphi$ and $\psi$. That is, the kernel of $\mu$ is

$$\langle [3] \rangle \cap \langle [2] \rangle = \{[0], [3]\} \cap \{[0], [2], [4]\} = \{[0]\}.$$

Thus, $\mu$ is actually injective (Corollary 12.4). Let's write out the homomorphism $\mu$ explicitly:

$$\begin{aligned}
[0]_6 &\longmapsto ([0]_2, [0]_3) \\
[1]_6 &\longmapsto ([1]_2, [1]_3) \\
[2]_6 &\longmapsto ([0]_2, [2]_3) \\
[3]_6 &\longmapsto ([1]_2, [0]_3) \\
[4]_6 &\longmapsto ([0]_2, [1]_3) \\
[5]_6 &\longmapsto ([1]_2, [2]_3)
\end{aligned}$$

Notice that this function is actually surjective, and so $\mathbb{Z}_6$ is isomorphic to $\mathbb{Z}_3 \times \mathbb{Z}_2$. We have represented $\mathbb{Z}_6$ as a direct product of simpler pieces. And because those simpler pieces $\mathbb{Z}_3$ and $\mathbb{Z}_2$ are domains (and in fact fields), the zero divisors in this ring are easy to determine: They are those elements with a zero in at least one component.

▷ **Quick Exercise.** Check that this analysis coincides with what you already know about which elements of $\mathbb{Z}_6$ are zero divisors. ◁

Why did $\mu$ turn out to be surjective? Here the domain and range of $\mu$ are both finite sets with 6 elements: Any injection from one such set to another must be onto – this is an example of what is called the **pigeonhole principle**: If you have $n$ pigeons to put into $n$ pigeon holes, and you never put two birds in one hole, all the holes must be filled!

We would like to apply this sort of analysis to any ring of the form $\mathbb{Z}_m$. Before we do this in general, let's examine one more example.

**Example 16.2**

Now consider $\mathbb{Z}_{12}$. Let's consider the residue homomorphisms $\mathbb{Z}_{12} \to \mathbb{Z}_4$ and $\mathbb{Z}_{12} \to \mathbb{Z}_3$: They have kernels $\{[0], [4], [8]\}$ and $\{[0], [3], [6], [9]\}$, and the intersection of these ideals is zero.

▷ **Quick Exercise.** Write down the two residue homomorphisms explicitly, verify that their kernels are as stated, and then write out the isomorphism between $\mathbb{Z}_{12}$ and $\mathbb{Z}_4 \times \mathbb{Z}_3$ explicitly. ◁

Now in this case, the representation is not quite so nice: $\mathbb{Z}_4$ is not a domain. Equivalently the kernel $\{[0], [4], [8]\}$ is not prime. The problem here is a repeated prime factor in the prime factorization of 12. In Exercise 3 you will find that the only prime ideals of $\mathbb{Z}_{12}$ are

$$\langle [2] \rangle = \{[0], [2], [4], [6], [8], [10]\} \text{ and } \langle [3] \rangle = \{[0], [3], [6], [9]\}.$$

Notice that
$$\langle [2] \rangle \cap \langle [3] \rangle = \{[0], [6]\} \supset \{[0]\}.$$

Now, what are the zero divisors of $\mathbb{Z}_4 \times \mathbb{Z}_3$? They must consist of elements that are either zero in exactly one component, or else zero in the second component and a zero divisor of $\mathbb{Z}_4$ in the first component. These elements must correspond under our isomorphism to the zero divisors of $\mathbb{Z}_{12}$.

▷ **Quick Exercise.** Check this explicitly. ◁

---

## 16.2   Chinese Remainder Theorem

We now prove the general theorem that explains the representation of $\mathbb{Z}_6$ and $\mathbb{Z}_{12}$ as direct products, which we obtained in Examples 16.1 and 16.2 above.

**Theorem 16.1** *Let* $p_1, p_2, \cdots, p_n$ *be distinct prime integers and* $m = p_1^{k_1} p_2^{k_2} \cdots p_n^{k_n}$. *Then* $\mathbb{Z}_m$ *is isomorphic to*
$$\mathbb{Z}_{p_1^{k_1}} \times \mathbb{Z}_{p_2^{k_2}} \times \cdots \times \mathbb{Z}_{p_n^{k_n}}.$$

**Proof:** Consider the function $\varphi_i : \mathbb{Z}_m \to \mathbb{Z}_{p_i^{k_i}}$ defined by $\varphi_i([a]_m) = [a]_{p_i^{k_i}}$; this is well-defined because if $[a]_m = [b]_m$, then $m|(a-b)$, and so $p_i^{k_i}|(a-b)$. It is clearly a homomorphism with kernel $\langle[p_i^{k_i}]\rangle$. Now define the function

$$\mu : \mathbb{Z}_m \to \mathbb{Z}_{p_1^{k_1}} \times \mathbb{Z}_{p_2^{k_2}} \times \cdots \times \mathbb{Z}_{p_n^{k_n}}$$

by setting

$$\mu([a]) = (\varphi_1([a]), \cdots, \varphi_n([a])).$$

This is evidently a ring homomorphism.

▷ **Quick Exercise.** Check this. ◁

Now, its kernel is

$$\bigcap_{i=1}^{n}\langle[p_i^{k_i}]\rangle = \{[a] : p_i^{k_i}|a, \text{for all } i\} = \{[0]\}.$$

This means that the homomorphism $\mu$ is injective.

To show that $\mu$ is an isomorphism, it remains to check that it is surjective. We will actually do this twice, because both proofs are illuminating:

*Existential Proof:* Now $\mathbb{Z}_m$ and $\mathbb{Z}_{p_1^{k_1}} \times \mathbb{Z}_{p_2^{k_2}} \times \cdots \times \mathbb{Z}_{p_n^{k_n}}$ are both finite sets with $m$ elements; thus, by the pigeonhole principle, because $\mu$ is injective, it *must* be surjective.

*Constructive Proof:* Given

$$([a_1], [a_2], \cdots, [a_n]) \in \mathbb{Z}_{p_1^{k_1}} \times \mathbb{Z}_{p_2^{k_2}} \times \cdots \times \mathbb{Z}_{p_n^{k_n}},$$

we must find $[a] \in \mathbb{Z}_m$ such that $\varphi_i([a]) = [a_i]$, for all $i$. Let

$$m_i = \frac{m}{p_i^{k_i}};$$

because $m_i$ and $p_i$ are relatively prime, by the GCD identity there exist integers $x_i$ and $y_i$ with $x_i m_i + y_i p_i^{k_i} = 1$. But then

$$[x_i m_i]_{p_i^{k_i}} = [1].$$

Now let

$$a = x_1 m_1 a_1 + x_2 m_2 a_2 + \cdots + x_n m_n a_n.$$

We claim that $\varphi_i([a]) = [a_i]$, for all $i$. But $p_i^{k_i}|m_j$ for all $j \neq i$, and so

$$\varphi_i([a]) = [a]_{p_i^{k_i}} = [x_i m_i a_i] = [x_i m_i][a_i] = [1][a_i] = [a_i],$$

as required. □

The number theory version of this theorem is known as the Chinese Remainder Theorem, because an example of the sort of number theory problem it solves first appears in a work by the Chinese mathematician Sun Tsu, in about the third century AD. We now restate the theorem in its number theoretic guise:

**Theorem 16.2   The Chinese Remainder Theorem**   *Let $p_1, p_2, \cdots, p_n$ be distinct prime integers and $m = p_1^{k_1} p_2^{k_2} \cdots p_n^{k_n}$. Then the set of congruences*

$$x \equiv a_1 \,(\mathrm{mod}\ p_1^{k_1})$$
$$x \equiv a_2 \,(\mathrm{mod}\ p_2^{k_2})$$
$$\cdots$$
$$x \equiv a_n \,(\mathrm{mod}\ p_n^{k_n})$$

*has a simultaneous solution, which is unique modulo $m$.*

**Proof:**   Consider the element

$$([a_1], [a_2], \cdots, [a_n]) \in \mathbb{Z}_{p_1^{k_1}} \times \mathbb{Z}_{p_2^{k_2}} \times \cdots \times \mathbb{Z}_{p_n^{k_n}}.$$

Because the function $\mu$ in the previous proof is surjective, there exists $[a] \in \mathbb{Z}_m$ so that $\varphi_i([a]) = [a_i]$, for all $i$. Then $a$ simultaneously satisfies the $n$ congruences in the theorem.

▷ **Quick Exercise.**   Why? ◁   □

### Example 16.3

Consider the three congruences $x \equiv 4 \,(\mathrm{mod}\ 7)$, $x \equiv 1 \,(\mathrm{mod}\ 2)$, and $x \equiv 3 \,(\mathrm{mod}\ 5)$; the theorem asserts that they have a simultaneous solution modulo 70. In fact, we can construct the solution by following the constructive proof above. In this case $m_1 = 10$, $m_2 = 35$, and $m_3 = 14$. Then $x_1 = 5$ because

$$[5]_7 [10]_7 = [5]_7 [3]_7 = [1]_7.$$

In a similar fashion $x_2 = 1$ and $x_3 = 4$. Then

$$a = (5)(10)(4) + (1)(35)(1) + (4)(14)(3) = 403.$$

But $403 \equiv 53 \,(\mathrm{mod}\ 70)$, and so our simultaneous solution of the three congruences is 53.

▷ **Quick Exercise.**   Check directly that 53 satisfies these three congruences. ◁

Alternatively, we can view this example in light of our first version of the Chinese Remainder Theorem (Theorem 16.1). From that version of the theorem, we know that $\mathbb{Z}_7 \times \mathbb{Z}_2 \times \mathbb{Z}_5$ is isomorphic to $\mathbb{Z}_{70}$. The isomorphism takes the element $(4, 1, 3)$ to 53.

Note that if all the primes in the factorization of $m$ occur only once, we then have that $\mathbb{Z}_m$ is a direct product of finitely many domains (in fact, fields). But suppose that at least one prime divisor of $m$ occurs with degree at least two. We claim that $\mathbb{Z}_m$ cannot be a direct product of domains; this amounts to claiming that the intersection of all the prime ideals of $\mathbb{Z}_m$ is not zero.

To show that this is true, we must first convince ourselves that $\mathbb{Z}_m$ even has any prime ideals. But because $\mathbb{Z}_m$ is a finite ring, it has only finitely many ideals; consequently, it certainly has at least one maximal ideal (which is necessarily prime).

To show that the intersection of all the prime ideals of $\mathbb{Z}_m$ is not zero, choose an arbitrary prime ideal $P$. Now $\mathbb{Z}_m$ is isomorphic to $\mathbb{Z}_{p_1^{k_1}} \times \mathbb{Z}_{p_2^{k_2}} \times \cdots \times \mathbb{Z}_{p_n^{k_n}}$ under the "usual" isomorphism given in the proof of Theorem 16.1, where we may as well assume that $k_1 > 1$. Let $[x]_m$ be the element of $\mathbb{Z}_m$ that gets mapped to the element of $\mathbb{Z}_{p_1^{k_1}} \times \mathbb{Z}_{p_2^{k_2}} \times \cdots \times \mathbb{Z}_{p_n^{k_n}}$ that has $[p_1]_{p_1^{k_1}}$ in the first component and $[0]$ in all the other components. Because $k_1 > 1$, the element $[x]_m$ is non-zero. Furthermore,

$$[x]_m^{k_1} = [0]_m \in P.$$

But because $P$ is prime, this implies that $[x] \in P$.

▷ **Quick Exercise.**   Why? ◁

Hence, the intersection of all prime ideals of $\mathbb{Z}_m$ contains $[x]_m$, and so cannot be $\{[0]_m\}$; thus, $\mathbb{Z}_m$ in this case cannot be a direct product of domains. This is the general explanation of the case $m = 12$.

▷ **Quick Exercise.**   What would be the value for $[x]_m$ in the case $m = 12$? ◁

## 16.3 A General Chinese Remainder Theorem

The proof of Theorem 16.1 depends essentially on the fact that the separate prime power factors are relatively prime to one another, and consequently we can use the GCD identity. We can make a more general definition for this for any commutative ring with unity. We will begin (as we did in Example 16.1) with only two ideals.

So let $R$ be a commutative ring with unity, and suppose that $I, J$ are proper ideals. We say that the ideals $I, J$ are **relatively prime** if for all $r \in R$ we have $a \in I$, $b \in J$ so that $r = a + b$.

Note that because $I$ and $J$ are ideals, we really only need to be able to obtain $1 = a + b$, where $a \in I, b \in J$.

▷ **Quick Exercise.** Why? ◁

Consider how this works in Example 16.1. There 2 and 3 are relatively prime, and so we get $1 = 2a + 3b$. This is the GCD identity!

We can now state and prove a theorem generalizing Theorem 16.1:

**Theorem 16.3** *Let $R$ be a commutative ring with unity, and suppose that $I, J$ are relatively prime proper ideals. Then $R/(I \cap J)$ is isomorphic to the ring $R/I \times R/J$.*

**Proof**:

Consider the natural homomorphisms $\nu_1, \nu_2$ from $R$ onto $R/I$ and $R/J$. Define the function $\mu : R \to R/I \times R/J$ by $\mu(t) = (\nu_1(t), \nu_2(t))$; this is clearly a ring homomorphism.

We claim this homomorphism is surjective. For that purpose, choose an arbitrary

$$(I + r, J + s) \in R/I \times R/J.$$

Because the ideals are relatively prime, we can find elements $a \in I, b \in J$ for which $a + b = r - s$. Then let $x = r - a = b + s$. Then

$$\mu(x) = (I + r - a, J + b + s) = (I + r, J + s),$$

as claimed.

Now the kernel of this function is $\text{ker}(\nu_1) \cap \text{ker}(\nu_2) = I \cap J$. Therefore, by the Fundamental Isomorphism Theorem 14.1 we have that $R/(I \cap J)$ is isomorphic to $R/I \times R/J$, as we claim. □

Notice that in Examples 16.1 and 16.2 (and also in the general theorem), we chose ideals whose intersection is zero, while in the general theorem we allow the intersection to be a non-trivial ideal. Recall also that in the Chinese Remainder Theorem we need surjectivity to obtain an isomorphism; we either had to use the GCD identity in the constructive proof, or else rely on the finiteness of the rings. Here we obtain the surjectivity from the abstract property of being relatively prime.

To provide a full generalization of the Chinese Remainder Theorem, we should really generalize Theorem 16.3 to the case where we have $n$ ideals, rather than just two. You can follow up on these ideas in Exercise 10.

**Example 16.4**

Consider the ideals $I = \langle x - 2 \rangle$ and $J = \langle x^2 \rangle$ in $\mathbb{Q}[x]$. Here a GCD of these two polynomials is 1, and so clearly there exists a linear combination of them equalling 1; that is, $I$ and $J$ are relatively prime. Note that $I \cap J = \langle (x-2)x^2 \rangle$. We of course know that $\mathbb{Q}[x]/I$ is isomorphic to the field of rational numbers. By Exercise 14.12, we know that $\mathbb{Q}[x]/J$ is isomorphic to the ring of matrices

$$M = \left\{ \begin{pmatrix} a & b \\ 0 & a \end{pmatrix} \in M_2(\mathbb{Q}) : a, \ b \in \mathbb{Q} \right\}.$$

We can thus conclude by Theorem 15.3 that $\mathbb{Q}[x]/(I \cap J)$ is isomorphic to $\mathbb{Q} \times M$.

## Historical Remarks

Sun Tsu's problem was phrased in this way: "We have things of which we do not know the number. If we count them by threes, the remainder is 2; if we count them by fives, the remainder is 3; if we count them by sevens, the remainder is 2. How many things are there?" This is clearly an example of the Chinese Remainder Theorem, which you will solve in Warmup Exercise e below. Because this is the only such problem treated by him, it is unclear whether he had available a general method of solving a system of linear congruences. However, in the 13th-century, another Chinese mathematician named Qin Jiushao published a mathematical text that includes a number of examples of such problems. These examples provide in essence our algorithmic proof of the Chinese Remainder Theorem. In particular, to solve such congruences as $[x_i m_i] = [1]$, he uses a version of Euclid's Algorithm, where operations are carried out on a counting board. This medieval Chinese mathematics is much more sophisticated than anything happening in Europe at the same time.

---

## Chapter Summary

In this chapter we proved that any ring of the form $\mathbb{Z}_m$ is isomorphic to a direct product of rings of the form $\mathbb{Z}_{p^k}$, where $p$ is prime. In its number-theoretic guise, this is known as the *Chinese Remainder Theorem*.

---

## Warm-up Exercises

a. According to the Chinese Remainder Theorem, what direct product of rings is $\mathbb{Z}_{24}$ isomorphic to? What about $\mathbb{Z}_{60}$? $\mathbb{Z}_{11}$? $\mathbb{Z}_9$?

b. Note that $[2]^4 = [0]$ in $\mathbb{Z}_{16}$. Why does this mean that $[2]$ belongs to *all* prime ideals of this ring?

c. What's the relationship between $m$ and $n$, if there exists an onto ring homomorphism $\varphi : \mathbb{Z}_m \to \mathbb{Z}_n$?

d. Solve the simultaneous congruences

$$\begin{aligned} x &\equiv 1 \,(\mathrm{mod}\ 2) \\ x &\equiv 2 \,(\mathrm{mod}\ 3). \end{aligned}$$

e. Express Sun Zi's problem in modern notation, and solve it.

# Exercises

1. Show directly that $\mathbb{Z}_{12}$ is isomorphic to $\mathbb{Z}_3 \times \mathbb{Z}_4$ by defining a homomorphism from $\mathbb{Z}_{12}$ onto $\mathbb{Z}_3 \times \mathbb{Z}_4$ as $\mu$ was defined in the proof of Theorem 16.2. Write out an explicit element-by-element description of this isomorphism, as we did in Section 16.1 for $\mathbb{Z}_6$.

2. Repeat Exercise 1 for the ring $\mathbb{Z}_{30}$. (First of all you must determine the direct product to which it is isomorphic.)

3. Determine the prime ideals of $\mathbb{Z}_{12}$; determine the maximal ideals of $\mathbb{Z}_{12}$.

4. Repeat Exercise 3 for $\mathbb{Z}_{30}$.

5. Solve the simultaneous congruences

$$
\begin{aligned}
x &\equiv 1 \,(\text{mod } 4) \\
x &\equiv 5 \,(\text{mod } 7) \\
x &\equiv 3 \,(\text{mod } 9).
\end{aligned}
$$

6. Solve the simultaneous congruences

$$
\begin{aligned}
x &\equiv 1 \,(\text{mod } 2) \\
x &\equiv 6 \,(\text{mod } 7) \\
x &\equiv 2 \,(\text{mod } 27) \\
x &\equiv 6 \,(\text{mod } 11).
\end{aligned}
$$

7. One can solve a problem like Exercises 5 or 6 algorithmically, without remembering the constructive proof as in Example 16.3. For example, in Exercise 5 one could rewrite the first congruence as $x = 4k + 1$, for some integer $k$, and put that into the second congruence, and continue from there. Try this approach.

8. Find $a, b \in \mathbb{Z}$ so that there is no simultaneous solution to

$$x \equiv a (\text{mod } 6) \text{ and } x \equiv b (\text{mod } 4).$$

Why does this not contradict the Chinese Remainder Theorem?

9. The rings $\mathbb{Z}_9$ and $\mathbb{Z}_3 \times \mathbb{Z}_3$ both have nine elements; indeed, a careless reading of Theorem 16.1 might lead one to suppose that these rings are isomorphic. Show that this is false.

10. In the text we formulate and prove Theorem 16.3 which gives criteria for which a commutative ring with unity is isomorphic to $R/I \times R/J$ for two proper ideals $I, J$. Generalize this to the case of finitely many ideals $I_1, I_2, \ldots I_n$.

11. From the Chinese Remainder Theorem, we know that if $m$ is an integer which has no repeated primes in its prime factorization, then $\mathbb{Z}_m$ is isomorphic to a direct product of domains. Here is a generalization you can prove: A commutative ring $R$ with unity is isomorphic to a direct product of finitely many integral domains if and only if it has a finite collection of prime ideals $\{P_i\}$ whose intersection is zero, and for each $r \in R$, there exist $r_i \in \cap \{P_j : j \neq i\}$ for which $r = \Sigma r_i$.

12. We call a ring $R$ a **finite subdirect product of domains** if there exist finitely many integral domains $D_1, D_2, \cdots, D_n$ and an injective homomorphism

$$\mu : R \to \prod D_i$$

such that $\pi_i(\mu(R)) = D_i$, for all $i$, where $\pi_i$ is the projection homomorphism. Prove that $R$ is a finite subdirect product of domains if and only if it has a finite collection of prime ideals whose intersection is zero.

13. Consider

$$R = \{(n, n + 3m) : n, m \in \mathbb{Z}\} \subseteq \mathbb{Z} \times \mathbb{Z}.$$

In this exercise you will show that $R$ is a finite subdirect product of domains, but it is not a direct product of domains.

   (a) Prove that $R$ is a subring of $\mathbb{Z} \times \mathbb{Z}$. Show that $R$ is not an integral domain.

   (b) Show by specific example that $R$ does not include all elements of $\mathbb{Z} \times \mathbb{Z}$. (That is, $R$ is not the entire direct product.)

   (c) Consider the kernels $P_1$ and $P_2$ of the projection homomorphisms $\pi_1$ and $\pi_2$ that project onto the first and second coordinates, respectively. Prove that $P_1 = \langle(0,3)\rangle$ and that $P_2 = \langle(3,0)\rangle$.

   (d) Show from the definition given in Exercise 12 that $R$ is a subdirect product of the domains $\mathbb{Z}$.

   (e) Suppose that $P$ is any other prime ideal of $R$. Show that $P$ necessarily contains either $P_1$ or $P_2$.

   (f) Now use the previous result and Exercise 12 to argue that $R$ cannot be a finite direct product of domains.

14. Show that $\mathbb{Z}_6[x]$ is a finite subdirect product of domains.

15. Make a definition for a *finite subdirect product of fields*, on the model of the definition in Exercise 12. Then state and prove the theorem analogous to the result proved in Exercise 12.

16. In this exercise we construct a ring that is a domain (and so is trivially a finite subdirect product of domains) but is not a finite subdirect product of fields; to do this exercise you need to understand Exercises 12 and 15. Let $p$ be a prime integer, and define

$$\mathbb{Z}_{\langle p \rangle} = \left\{q \in \mathbb{Q} : q = \frac{a}{b}, \ a, b \in \mathbb{Z}, \quad p \text{ does not divide } b\right\}.$$

(We considered this ring in Exercise 7.11 for $p = 2$.)

   (a) Show that $\mathbb{Z}_{\langle p \rangle}$ is a subring of $\mathbb{Q}$, and so is an integral domain.

   (b) Define

$$\varphi : \mathbb{Z}_{\langle p \rangle} \to \mathbb{Z}_p$$

   by setting $\varphi(\frac{a}{b}) = [a]_p [b]_p^{-1}$. Prove that $\varphi$ is an onto ring homomorphism.

   (c) Prove that the kernel of the homomorphism $\varphi$ from part (b) is $\langle p \rangle$. This means that $\langle p \rangle$ is a maximal ideal of $\mathbb{Z}_{\langle p \rangle}$.

   (d) Prove that $\frac{a}{b} \in \mathbb{Z}_{\langle p \rangle}$ is a unit if and only if $\frac{a}{b} \notin \langle p \rangle$.

(e) Use part (d) to argue that $\langle p \rangle$ contains *every* proper ideal of $\mathbb{Z}_{\langle p \rangle}$, and so this ring has a unique maximal ideal.

(f) Why does part (e) mean that $\mathbb{Z}_{\langle p \rangle}$ is not a finite subdirect product of fields?

(g) Verify explcitly that $\langle p \rangle$ consists exactly of the non-units of $\mathbb{Z}_{\langle p \rangle}$ (see Exercise 15.31b).

# Section III in a Nutshell

This section considers functions from one ring $R$ to another ring $S$ that preserve certain algebraic properties: Consider $\varphi : R \to S$ such that

$$\varphi(a + b) = \varphi(a) + \varphi(b) \quad \text{and} \quad \varphi(ab) = \varphi(a)\varphi(b),$$

for all $a, b \in R$. We call $\varphi$ a *ring homomorphism*.

A ring homomorphism always preserves the zero of the ring, additive inverses, unity, and multiplicative inverses (Theorem 11.1). While a ring homomorphism $\varphi : R \to S$ need not be onto, it is onto the image of $R$ in $S$ ($\varphi(R)$) which is itself a subring of $S$ (Theorem 11.2).

The *kernel* of $\varphi$ is defined by

$$\ker(\varphi) = \varphi^{-1}(0) = \{r \in R : \varphi(r) = 0\}.$$

The kernel is always an ideal of $R$ (Theorem 12.1). Furthermore, if $s \in \varphi(R) \subseteq S$, then $\varphi^{-1}(s)$ is the coset $\ker(\varphi) + r$, for any $r \in \varphi^{-1}(s)$. $\ker(\varphi) = 0$ if and only if $\varphi$ is one-to-one (Theorem 12.4).

The *cosets* of any ideal $I$ of $R$ partitions $R$ into pairwise disjoint sets (Theorem 13.1). The set of cosets $R/I$ is itself a ring (Theorem 13.2), called the *ring of cosets* or the *quotient ring of $R$ mod $I$*. There is a *natural homomorphism* from $R$ onto $R/I$ given by

$$\nu(a) = I + a.$$

The kernel of $\nu$ is $I$ (Theorem 13.3).

Recall from Chapter 7 that if $\varphi : R \to S$ is a one-to-one onto homomorphism, we call $\varphi$ an *isomorphism*; in this case we say the $R$ and $S$ are *isomorphic*. The *Fundamental Isomorphism Theorem* (Theorem 14.1) states that if $\varphi : R \to S$ is an onto homomorphism and $\nu : R \to R/\ker(\varphi)$ is the natural homomorphism, then $R/\ker(\varphi)$ is isomorphic to $S$. Furthermore, if we define $\mu : R/\ker(\varphi) \to S$ by $\mu(\ker(\varphi) + r) = \varphi(r)$, then $\mu$ is an isomorphism and $\mu \circ \nu = \varphi$. The essential content of this important theorem is that the output of a ring homomorphism can be obtained by forming the ring of cosets of the appropriate ideal. Chapter 14 gives many examples of this theorem.

An ideal $I$ of $R$ is *maximal* if the only ideal of $R$ properly containing $I$ is $R$ itself. An ideal $I$ is maximal if and only if $R/I$ is a field (Theorem 15.2).

An ideal $I$ of $R$ is *prime* if whenever $ab \in I$ then either $a \in I$ or $b \in I$. An ideal $I$ is prime if and only if $R/I$ is a domain (Theorem 15.3). It follows that every maximal ideal is a prime ideal, since every field is a domain.

Finally, the section closes with a chapter on the famous *Chinese Remainder Theorem*. This theorem asserts that any ring $\mathbb{Z}_m$ is isomorphic to a direct product of rings of the form $\mathbb{Z}_{p^k}$, where $p^k$ is the largest power of the prime $p$ which divides $m$. This can be rephrased as an assertion that any finite set of congruences modulo integers of the form $p^k$, for distinct primes $p$ has a common solution. The theorem can also be placed in a more abstract context using the idea of *relatively prime ideals*.

# Part IV

# Groups

# Chapter 17

## Symmetries of Geometric Figures

Many geometric objects possess a large amount of symmetry. Roughly speaking, this means that a change of the viewer's perspective does not change what is seen. Equivalently, we can move the object instead. In this case we want to consider motions of the object that leave it apparently unchanged.

### 17.1 Symmetries of the Equilateral Triangle

For example, consider an equilateral triangle with vertices labelled 1, 2, and 3:

Notice that a counterclockwise **rotation** through 120° moves vertex 1 to the location vertex 2 has just vacated, and moves vertex 2 to location 3, and vertex 3 to location 1. We shall denote this rotation by $\rho$. Notice that if we ignore the labels of the vertices, after applying $\rho$ the triangle is in the same position as it was before the motion. Note that if we apply $\rho$ twice, that is, we rotate the triangle through 240°, the triangle again appears unchanged. We shall denote this rotation by $\rho\rho$, or $\rho^2$ for short.

Perhaps we need a more precise definition: A **rigid motion** of the plane is a bijection from the plane onto itself that preserves distance. We call these rigid *motions* because they can be realized by moving the plane in three-dimensional space. If $S$ is a subset of the plane (that is, a figure in the plane like our equilateral triangle), a **symmetry** of $S$ is a rigid motion of the plane that takes $S$ onto itself.

So, when we talk of a rotation of the equilateral triangle, we can just as well think of rotating the entire plane. Thus, $\rho$ and $\rho^2$ are symmetries of the triangle.

Are there any other symmetries of the triangle? First of all, there is certainly the identity: The bijection that takes each point to itself; this function clearly preserves distance. This corresponds to *no motion at all*, and we will denote it by $\iota$.

But are there any more interesting symmetries? Yes! Consider the **reflection** through the line $\ell_1$ which assigns to each point $P$ the point $P'$ which is the same perpendicular distance from $\ell_1$ but on the other side. (If $P$ is actually on the line, it is sent to itself.)

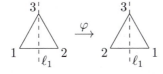

We could think of this as rotating the plane on axis $\ell_1$ through 180°. (Notice this motion happens in three-dimensional space.) This rigid motion takes the triangle onto itself, and so is a symmetry of the triangle that we will call $\varphi$ (for *flip*).

Now, obviously one symmetry followed by another is still a symmetry; note that 'followed by' here means *functional composition*, if we think of these rigid motions as functions from the plane onto itself.

So we have

$$\rho, \rho^2, \rho^3 = \text{no motion at all} = \iota.$$
$$\varphi, \varphi^2 = \iota$$

in our list of symmetries of the triangle. But we also have $\rho\varphi$ (where we mean by this juxtaposition the composition of these two functions, first $\varphi$ and then $\rho$).

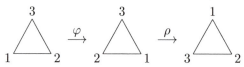

Now this is just a reflection about the line $\ell_2$:

Similarly, $\varphi\rho$ looks like a reflection about the line $\ell_3$:

▷ **Quick Exercise.**  Verify that $\varphi\rho$ is indeed equivalent to the reflection about the line $\ell_3$. ◁

Note that the order in which these compositions are done makes a big difference! These six turn out to be a list of *all* the symmetries of the equilateral triangle (we'll prove this later).

We can now obtain a *multiplication table* of these symmetries as follows. The entry in the $\alpha$ row and $\beta$ column consists of the composition $\alpha\beta$:

|            | $\iota$ | $\rho$ | $\rho^2$ | $\varphi$ | $\rho\varphi$ | $\varphi\rho$ |
|------------|---------|--------|----------|-----------|---------------|---------------|
| $\iota$    | $\iota$ | $\rho$ | $\rho^2$ | $\varphi$ | $\rho\varphi$ | $\varphi\rho$ |
| $\rho$     | $\rho$  | $\rho^2$ | $\iota$ | $\rho\varphi$ | $\varphi\rho$ | $\varphi$ |
| $\rho^2$   | $\rho^2$ | $\iota$ | $\rho$ | $\varphi\rho$ | $\varphi$ | $\rho\varphi$ |
| $\varphi$  | $\varphi$ | $\varphi\rho$ | $\rho\varphi$ | $\iota$ | $\rho^2$ | $\rho$ |
| $\rho\varphi$ | $\rho\varphi$ | $\varphi$ | $\varphi\rho$ | $\rho$ | $\iota$ | $\rho^2$ |
| $\varphi\rho$ | $\varphi\rho$ | $\rho\varphi$ | $\varphi$ | $\rho^2$ | $\rho$ | $\iota$ |

For instance, let's compute $\rho\varphi\rho$. We can picture this composition as follows:

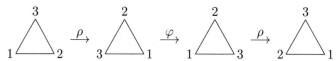

We see that this is indeed equivalent to $\varphi$.

▷ **Quick Exercise.** Try generating the remainder of this multiplication table yourself, thinking geometrically about composing movements of the triangle. ◁

Notice that this 'multiplication' obeys the associative law because it is simply the composition of functions, which is associative. Note also that there seems to be a block substructure to this table. Namely, if we denote the three rotations by $R$ and the three flips by $F$, we have

$$
\begin{array}{c|cc}
 & R & F \\
\hline
R & R & F \\
F & F & R
\end{array}
$$

We will return to this substructure of the table in Example 22.6.

---

## 17.2   Permutation Notation

Now, how do we know that this comprises the complete list of the symmetries of the equilateral triangle?

To answer this question, we make the following observation: If we specify where each of the vertices goes, then we have determined the symmetry. We introduce some notation here, to describe what a symmetry does to the vertices of the triangle. For example, to describe $\rho$, we put beneath the integers 1, 2, and 3 the names of the vertex locations they're taken to. Thus, we should put 2 under 1 to indicate that the counterclockwise rotation of $\rho$ takes vertex 1 to the location vertex 2 has just vacated. This is an example of a general notation for *permutations*, which we will study further in the next chapter. Thus, the permutation of the vertices $1, 2, 3$ accomplished by the rotation $\rho$ is just

$$
\begin{pmatrix} 1 & 2 & 3 \\ 2 & 3 & 1 \end{pmatrix}.
$$

This then means that the vertex in location 1 is moved to location 2, and likewise for the other two columns.

This latter notation gives us a function from $\{1, 2, 3\}$ onto $\{1, 2, 3\}$. For example,

$$
\begin{pmatrix} 1 & 2 & 3 \\ 2 & 3 & 1 \end{pmatrix} (2) = 3.
$$

Explicitly, we have the following correspondence between symmetries of the triangle and permutations of their vertex locations:

$$
\iota \longleftrightarrow \begin{pmatrix} 1 & 2 & 3 \\ 1 & 2 & 3 \end{pmatrix}
$$

$$
\rho \longleftrightarrow \begin{pmatrix} 1 & 2 & 3 \\ 2 & 3 & 1 \end{pmatrix}
$$

$$
\rho^2 \longleftrightarrow \begin{pmatrix} 1 & 2 & 3 \\ 3 & 1 & 2 \end{pmatrix}
$$

$$\varphi \longleftrightarrow \begin{pmatrix} 1 & 2 & 3 \\ 2 & 1 & 3 \end{pmatrix}$$

$$\varphi\rho \longleftrightarrow \begin{pmatrix} 1 & 2 & 3 \\ 1 & 3 & 2 \end{pmatrix}$$

$$\rho\varphi \longleftrightarrow \begin{pmatrix} 1 & 2 & 3 \\ 3 & 2 & 1 \end{pmatrix}$$

Now, there are only six ways to rearrange the list $1, 2, 3$: We have three choices for the first element in the list, two remaining choices for the second element, and only one remaining choice for the third element, making $3 \cdot 2 \cdot 1 = 3! = 6$ altogether. So we have the complete list of the symmetries of the triangle.

We call the set of these six symmetries the **group of symmetries** of the equilateral triangle; the table above is called its **group table** or **multiplication table**. The group of symmetries of the equilateral triangle is sometimes called the **3rd dihedral group** and denoted $D_3$. In general, the group of symmetries of a regular $n$-sided polygon is called the **nth dihedral group** and is denoted $D_n$. In the future, we will call these groups either dihedral groups or the groups of symmetries of appropriate regular polygons. In Exercise 8, you will show that $D_n$ has $2n$ elements.

From the group table, we know that the composition of symmetries $\rho(\varphi\rho)$ gives us the symmetry $\varphi$. But consider the two corresponding permutations of the vertices:

$$\rho = \begin{pmatrix} 1 & 2 & 3 \\ 2 & 3 & 1 \end{pmatrix} \quad \text{and} \quad \varphi\rho = \begin{pmatrix} 1 & 2 & 3 \\ 1 & 3 & 2 \end{pmatrix}.$$

These are both functions, which can be composed together. Let's see what $\rho(\varphi\rho)$ does to vertex 1. First, the symmetry $\varphi\rho$ sends 1 to 1, then $\rho$ sends 1 to 2. Thus, the composition function sends 1 to 2.

▷ **Quick Exercise.**   Compute what this composition does to 2 and 3. ◁

We thus have

$$\begin{pmatrix} 1 & 2 & 3 \\ 2 & 3 & 1 \end{pmatrix} \circ \begin{pmatrix} 1 & 2 & 3 \\ 1 & 3 & 2 \end{pmatrix} = \begin{pmatrix} 1 & 2 & 3 \\ 2 & 1 & 3 \end{pmatrix}.$$

(Remember, this composition is done right-to-left.) This is of course exactly the permutation describing what the corresponding symmetry $\varphi$ does to the vertex locations! Thus, $\rho(\varphi\rho) = \varphi$.

---

## 17.3   Matrix Notation

If you've had a bit of linear algebra, we can describe the symmetries of the equilateral triangle using matrix notation. Let's now look again at the rigid motions of the plane we've used in describing the symmetries of the equilateral triangle. Remember a rigid motion of the plane is really a bijection from the plane onto the plane, which preserves distance.

Now the plane can be considered algebraically as the set $\mathbb{R}^2$ of all ordered pairs of real numbers. We denote this by writing $P(x, y)$ for the point $P$ in the plane with coordinates $(x, y)$.

How then can we represent rotation (about the origin, say) through angle $\theta$? To answer this, suppose that $P(x, y)$ is rotated through angle $\theta$ to point $P'(x', y')$. If we represent $P$ by the polar coordinates $(r, \varphi)$ then

$$x = r \cos \varphi \quad \text{and} \quad y = r \sin \varphi,$$

and so

$$x' = r \cos(\theta + \varphi) \quad \text{and} \quad y' = r \sin(\theta + \varphi).$$

By using the trigonometric sum formulas, we obtain

$$
\begin{aligned}
x' &= r \cos \varphi \cos \theta - r \sin \varphi \sin \theta = x \cos \theta - y \sin \theta \\
y' &= r \cos \varphi \sin \theta + r \sin \varphi \cos \theta = x \sin \theta + y \cos \theta.
\end{aligned}
$$

These two equations describe the transformation $P(x, y) \to P'(x', y')$, which is the rotation through the angle $\theta$ about the origin.

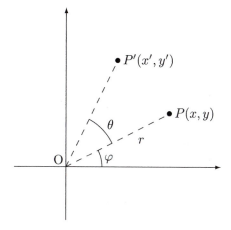

A particularly nice way to look at this transformation is as a matrix multiplication. Consider the $2 \times 2$ matrix

$$R = \begin{pmatrix} \cos \theta & -\sin \theta \\ \sin \theta & \cos \theta \end{pmatrix}$$

and the $2 \times 1$ arrays $P = \begin{pmatrix} x \\ y \end{pmatrix}$ and $P' = \begin{pmatrix} x' \\ y' \end{pmatrix}$. Then we have that $RP = P'$. Note that we think of $P$ and $P'$ as columns in order to make this matrix multiplication work. Thus, we can represent rotation of the plane by means of a *matrix multiplication*. (See Exercise 6.7.)

What about the reflection through lines? Consider the matrix

$$F = \begin{pmatrix} -1 & 0 \\ 0 & 1 \end{pmatrix}.$$

Note that $FP = \begin{pmatrix} -x \\ y \end{pmatrix}$, and so we see that multiplication by $F$ exactly describes reflection of the plane through the $y$-axis.

So now consider the equilateral triangle centered about the origin $(0, 0)$, so that vertex 1 is $(-\frac{\sqrt{3}}{2}, -\frac{1}{2})$, vertex 2 is $(\frac{\sqrt{3}}{2}, -\frac{1}{2})$, and vertex 3 is $(0, 1)$. (Note that the vertices of this triangle are all distance 1 from the origin.) We can then explicitly describe the symmetries $\rho$ and $\varphi$ by multiplication by $R$ and $F$, respectively. We then obtain the following correspondence between symmetries and matrices:

$$\iota \longleftrightarrow I = \begin{pmatrix} 1 & 0 \\ 0 & 1 \end{pmatrix}$$

$$\rho \longleftrightarrow R = \begin{pmatrix} -\frac{1}{2} & -\frac{\sqrt{3}}{2} \\ \frac{\sqrt{3}}{2} & -\frac{1}{2} \end{pmatrix}$$

$$\rho^2 \longleftrightarrow R^2 = \begin{pmatrix} -\frac{1}{2} & \frac{\sqrt{3}}{2} \\ -\frac{\sqrt{3}}{2} & -\frac{1}{2} \end{pmatrix}$$

$$\varphi \longleftrightarrow F = \begin{pmatrix} -1 & 0 \\ 0 & 1 \end{pmatrix}$$

$$\varphi\rho \longleftrightarrow FR = \begin{pmatrix} \frac{1}{2} & \frac{\sqrt{3}}{2} \\ \frac{\sqrt{3}}{2} & -\frac{1}{2} \end{pmatrix}$$

$$\rho\varphi \longleftrightarrow RF = \begin{pmatrix} \frac{1}{2} & -\frac{\sqrt{3}}{2} \\ -\frac{\sqrt{3}}{2} & -\frac{1}{2} \end{pmatrix}$$

Note the interconnection among the composition of symmetries, the composition of permutations of the vertex locations, and the matrix multiplication! In other words, the matrix corresponding to the composition of two symmetries is exactly the product of the matrices corresponding to those symmetries.

▷ **Quick Exercise.**  Pick a couple of symmetries and verify that the product of the symmetries has as its corresponding matrix exactly the product of the corresponding matrices of the original symmetries.  ◁

(We will explore this correspondence more carefully later.)

---

## 17.4   Symmetries of the Square

Let's now examine the symmetries of the square. We consider the square pictured below, with vertices at

$$(-1, -1), \quad (1, -1), \quad (1, 1), \quad (-1, 1).$$

(Call these vertices 1, 2, 3, and 4.)

Because vertices must be moved to vertices, there can be no more than $4! = 24$ symmetries of the square. However, consider the permutation of the vertex locations

$$\begin{pmatrix} 1 & 2 & 3 & 4 \\ 1 & 2 & 4 & 3 \end{pmatrix}.$$

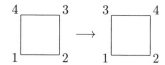

There is no way to move the square in this way, because vertex 1 (wherever it is moved) must stay adjacent to vertices 2 and 4. Let's count how many such permutations are possible. We can move vertex 1 to any of four places. But because vertex 2 must be adjacent, this means that there remain only two possibilities for 2. And once we've established where the 1-2 edge goes, the entire square's motion is accounted for. This means there are at most 8 symmetries of the square. In Exercise 1, you will find all 8, and analyze them as we have analyzed the symmetries of the equilateral triangle. As mentioned before, we denote the group of the symmetries of the square by $D_4$, the 4th dihedral group.

## 17.5   Symmetries of Figures in Space

We'd now like to turn to symmetries of three-dimensional objects, in particular of the regular tetrahedron. By way of analogy with plane figures, it should be clear that we should be concerned with rigid motions of three-dimensional space ($\mathbb{R}^3$) taking the given object onto itself. However, we should make clear what is meant by a rigid motion of $\mathbb{R}^3$. We mean more than just a bijective distance-preserving function. Consider $\mathbb{R}^3$ equipped with a coordinate system with the usual right-hand rule orientation as depicted below (where $\vec{i} \times \vec{j} = \vec{k}$):

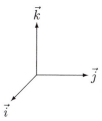

The function $\beta : \mathbb{R}^3 \to \mathbb{R}^3$ defined by $\beta(x, y, z) = (x, -y, z)$ (that is, reflection through the $xz$-plane) is a bijective distance-preserving function that *cannot* be accomplished by moving $\mathbb{R}^3$ *in* $\mathbb{R}^3$. We can see this because it changes the set of vectors $\vec{i}, \vec{j}, \vec{k}$, which has *right-handed* orientation, into the set $\vec{i}, -\vec{j}, \vec{k}$, which has *left-handed* orientation.

The pertinent comparison is to the reflection $(x, y) \mapsto (x, -y)$ in the plane, which can't be accomplished in the plane. It is an accident of physics that we inhabit three-dimensional space. Hence, we are happy with reflections through a line in the plane, because they can be accomplished in three-dimensional space. We are not so happy with reflections through a plane in space (even though it turns out they can be accomplished as a motion in four dimensions). The key here is to restrict ourselves to bijective distance-preserving functions that preserve the right-handed orientation of our coordinate system; we call such functions **rigid motions of space**. (This should jibe with your intuition of rigid motion.) This definition of symmetry is quite restrictive because it excludes functions like $\beta$, which can be realized as mirror reflections. Many people would call such functions symmetries too, but we will not do so here.

## 17.6 Symmetries of the Regular Tetrahedron

We're now ready to determine the group of symmetries of a regular tetrahedron. Because it has four vertices, there are no more than $4! = 24$ such symmetries.

Let's first consider symmetries leaving one vertex (say, number 1) fixed. We have the rotations

$$\rho_1 \longmapsto \begin{pmatrix} 1 & 2 & 3 & 4 \\ 1 & 3 & 4 & 2 \end{pmatrix}$$

and

$$\rho_1^2 \longmapsto \begin{pmatrix} 1 & 2 & 3 & 4 \\ 1 & 4 & 2 & 3 \end{pmatrix},$$

where, of course, $\rho_1^3 = \iota$, the identity.

But consider another permutation of the vertices leaving number 1 fixed, say,

$$\begin{pmatrix} 1 & 2 & 3 & 4 \\ 1 & 4 & 3 & 2 \end{pmatrix}.$$

This corresponds to a reflection through a plane.

▷ **Quick Exercise.** What plane? ◁

But this *can't* be accomplished with a rigid motion! To see this, paint the exterior of face 234 white and the interior of face 234 red. Now apply the above permutation. Clearly, for this to be a symmetry of the regular tetrahedron, a face must be mapped to another face. Here, because vertex 1 is fixed, face 234 is clearly mapped to itself. But after the permutation, what color are the exterior and interior of face 234? They've switched colors! This permutation has caused the tetrahedron to turn 'inside itself'. This doesn't jibe with our intuition of rigid motion. Indeed, this is the sort of thing that happens when a distance-preserving permutation is applied that changes the right-handed orientation.

To see this change in orientation another way, place vectors $\vec{i}$ and $\vec{j}$ in the plane of the face 234, with $\vec{j}$ parallel to the edge 24. Then the vector cross product $\vec{i} \times \vec{j}$ points toward the tetrahedron. Suppose $\vec{i}'$ and $\vec{j}'$ are the images of $\vec{i}$ and $\vec{j}$, respectively. Note that $\vec{i}' \times \vec{j}'$ now points away from the tetrahedron. That is, the side of the face that was formerly in the interior of the tetrahedron is now on the exterior.

So we lose altogether three of the symmetries of the triangle 234: namely,

$$\begin{pmatrix} 1 & 2 & 3 & 4 \\ 1 & 4 & 3 & 2 \end{pmatrix}, \begin{pmatrix} 1 & 2 & 3 & 4 \\ 1 & 2 & 4 & 3 \end{pmatrix}, \text{ and } \begin{pmatrix} 1 & 2 & 3 & 4 \\ 1 & 3 & 2 & 4 \end{pmatrix}.$$

Similarly, we have rotations corresponding to the other three vertices:

$$\rho_2 \rightarrow \begin{pmatrix} 1 & 2 & 3 & 4 \\ 4 & 2 & 1 & 3 \end{pmatrix} \qquad \rho_2^2 \rightarrow \begin{pmatrix} 1 & 2 & 3 & 4 \\ 3 & 2 & 4 & 1 \end{pmatrix}$$

$$\rho_3 \rightarrow \begin{pmatrix} 1 & 2 & 3 & 4 \\ 2 & 4 & 3 & 1 \end{pmatrix} \qquad \rho_3^2 \rightarrow \begin{pmatrix} 1 & 2 & 3 & 4 \\ 4 & 1 & 3 & 2 \end{pmatrix}$$

$$\rho_4 \rightarrow \begin{pmatrix} 1 & 2 & 3 & 4 \\ 3 & 1 & 2 & 4 \end{pmatrix} \qquad \rho_4^2 \rightarrow \begin{pmatrix} 1 & 2 & 3 & 4 \\ 2 & 3 & 1 & 4 \end{pmatrix}$$

These give us 8 symmetries fixing exactly one vertex, together with one fixing all four, while we have banned 6.

▷ **Quick Exercise.** For each fixed vertex, we found 3 permutations that are not allowed. There are 4 vertices, so why are there only 6 banned permutations and not 12? ◁

What about the $24 - 15 = 9$ permutations that leave no vertex fixed? How many of these lead to symmetries? Let's try to consider a typical one. We may suppose without loss of generality that vertex 1 is sent to vertex 2. So, the permutation looks like this:

$$\begin{pmatrix} 1 & 2 & 3 & 4 \\ 2 & ? & ? & ? \end{pmatrix}.$$

If $2 \mapsto 1$, then (because we've assumed that it fixes no vertex) it must look like this:

$$\begin{pmatrix} 1 & 2 & 3 & 4 \\ 2 & 1 & 4 & 3 \end{pmatrix}.$$

This is a symmetry! Just let $\ell$ be the line connecting the midpoints of the segments 12 and 34, respectively. Then this symmetry can be accomplished by rotating 180° about the line $\ell$.

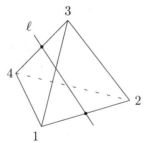

We will call this symmetry $\varphi_1$; note that $\varphi_1^2 = \iota$. In a similar fashion we have

$$\varphi_2 \rightarrow \begin{pmatrix} 1 & 2 & 3 & 4 \\ 3 & 4 & 1 & 2 \end{pmatrix}, \qquad \varphi_3 \rightarrow \begin{pmatrix} 1 & 2 & 3 & 4 \\ 4 & 3 & 2 & 1 \end{pmatrix}.$$

But what if 2 is not mapped to 1? Suppose without loss of generality that $2 \mapsto 3$. Then, because we have no fixed points, we must have

$$\begin{pmatrix} 1 & 2 & 3 & 4 \\ 2 & 3 & 4 & 1 \end{pmatrix}.$$

We claim this *can't* be accomplished by a rigid motion of $\mathbb{R}^3$. Why? Once again, this motion would not preserve the orientation of our coordinate system. Consider any face of the tetrahedron. Note that the permutation will flip this face around—the exterior side becomes the interior side. This cannot be accomplished by a rigid motion.

▷ **Quick Exercise.** Pick a particular face of the tetrahedron and check explicitly that this permutation interchanges the interior and exterior sides of the face. ◁

Thus, there are six more permutations that are forbidden:

$$\begin{pmatrix} 1 & 2 & 3 & 4 \\ 2 & 3 & 4 & 1 \end{pmatrix} \quad \begin{pmatrix} 1 & 2 & 3 & 4 \\ 3 & 4 & 2 & 1 \end{pmatrix} \begin{pmatrix} 1 & 2 & 3 & 4 \\ 4 & 1 & 2 & 3 \end{pmatrix}$$

$$\begin{pmatrix} 1 & 2 & 3 & 4 \\ 2 & 4 & 1 & 3 \end{pmatrix} \quad \begin{pmatrix} 1 & 2 & 3 & 4 \\ 4 & 3 & 1 & 2 \end{pmatrix} \begin{pmatrix} 1 & 2 & 3 & 4 \\ 3 & 1 & 4 & 2 \end{pmatrix}$$

We thus have 12 symmetries of the regular tetrahedron: 8 that fix exactly one vertex, 3 that fix no vertex, and the one that fixes all four. We then obtain the following group table for the symmetries of the regular tetrahedron:

| | $\iota$ | $\varphi_1$ | $\varphi_2$ | $\varphi_3$ | $\rho_1$ | $\rho_2$ | $\rho_3$ | $\rho_4$ | $\rho_1^2$ | $\rho_2^2$ | $\rho_3^2$ | $\rho_4^2$ |
|---|---|---|---|---|---|---|---|---|---|---|---|---|
| $\iota$ | $\iota$ | $\varphi_1$ | $\varphi_2$ | $\varphi_3$ | $\rho_1$ | $\rho_2$ | $\rho_3$ | $\rho_4$ | $\rho_1^2$ | $\rho_2^2$ | $\rho_3^2$ | $\rho_4^2$ |
| $\varphi_1$ | $\varphi_1$ | $\iota$ | $\varphi_3$ | $\varphi_2$ | $\rho_3$ | $\rho_4$ | $\rho_1$ | $\rho_2$ | $\rho_4^2$ | $\rho_3^2$ | $\rho_2^2$ | $\rho_1^2$ |
| $\varphi_2$ | $\varphi_2$ | $\varphi_3$ | $\iota$ | $\varphi_1$ | $\rho_4$ | $\rho_3$ | $\rho_2$ | $\rho_1$ | $\rho_2^2$ | $\rho_1^2$ | $\rho_4^2$ | $\rho_3^2$ |
| $\varphi_3$ | $\varphi_3$ | $\varphi_2$ | $\varphi_1$ | $\iota$ | $\rho_2$ | $\rho_1$ | $\rho_4$ | $\rho_3$ | $\rho_3^2$ | $\rho_4^2$ | $\rho_1^2$ | $\rho_2^2$ |
| $\rho_1$ | $\rho_1$ | $\rho_4$ | $\rho_2$ | $\rho_3$ | $\rho_1^2$ | $\rho_4^2$ | $\rho_2^2$ | $\rho_3^2$ | $\iota$ | $\varphi_3$ | $\varphi_1$ | $\varphi_2$ |
| $\rho_2$ | $\rho_2$ | $\rho_3$ | $\rho_1$ | $\rho_4$ | $\rho_3^2$ | $\rho_2^2$ | $\rho_4^2$ | $\rho_1^2$ | $\varphi_3$ | $\iota$ | $\varphi_2$ | $\varphi_1$ |
| $\rho_3$ | $\rho_3$ | $\rho_2$ | $\rho_4$ | $\rho_1$ | $\rho_4^2$ | $\rho_1^2$ | $\rho_3^2$ | $\rho_2^2$ | $\varphi_1$ | $\varphi_2$ | $\iota$ | $\varphi_3$ |
| $\rho_4$ | $\rho_4$ | $\rho_1$ | $\rho_3$ | $\rho_2$ | $\rho_2^2$ | $\rho_3^2$ | $\rho_1^2$ | $\rho_4^2$ | $\varphi_2$ | $\varphi_1$ | $\varphi_3$ | $\iota$ |
| $\rho_1^2$ | $\rho_1^2$ | $\rho_3^2$ | $\rho_4^2$ | $\rho_2^2$ | $\iota$ | $\varphi_2$ | $\varphi_3$ | $\varphi_1$ | $\rho_1$ | $\rho_3$ | $\rho_4$ | $\rho_2$ |
| $\rho_2^2$ | $\rho_2^2$ | $\rho_4^2$ | $\rho_3^2$ | $\rho_1^2$ | $\varphi_2$ | $\iota$ | $\varphi_1$ | $\varphi_3$ | $\rho_4$ | $\rho_2$ | $\rho_1$ | $\rho_3$ |
| $\rho_3^2$ | $\rho_3^2$ | $\rho_1^2$ | $\rho_2^2$ | $\rho_4^2$ | $\varphi_3$ | $\varphi_1$ | $\iota$ | $\varphi_2$ | $\rho_2$ | $\rho_4$ | $\rho_3$ | $\rho_1$ |
| $\rho_4^2$ | $\rho_4^2$ | $\rho_2^2$ | $\rho_1^2$ | $\rho_3^2$ | $\varphi_1$ | $\varphi_3$ | $\varphi_2$ | $\iota$ | $\rho_3$ | $\rho_1$ | $\rho_2$ | $\rho_4$ |

In examining this group table we can detect a block substructure, similar to that we discovered in the group table for $D_3$. We can highlight this by labelling the symmetries

$$\{\iota, \varphi_1, \varphi_2, \varphi_3\}$$

by $F$, the symmetries

$$\{\rho_1, \rho_2, \rho_3, \rho_4\}$$

by $R$, and the symmetries

$$\{\rho_1^2, \rho_2^2, \rho_3^2, \rho_4^2\}$$

by $R^2$. Then the block structure looks like this:

| | $F$ | $R$ | $R^2$ |
|---|---|---|---|
| $F$ | $F$ | $R$ | $R^2$ |
| $R$ | $R$ | $R^2$ | $F$ |
| $R^2$ | $R^2$ | $F$ | $R$ |

We will examine this block structure further in Example 22.7.

In Exercises 17 and 18 we will lead you through a determination of the symmetries of a cube, discovering that there are 24 such symmetries.

## Warm-up Exercises

a. Explain geometrically why $\iota$ appears in each row and in each column of the group table for $D_3$.

b. We pointed out a block substructure in the group table for $D_3$. What does this sub-structure mean geometrically?

c. Give the matrices that accomplish rotations through $45°$ and $30°$. Check that these matrices work for $(1, 0)$ and $(0, 1)$.

d. Compute $\rho\varphi\rho$ three times: **(1)** by making an equilateral triangle out of cardboard, and actually performing the motions; **(2)** by composing the permutation functions which describe what locations the vertices are taken to; **(3)** by multiplying the corresponding matrices. Did you get the same answer all three times?

e. Consider an *isosceles* triangle, which is not equilateral. How many symmetries does it have?

f. Consider a *scalene* triangle (all sides have different lengths). How many symmetries does it have?

g. Compute the following elements twice in the group of symmetries of the tetrahedron, once using the group table, and once using the representation as permutations:

$$\varphi_1\rho_2{}^2\varphi_2, \quad \varphi_2\varphi_1\varphi_2, \quad \rho_1\rho_3.$$

h. Explain the geometric meaning of the block substructure in the group table for the symmetries of the regular tetrahedron.

## Exercises

1. Complete the analysis of the symmetries of the square, which we began in the text. Some will be rotations, and some will be flips. Determine matrix and permutation representations for them, draw a table of correspondence, and compute the group table for your symmetries.

2. Repeat Exercise 1 for $D_5$, the group of symmetries of a regular pentagon.

3. Determine all symmetries of a non-square rectangle, and represent them with matrices and permutations. How many are rotations, and how many are flips?

4. Repeat Exercise 3 for a rhombus (that is, an equilateral parallelogram, which is not a square).

5. Show algebraically that the rotation transformation preserves distance: Consider the points
$$P_1(x_1, y_1) \text{ and } P_2(x_2, y_2).$$

   (a) What is the square of the distance between $P_1$ and $P_2$?

(b) Now rotate through the angle $\theta$, by multiplying by the appropriate matrix, to obtain the points

$$P_1'(x_1', y_1') \text{ and } P_2'(x_2', y_2').$$

Compute the square of the distance between these points. Use trig identities to show that this is the same as in part (a).

6. Verify by multiplying two matrices together that a rotation through angle $\theta$, followed by a rotation through angle $\varphi$, gives a rotation through angle $\theta + \varphi$.

7. How many symmetries can you find for the unit circle? Which rotations are possible? Which flips?

8. Find out how many elements there are in $D_n$, the group of symmetries of a regular $n$-sided polygon.

9. You can check that all of the matrices of the symmetries of the equilateral triangle and the square have the property that their determinants are always $\pm 1$. (See Exercise 8.2 for a definition of the determinant of a $2 \times 2$ matrix.) In this exercise you will show that if a matrix preserves distance, then its determinant must be $\pm 1$.

   (a) Suppose that $A \in M_2(\mathbb{R})$, and $\det(A) = 0$. Show that multiplication by $A$ cannot preserve distance. Do this by showing that multiplication by $A$ takes some point in the plane to the origin, and hence cannot preserve distance.

   (b) Suppose next that

   $$\begin{pmatrix} a & b \\ c & d \end{pmatrix} = A \in M_2(\mathbb{R}),$$

   but $\det(A) \neq 0$. Suppose that multiplication by $A$ does preserve distance, and consider successively what happens to

   $$\begin{pmatrix} 1 \\ 0 \end{pmatrix}, \quad \begin{pmatrix} 0 \\ 1 \end{pmatrix}, \quad \begin{pmatrix} d \\ -c \end{pmatrix}, \quad \begin{pmatrix} -b \\ a \end{pmatrix}.$$

   You will be able to infer that $\det(A) = \pm 1$.

10. Our description of the symmetries of the equilateral triangle can be elegantly rephrased using the arithmetic of the complex numbers $\mathbb{C}$, described in Chapter 8.

    (a) Argue that the three vertices of the triangle can be thought of as numbers of the form $e^{i\alpha_k}$ in the complex plane, for appropriate angles $\alpha_k$, where $k = 1, 2, 3$.

    (b) Show that you can represent the rotations of the triangle in the symmetry group by complex multiplication by a number of the form $e^{i\theta}$, for an appropriate choice of $\theta$.

    (c) What operation on the complex numbers performs the flip $\varphi$?

11. Find all symmetries of a pyramid as drawn below (the base is a square, and the four sides are congruent):

12. Show that $\rho_1$ and $\varphi_1$ *generate* the group of symmetries of the tetrahedron. This means that you can find a formula for each of the other symmetries, in terms only of $\rho_1$ and $\varphi_1$. (By all means, use our group table!)

13. Consider the *flatlanders*, who live in the plane and consequently cannot conceive of motion in three dimensions. Formulate a definition for symmetry in $\mathbb{R}^2$ for flatlanders, and then determine for them the group of symmetries of the equilateral triangle and the square.

14. In our discussion of the symmetries of the tetrahedron, we excluded the symmetry corresponding to the permutation

$$\begin{pmatrix} 1 & 2 & 3 & 4 \\ 1 & 4 & 3 & 2 \end{pmatrix},$$

because it could not be accomplished by a rigid motion of space. However, we did observe that it could be accomplished by a reflection through a plane; this is a called a *mirror reflection*. Figure out which of the excluded permutations for the group of symmetries of the tetrahedron are mirror reflections. Specify the plane for each of these.

15. Find all symmetries of the 'tent':

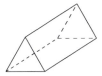

The ends are equilateral triangles, and the sides are congruent non-square rectangles. Give the group table. Comment on its relationship to the group table for the triangle.

16. (For those comfortable with multiplying $3 \times 3$ matrices.) Consider the tetrahedron in $\mathbb{R}^3$ with vertices at

$$(1, 1, -1), (-1, -1, -1), (1, -1, 1), \text{ and } (-1, 1, 1).$$

If we label these vertices 1, 2, 3, 4, respectively, then this corresponds to the labeling of the tetrahedron we used in the text.

Notice that the matrix

$$F_1 = \begin{pmatrix} -1 & 0 & 0 \\ 0 & -1 & 0 \\ 0 & 0 & 1 \end{pmatrix}$$

corresponds to the symmetry $\varphi_1$. Find a matrix which corresponds to the symmetry $\rho_1$, and then determine matrices for all the other symmetries, using Exercise 12 above.

17. In this exercise we will count the number of symmetries of a cube. We here introduce some labeling and terminology which will be helpful for what follows. We shall identify the vertices and the faces of the cube as follows:

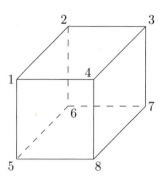

| | | |
|---|---|---|
| face 1485 | F | (front) |
| face 3267 | B | (back) |
| face 4378 | R | (right) |
| face 2156 | L | (left) |
| face 2341 | U | (up) |
| face 5876 | D | (down) |

Furthermore, we shall paint each of the six faces of the cube, according to the following scheme:

| | | |
|---|---|---|
| front | — W (white) | left — O (orange) |
| back — Y (yellow) | | up — B (black) |
| right — R (red) | | down — G (green) |

(a) Argue that a symmetry of the cube is entirely determined when we specify the colors of the front and right faces, once the symmetry has been performed.

(b) Now count the number of such symmetries, taking advantage of part (a).

18. In this exercise, we we will consider the symmetries of the cube in terms of permutations of various geometric parts of the cube.

(a) We could think of symmetries of the cube as permutations of the vertices. But how many such permutations are there? Find a specific such permutation that cannot be realized as a rigid motion of space.

(b) We could instead think of symmetries of the cube as permutations of the faces. But how many such permutations are there? Find a specific example of such a permutation that cannot be realized as a rigid motion of space.

(c) Now consider the diagonal segments of the cube, labeled as follows:

$$a(1-7), \quad b(2-8), \quad c(3-5), \quad d(4-6).$$

We can now think of symmetries of the cube as permutations of the diagonals. But how many such permutations are there? How does this compare to your count of the possible symmetries of the cube in Exercise 17? Is each permutation of diagonals a symmetry of the cube?

(d) One set of rigid motions which are symmetries of the cube is to do $90°, 180°, 270°$ rotations about a line through the middle of two opposite faces. Argue that there are 9 distinct symmetries of this form.

(e) Another set of rigid motions which are symmetries of the cube is to do a 180°
rotation of the cube on lines connecting the midpoints of opposite edges; for
example, using the picture below, rotate about the line connecting points 18 and
12:

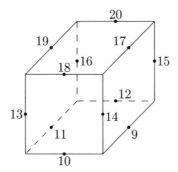

Argue that there are 6 distinct symmetries of this form.

(f) Consider now rotating about a fixed diagonal like *a*. Argue that we can rotate
120° and 240° about these. Then argue that there are 8 distinct symmetries of
this form.

(g) Have we accounted for all of the symmetries with the classification we've given
in parts (d), (e), and (f)?

(h) This is challenging! Build a chart which describes each of the symmetries of the
cube in terms of a permutation of the 4 diagonals.

# Chapter 18

## *Permutations*

In the previous chapter we used the idea that a symmetry of a geometric object like a triangle or tetrahedron must take vertices to vertices. For example, specifying where the vertices of a tetrahedron are sent completely determines the symmetry function. We used this reasoning to determine a complete list of symmetries of the tetrahedron. We called a specification of how the vertices are moved a *permutation* of the vertices. We found that using permutation notation was one way to compute the composition of symmetries of the tetrahedron (or some other geometric figure). We will now consider the notion of permutations in an abstract setting. This will provide us with a great source for examples and computational techniques when we move to abstract groups in the next chapter.

---

## 18.1  Permutations

Consider the list $1, 2, 3, 4, \cdots, n$ of the first $n$ positive integers. We wish to rearrange or *permute* this list. To do this, we must tell ourselves which slot the integer 1 should be placed in, which slot the integer 2 should be placed in, and so forth. What this means is that a permutation of this list amounts to a function

$$\alpha : \{1, 2, 3, \cdots, n\} \to \{1, 2, 3, \cdots, n\}$$

that is bijective. It is one-to-one, because no two integers can be placed in the same slot. It is onto, because each slot must be filled. Formally then, a **permutation of the set** $\{1, 2, 3, \cdots, n\}$ is a bijection from this set to itself.

▷ **Quick Exercise.** Recall that if a function from *finite* set to itself is injective, it is automatically surjective. What is the explanation for this? Give an example to show that this is false in the infinite case. (You might want to consult Section 16.1, where we discuss the pigeonhole principle. ◁

As we did in Chapter 17, we can denote such functions by a $2 \times n$ array: The top row tells the names of the elements of the set $\{1, 2, \cdots, n\}$, and the bottom row records the slots to which they are sent. Thus,

$$\begin{pmatrix} 1 & 2 & 3 & 4 & 5 \\ 3 & 2 & 5 & 4 & 1 \end{pmatrix}$$

is a permutation of the set $\{1, 2, 3, 4, 5\}$, which does not move 4 or 2, sends 1 to 3, 3 to 5, and 5 to 1. We could use functional notation to express the above facts. For instance, sending 1 to 3 could be written as

$$\begin{pmatrix} 1 & 2 & 3 & 4 & 5 \\ 3 & 2 & 5 & 4 & 1 \end{pmatrix} (1) = 3.$$

We denote the set of all permutations on the set $\{1, 2, \cdots n\}$ by $S_n$. This set has $n!$

elements because for 1 there are $n$ slots available, for 2 there are $n - 1$ slots available (because 1 has taken one of them), for 3 there are $n - 2$ slots available, and so on. This gives us

$$n(n - 1)(n - 2) \cdots (3)(2)(1) = n!$$

possibilities altogether.

## 18.2    The Symmetric Groups

Since two elements $\alpha, \beta \in S_n$ are functions with the same domain and range, we can combine them together with *functional composition*. We denote that operation by $\circ$, or by juxtaposition, writing either $\alpha \circ \beta$ or $\alpha\beta$. Note that composition of two elements of $S_n$ gives another element of the set: The composition is still a bijection.

▷ **Quick Exercise.**    Prove this. ◁

As we observed in Chapter 17, functional composition is an associative operation, and so we can write down a string of compositions like $\alpha\beta\gamma$ without inserting any unnecessary parentheses.

A trivial but important example of a permutation is the **identity function** $\iota$, where $\iota(k) = k$, for $k = 1, 2, \cdots, n$. We would write this as

$$\iota = \begin{pmatrix} 1 & 2 & 3 & \cdots & n \\ 1 & 2 & 3 & \cdots & n \end{pmatrix},$$

using our matrix notation. Note that if $\alpha$ is any element of $S_n$ and $1 \leq k \leq n$, then

$$\iota \circ \alpha(k) = \iota(\alpha(k)) = \alpha(k),$$

and so $\iota \circ \alpha = \alpha$.

▷ **Quick Exercise.**    Write down the corresponding proof for $\alpha \circ \iota$. ◁

From our discussion of bijections in Section 7.3, it follows that for $\alpha \in S_n$, there exists a unique **inverse function** $\alpha^{-1}$:

$$\alpha^{-1}(n) = m \text{ is true exactly if } \alpha(m) = n.$$

This is exactly the function that composes with $\alpha$ to give the identity function $\iota$. Informally, the existence of $\alpha^{-1}$ means that we can 'un-do' any rearrangement of our list.

We call the set $S_n$ the **symmetric group** on $n$. We will also call it a **group of permutations**. Notice that although for convenience's sake we think of $S_n$ as permutations of the set $\{1, 2, \cdots, n\}$, it really doesn't matter what finite set of objects we're permuting (as long as they are distinguishable). This is essentially what we did in Chapter 17, where we were really permuting vertices of geometric figures, rather than the integer labels.

Let's look at some particular computations in $S_n$:

## Example 18.1

What is the inverse of the permutation $\begin{pmatrix} 1 & 2 & 3 & 4 & 5 \\ 3 & 2 & 5 & 4 & 1 \end{pmatrix}$ in $S_5$? It is
$\begin{pmatrix} 1 & 2 & 3 & 4 & 5 \\ 5 & 2 & 1 & 4 & 3 \end{pmatrix}$.

▷ **Quick Exercise.** Verify this. ◁

## Example 18.2

Consider the permutations

$$\alpha = \begin{pmatrix} 1 & 2 & 3 & 4 \\ 3 & 2 & 1 & 4 \end{pmatrix} \quad \text{and} \quad \beta = \begin{pmatrix} 1 & 2 & 3 & 4 \\ 4 & 3 & 2 & 1 \end{pmatrix}.$$

What is the permutation $\alpha \circ \beta$? Because we read functional composition from right to left, we have

$$\alpha \circ \beta(1) = \alpha(\beta(1)) = \alpha(4) = 4.$$

By doing the three other necessary computations, we obtain that

$$\alpha \circ \beta = \begin{pmatrix} 1 & 2 & 3 & 4 \\ 4 & 1 & 2 & 3 \end{pmatrix}.$$

▷ **Quick Exercise.** Verify the calculation for $\alpha \circ \beta$ given in Example 18.2. Then compute $\beta \circ \alpha, \alpha^2$, and $\beta^{-1}$. ◁

One of your results from the previous Quick Exercise drives home a point we had already encountered in Chapter 17. Order matters when composing permutations.

▷ **Quick Exercise.** Choose two elements at random from $S_6$ and compute their products in both orders. It is highly likely that your answers will not be the same. ◁

We now have a nice way of thinking of our symmetry calculations in Chapter 17. While the symmetries, thought of geometrically, are quite concrete, they are rather difficult to compute with (especially if we don't have a cube or tetrahedron around to handle!). But representing them as permutations, while more abstract, provide us with a really easy way to compute explicitly. We also represented some of our symmetries using matrices, because multiplying matrices is also computationally straightforward; this approach is also important, and we will explore these ideas more later.

We first make a few notational remarks about the symmetry groups $S_n$. Note that if $m < n$, we can think of $S_m$ as a subset of $S_n$ because any permutation of $\{1, 2, \cdots, m\}$ can be thought of as a permutation of the larger set $\{1, 2, \cdots, n\}$, which leaves the elements $m+1, m+2, \cdots, n$ fixed. Consequently, in what follows we will not bother to be too specific about which symmetric group a given permutation belongs to.

Although we combine elements of $S_n$ together by means of functional composition, we will for the most part in this (and future chapters) speak less formally about this operation; since we will often use juxtaposition to indicate composition, we will tend to speak of the *product* of two permutations, rather than their composition.

## 18.3   Cycles

While the composition of two permutations using the double row notation above is straightforward, it does seem a bit cumbersome and awkward to use. Our immediate goal is to introduce a much more efficient way of describing permutations; this notation will depend on some important facts about permutations that we will now explore.

Consider first the permutation

$$\alpha = \begin{pmatrix} 1 & 2 & 3 & 4 \\ 2 & 3 & 4 & 1 \end{pmatrix}.$$

It permutes the elements 1, 2, 3, and 4 *cyclically,* as the following picture suggests:

In general a permutation $\alpha \in S_n$ is a **cycle of length** $k$, if there exist integers $a_1, a_2, \cdots, a_k$ so that

$$\alpha(a_1) = a_2, \ \alpha(a_2) = a_3, \ \cdots, \ \alpha(a_{k-1}) = a_k, \ \alpha(a_k) = a_1,$$

and $\alpha$ leaves fixed the remaining $n - k$ elements in its domain. We will denote this cycle by the notation

$$(a_1 a_2 a_3 \cdots a_k).$$

**Example 18.3**

Consider the cycle (125). It is short-hand for the permutation

$$\begin{pmatrix} 1 & 2 & 3 & 4 & 5 \\ 2 & 5 & 3 & 4 & 1 \end{pmatrix}.$$

Notice that we could just as well have represented this cycle by (251) or (512). It is clear that (125) represents an element in $S_5$, at least. But in a particular situation, we might use the same notation to denote the corresponding element

$$\begin{pmatrix} 1 & 2 & 3 & 4 & 5 & 6 & 7 \\ 2 & 5 & 3 & 4 & 1 & 6 & 7 \end{pmatrix}$$

in $S_7$. Notice that the inverse permutation cycles elements in the opposite direction, which in this case is (152).

▷ **Quick Exercise.**   Express the cycle (3251) in our earlier notation. Write this cycle three other ways in cycle notation. What is its inverse? ◁

Of course, not all permutations are cycles: Consider the permutation

$$\beta = \begin{pmatrix} 1 & 2 & 3 & 4 & 5 \\ 2 & 5 & 4 & 3 & 1 \end{pmatrix}.$$

While it behaves as a cycle on the set $\{1, 2, 5\}$, it also behaves cyclically on the set $\{3, 4\}$. What this really means is that we can factor $\beta$ as a composition of two cycles:

$$\beta = (125)(34) = (34)(125).$$

▷ **Quick Exercise.** Convince yourself that $\beta$ really equals the composition of the two-cycle permutations $(125)$ and $(34)$, composed in either order. What is the inverse of $\beta$? ◁

## 18.4 Cycle Factorization of Permutations

This suggests that perhaps we could factor *any* permutation into a product of cycles. To describe this, we need some terminology. The **support** of a permutation $\alpha$ is the set of all integers $k$ so that $\alpha(k) \neq k$. Speaking more colloquially, the support of a permutation is the set of elements in its domain *that it moves*. We use the notation $\mathrm{Supp}(\alpha)$ for this set, which is always a subset of $\{1, 2, 3, \ldots, n\}$. We have that $\mathrm{Supp}((125)) = \{1, 2, 5\}$, while $\mathrm{Supp}(\iota) = \emptyset$, the empty set.

▷ **Quick Exercise.** What is $\mathrm{Supp}((125)(34))$? Give an element in $S_7$ whose support is $\{2, 5, 6\}$. ◁

Now if $\alpha(k) = k$, then $k$ is not in the support of $\alpha$; we say that $\alpha$ **fixes** $k$. The set of all elements that an element fixes is its **fixed set**, which we denote by $\mathrm{Fix}(\alpha)$.

**Example 18.4**

Consider the permutation $(3251) \in S_5$. What is its support? What is its fixed set? In general, what can you say about $\mathrm{Supp}(\alpha) \cap \mathrm{Fix}(\alpha)$ and $\mathrm{Supp}(\alpha) \cup \mathrm{Fix}(\alpha)$?

Let's make a technical observation about the support of a permutation, which we will repeatedly find of use in the arguments which follow.

**Lemma 18.1** *Suppose that $\alpha$ is a permutation, and $k \in \mathrm{Supp}(\alpha)$. Then $\alpha(k) \in \mathrm{Supp}(\alpha)$.*

**Proof:** Because $k$ is in the support of $\alpha$, $\alpha(k) \neq k$. Now apply the function $\alpha$ to these two distinct integers. Because $\alpha$ is injective, we must get distinct integers:

$$\alpha(\alpha(k)) \neq \alpha(k).$$

But this means exactly that $\alpha(k)$ is in the support of $\alpha$. □

Two permutations are **disjoint** if their supports contain no element in common; using our notation, permutations $\alpha$ and $\beta$ are disjoint exactly if $\mathrm{Supp}(\alpha) \cap \mathrm{Supp}(\beta) = \emptyset$. In a previous Quick Exercise it was the disjointness of $(125)$ and $(34)$ that allowed them to commute. Let's prove this in general:

**Theorem 18.2** *Suppose that $\alpha$ and $\beta$ are disjoint permutations. Then*

$$\alpha\beta = \beta\alpha.$$

**Proof:**   Suppose that $\alpha$ and $\beta$ are disjoint. We must show that the two functions $\alpha\beta$ and $\beta\alpha$ are the same: So we show they do the same thing to typical elements of their domain.

If we choose an integer $k$ belonging to neither $\text{Supp}(\alpha)$ nor $\text{Supp}(\beta)$, then clearly

$$\alpha\beta(k) = \alpha(k) = k = \beta(k) = \beta\alpha(k).$$

Any other integer belongs to the support of exactly one of $\alpha$ and $\beta$, because they are disjoint. Without loss of generality, let's suppose that integer $k \in \text{Supp}(\alpha)$, but $k \notin \text{Supp}(\beta)$. Now by the previous lemma, $\alpha(k)$ also belongs to the support of $\alpha$, and hence not to that of $\beta$. Thus, we have

$$\alpha\beta(k) = \alpha(k)$$

and

$$\beta\alpha(k) = \alpha(k).$$

Hence, the two functions $\alpha\beta$ and $\beta\alpha$ are the same, as claimed.       $\square$

We're now ready to state and prove the factorization theorem for permutations. Note that this theorem is very similar in flavor to the Fundamental Theorem of Arithmetic 2.8 and 2.9: It asserts that every permutation can be factored uniquely into simpler pieces (cycles), just as the Fundamental Theorem of Arithmetic asserts that integers can be factored uniquely into simpler pieces (primes).

**Theorem 18.3    Cycle Factorization Theorem for Permutations**       *Every non-identity permutation is either a cycle or can be uniquely factored (up to order) as a product of pairwise disjoint cycles.*

**Proof:**   We first prove the *existence* of the required factorization by using induction on the size of the support of the permutation. Note first that if the size of the support is zero, the permutation moves no element, and is consequently the identity permutation. No permutation can have exactly one element in its support.

▷ **Quick Exercise.**   Why can no permutation have exactly one element in its support? ◁

Thus, the base case for the induction is support size two. Suppose that the support of permutation $\alpha$ has two elements. If $k$ is an element of this support, by Lemma 18.1 $\alpha(k)$ must be the other element in the support. But $\alpha(k) \neq \alpha^2(k)$, and so the only choice is that $\alpha^2(k) = k$. That is, $\alpha$ must be a cycle of length two.

Now suppose that $\alpha$ is a permutation with $n$ elements in its support. Our induction hypothesis says that all non-identity permutations with support size less than $n$ are either cycles, or products of disjoint cycles. Pick $a_1 \in \text{Supp}(\alpha)$. Then

$$a_1, \ a_2 = \alpha(a_1), \ a_3 = \alpha(a_2), \ \cdots$$

are all elements of the support. Because the support is finite, sooner or later we will have a duplication in this list. Suppose the first duplication in the list occurs at $a_{k+1} = \alpha(a_k)$. Then because $\alpha$ is injective, the only duplication possible is if $\alpha(a_k) = a_1$ (any other duplication would give different elements with the same $\alpha$ value). Now consider the cycle $\beta = (a_1 a_2 a_3 \cdots a_k)$. If $k = n$, then $\alpha = \beta$, which is a cycle, as required. Otherwise, consider the element $\beta^{-1}\alpha$. This function equals $\alpha$ on any integer other than $a_1, a_2, \cdots, a_k$, but does not move the $a_i$'s. Thus, the support of $\beta^{-1}\alpha$ is disjoint from $\{a_1, a_2, \cdots, a_k\}$ and has $n-k$ elements, which is fewer than $n$. Consequently, by the induction hypothesis, we can factor $\beta^{-1}\alpha$ as a product of disjoint cycles. Then $\alpha$ is the product of these cycles, times $\beta$. This completes the proof, by the Principle of Mathematical Induction.

It remains to prove that the factorization we have obtained is unique. This is a similar induction proof, left as Exercise 18.2. □

Henceforth, when we compute with permutations, we will invariably use the disjoint cycle representation guaranteed by this theorem.

## Example 18.5

Suppose that
$$\alpha = (154)(23)(689) \quad \text{and} \quad \beta = (14895)(27)$$
are two permutations, written as products of disjoint cycles. Let's compute the product $\alpha\beta$. This is an entirely straightforward matter, if we remember that the operation here is functional composition, and so we operate from right to left. Start with any element of the domain (say, 1):
$$\alpha\beta(1) = \alpha(4) = 1.$$
Thus, the product fixes 1. Let's try 2:
$$\begin{aligned}\alpha\beta(2) &= \alpha(7) = 7; \\ \alpha\beta(7) &= \alpha(2) = 3; \\ \alpha\beta(3) &= \alpha(3) = 2.\end{aligned}$$

Thus, $(273)$ is one of the cycle factors of the product $\alpha\beta$. We leave it to you to complete the computation to obtain:
$$\alpha\beta = (273)(49)(68).$$

▷ **Quick Exercise.** Complete this computation. Then compute $\beta\alpha$ and $\beta^2\alpha^{-1}$ in the same way. ◁

Once a permutation has been factored as a product of disjoint cycles, we call the set of elements moved by each of its constituent cycle factors its **orbits**. Thus, the permutation

$$(157)(2389)$$

has orbits $\{1, 5, 7\}$ and $\{2, 3, 8, 9\}$. In addition, it has the **trivial orbits** $\{4\}$ and $\{6\}$. We therefore have that every element of the domain of the permutation belongs to exactly one orbit, and the other elements of the orbit it belongs to are exactly the locations we can move the element to, by repeatedly applying the permutation. Given $\alpha \in S_n$ and $1 \leq k \leq n$, we will then let $\text{Orb}(k)$ be the orbit of $\alpha$ that contains $k$. Formally, we can say that

$$\text{Orb}(k) = \{k, \alpha(k), \alpha^2(k), \alpha^3(k), \ldots\}.$$

Note that this is of course a finite set, because the list of elements eventually starts to repeat, as we have seen in the proof of Theorem 18.3 above.

▷ **Quick Exercise.** Give examples in $S_7$ of an element with one orbit of seven elements and also an element with 3 orbits: one orbit of 4 elements, one orbit of 2 elements, and a trivial orbit. ◁

Let's now look carefully at the disjoint cycle and orbit structure for the elements of $S_3$. The identity of course has three orbits $\{1\}, \{2\}, \{3\}$. We then also have three 2-cycles $(12), (13), (23)$, each of which has two orbits, one with two elements and the other with one element. And finally, we have two 3-cycles $(123), (132)$, each of which has one orbit, which consists of the entire set $\{1, 2, 3\}$. Since 3 is so small, all non-identity elements of $S_3$ are cycles.

If we interpret these 6 permutations as permutations of the vertices of an equilateral triangle, we then see that the different cycle structures have geometric meaning. The 2-cycles are exactly the flips across a line, while the 3-cycles are the two rotations through 120° and 240°.

Consider now $S_4$. As before, we have the identity as a special case. We have 6 distinct 2-cycles, each of which has 3 orbits:

$$(12), (13), (14), (23), (24), (34).$$

We have 8 distinct 3-cycles in $S_4$, each of which has 2 orbits:

$$(123), (132), (124), (142), (134), (143), (234), (243).$$

We have 6 distinct 4-cycles, each of which has 1 orbit:

$$(1234), (1243), (1324), (1342), (1423), (1432).$$

But what is different in $S_4$ is that we have some elements that are not cycles. There are 3 of these, each with 2 orbits, each with 2 elements:

$$(12)(34), (13)(24), (14)(23).$$

Notice that we have accounted for all 24 elements of $S_4$.

If we interpret our elements of $S_4$ as permutations of the vertices of the tetrahedron (as in Section 17.6), we have the following geometric interpretations. The 2-cycles cannot be accomplished by a rigid motion of three dimensional space. The 3-cycles amount to rotations about a line through one of the vertices and the centroid of the opposite triangle. The elements which are products of two disjoint 2-cycles are the 180° rotations about the line through the midpoints of two of the edges. And the 4-cycles cannot be accomplished by a rigid motion of three dimensional space. This leaves us with the 12 symmetries of the tetrahedron.

---

## Chapter Summary

In this chapter we considered the symmetric group $S_n$, which consists of all permutations of the first $n$ positive integers; there are $n!$ elements of $S_n$. We introduced an efficient notation for such permutations, which depends on the theorem that any permutation can be factored as a product of disjoint cycles.

---

## Warm-up Exercises

a. Suppose that $\alpha$ is the element of $S_7$ specified by

$$\begin{pmatrix} 1 & 2 & 3 & 4 & 5 & 6 & 7 \\ 3 & 7 & 2 & 4 & 1 & 6 & 5 \end{pmatrix}.$$

What is $\alpha(5)$? What is $\alpha(4)$? What is the inverse of $\alpha$?

b. Express the element $\alpha$ from the previous exercise in disjoint cycle notation. What is its inverse, using this notation? What is Fix($\alpha$)? What is Supp($\alpha$)? What is Orb(3)? Orb(6)?

c. Suppose in addition that $\beta$ is the element of $S_7$ specified by

$$\begin{pmatrix} 1 & 2 & 3 & 4 & 5 & 6 & 7 \\ 4 & 6 & 3 & 1 & 5 & 7 & 2 \end{pmatrix}.$$

Express this element in disjoint cycle notation. Then compute

$$\beta^2, \ \beta\alpha, \ \alpha\beta, \ \beta\alpha^{-1}, \ \alpha\beta\alpha.$$

d. Using the elements $\alpha$ and $\beta$ from the previous exercises, calculate $(\alpha\beta)^{-1}$ twice, once by using your computation of $\alpha\beta$ in Exercise c, and once by computing $\alpha^{-1}$ and $\beta^{-1}$, and then composing them (in the proper order!).

e. How many elements are there in $S_4$? $S_5$? $S_6$?

f. What kind of element in $S_8$ has 8 orbits? What kind of element in $S_8$ has one orbit?

g. Two non-identity permutations with disjoint supports necessarily commute. However, the converse is false. Find an example to show this.

h. In what ways are the Cycle Factorization Theorem 18.3 and the Fundamental Theorem of Arithmetic 2.8 and 2.9 similar? In what ways are they different? (For the latter, consider whether cycles are *irreducible*.)

## Exercises

1. If $\alpha \in S_n$ is a cycle, is $\alpha^2$ a cycle also? Give an example where this is true, and an example where this is false. Then characterize cycles where this is true, proving your assertion.

2. Use induction to complete the proof of Theorem 18.3, by showing that the factorization into disjoint cycles of a permutation is unique, up to order.

3. What are the possible disjoint cycle structures for elements of $S_5$? Now count the number of distinct elements for each possibility. You should of course account for $120 = 5!$ elements altogether. Try to perform these counts without writing down all 120 elements explicitly!

4. A permutation $\alpha \in S_n$ is called a **derangement** if $\text{Fix}(\alpha)$ is the empty set. How many of the elements of $S_5$ are derangements? Hint: Use the previous exercise.

5. In $S_n$ (where $n \geq 6$), determine all possible disjoint cycle structures for a product of two 3-cycles. Clearly this will depend on a case by case analysis, which depends on the extent to which the supports of the two 3-cycles overlap.

6. This exercise depends on Exercise 17.18, where in part (c) you saw you could classify the symmetries of the cube in terms of the permutations of the four diagonals of the cube. What is the geometric meaning (in terms of rigid motions of the types given in Exercise 17.18 (d), (e), and (f)) of the distinct disjoint cycle structures for elements of $S_4$ that we discussed at the end of this chapter?

7. How many distinct 3-cycles are there in $S_3, S_4, S_5$? How about in $S_n$?

8. What is the cycle structure of $\alpha \in S_n$, if $\alpha^{-1} = \alpha$?

9. Suppose that $\mathrm{Fix}(\alpha) = \mathrm{Fix}(\beta)$, where $\alpha, \beta \in S_n$. Prove that

$$\mathrm{Fix}(\alpha) \subseteq \mathrm{Fix}(\alpha\beta).$$

Give examples where equality holds, and where the first set is a proper subset of the second.

# Chapter 19

## Abstract Groups

In this chapter we intend to generalize the examples of the previous two chapters, to obtain an abstract notion of *group*, quite similar in flavor to our abstract definition of *ring* in Chapter 6. We have a set of objects (which up to now have been symmetries or permutations), equipped with a binary operation (which up to now has been functional composition). What abstract properties should this binary operation satisfy?

The composition of functions (whether symmetries or permutations) is *associative*, and we will include this in our abstract definition.

Notice that we always included the *identity* function: As a symmetry, this consists of no motion at all; as a permutation, this is again a function that moves nothing. When composed with any other function, we then obtain the second function. This clearly seems similar to our definitions of additive and multiplicative identities. We will include an identity element in our abstract definition.

Any symmetry or permutation can be undone: For symmetries, there is a motion that restores the object in question to its original orientation. For permutations, we can move the elements of the set $\{1, 2, \ldots\}$ back where they came from. What does this mean in terms of the group tables we constructed for the symmetries of the equilateral triangle, and the tetrahedron? For example, in the group $D_3$ of symmetries of the equilateral triangle, the motion that undoes the rotation $\rho$ is simply $\rho^2$, and this fact is reflected in the group table by the fact that

$$\rho\rho^2 = \iota \quad \text{and} \quad \rho^2\rho = \iota.$$

So, the existence of *inverses* is the third requirement of our definition.

▷ **Quick Exercise.** Choose several elements in the groups of symmetries for the triangle and tetrahedron. What are their inverses? Does this make sense geometrically? ◁

## 19.1 Definition of Group

We now state our definition formally. A **group** $G$ is a set of elements on which one binary operation $(\circ)$ is defined that satisfies the following properties: (The symbols $g, h$, and $k$ represent any elements from $G$.)

**(Rule 1)** $(g \circ h) \circ k = g \circ (h \circ k)$

**(Rule 2)** There exists an element $e$ in $G$ such that

$$g \circ e = e \circ g = g$$

**(Rule 3)** For each element $g$ in $G$ there exists an element $x$ so that

$$g \circ x = x \circ g = e$$

We introduce some terminology to describe these rules (which will seem familiar after our experience with rings): Rule 1 says that the operation ∘ is **associative**, Rule 2 says that an **identity** exists, and Rule 3 says that each element of the group has an **inverse**. We can thus paraphrase our definition by saying this: A group is a set with an associative binary operation with an identity, where all elements have inverses.

---

## 19.2   Examples of Groups

What are some examples of groups?

### Example 19.1

The groups of symmetries for the triangle, square, and tetrahedron, where the operation ∘ is functional composition. The operation (being functional composition) is associative, and the identity is the symmetry consisting of no motion (the *identity function*). The inverse of each element is also clearly a symmetry and consists geometrically of 'undoing' the symmetry. Function-theoretically, it is exactly the *inverse function* of the given symmetry.

### Example 19.2

The symmetric groups $S_n$ of permutations of the set $\{1, 2, \ldots, n\}$, where the operation is functional composition. The discussion in Section 18.2 makes clear that this set satisfies the three rules for groups.

Now let $R$ be a ring, equipped with operations $+$ and $\cdot$. Let's forget the multiplication. We claim that $R$ equipped with $+$ is a group. Obviously, $+$ is associative (Rule 2 in the definition of ring). Clearly, 0 plays the role of the identity for $+$ (Rule 3 in the definition of ring). And every element in a ring has an (additive) inverse, by Rule 4 in the definition of a ring.

We have thus provided ourselves with an extremely large fund of group examples, because we know about so many rings. It's probably worthwhile listing a few of these explicitly, so that you can better appreciate the ground covered by this observation.

### Example 19.3

The integers $\mathbb{Z}$, equipped with $+$.

### Example 19.4

The integers modulo $m$ (that is, $\mathbb{Z}_m$), equipped with $+$, as defined in Chapter 3.

### Example 19.5

The set $M_2(\mathbb{Z})$ of $2 \times 2$ matrices, with integer entries, equipped with matrix addition.

**Example 19.6**

The Gaussian integers $\mathbb{Z}[i]$, under addition.

When we look at the additive structure of a ring, and consider it as a group, we call it the **additive group of a ring**. If the only groups around occurred as the additive groups of rings, the abstract concept of group would clearly be superfluous. But by Rule 1 from the definition of ring, the binary operation addition for a ring is always *commutative*. That is, it satisfies the rule

$$a + b = b + a.$$

Note that we do *not* include such a requirement on the abstract operation ∘ in our definition of group above, because we wish to include those groups in Example 19.1 and 19.2.

Thus, the notion of group is a genuine generalization of the notion of additive group of a ring because it includes more structures. As a more general concept, we will in the next chapters discuss quite a distinct algebraic theory regarding groups.

If a group satisfies the additional rule

**Rule 4**     $a \circ b = b \circ a$

we say that the group is **abelian**. That is, a group with a commutative operation is abelian. The word 'abelian' honors the early 19th-century mathematician Abel, whom we mentioned in Chapter 10. (Although 'commutative group' might seem a reasonable term to use, we will follow long-standing practice and say 'abelian' instead.)

---

## 19.3   Multiplicative Groups

We need some more examples:

**Example 19.7**

The set $\mathbb{Q}^*$ of non-zero rational numbers, under multiplication, is a group. Multiplication is clearly an associative operation, and its identity is 1. Note that every such rational number has a (multiplicative) inverse: The multiplicative inverse of rational number $a/b$ (where $a, b \neq 0$) is $b/a$.

Examples 19.3 through 19.6 might have led you to infer that we should invariably associate 'group operation' with addition, rather than multiplication. This is certainly false, as Example 19.7 and several examples below show.

We can generalize Example 19.7. Let $F$ be any field. Then let $F^*$ be the set of non-zero elements of $F$. Equip this set with its usual multiplication. Then this is a group.

▷ **Quick Exercise.**   Why? ◁

**Example 19.8**

Let's create this group for a particular field. Consider the field $\mathbb{Z}_7$. We are then claiming that the set

$$\mathbb{Z}_7^* = \{1, 2, 3, 4, 5, 6\}$$

of residue classes modulo 7 forms a group under multiplication. Because $\mathbb{Z}_7$ is a field, we know that all these elements have (multiplicative) inverses, but let's compute them explicitly anyway: Because it is the identity, clearly the inverse of 1 is 1. Because $2 \cdot 4 = 1$, 2 and 4 are inverses of one another; because $3 \cdot 5 = 1$, 3 and 5 are inverses of one another. And because $6^2 = 1$, 6 is its own inverse.

▷ **Quick Exercise.** Compute the inverses in the multiplicative group $\mathbb{Z}_{11}^*$. ◁

We can generalize even further. Suppose that $R$ is any ring with unity. In Chapter 8 we defined $U(R)$, the set of those elements of $R$ that are *units*; that is, $U(R)$ is the set of those elements of $R$ that have multiplicative inverses. For a field $F$, $U(F)$ is just $F^*$, its set of non-zero elements.

But we claim that for *any* ring with unity, $U(R)$ is a group under multiplication. Because $R$ has unity, the set $U(R)$ evidently possesses an identity. And clearly (by definition) every element of $U(R)$ has a (multiplicative) inverse. There remains a subtle point to verify, before we can claim that $U(R)$ is a group: Is in fact $U(R)$ *closed* under multiplication? That is, is multiplication on $U(R)$ a binary operation? The answer is yes, and we proved exactly this in Chapter 8.

▷ **Quick Exercise.** Carefully re-read this proof in Section 8.2. ◁

We will consequently call $U(R)$ the **group of units** for the ring (with unity) $R$.

▷ **Quick Exercise.** Why are we restricting ourselves to rings *with unity*? ◁

Thus, each ring with unity has associated with it *two* groups: its additive group, and its group of units. Let's look concretely at a few groups of units:

**Example 19.9**

The set $U(\mathbb{Z}_6) = \{1, 5\}$ is a group under multiplication. The multiplicative inverse of 5 is itself.

**Example 19.10**

Consider the group of units of the Gaussian integers:

$$U(\mathbb{Z}[i]) = \{1, -1, i, -i\}.$$

Note that $-1$ is its own inverse and $i$ and $-i$ are inverses of each other.

## Example 19.11

Consider the sets $U(M_2(\mathbb{Z}))$ and $U(M_2(\mathbb{R}))$; they are both groups, where the operation is matrix multiplication. Recall that the units of $M_2(\mathbb{Z})$ and $M_2(\mathbb{R})$ are those matrices with non-zero determinant. (See Exercise 8.2.) Both groups have infinitely many elements. For example,

$$\begin{pmatrix} 3 & 1 \\ 5 & 2 \end{pmatrix} \quad \text{and} \quad \begin{pmatrix} 3 & 4 \\ 4 & 5 \end{pmatrix}$$

are both elements of $U(M_2(\mathbb{Z}))$.

▷ **Quick Exercise.** What are their inverses? ◁

▷ **Quick Exercise.** Show that $U(M_2(\mathbb{Z}))$ is a non-abelian group. (The examples we've just given will do!) ◁

## Example 19.12

Consider $U(\mathbb{Z}[\sqrt{2}])$. In Exercise 8.6 we computed infinitely many elements of this multiplicative group. Depending on how you solved this exercise, you might have even obtained the complete list of all elements of this group; it is a bit difficult to prove that such a list is complete, however.

▷ **Quick Exercise.** Verify that $7 + 5\sqrt{2}$ is an element of this group. (That is, compute its inverse.) ◁

We can have multiplicative groups living inside previously studied rings, which do not consist of the entire group of units. For a couple of such examples, consider the following.

## Example 19.13

Let's work inside the field $\mathbb{Z}_{13}$. Take the element 4. We now compute the (multiplicative) powers of this element; to make our computations clear, we will this time make explicit use of modular notation:

$$[4]^1 = [4], \quad [4]^2 = [16] = [3], \quad [4]^3 = [4][3] = [12],$$
$$[4]^4 = [4][12] = [48] = [9], \quad [4]^5 = [4][9] = [10],$$
$$[4]^6 = [4][10] = [40] = [1].$$

We have stopped here, because if we continue, we will repeat elements we have already obtained. We now claim that

$$\{[1], [4], [3], [12], [9], [10]\} =$$
$$\{[1], [4]^1, [4]^2, [4]^3, [4]^4, [4]^5\}$$

is a group under multiplication. You can check this by brute force, but the fact that all elements are powers of $[4]$, and that $[4]^6 = [1]$, means that computation of products and inverses is easy! For example,

$$[4]^3 [4]^4 = [4]^7 = [4]^6 [4] = [4].$$

And what's the inverse of $[4]^2$? We claim it is $[4]^{6-2}$; the computation

$$[4]^2 [4]^4 = [4]^6 = [1]$$

makes this evident. We will return to this example (and generalizations) in Chapter 21.

**Example 19.14**

Let's work inside the field $\mathbb{C}$. Consider the set of all complex numbers of modulus 1. We call this set $\mathbb{S}$:

$$\mathbb{S} = \{\alpha \in \mathbb{C} : |\alpha| = 1\} = \{a + bi \in \mathbb{C} : a^2 + b^2 = 1\}.$$

Considered within the complex plane, this consists of exactly the points on the circle centered at the origin, of radius one. (You should certainly review our discussion of complex numbers in Chapter 8, if necessary.) We consequently call $\mathbb{S}$ the **unit circle**. We claim that $\mathbb{S}$ is a group under complex multiplication. This set is certainly closed under multiplication because by Theorem 8.3

$$|\alpha\beta| = |\alpha||\beta| = 1 \cdot 1 = 1.$$

And clearly, the multiplicative identity 1 belongs to $S$. But what about multiplicative inverses? If we express an element of $\mathbb{S}$ in trigonometric form, it looks like

$$e^{i\theta} = \cos\theta + i\sin\theta.$$

And by DeMoivre's Theorem 8.4, its multiplicative inverse is exactly

$$e^{-i\theta} = \cos(-\theta) + i\sin(-\theta) = \cos\theta - i\sin\theta,$$

which still belongs to $\mathbb{S}$, as you can easily verify.

     The diagram below shows graphically the values of $\alpha$, $\alpha^2$, and $\alpha^{-1}$ for an arbitrary complex number $\alpha \in \mathbb{S}$.

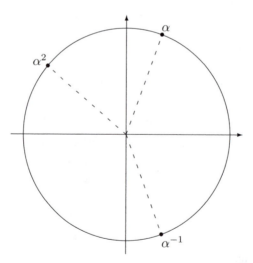

**Example 19.15**

We now work inside $M_2(\mathbb{R})$. Consider the following set of $2 \times 2$ matrices:

$$\begin{pmatrix} 1 & 0 \\ 0 & 1 \end{pmatrix} \quad \begin{pmatrix} -\frac{1}{2} & -\frac{\sqrt{3}}{2} \\ \frac{\sqrt{3}}{2} & -\frac{1}{2} \end{pmatrix} \quad \begin{pmatrix} -\frac{1}{2} & \frac{\sqrt{3}}{2} \\ -\frac{\sqrt{3}}{2} & -\frac{1}{2} \end{pmatrix}$$

$$\begin{pmatrix} -1 & 0 \\ 0 & 1 \end{pmatrix} \quad \begin{pmatrix} \frac{1}{2} & \frac{\sqrt{3}}{2} \\ \frac{\sqrt{3}}{2} & -\frac{1}{2} \end{pmatrix} \quad \begin{pmatrix} \frac{1}{2} & -\frac{\sqrt{3}}{2} \\ -\frac{\sqrt{3}}{2} & -\frac{1}{2} \end{pmatrix}$$

Then this is a group under matrix multiplication. To verify this explicitly, we would need to make sure each element has an inverse in the set (this is not too difficult), and also check that this set is closed under multiplication. If we were to do this by brute force, there would be a lot of multiplications to check! (How many?) However, this set of matrices should look familiar: It was a geometrically explicit realization of $D_3$, the group of symmetries of an equilateral triangle. (See Chapter 17.)

▷ **Quick Exercise.** Recall from Chapter 17 that multiplication of these matrices corresponds to composition of symmetries. With this observation, why must this set of matrices be closed under multiplication? ◁

## Example 19.16

We now work inside $M_2(\mathbb{C})$, the ring of $2 \times 2$ matrices with complex entries. Consider the following set of $2 \times 2$ matrices:

$$\begin{pmatrix} 1 & 0 \\ 0 & 1 \end{pmatrix}, \begin{pmatrix} 0 & 1 \\ -1 & 0 \end{pmatrix}, \begin{pmatrix} 0 & i \\ i & 0 \end{pmatrix}, \begin{pmatrix} i & 0 \\ 0 & -i \end{pmatrix}$$

$$\begin{pmatrix} -1 & 0 \\ 0 & -1 \end{pmatrix}, \begin{pmatrix} 0 & -1 \\ 1 & 0 \end{pmatrix}, \begin{pmatrix} 0 & -i \\ -i & 0 \end{pmatrix}, \begin{pmatrix} -i & 0 \\ 0 & i \end{pmatrix}.$$

Then this is a group under matrix multiplication. It is a tedious matter to verify that this set is closed under multiplication, and we will not carry out this project. However, it would be worthwhile for you to try a few sample multiplications, to get a feel for how this group works:

▷ **Quick Exercise.** Do this. ◁

It is important to note that every element in this set has an inverse; you will verify that this is the case in Exercise 19.3. This group is non-abelian; it is known as the group of **quaternions**. We will denote it by $Q_8$. In Exercise 21.12 we will introduce a more compact notation for the elements of $Q_8$.

## Example 19.17

We now work inside the group $S_3$. Consider the set of permutations $\{\iota, (123), (132)\}$. Since the two permutations $(123)$ and $(132)$ are inverses of one another, it is clear that this set is closed under multiplication (which is really functional composition), and so this is a group.

## Example 19.18

Let's work inside $S_4$. We claim that the set

$$\iota, \quad (12)(34), \quad (13)(24), \quad (14)(23)$$

is a group. It is evident that each of these permutations is its own inverse. But we must check that this set is closed under multiplication. There are only three interesting products to compute.

▷ **Quick Exercise.** Check this. For example, $(12)(34) \ (13)(24) = (14)(23)$. ◁

It's worth noting that many sets equipped with a single operation do *not* satisfy the group axioms.

**Example 19.19**

Consider the integers $\mathbb{Z}$ equipped with multiplication. This is not a group. Although the operation is associative, and there is an identity (namely, 1), almost none of the elements have inverses (with respect to this operation).

More generally, if we take a ring $R$ and forget its addition, we will never obtain a group! To begin with, we wouldn't get a group unless we had a multiplicative identity (unity). But even if $R$ has unity, we would then require that every element of $R$ have a multiplicative inverse; yet 0 never has a multiplicative inverse! (The only exception here is the trivial example of the *zero ring*.)

Of course, this isn't the only way a set with a binary operation can fail to be a group.

**Example 19.20**

The non-negative integers $\{0, 1, 2, \cdots\}$ is not a group under addition: Although it has an additive identity, all elements (except 0 itself) lack an inverse.

**Example 19.21**

Let $R$ be an arbitrary ring. We consider the set $\mathrm{Aut}(R)$ of all ring isomorphisms from $R$ onto itself; such isomorphisms are called **automorphisms**. We claim that $\mathrm{Aut}(R)$ is a group, under functional composition. We need to check that this operation is well defined; that is, is the composition of two such isomorphisms itself an isomorphism? You will do this in Exercise 19.10, where you check that such a composition still preserves addition and multiplication and is a bijection. Next, note that for any ring $R$, the identity function $\iota : R \to R$ defined by $\iota(r) = r$ is in fact a ring isomorphism (see Example 11.8). This element will be the identity element in this group, because when we compose it with any other automorphism, we get the same function back. And in Section 7.4 we described how the inverse function of a ring isomorphism is itself a ring isomorphism, and so each element of $\mathrm{Aut}(R)$ has a inverse. In Exercises 19.11–19.15 you will determine the elements of $\mathrm{Aut}(R)$, for certain specific rings.

---

# Chapter Summary

In this chapter we have generalized the notion of group of symmetries and the symmetric groups $S_n$, to obtain the concept of abstract *group*. We looked at many examples of groups, including two important classes arising from rings: the *additive group of a ring*, and the *group of units* of a ring with unity.

---

# Warm-up Exercises

a. Give examples of the following:

(a) An abelian group with infinitely many elements.

(b) An abelian group with finitely many elements.

(c) A non-abelian group with infinitely many elements.

(d) A non-abelian group with finitely many elements.

b. Determine the inverses of the following elements, specifying the group operation in each case:

  (a) $[3] \in \mathbb{Z}_5$.

  (b) $[3] \in \mathbb{Z}_5^*$.

  (c) $\begin{pmatrix} 3 & 4 \\ -4 & -5 \end{pmatrix} \in U(M_2(\mathbb{Z}))$.

  (d) $\varphi\rho \in D_3$.

  (e) $(156)(234) \in S_6$.

  (f) $(2, -4) \in \mathbb{Q} \times \mathbb{Q}$.

  (g) $(2, -4) \in U(\mathbb{Q} \times \mathbb{Q})$.

  (h) $\frac{3}{5} + \frac{4}{5}i \in \mathbb{S}$.

c. Give examples of the following (you need to specify both the group in which you are computing, as well as specific elements):

  (a) A non-identity element that is its own inverse.

  (b) Two elements $a$ and $b$, so that $(a \circ b)^{-1} \neq a^{-1} \circ b^{-1}$.

  (c) A non-identity element $a$ so that $a \circ a \circ a \circ a = 1$.

d. Let $R$ be a commutative ring with unity. What two groups are associated with $R$?

e. Consider the element
$$A = \begin{pmatrix} 2 & 0 \\ 0 & 2 \end{pmatrix}$$
in the group $U(M_2(\mathbb{R}))$. Why is it true that $AB = BA$, for all other elements $B$ in this group? Does this mean that this group is abelian?

f. Consider the following sets, equipped with an operation. Are they groups, or not?

  (a) $\mathbb{Z}$, with subtraction.

  (b) $\mathbb{Z}$, with multiplication.

  (c) $M_2(\mathbb{Z})$, with matrix addition.

  (d) $M_2(\mathbb{Z})$, with matrix multiplication.

  (e) $\mathbb{R}^*$, with multiplication.

  (f) $\mathbb{R}^*$, with division.

  (g) $\mathbb{R}^*$, with addition.

  (h) $\mathbb{Z} \times \mathbb{Z}$, with the operation $*$, defined by $(a, b) * (c, d) = (a, d)$.

  (i) The set of vectors in $\mathbb{R}^3$, with cross product.

  (j) The set $\mathbb{R}^+$ of strictly positive real numbers, with multiplication.

  (k) The set of 3-cycles in $S_4$, with functional composition.

g. The following table represents a binary operation on the set $\{a, b, c, d\}$. Argue that this set with this operation is not a group. (This fails to be a group for more than one reason.)

|   | $a$ | $b$ | $c$ | $d$ |
|---|---|---|---|---|
| $a$ | $a$ | $b$ | $d$ | $c$ |
| $b$ | $b$ | $a$ | $c$ | $d$ |
| $c$ | $c$ | $d$ | $b$ | $a$ |
| $d$ | $d$ | $c$ | $a$ | $b$ |

## Exercises

1. Suppose you have a set of $n$ elements with a binary operation that you think might be a group. You easily check that there is an identity and that every element has an inverse, but you are now faced with showing the operation is associative. So, you check the equation $a \circ (b \circ c) = (a \circ b) \circ c$ for all possible $a$, $b$, and $c$. How many equations must you check? If $n = 5$, how many equations is this? If $n = 10$, how many?

2. Specify the elements of the groups of units of the following commutative rings:

$$\mathbb{Z}_{15}, \quad \mathbb{Q}[x], \quad \mathbb{Z}[x], \quad \mathbb{Z} \times \mathbb{Q}.$$

3. Consider the group of quaternions, described in Example 19.16. Determine explicitly which of these eight elements are inverses of one another. Also, show by example that this is not an abelian group.

4. Consider the set $\mathbb{R} \setminus \{-1\}$ of all real numbers except $-1$. Define the operation $*$ by

$$a * b = a + b + ab.$$

Prove that this is a group.

5. In this problem we consider permutations of the set $\mathbb{R}$.

   (a) Let $S(\mathbb{R})$ denote the set of all real-valued functions $f : \mathbb{R} \to \mathbb{R}$, such that $f$ is a bijection. Prove that $S(\mathbb{R})$ is a group, where the operation is functional composition.

   (b) Now let $A(\mathbb{R})$ be the set of functions from $S(\mathbb{R})$ that are also **order-preserving**: By this we mean that if $x < y$, then $f(x) < f(y)$. Prove that $A(\mathbb{R})$ is a group under functional composition. (Note that $S(\mathbb{R})$ includes many strange functions; however, the functions in $A(\mathbb{R})$ are all continuous. You could try to prove this, if you have had a course in real analysis.)

6. Consider the set of matrices of the form

$$\begin{pmatrix} 1 & a & b \\ 0 & 1 & c \\ 0 & 0 & 1 \end{pmatrix},$$

   where $a, b, c$ are arbitrary real numbers. Show that this set forms a group under matrix multiplication.

7. Let $n$ be a positive integer and $\mathcal{C}$ be a circle. Now for $i = 0, 1, \ldots, n - 1$, let $\rho_i$ be the rotation of $\mathcal{C}$ counterclockwise through the angle $2\pi i / n$ radians. Show that this set of rotations is a group under the operation of composition. How many elements are in this group?

8. Let $G$ be a group with operation $\circ$. Suppose that $x \circ x = 1$, for all $x \in G$. Prove that $G$ is abelian.

9. Suppose that $G$ is a group with operation $\circ$; suppose that $x, y \in G$. Show that if

$$(x \circ y) \circ (x \circ y) = (x \circ x) \circ (y \circ y),$$

then $x \circ y = y \circ x$.

10. Let $R$ be any ring, and suppose that $\phi, \psi \in \operatorname{Aut}(R)$. Show that the composition $\phi\psi \in \operatorname{Aut}(R)$, by checking that this function has the appropriate domain and range, is a bijection, and preserves addition and multiplication. (This exercise verifies that $\operatorname{Aut}(R)$ is closed under functional composition; in Example 19.21 we complete the verification that $\operatorname{Aut}(R)$ is a group under this operation.)

11. Prove that $\operatorname{Aut}(\mathbb{Z})$ is a group with only a single element.

12. Show that $\operatorname{Aut}(\mathbb{Q})$ is a group with only a single element.

13. In this problem you will sketch the proof that $\operatorname{Aut}(\mathbb{R})$ is a group with only a single element. You will use the fact that all positive real numbers have exactly two square roots.

    (a) Let $a, b \in \mathbb{R}$. Show that $a \geq b$ if and only if $a - b = x^2$, for some $x \in \mathbb{R}$.

    (b) Use part (a) to show that if $\varphi \in \operatorname{Aut}(\mathbb{R})$, then $a \geq b$ if and only if $\varphi(a) \geq \varphi(b)$.

    (c) Argue that any automorphism of $\mathbb{R}$ is fixed on the rational numbers $\mathbb{Q}$. (See Exercise 12.)

    (d) You may assume that between any two real numbers is a rational number. Use this to prove that any automorphism of $\mathbb{R}$ is fixed on all real numbers, and so $\operatorname{Aut}(\mathbb{R})$ has only a single element.

14. Consider the field of complex numbers $\mathbb{C}$. Show that there are only two two automorphisms of $\mathbb{C}$ that leave $\mathbb{R}$ fixed, namely the identity automorphism $\iota$, and the complex conjugation map $\varphi$ defined by $\varphi(a + bi) = a - bi$. (See Example 11.5.)

15. Consider the field $F = \{0, 1, \alpha, \alpha + 1\}$ introduced in Exercise 8.15. Show that $\operatorname{Aut}(F)$ has only two elements, the identity automorphism $\iota$ and the Frobenius automorphism discussed in Exercise 12.17.

16. Let $n$ be a positive integer; consider the set

$$G_n = \{1, -1\} \times \mathbb{Z}_n.$$

We define an operation on this set by

$$(i, [k]) * (j, [m]) = (ij, [m + jk]).$$

Prove that this makes $G_n$ a group. *Note:* Do not neglect to show that this operation is associative! In Exercise 20.29 we will provide a more compact notation for this group, and show rigorously that it is 'essentially the same' as the dihedral group $D_n$ we introduced in Chapter 17—the group of symmetries of a regular $n$-sided polygon.

# Chapter 20

## Subgroups

In this chapter we will closely follow our exposition of abstract rings in Chapters 6 and 7. We first prove a theorem about arithmetic in an abstract group, quite similar to Theorem 6.1. We then introduce the idea of *subgroup*, which is quite analogous to the corresponding idea *subring*. Finally, we will introduce *group isomorphisms*.

## 20.1 Arithmetic in an Abstract Group

**Theorem 20.1** *Suppose that $G$ is a group with operation $\circ$, and $g, h, k \in G$.*

a. *(Cancellation on the right) If $g \circ k = h \circ k$, then $g = h$.*

b. *(Cancellation on the left) If $k \circ g = k \circ h$, then $g = h$.*

c. *(Solution of Equations) The equation $g \circ x = h$ always has a unique solution in $G$; likewise, the equation $x \circ g = h$ always has a unique solution in $G$.*

d. *(Uniqueness of inverse) Every element of $G$ has exactly one inverse.*

e. *(Uniqueness of identity) There is only one element of $G$ which satisfies the equations $z \circ g = g \circ z = g$ for all $g$; namely, the element $e$.*

f. *(Inverse of a product) The inverse of a product is the product of the inverses, in reversed order: $(g \circ h)^{-1} = h^{-1} \circ g^{-1}$.*

Notice that we need to state both parts (a) and (b), because in an arbitrary group the operation is not commutative: Hence, one of these statements does not immediately follow from the other. Likewise, the solutions guaranteed for the two equations in part (c) need not be the same.

▷ **Quick Exercise.**   Consider these equations in the group of symmetries of the tetrahedron:

$$\varphi_1 x = \rho_3 \quad \text{and} \quad x\varphi_1 = \rho_3.$$

Discover the solutions to these equations (by examining the group table in Section 17.6); note that they are not the same. ◁

Notice that we have already encountered the rule in part (f) for computing inverses of products, when we discussed the multiplicative inverse of a product in a non-commutative ring; it might be worthwhile to re-read our discussion of this rule in Section 8.2.

**Proof:**   (a) & (b): Merely operate on the appropriate side of the equation by an inverse of $k$.

(c): This is Exercise 2; your proof will be similar to part (b) of Theorem 6.1.

(d): This is Exercise 3.

(e): Suppose that $e$ and $z$ both serve as identities in $G$. Then $e = e \circ z$ (because $z$ is an identity); but $e \circ z = z$, (because $e$ is an identity). Thus $e = z$.

(f): Consider that

$$(g \circ h) \circ (h^{-1} \circ g^{-1}) = g \circ (h \circ h^{-1}) \circ g^{-1} = g \circ e \circ g^{-1} = g \circ g^{-1} = e.$$

▷ **Quick Exercise.**   Similarly, show that $(h^{-1} \circ g^{-1}) \circ (g \circ h) = e.$ ◁

Because inverses are unique, $(g \circ h)^{-1} = h^{-1} \circ g^{-1}$.                            □

## 20.2    Notation

Henceforth, when we discuss a generic abstract group, we will tend to denote its operation by juxtaposition, rather than by using the symbol $\circ$ for the operation; we will say that the group is being written *multiplicatively*. In this case, we will usually denote the (unique) identity by 1 and will denote the (unique) inverse of element $g$ by $g^{-1}$. Note that we have already used juxtaposition for the group operation for symmetry groups and permutations groups (this despite the fact that the operations there are really *functional composition*). As in those groups, $g^2 = g \circ g$, $g^3 = g \circ g \circ g$, and so on. In fact, when we use this multiplicative notation, we will often refer informally to the operation as multiplication.

If a generic abstract group is abelian, we will tend to denote its operation by $+$; we will say that the group is being written *additively*. In this case, we will usually denote the identity by 0 and will denote the inverse of element $g$ by $-g$. We can in this situation talk about **subtraction**:

$$g - h = g + (-h).$$

However, for some abelian groups, it is still most natural to use multiplicative notation; for example, we will still use multiplicative notation for the abelian multiplicative group $\mathbb{Q}^*$ of non-zero rational numbers.

You should exercise extreme care in sorting out which sort of notation we use for a given group. General remarks, definitions, and theorems may be expressed multiplicatively, but they still apply for groups written additively. Note that for many sets equipped naturally with more than one operation (such as rings), only one of those operations makes the set a group. Be sure you know which!

## 20.3    Subgroups

It is often easier to check whether a given set is a group, if it is a subset of a larger group with the same operation. This is directly analogous to the ring theoretic situation, where we were led to a definition of *subring*. Similarly, we say that a subset $H$ of a group $G$ is a **subgroup** if $H$ is itself a group under the operation induced from $G$.

We can immediately list some examples:

## Example 20.1

The additive group of integers $\mathbb{Z}$ is a subgroup of the additive group of reals $\mathbb{R}$.

We can generalize Example 20.1. If $R$ is a subring of $S$, and we ignore the multiplication, then clearly the additive group $R$ is a subgroup of the additive group $S$. This obviously provides us with a raft of additional examples of subgroups.

▷ **Quick Exercise.** List for yourself at least six interesting examples of subgroups arising in this way. ◁

## Example 20.2

The unit circle $\mathbb{S}$ is a subgroup of the multiplicative group $\mathbb{C}^*$ of non-zero complex numbers. (See Example 19.14.)

## Example 20.3

The set $\{1, 4, 3, 12, 9, 10\}$ is a subgroup of the multiplicative group of units $U(\mathbb{Z}_{13}) = \mathbb{Z}_{13}^*$. (See Example 19.13.)

## Example 20.4

The set $\{\iota, (123), (132)\}$ is a subgroup of the symmetric group $S_3$. (See Example 19.17.)

## Example 20.5

Let $G$ be any group (written multiplicatively). Then $\{1\}$ and $G$ itself are always subgroups of $G$. We call $\{1\}$ the **trivial subgroup** of $G$. $G$ is the **improper subgroup** of $G$. All subgroups of $G$ other than the improper subgroup $G$ are **proper subgroups**.

## Example 20.6

Note that because $\mathbb{Z}$ is an additive group, the trivial subgroup of this group is $\{0\}$.

▷ **Quick Exercise.** Is $\{1\}$ a subgroup of $\mathbb{Z}$? (Under what operation? Remember that when we say that $H$ is a subgroup of $G$, we are implying that $G$ is also a group.) ◁

## Example 20.7

Consider the subset $\{\iota, \varphi\}$ of $D_3$. Because $\varphi$ is its own inverse, it is easy to see that this is a subgroup.

## 20.4    Characterization of Subgroups

As with rings, we have a theorem that makes it easy to check whether a given set is a subgroup:

**Theorem 20.2    The Subgroup Theorem**    *A non-empty subset $H$ of a group $G$ is a subgroup if and only if whenever $h, k \in H$, then $hk^{-1} \in H$.*

Note that in an abelian group (written additively), this says just that a non-empty subset of a group is a subgroup if and only if it is closed under *subtraction*.

**Proof:**    This theorem is proved just like the corresponding theorem for rings (Theorem 7.1) and is Exercise 4.                                                             □

Note that Examples 7.6 through 7.10 are examples of this theorem in action (if you just ignore multiplication).

▷ **Quick Exercise.**    Review these examples from Chapter 7. ◁

Here are some examples of subgroups which are not additive groups of rings:

**Example 20.8**

We work in the multiplicative group $\mathbb{Q}^*$. Let

$$H = \left\{ \frac{m}{n} : m, n \text{ are odd integers} \right\}.$$

We claim that $H$ is a subgroup. Pick two typical elements of $H$, $m/n$ and $r/s$, where $m, n, r,$ and $s$ are all odd integers. Then

$$\frac{m}{n} \left( \frac{r}{s} \right)^{-1} = \frac{m}{n} \frac{s}{r} = \frac{ms}{nr}.$$

But $ms$ and $nr$ are clearly odd integers, and so this is an element of $H$. By the theorem, $H$ is a subgroup. (The most interesting thing about this conclusion is that $H$ is a group at all!)

**Example 20.9**

We work in $D_4$, the group of symmetries of the square. Consider the set

$$\{\iota, \rho, \rho^2, \rho^3\}.$$

Note that $\rho^{-1} = \rho^3$, and $(\rho^2)^{-1} = \rho^2$. This means that any substitution into $hk^{-1}$ by elements from this set will turn into a power of $\rho$ and hence will belong to the set. We will generalize this example in the next chapter.

## 20.5    Group Isomorphisms

Just as we did for rings, we now make sense of what it means to say that two groups are essentially the same. Suppose that $G$ is a group with operation $\circ$, and $H$ is a group with operation $*$. We then say that $G$ and $H$ are **isomorphic** if there is a bijection $\varphi : G \to H$ which preserves the group operations; that is, for any $g, k \in G$ we have that $\varphi(g \circ k) = \varphi(g) * \varphi(k)$. Notice that $g$ and $k$ are elements of $G$, and so we are combining them with the $G$ operation $\circ$, while $\varphi(g), \varphi(k)$ are elements of $H$, and so we are combining them with the $H$ operation $*$. We call the function $\varphi$ a **group isomorphism**.

Just as with rings, group isomorphisms preserve the other essential group structures, and the proofs are essentially the same. If $e_G, e_H$ are the identity elements for $G$ and $H$ respectively, then $\varphi(e_G) = e_H$ (this is Exercise 21). If $g$ and $k$ are inverses in group $G$, then $\varphi(g)$ and $\varphi(k)$ are inverses in the group $H$ (this is Exercise 22). Furthermore, if $\psi$ is the inverse function of the bijection $\varphi$ (as discussed in Section 7.4), then $\psi : H \to G$ is a group isomorphism too (this is Exercise 23). That is, two groups being isomorphic is a symmetric relationship.

Let's now turn to some examples of group isomorphisms. These examples will illuminate and clarify a lot of what we have done so far. But we first note that if we have a ring isomorphism, then we automatically have a group isomorphism between the additive groups of the rings. This applies to such Examples as 7.16, 7.17, and 11.5, among others.

## Example 20.10

Consider the groups $D_3$ (symmetries of an equilateral triangle), $S_3$ (symmetric group on the set $\{1, 2, 3\}$, and the group of six matrices given in Example 19.15. All of these groups are isomorphic to one another, and the bijections between them may be found in Sections 17.2 and 17.3. In Chapter 17 we argued in some detail that these bijections preserve the group operations (although in that chapter we did not use that language).

## Example 20.11

Our discussion at the end of Section 18.4 regarding the cycle structure of elements of $S_4$, and the determination of those which correspond to elements of the symmetry group of the regular tetrahedron can now be rephrased like this: The symmetry group for the tetrahedron is isomorphic to the subgroup of $S_4$ consisting of the twelve elements

$$\iota, (123), (132), (124), (142), (134), (143), (234), (243), (12)(34), (13)(24), (14)(23).$$

We will henceforth denote this subgroup of $S_4$ by $A_4$, and call it the **alternating group on 4 elements**; we will see in Chapter 27 the origin of this terminology.

## Example 20.12

Consider now the additive group $\mathbb{Z}_2$ and the multiplicative group $U(\mathbb{Z}) = \{1, -1\}$. Consider the bijection that assigns $0 \in \mathbb{Z}_2$ to $1 \in U(\mathbb{Z})$; we have to do this because we know that a group isomorphism must send one group identity to the other. We then must send $1$ to $-1$. This is a group isomorphism $\varphi$ that sends addition to multiplication; for example,

$$1 = \varphi(0) = \varphi(1 + 1) = \varphi(1)\varphi(1) = (-1)(-1) = 1.$$

There aren't that many more cases to consider for a brute force verification that this bijection takes addition to multiplication in all cases. This example is very close to a proof that *all* groups with two elements are isomorphic to one another. We will prove this (and much more) carefully in the next chapter.

▷ **Quick Exercise.** Argue that the subgroup $\{\iota, (12)\}$ of $S_3$ is isomorphic to the two groups we discuss in this example, by providing an explicit bijection that preserves the group operation. ◁

To understand any given group, we thus are not so concerned with the particular notation or type of operation we use. All three groups with two elements in the previous example behave in essentially the same way!

---

# Chapter Summary

In this chapter we proved some elementary properties of the arithmetic of an abstract group. We then introduced the concept of *subgroup*, listed many examples, and proved a theorem which characterizes subgroups. We then introduced the idea of *group isomorphism*, which we used to illuminate some of our earlier examples.

---

# Warm-up Exercises

a. Which of the following are subgroups? (The operation on the larger group is always the obvious one.)

 (a) The subset of even integers in $\mathbb{Z}$.

 (b) The subset $\{0, 2, 4\}$ in $\mathbb{Z}_7$.

 (c) The subset $\{2^n : n \in \mathbb{Z}\}$ in $\mathbb{Q}$.

 (d) The subset $\{2^n : n \in \mathbb{Z}\}$ in $\mathbb{Q}^*$.

 (e) The subset $\{\iota, \rho\}$ in $D_3$.

 (f) The subset $\mathbb{N}$ of $\mathbb{Z}$.

b. Is every subring also a subgroup?

c. Is the group of units of a ring a subgroup of the ring?

d. How many identity elements can a group have? How many inverses can a given element of a group have?

e. Provide the (unique) solution to the following group equations:

 (a) $\varphi_2 x = \rho_1^2$, in the group of symmetries of the tetrahedron.

 (b) $x\varphi_2 = \rho_1^2$, in the group of symmetries of the tetrahedron.

 (c) $11 + x = 4$, in $\mathbb{Z}_{15}$.

 (d) $4x = 11$, in $U(\mathbb{Z}_{15})$.

 (e) $(24)(176)x = (145)$, in $S_7$.

f. Is a ring isomorphism always a group isomorphism?

g. Consider the bijection between the groups $\mathbb{Z}_3$ and $\{\iota, (123), (132)\} \subset S_3$, given by $\varphi(1) = (123), \varphi(2) = (132), \varphi(0) = \iota$. Is this a group isomorphism?

## Exercises

1. Consider the set

$$iR = \{ai : a \in \mathbb{R}\} \subseteq \mathbb{C};$$

   these are the **imaginary** numbers. Prove that this is a subgroup of the additive group of $\mathbb{C}$. Is $i\mathbb{R}$ a subring of the ring $\mathbb{C}$? Similarly, show that $i\mathbb{Z} = \{ni : n \in \mathbb{Z}\}$ is a subgroup of the additive group of the Gaussian integers $\mathbb{Z}[i]$.

2. Prove Theorem 20.1c. That is, suppose that $G$ is a group and $g, h \in G$. Prove that $gx = h$ has a unique solution; likewise, prove that $xg = h$ has a unique solution. (We have written the equations multiplicatively.)

3. Prove Theorem 20.1d. That is, prove that in a group, every element has exactly one inverse.

4. Prove the Subgroup Theorem 20.2: A non-empty subset $H$ of a group $G$ is a subgroup if and only if whenever $h, k \in H$, then $hk^{-1} \in H$.

5. Show that if $H$ and $K$ are subgroups of the group $G$, then $H \cap K$ is also a subgroup of $G$. Show by example that $H \cup K$ need not be a subgroup. (This exercise can and should be compared to Exercises 7.9 and 7.10.)

6. Suppose that $G$ is a group, written multiplicatively. Let $g \in G$, and suppose that $g^2 = g$. Prove that $g$ is the identity.

7. Let $G$ be a group, and $a, b, c \in G$. Prove that the equation $axc = b$ has a unique solution in $G$.

8. Suppose that $G$ is equipped with an associative operation $*$. Suppose that $G$ has an element $e$ so that $g * e = g$, for all $g \in G$; furthermore, for all $g \in G$, there exists an element $g' \in G$ so that $g * g' = e$. Why are these assumptions apparently weaker than decreeing that $G$ be a group? Prove, however, that these assumptions are sufficient to force $G$ to be a group.

9. Show that if $(xy)^{-1} = x^{-1}y^{-1}$ for all $x$ and $y$ in the group $G$, then $G$ is abelian.

10. Complete the following multiplication table so the following will be a group.

|   | $a$ | $b$ | $c$ | $d$ |
|---|---|---|---|---|
| $a$ |   |   |   |   |
| $b$ |   |   |   | $d$ |
| $c$ |   |   | $d$ |   |
| $d$ |   |   |   |   |

11. Find all subgroups of $U(\mathbb{Z}_8)$; of $\mathbb{Z}_7^*$; of $U(\mathbb{Z}_{15})$; of $S_3$.

12. Show that $n\mathbb{Z}$ is a subgroup of the additive group of integers $\mathbb{Z}$, for all integers $n$.

13. Find all finite subgroups of the additive group $\mathbb{C}$. What can you say about all finite subgroups of the multiplicative group $\mathbb{C}^*$?

14. Argue *geometrically* that the dihedral group, $D_n$, has a subgroup of order $n$.

15. Let $G$ be a group and $a \in G$. Define the **centralizer of** $a$ to be

$$C(a) = \{g \in G : ga = ag\}.$$

That is, $C(a)$ consists of all the elements that commute with $a$.

(a) Find $C(\rho)$ in $D_3$.

(b) Find $C(4)$ in $\mathbb{Z}_7$.

(c) Show that $C(a)$ is a subgroup of $G$.

(d) Let $H$ be a subgroup of $G$, and let

$$C(H) = \{g \in G : gh = hg \text{ for all } h \in H\};$$

call $C(H)$ the **centralizer of** $H$. Show that $C(H)$ is a subgroup of $G$.

16. Let $Z(G)$, the **center of** $G$, be the set of elements of $G$ that commute with *all* elements of $G$.

(a) Find the center of the quaternions, defined in Example 19.16.

(b) Find the center of $\mathbb{Z}_5$.

(c) Show that $Z(G)$ is a subgroup of $G$.

(d) If $Z(G) = G$, what can you say about the group $G$?

17. If $H$ is a subgroup of $G$, then show that $Z(G) \cap H$ is a subgroup of $Z(H)$.

18. Recall that the elements of $U(M_2(\mathbb{R}))$ are the real-valued $2 \times 2$ matrices with non-zero determinants. (See Exercise 8.2.) Show that the collection of such matrices with determinants equal to one is a subgroup of $U(M_2(\mathbb{R}))$.

19. Consider the elements of $U(M_2(\mathbb{R}))$ of the form

$$\begin{pmatrix} a & 0 \\ b & 1 \end{pmatrix},$$

where $a \neq 0$. Prove that this is a subgroup.

20. Show that the group given in Exercise 19.7 is a subgroup of $\mathbb{S}$, the group given in Example 19.14.

21. Suppose that $G$ and $H$ are groups with operations $\circ$ and $*$ and identities $e_G$ and $e_H$. If $\varphi : G \to H$ is a group isomorphism, prove that $\varphi(e_G) = e_H$.

22. Suppose that $G$ and $H$ are groups with operations $\circ$ and $*$ and suppose $g, k \in G$ are inverses; that is, $g \circ k = e_G$. If $\varphi : G \to H$ is a group isomorphism, prove that $\varphi(g)$ and $\varphi(k)$ are inverses in $H$.

23. Suppose that $\varphi : G \to H$ is a group isomorphism and that $\psi$ is the inverse function of $\varphi$, which exists because $\varphi$ is a bijection. Prove that $\psi$ is a group isomorphism too.

24. Prove that the isomorphic relation between groups is an *equivalence relation*. That is, show that it is *reflexive* (every group is isomorphic to itself), *symmetric* (if $G$ is isomorphic to $H$, then $H$ is isomorphic to $G$), and *transitive* (if $G$ is isomorphic to $H$ and $H$ is isomorphic to $K$, then $G$ is isomorphic to $K$).

25. Let $G$ be any (multiplicative) group, and consider the function $\varphi : G \to G$ defined by $\varphi(g) = g^{-1}$. Show that $\varphi$ is a bijection. Prove that $\varphi$ is a group isomorphism if and only if $G$ is abelian.

26. Let $G$ be a group, and let $H = \{g \in G : g^2 = 1\}$. Show that $H$ is a subgroup of $G$ if $G$ is abelian. Give an example where this is false, if $G$ is not abelian.

27. We give the group version of Example 19.21. Let $G$ be a group, and $\varphi : G \to G$ a group isomorphism from $G$ onto $G$. We call $\varphi$ a **group automorphism**. Let $\text{Aut}(G)$ be the set of all such automorphisms. Prove that $\text{Aut}(G)$ is a group under functional composition.

28. Show that for group $G$, if $a, b \in G$ and $ab = 1$, then $ba = 1$.

29. Let $n$ be a positive integer, and consider the group $G_n$ described in Exercise 19.16. We will relabel these elements in such a way that we can consider $G_n$ as the list of objects
$$G_n = \{I, R, R^2, \cdots, R^{n-1}, F, FR, FR^2, \cdots FR^{n-1}\},$$
where $I$ is the label we assign to the identity, and the elements $(1, [k])$ are labeled as $R^k$, and the elements $(-1, [k])$ are labeled as $FR^k$.

(a) Show that the elements of $G_n$ under this labeling do satisfy the identities
$$RF = FR^{n-1}, F^2 = 1, \text{ and } R^n = 1.$$

In fact, it is possible to prove that our list above of the elements $R^k, FR^k$, together with these identities, abstractly characterizes the group $G_n$. We will not rigorously prove this here, however.

(b) Using the abstract characterization of $G_n$ described above, prove that the dihedral group $D_n$ as discussed in Chapter 17, is isomorphic to $G_n$. (You need only establish a one-to-one correspondence that preserves the identities given in part a.)

# Chapter 21

## Cyclic Groups

Suppose that $G$ is a group (written multiplicatively), and $g \in G$. If we repeatedly multiply $g$ by itself, we get the **powers** of $g$:

$$g^1 = g, \quad g^2 = gg, \quad g^3 = g(g^2), \quad g^4 = g(g^3), \quad \cdots.$$

Given an element $g^n$ of this form, we call $n$ an **exponent**; for now, we are restricting ourselves to exponents that are positive integers.

### Example 21.1

Choose the element 2 in $\mathbb{Q}^*$. Then the powers of 2 are the distinct elements

$$2, \quad 4, \quad 8, \quad 16, \quad \cdots.$$

Note that there are infinitely many distinct elements in this list.

### Example 21.2

Choose element $(123)$ in the symmetric group $S_3$. Then the powers of $(123)$ are the repeating list of elements

$$(123), \quad (123)^2 = (132), \quad (123)^3 = \iota, \quad (123)^4 = (123), \quad \cdots.$$

Note that there are exactly three distinct elements in this list.

## 21.1 The Order of an Element

It turns out that all elements in a group behave in one or the other of the two ways illustrated by our examples above. We make this precise in the following theorem:

**Theorem 21.1** *Suppose that $G$ is a group and let $g \in G$. Then exactly one of the following two cases holds:*

a. *(The Torsion-free Case) All the powers*

$$g, \quad g^2, \quad g^3, \quad g^4, \quad \cdots$$

*of $g$ are distinct elements of $G$.*

b. *(The Torsion Case) There exists a least positive integer $n$ for which $g^n = 1$. In this case, the elements*

$$g, \quad g^2, \quad g^3, \quad \cdots, \quad g^{n-1}, \quad g^n = 1$$

*are all distinct.*

If an element falls into the second case, we say that it is a **torsion** element. The integer $n$ we call its **order**. In this case, we denote the (finite) order of an element $g$ by $o(g)$. Thus, in Example 21.2 above, $o((123)) = 3$. If an element falls into the first case, we say that it is a **torsion-free** element and that it has **infinite order**. The word 'torsion' is intended to reflect the fact that in the second case the powers of the element cycle back on themselves:

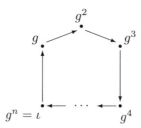

**Proof**:    Let $G$ be a group, and $g \in G$. If all the powers of $g$ are distinct, we obviously have the first case. Suppose, then, that not all powers of $g$ are distinct. Then there exist positive integers $r$ and $s$ with $g^r = g^s$. We may as well assume that $r < s$. But now multiply both sides of this equation by $r$ copies of $g^{-1}$:

$$1 = (g^{-1})^r (g^r) = (g^{-1})^r g^s = g^{s-r}.$$

This tells us that there exists some positive integer $n$ (in this case, $s - r$) so that $1 = g^n$. By the Well-ordering Principle choose the smallest such $n$. We then claim that

$$g, \quad g^2, \quad g^3, \quad \cdots, \quad g^{n-1}, \quad g^n = 1$$

are all distinct elements. If not, then $g^r = g^s$, where $1 \leq r < s \leq n$. By the same argument as above, $g^{s-r} = 1$. But this is impossible because $s - r < n$, and we had chosen $n$ to be the *least* such exponent. We have thus concluded that the second case holds, if the first fails.□

### Example 21.3

The order of 3 in the group $U(\mathbb{Z}_8)$ is 2. Of course, we could then write $o(3) = 2$.

### Example 21.4

The order of the identity in any group is 1.

▷ **Quick Exercise.**    What are the orders of the elements 1, 2, and 4 in $\mathbb{Z}_{13}^*$?    ◁

Note that you have just illustrated the fact that non-identity elements in a given group need not have the same order.

▷ **Quick Exercise.**    What is the order of the element $-1$ in the group $U(\mathbb{Z}[\sqrt{2}])$? What about the element $1 + \sqrt{2}$?    ◁

The point of the previous Quick Exercise is this: It is possible that a group possess some elements with infinite order, and some elements (other than the identity) with finite order. The next two examples provide a somewhat more complicated illustration of this.

**Example 21.5**

The element $e^{\frac{\pi i}{5}} = \cos(\pi/5) + i\sin(\pi/5)$ in the unit circle $\mathbb{S}$ has order 10. (Use DeMoivre's Theorem 8.4.)

**Example 21.6**

Consider the element $\cos 1 + i\sin 1$ in the unit circle $\mathbb{S}$. (Here, the angle in question is 1 radian, or about 57 degrees.) We claim that this element is torsion-free. If not, there must exist a positive integer $n$ for which

$$1 = (\cos 1 + i\sin 1)^n = \cos n + i\sin n.$$

Thus, $n$ is an angle whose cosine is 1. Elementary trigonometry tells us that $n$ must be an integer multiple of $2\pi$: $n = 2\pi k$. But then $\pi = n/(2k)$. That is, $\pi$ must be a rational number! This is certainly false, and is known as Lambert's Theorem; we will discuss this fact further in Section 38.3. Thus, the element $\cos 1 + i\sin 1$ has infinite order. (You may have already worked with this example in Exercise 8.8.)

Let's consider now a cycle in a symmetric group. For example, the cycle (1234) has order four:

$$(1234), \quad (1234)^2 = (13)(24),$$
$$(1234)^3 = (1234)(13)(24) = (1432), \quad (1234)^4 = \iota.$$

We can generalize this: A cycle of length $n$ has order $n$. For if $\alpha$ is such an element, obviously we must apply it to itself exactly $n$ times to bring any element of its support back to itself.

But now consider an arbitrary permutation $\alpha$. We know by Theorem 18.3 that it can be factored as a product $k$ disjoint cycles

$$\alpha = \beta_1\beta_2 \cdots \beta_k,$$

which necessarily commute (Theorem 18.2). Therefore, we can easily compute the $m$th power of $\alpha$:

$$\alpha^m = \beta_1^m \beta_2^m \cdots \beta_k^m.$$

To make such a product the identity, we must guarantee that simultaneously all the terms $(\beta_i)^m$ are the identity (for otherwise, the element $\alpha^m$ would not be the identity on elements in the orbit corresponding to $\beta_i$). To accomplish this, we need $m$ to be a common multiple of the orders of all the $\beta_i$'s. The least common multiple will do the job. In fact, the least common multiple of the orders of all the $\beta_i$'s is the order of $\alpha$. (See Exercise 5.)

**Example 21.7**

What is the order of the element (1234)(567)(89)? The least common multiple of 4, 3, and 2 is 12, so this must be the order of this permutation.

## 21.2 Rule of Exponents

We will now define what we mean by **negative exponents** on an element in a multiplicative group. If $n$ is a positive integer, we define $g^{-n}$ to mean $(g^{-1})^n$. And by $g^0$, we mean the identity 1. The reason we make these definitions is exactly so that the ordinary rule of exponents works:

$$(g^r)(g^s) = g^{r+s},$$

for all integers positive, negative, or zero. In ordinary arithmetic, preserving the rule of exponents is exactly the reason why such computations as $3^{-2} = 1/9$ and $3^0 = 1$ are defined in the way they are.

**Proof of Rule of Exponents**:   If $r$ and $s$ are both positive, this is clear, by the definition of positive exponents. If $r$ and $s$ are both negative, then

$$g^r g^s = (g^{-1})^{-r}(g^{-1})^{-s},$$

where the exponents $-r$ and $-s$ are both positive. This then reduces to the previous case. What if one exponent is positive and one is negative? Suppose that $r, s > 0$; then

$$g^r g^{-s} = g^r (g^{-1})^s.$$

We then cancel out terms until we run out of either $g$'s or $g^{-1}$'s. In either case, we obtain $g^{r-s}$, as we require.

▷ **Quick Exercise.**   There remains the case when $r$ or $s$ (or both) is 0; handle this. ◁

□

Suppose that $g$ has finite order $n$; then

$$1 = g^n = g^{n-1}g.$$

Because the inverse of $g$ is unique, $g^{-1} = g^{n-1}$. Consequently, any negative power of $g$ can be expressed as a positive power, if $g$ is a torsion element.

**Example 21.8**

Consider the element $[3]$ in

$$U(\mathbb{Z}_{14}) = \{1, 3, 5, 9, 11, 13\}.$$

The order of this element is 6.

▷ **Quick Exercise.**   Check that the order of $[3]$ is 6. ◁

Thus, $[3]^{-1} = [3]^5$. We can check this directly:

$$[3]^5 = [243] = [14 \cdot 17 + 5] = [5],$$

and $[3][5] = [15] = [1]$.
Let's express $[3]^{-4}$ as a positive power of $[3]$:

$$[3]^{-4} = ([3]^{-1})^4 = ([3]^5)^4 = [3]^{20} = ([3]^6)^3[3]^2 = [1][3]^2 = [3]^2.$$

On the other hand, if an element $g$ has infinite order, its inverse cannot be expressed as a positive power of $g$. Indeed, all the elements

$$\cdots, \; g^{-2}, \; g^{-1}, \; g^0 = 1, \; g^1 = g, \; g^2, \; g^3, \; \cdots$$

are distinct. The proof of this is a slight variation of the proof regarding the torsion-free case in Theorem 21.1; we leave it as Exercise 4.

▷ **Quick Exercise.**   What is the complete list of all powers (positive, negative, or zero) of the element 2 in $\mathbb{Q}^*$? List all powers for $-4 < n < 4$ of the element $1 + \sqrt{2}$ in $U(\mathbb{Z}[\sqrt{2}])$. ◁

In case an element $g$ has finite order, there is a simple relationship between $o(g)$ and *any* integer $m$ for which $g^m = 1$. We use the Division Theorem 2.1 for $\mathbb{Z}$ to see this:

**Theorem 21.2** *Suppose that $g$ is an element of a group, with order $n$. Suppose also that $g^m = 1$, where $m$ is some positive integer. Then $n$ divides $m$.*

**Proof:** Let $g$ be a group element, and $n = o(g)$. Suppose also that $g^m = 1$. Now $n$ is the least positive integer so that $g^n = 1$. Thus, $n \leq m$. By the Division Theorem for the integers (Theorem 2.1), we can write $m = qn + r$, where $q$ is a positive integer, and $r$ is an integer with $0 \leq r < n$. But then

$$1 = g^m = g^{qn+r} = (g^n)^q g^r = g^r.$$

But because $n$ is the least positive integer for which $g^n = 1$, we must have that $r = 0$. In other words, $n$ divides $m$. □

In our discussion in this chapter we have up to now restricted ourselves to groups written multiplicatively. Of course, all our definitions and results apply to groups written additively as well, but the notation looks quite different.

To begin with, consider the positive powers of an element $g$ of a group $G$, which is written additively. Our previous experience with rings supplies us with the appropriate notation:

$$g, \; g + g = 2g, \; g + g + g = 3g, \; \cdots, \; g + g + \cdots g = ng, \; \cdots.$$

If the order of $g$ is infinite, then these elements are all distinct.

## Example 21.9

The element 2 in the additive group $\mathbb{Z}$ has infinite order, because

$$2, \quad 4, \quad 6, \quad 8, \cdots$$

are all distinct.

▷ **Quick Exercise.** What is the order of $-1$ in the additive group $\mathbb{Z}[\sqrt{2}]$? Notice how different this question is than our earlier one about the order of $-1$ in the multiplicative group $U(\mathbb{Z}[\sqrt{2}])$. ◁

If the order of $g$ is finite, adding $g$ to itself finitely many times yields the identity, which in this case we denote by 0.

## Example 21.10

The element 2 in the additive group $\mathbb{Z}_6$ has order 3, because the least $n$ for which $n2 = 0$ is 3: $2 + 2 + 2 = 0$.

▷ **Quick Exercise.** Determine the orders of all elements in $\mathbb{Z}_6$. ◁

Note that if $g$ has finite order in an additive group (say, $o(g) = n$), then $-g = (n - 1)g$. (This, of course, is just the observation we made in multiplicative groups that $g^{-1} = g^{n-1}$.) That is, $-g$ is an integer multiple of $g$. For example, the order of $[2]$ in $\mathbb{Z}_{24}$ is 12. Then $[(12 - 1)2] = [22] = -[2]$, the inverse of $[2]$.

▷ **Quick Exercise.** What is the order of $3 + 2x$ in the additive group $\mathbb{Z}_6[x]$? Express its inverse as an integer multiple of itself, and check directly that this works. ◁

## 21.3   Cyclic Subgroups

Suppose now that $g$ is an element of order $n$ in the group $G$ (written multiplicatively). Now consider the subset

$$\{1, g, g^2, \cdots, g^{n-1}\}$$

of $G$. Because all the negative powers of $g$ can be expressed as a positive power, it is quite clear that this is a subgroup. Furthermore, it is evidently *the smallest subgroup of* G *that contains* g, because by closure of the operation all powers of $g$ belong to any group containing $g$. We call this subgroup the **cyclic subgroup generated by** $g$, and denote it by $\langle g \rangle$.

**Example 21.11**

> The cyclic subgroup generated by $(123)$ in $S_3$ is $\langle (123) \rangle = \{\iota, (123), (132)\}$ and the cyclic subgroup generated by $(12)$ in this group is $\langle (12) \rangle = \{\iota, (12)\}$.

**Example 21.12**

> The cyclic subgroup of $\mathbb{Z}_7^*$ generated by 2 is $\langle 2 \rangle = \{1, 2, 4\}$ and that generated by 3 is the entire group.

**Example 21.13**

> The cyclic subgroup of $\mathbb{C}^*$ generated by $i$ is
>
> $$\langle i \rangle = \{1, i, -1, -i\}.$$

**Example 21.14**

> The cyclic subgroup of $\mathbb{Z}_{12}$ generated by 4 is $\langle 4 \rangle = \{0, 4, 8\}$ and that generated by 0 is the trivial subgroup $\{0\}$.

What if $g$ is an element of infinite order in $G$? Then the smallest subgroup of $G$ containing $g$ must at least include all integer powers of $g$. In fact, in this case we set

$$\langle g \rangle = \{\cdots, g^{-2}, g^{-1}, 1, g, g^2, \cdots\}.$$

▷ **Quick Exercise.**   Show that $\langle g \rangle$ as just defined is indeed a subgroup of $G$. ◁

**Example 21.15**

> The cyclic subgroup generated by 2 in the group $\mathbb{Z}$ is the subgroup $\langle 2 \rangle = 2\mathbb{Z}$.

## Example 21.16

The cyclic subgroup generated by $\pi$ in the group $\mathbb{R}^*$ is the infinite set

$$\langle \pi \rangle = \left\{ \cdots, \frac{1}{\pi^2}, \frac{1}{\pi}, 1, \pi, \pi^2, \cdots \right\}.$$

We can subsume these two definitions of cyclic subgroup under a single expression: For any element $g$ in the group $G$,

$$\langle g \rangle = \{ g^m : m \in \mathbb{Z} \}.$$

Of course, if $g$ is torsion-free, we obtain distinct elements for each choice of integer $m$. If $g$ has finite order, each element of the cyclic subgroup is obtained by infinitely many choices of $m$.

Notice that we have used the same notation for cyclic subgroup that we have earlier used for principal ideal. This should cause no confusion, as long as you are certain whether we are working with groups or rings. This notational coincidence underlies the similarity of the concepts: The cyclic subgroup generated by an element is the smallest subgroup containing the element, while the principal ideal for an element is the smallest ideal containing the element.

## 21.4 Cyclic Groups

We say that a group $G$ is itself **cyclic** if there exists some element $g \in G$ so that $\langle g \rangle = G$; that is, the cyclic subgroup of $G$ generated by $g$ is the entire group. A cyclic group with infinitely many elements is necessarily generated by an element of infinite order. A cyclic group with finitely many elements (say, $n$) is necessarily generated by an element with order $n$.

## Example 21.17

The integers $\mathbb{Z}$ is a cyclic group, because $\mathbb{Z} = \langle 1 \rangle$.

## Example 21.18

The group $\mathbb{Z}_m$ is cyclic, for any integer $m > 1$, because $\mathbb{Z}_m = \langle 1 \rangle$.

## Example 21.19

The group $\mathbb{Z}_{13}^*$ is cyclic, with generator 2.

Note that in a cyclic group not all non-identity elements serve as generators.

▷ **Quick Exercise.** Give non-identity elements in the groups $\mathbb{Z}$ and in $\mathbb{Z}_{13}^*$ that do *not* generate the entire group. Can you do this in $\mathbb{Z}_5^*$? ◁

## Example 21.20

The group $S_3$ is not cyclic because it has 6 elements, but it contains no element of order 6.

**Example 21.21**

> The group $U(\mathbb{Z}_8) = \{1, 3, 5, 7\}$ is not cyclic either; all non-identity elements have order
> 2. That is, there is no element of order 4.

Except for notational differences, it may by now seem that any cyclic group of a given
order is essentially like any other of the same order and this is in fact the case, as we show
in the next theorem.

**Theorem 21.3** *All infinite cyclic groups are isomorphic to the additive group $\mathbb{Z}$. All cyclic
groups of order $n$ (where $n$ is an integer greater than 1) are isomorphic to the additive group
$\mathbb{Z}_n$.*

**Proof**: Suppose that $G$ is an infinite cyclic group (which we may as well write multiplicatively). Then $G$ has a generator $g$. That is,

$$G = \{g^m : m \in \mathbb{Z}\},$$

where if $r \neq s$, then $g^r \neq g^s$. Define the function

$$\varphi : G \to \mathbb{Z}$$

by $\varphi(g^m) = m$. (This looks like a symbolic version of the logarithm function because it
picks off exponents!) The rule of exponents says that this function takes the operation in $G$
to addition in $\mathbb{Z}$. It is clearly a bijection.

▷ **Quick Exercise.**   Why is $\varphi$ one-to-one and onto? ◁

Suppose now that $G$ is a cyclic group of order $n$. Then we know that $G$ has a generator
$g$, so that
$$G = \{1, g, g^2, \cdots, g^{n-1}\},$$
and $g^n = 1$. Define the function
$$\varphi : G \to \mathbb{Z}_n$$
by $\varphi(g^m) = [m]$. Let's show that this preserves the group operation. So choose $g^r, g^s \in G$;
here, $0 \leq r, s \leq n - 1$. Then
$$\varphi(g^r g^s) = \varphi(g^{r+s}).$$
If $r + s \leq n - 1$, then the value of the function is
$$[r + s] = [r] + [s] = \varphi(g^r) + \varphi(g^s),$$
as required. Otherwise, $n \leq r + s \leq 2n - 2$, and so
$$\varphi(g^{r+s}) =$$
$$\varphi(g^n g^{r+s-n}) = \varphi(1 \cdot g^{r+s-n}) = \varphi(g^{r+s-n}) =$$
$$[r + s - n] = [r + s] = [r] + [s],$$
again as required.                                                                               □

This last theorem is a good example of a recurrent theme in algebra: We have characterized the concrete example $\mathbb{Z}$ as the apparently abstract construct 'infinite cyclic group'.
This provides us real insight into the essence of the group structure of $\mathbb{Z}$.

We now have a careful explication of Example 20.12: All groups with 2 elements are
cyclic, and isomorphic to the additive group $\mathbb{Z}_2$.

Let's now return to our examples of cyclic groups from earlier in this chapter.

For example, the cyclic subgroup of the multiplicative group $\mathbb{Z}_7^*$ generated by 2 considered in Example 21.12 has three elements. Likewise, so does the cyclic subgroup of the additive group $\mathbb{Z}_{12}$ generated by 4, considered in Example 21.14. Consequently, both these groups are isomorphic to $\mathbb{Z}_3$, and hence to each other.

$\triangleright$ **Quick Exercise.** Write down explicit isomorphisms between each pair of these three groups. $\triangleleft$

The cyclic subgroup of $\mathbb{Z}$ generated by 2, which we considered in Example 21.15, is infinite. Likewise, so is the cyclic subgroup of $\mathbb{R}^*$ generated by $\pi$ which we considered in Example 21.16. Therefore, both of these groups are isomorphic to $\mathbb{Z}$.

$\triangleright$ **Quick Exercise.** Write down explicit isomorphisms between each pair of these three groups. $\triangleleft$

As a consequence of this isomorphism theorem, when we wish to discuss a cyclic group in the abstract and wish to use multiplicative notation, we will tend to denote it by

$$\langle a \rangle = \{\cdots, a^{-2}, a^{-1}, 1, a, a^2, \cdots\}$$

(if it is infinite), or

$$\langle a \rangle = \{1, a, a^2, \cdots, a^{n-1}\}$$

(if it is finite). We won't worry concretely about what $a$ is, because *any* such cyclic group is essentially the same.

---

## Chapter Summary

In this chapter we analyzed the important notions of the *order* of an element of a group and saw that an element of a group can have either infinite or finite order. We also discussed the *cyclic subgroup* generated by an element. We showed that all infinite cyclic groups are isomorphic to the additive group $\mathbb{Z}$, and all finite cyclic groups are isomorphic to the additive group $\mathbb{Z}_n$, for some positive integer $n$.

---

## Warm-up Exercises

a. What are the orders of the following elements?

   (a) $5 \in \mathbb{Z}_{15}$.

   (b) $7 \in U(\mathbb{Z}_{15})$.

   (c) $\varphi\rho \in D_3$.

   (d) $6 \in \mathbb{Z}$.

   (e) $6 \in \mathbb{Q}^*$.

   (f) $i \in \mathbb{S}$.

   (g) $i \in \mathbb{C}$.

   (h) $\begin{pmatrix} \frac{1}{\sqrt{2}} & \frac{1}{\sqrt{2}} \\ -\frac{1}{\sqrt{2}} & \frac{1}{\sqrt{2}} \end{pmatrix} \in U(M_2(\mathbb{R}))$.

b. Are the following groups cyclic? Either explain why not or else specify a generator.

   (a) $\mathbb{Z}$.

   (b) $\mathbb{Q}$.

   (c) The group of symmetries of the cube.

   (d) $\mathbb{Z}_8$.

   (e) $U(\mathbb{Z}_8)$.

   (f) $M_2(\mathbb{Z})$.

   (g) $\mathbb{Z}[i]$.

   (h) $\{1, -1, i, -i\} \subseteq \mathbb{C}^*$.

   (i) $\mathbb{Z} \times \mathbb{Z}$.

   (j) $\mathbb{Z}_2 \times \mathbb{Z}_4$.

   (k) $\mathbb{Z}_2 \times \mathbb{Z}_3$.      (Be careful.)

c. Can a group possess (non-identity) elements of both infinite and finite order? Explain, or give an example.

d. Are all cyclic groups abelian?

e. Why do all non-abelian groups have (non-trivial) abelian subgroups?

f. How many different generators do the following cyclic groups possess? List all such generators.

   (a) $\mathbb{Z}$.

   (b) $\mathbb{Z}_{10}$.

   (c) $\mathbb{Z}_7$.

   (d) $U(\mathbb{Z}_9)$.

g. Explain why the order of $g^{-1}$ is the same as $g$.

h. Suppose that $g^{-14} = 1$. What can you say about the order of $g$?

i. Suppose that $g^7 = g^{15}$. What can you say about the order of $g$?

j. Explain why none of the following groups are isomorphic (even though all have eight elements):
$$\mathbb{Z}_8, \ \mathbb{Z}_4 \times \mathbb{Z}_2, \ D_4.$$

k. Explain why $\mathbb{Z}[i]$ and $\mathbb{Z}$ are not isomorphic as additive groups. That is, why isn't $\mathbb{Z}[i]$ cyclic?

l. Can a group be isomorphic to a proper subgroup of itself? What if the group is finite?

m. Let $R$ be a finite ring, and consider its additive group and its group of units. Could these two groups be isomorphic?

## Exercises

1. Determine the cyclic subgroups of $U(M_2(\mathbb{Z}))$ generated by

$$\begin{pmatrix} 1 & 1 \\ 0 & 1 \end{pmatrix} \text{ and } \begin{pmatrix} 0 & 1 \\ 1 & 0 \end{pmatrix}.$$

2. Prove that every subgroup of a cyclic group is cyclic.

3. Find an example to show that the converse of Exercise 2 is false: That is, give a non-cyclic group, each of whose proper subgroups is cyclic.

4. Suppose that $g$ is an element of infinite order in a group $G$. Prove that no two distinct powers of $g$ (with any integer exponent) are equal.

5. Prove that the order of a permutation is the least common multiple of the orders of the cycles that make up its disjoint cycle representation.

6. If $a$ and $b$ are elements of a group that commute and $\langle a \rangle \cap \langle b \rangle = \{1\}$, what is the order of $ab$ if the order of $a$ is $m$ and the order of $b$ is $n$? Prove your assertion. (This is really a generalization of the previous exercise.) Show by example that your assertion is false in general, in the case that $a$ and $b$ do not commute.

7. If $a$ and $b$ are elements of a group whose orders are relatively prime, what can you say about $\langle a \rangle \cap \langle b \rangle$? Prove your assertion.

8. How many generators does an infinite cyclic group have?

9. Prove that if $G$ is a finite cyclic group with more than two elements, then $G$ has more than one generator.

10. (a) Show that in a cyclic group, the equation $x^2 = 1$ has no more than two solutions. (Of course, the identity is always one of the solutions.)

    (b) Give an example of a non-cyclic group where $x^2 = 1$ has more than two solutions.

11. Consider the set of complex numbers $\{1, -1, i, -i\}$. Note this is a group under multiplication. Show that this group is isomorphic to $\mathbb{Z}_4$.

12. Consider the group of eight matrices we consider in Example 19.16, which we call $Q_8$, the group of quaternions. In this problem we will introduce the more usual notation for this group.

    (a) What are the orders of the elements of this group?

    (b) Argue that the group in the previous exercise is isomorphic to the cyclic subgroup

$$\left\langle \begin{pmatrix} 0 & 1 \\ -1 & 0 \end{pmatrix} \right\rangle$$

    of the group of quaternions.

    (c) Repeat this argument for the cyclic subgroups

$$\left\langle \begin{pmatrix} 0 & i \\ i & 0 \end{pmatrix} \right\rangle \text{ and } \left\langle \begin{pmatrix} i & 0 \\ 0 & -i \end{pmatrix} \right\rangle.$$

(d) This all suggests that we should label the three matrices above with the symbols $i, j, k$, and thus write $Q_8$ efficiently as $\{1, -1, i, -i, j, -j, k, -k\}$, where we can identify the first four elements with the usual complex numbers, and consider the next four as other elements that behave similarly. Write down the complete correspondence of matrices to these eight symbols.

(e) Now verify that the elements in this notation satisfy the following rules of computation, by checking with the corresponding matrix calculations. First the element 1 is the identity, and multiplication by $-1$ changes signs as the notation suggests.

(f) Check that $i^2 = j^2 = k^2 = -1$.

(g) Check that $ij = k, jk = i, ki = j$.

(h) Check that $ji = -k, kj = -i, ik = -j$.

13. (a) Show that if $G$ is a cyclic group of order $m$ and $n$ divides $m$, then $G$ has a subgroup of order $n$. (This subgroup will itself be cyclic.)

(b) If $G$ is an arbitrary group with order $m$ and $n$ divides $m$, then $G$ need not have a cyclic subgroup of order $n$. Find two such examples, one where $n = m$, and one where $n < m$.

14. Generalizing the group given in Exercise 19.7, let $\rho_{i,n}$ be the rotation of a circle counterclockwise through an angle of $2\pi i/n$, for $i = 1, 2, \ldots, n - 1$ and all $n \geq 2$. Let $e$ be the 'identity' rotation. Show that this set of rotations forms a group under composition. Show that all elements in this group are torsion (even though the group itself is infinite). Show that each finite subgroup is cyclic.

15. Show that all finite subgroups of the group $\mathbb{S}$, given in Example 19.14, are cyclic.

16. If $G$ is a finite group where every non-identity element is a generator of $G$, what can you say about the order of $G$? Prove your assertion.

17. If $\alpha \in S_n$, is $\alpha^2$ a cycle also? Give an example where this is true and an example where this is false. Now characterize cycles where this is true and where this is false.

18. Determine all possible orders for a product of two 3-cycles in $S_n$ (where $n \geq 6$).

19. Repeat the previous exercise for two 4-cycles.

20. (a) Show that if $G$ is a group and $H_1, H_2$ are proper subgroups, then it is impossible that $G = H_1 \cup H_2$.

(b) Find an example of a group $G$ and three proper subgroups $H_1, H_2, H_3$ where it is true that $G = H_1 \cup H_2 \cup H_3$.

(c) Show that it is impossible for $\mathbb{Z}$ to be the union of finitely many proper subgroups.

21. Suppose $a$ and $b$ are non-identity elements of a multiplicative group $G$, and that $ab = ba$ and $b^2 = 1$. Show that $\{a^n, ba^n : n \in \mathbb{Z}\}$ is a subgroup of $G$.

22. We generalize the previous exercise: Suppose that $a$ and $b$ are non-identity elements of a group $G$, and that $ab = ba$ and $b^3 = 1$. Show that $\{a^n, ba^n, b^2a^n : n \in \mathbb{Z}\}$ is a subgroup.

23. Generalize the situation in the previous two exercises, replacing 2 and 3 by some positive integer $m$.

# Section IV in a Nutshell

This section defines the abstract notion of group, after examining some important examples: symmetries of regular $n$-sided polygons (called the *nth dihedral groups*, $D_n$), symmetries of the regular tetrahedron in 3-dimensional space, and most importantly, the permutations of a set of $n$ objects, which we call the *symmetric group on n*, $S_n$. (We mostly confine ourselves to permutations of a finite set, but the collection of permutations of infinite sets are also groups.) The dihedral groups and the symmetries in space can all be thought of as subgroups of the permutations of the vertices of the corresponding geometric objects. The dihedral group $D_n$ has $2n$ elements, and the symmetric group $S_n$ has $n!$ elements.

Elements of $S_n$ can be uniquely factored into disjoint cycles (Theorem 18.3). Disjoint cycles will always commute (Theorem 18.2).

A *group* $G$ is a set of elements with one binary operation ($\circ$) that satisfies three rules:

1. $(g \circ h) \circ k = g \circ (h \circ k)$, for all $g, h, k \in G$,

2. There exists an element $e \in G$ (called the *identity* of $G$) such that $g \circ e = e \circ g = g$ for all $g \in G$, and

3. For each $g \in G$, there exists an element $x$ (called the *inverse* of $g$) such that $g \circ x = x \circ g = e$.

In addition to the groups mentioned above, other important examples of a group are the additive group of a ring (that is, the elements of a ring with only the addition considered) and the group of units of a ring with unity under multiplication. An *abelian* group is one in which the binary operation $\circ$ is commutative.

Groups enjoy cancellation on both the right and the left, and the solution of equations. Furthermore, the group identity is unique as is the inverse of each element. (All this is Theorem 20.1.)

Paralleling the idea of subrings is the idea of subgroup: If $G$ is a group then a subset $H$ of $G$ is a *subgroup* if it is itself a group under the operation induced from $G$. To determine if $H$ is a subgroup of a group $G$, we need only check that $hk^{-1} \in H$ for every $h, k \in H$ (Theorem 20.2).

If $G$ is a group with operation $\circ$ and $H$ is a group with operation $*$, a map $\varphi : G \to H$ is a *group isomorphism* if $\varphi$ is a bijection and $\varphi(g \circ h) = \varphi(g) * \varphi(h)$ for all $g, h \in G$. This parallels the idea of a ring isomorphism.

An important class of subgroup of any group $G$ is the *cyclic subgroup generated by $g$*, which we denote by $\langle g \rangle$. It consists of the powers of $g$. The subgroup $\langle g \rangle$ may be infinite or finite, depending on whether the order of $g$ is infinite or finite. If $G$ itself is generated by the powers of a single element, we say that $G$ is a *cyclic group*. All infinite cyclic groups are isomorphic to $\mathbb{Z}$; all finite cyclic groups with $n$ elements are isomorphic to $\mathbb{Z}_n$ (Theorem 21.3).

# Part V

# Group Homomorphisms

# Chapter 22

## Group Homomorphisms

We are now ready to make the appropriate definition of a *group homomorphism*, generalizing the idea of group isomorphism from Section 20.5. This definition is exactly what you should expect, given our earlier experience with *ring homomorphisms* in Chapter 11.

## 22.1  Homomorphisms

Following our definition of isomorphism, we will be careful to consider two groups whose operations are denoted by explicit symbols. So, let $G$ together with operation $\circ$, and $H$ together with operation $*$, be groups. A function $\varphi : G \to H$ such that

$$\varphi(g \circ k) = \varphi(g) * \varphi(k),$$

for all $g, k \in G$ is a **group homomorphism**. Speaking more colloquially, a group homomorphism is a function between groups that preserves the group operation. We are requiring neither that this function be injective nor surjective.

## 22.2  Examples

It is time to look at some examples of group homomorphisms. First, the group isomorphisms we discussed in Chapter 20 serve as examples.

### Example 22.1

Examine Examples 20.10, 20.11, and 20.12. These are functions between groups that preserve the group operation.

We can also obtain many examples from our work with rings; we will begin with a general class of examples.

Namely, let $R$ and $S$ be rings, and $\varphi : R \to S$ a *ring homomorphism*. Now just look at the additive groups of $R$ and $S$ (so the operations on $R$ and $S$ are both just $+$). This function preserves addition, and so is a group homomorphism too. We mention specifically one important example of this type.

### Example 22.2

The residue function $\varphi : \mathbb{Z} \to \mathbb{Z}_m$ is a ring homomorphism, and therefore a group homomorphism too.

▷ **Quick Exercise.** Now look at Examples 11.2, 11.3, 11.4, and 11.5. Because they are ring homomorphisms, they are certainly group homomorphisms too. ◁

## Example 22.3

Now consider Example 11.6. It is the function $\rho : \mathbb{Z} \to \mathbb{Z}$ defined by $\rho(n) = 3n$. Here, we are considering $\mathbb{Z}$ as an additive group. This function preserves addition:

$$\rho(n + m) = 3(n + m) = 3n + 3m = \rho(n) + \rho(m).$$

Thus, $\rho$ is a group homomorphism. In Example 11.6 we saw that this is not a ring homomorphism: It preserves addition but *not* multiplication. This group homomorphism is certainly injective, but it is not surjective: For the first assertion, if $3n = 3m$, then $n = m$. For the second, note that $\rho(n) \neq 4$, for all $n \in \mathbb{Z}$.

It's time to consider some examples further afield from rings:

## Example 22.4

Consider the groups $\mathbb{R}$ under addition, and $\mathbb{R}^+$ of positive real numbers, under multiplication. Recall the function $\log : \mathbb{R}^+ \to \mathbb{R}$, the *natural logarithm* function. (That is, $\log(r)$ is the exponent needed on the irrational number $e$ so that $e^{\log(r)} = r$.)

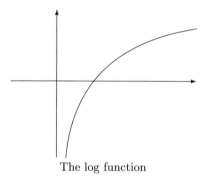

The log function

Recall first of all that this function is only defined for positive real numbers. The most important and useful property of the logarithm function is this:

$$\log(ab) = \log(a) + \log(b).$$

That is, the logarithm turns multiplication into addition. And this equation is exactly what is required to assert that log is a homomorphism! This example is well worth thinking about. It shows us that the group operations in two groups connected by a homomorphism can be quite different. Notice that this function is actually a bijection.

▷ **Quick Exercise.** Use what you know from calculus about $\log(x)$ and $e^x$ to check this. ◁

Thus the multiplicative group of positive real numbers is isomorphic to the additive group of real numbers! Now we know that a group isomorphism has an inverse function that also preserves the group operation (see Exercise 20.23). In this case the inverse function is the group isomorphism $\exp : \mathbb{R} \to \mathbb{R}^+$.

## Example 22.5

Consider another famous function, this time between the groups $U(M_2(\mathbb{R}))$, the group of units of the $2 \times 2$ real-valued matrices, and $\mathbb{R}^*$, the multiplicative group of non-zero reals. Recall that $U(M_2(\mathbb{R}))$ consists precisely of those matrices in $M_2(\mathbb{R})$ with non-zero determinant. (See Exercise 8.2.) Here, the operation in the first group is matrix multiplication, while in the second it is ordinary real number multiplication. The function is the *determinant* function det:

$$\det \begin{pmatrix} a & b \\ c & d \end{pmatrix} = ad - bc.$$

Let's show that this is a homomorphism. For that purpose, we need to choose two arbitrary matrices,

$$\begin{pmatrix} a & b \\ c & d \end{pmatrix} \quad \text{and} \quad \begin{pmatrix} r & s \\ t & u \end{pmatrix},$$

where the entries are all real numbers. The product of these two matrices is

$$\begin{pmatrix} ar + bt & as + bu \\ cr + dt & cs + du \end{pmatrix},$$

and the determinant of this matrix is

$$ardu + btcs - bucr - asdt.$$

▷ **Quick Exercise.** Check these computations. ◁

But the product of the determinants is

$$(ad - bc)(ru - ts),$$

which is the same. Thus, the determinant function preserves multiplication. To paraphrase, the determinant of a product is the product of the determinants. It is easy to see that this homomorphism is not injective (just find two matrices with the same determinant), but it is surjective (just find a matrix with determinant $r$, for any non-zero real number $r$).

▷ **Quick Exercise.** Carry out the details to verify the assertions in the last sentence. ◁

## Example 22.6

Consider the group

$$D_3 = \{\iota, \rho, \rho^2, \varphi, \rho\varphi, \varphi\rho\}$$

of symmetries of the equilateral triangle, whose operation is functional composition. Consider also the multiplicative subgroup $\{1, -1\}$ of the integers. Define the function

$$\Phi : D_3 \to \{1, -1\}$$

given as follows: $\Phi(\alpha) = 1$ if $\alpha$ is one of the rotations $\iota, \rho, \rho^2$; $\Phi(\alpha) = -1$ if $\alpha$ is one of the flips $\varphi, \varphi\rho, \rho\varphi$.

To see that this is a homomorphism, it is best to return to the group table we compiled for the symmetry group in Section 17.1. The pattern of $F$ and $R$ we observed there shows us that a rotation times a rotation is a rotation, a flip times a flip is a rotation, and a rotation times a flip (in either order) is a flip. Now replace 'rotation' by 1 and 'flip' by $-1$ in the previous sentence. This is just the way multiplication in the group $\{1, -1\}$ works! This example is obviously surjective but not injective.

**Example 22.7**

Consider the group $G$ of symmetries of the tetrahedron, discussed in Section 17.6. Consider also the additive group $\mathbb{Z}_3$. Define the function

$$\Phi : G \to \mathbb{Z}_3$$

given as follows: $\Phi(\alpha) = 0$ if $\alpha$ is one of the four symmetries $\iota, \varphi_1, \varphi_2$, or $\varphi_3$; $\Phi(\alpha) = 1$ if $\alpha$ is one of the four symmetries $\rho_1, \rho_2, \rho_3, \rho_4$; and $\Phi(\alpha) = 2$ if $\alpha$ is one of the four symmetries $\rho_1^2, \rho_2^2, \rho_3^2, \rho_4^2$.

To see that this is a homomorphism, it is best to return to the group table we compiled for the symmetry group in Section 17.6. The pattern we observed of $F, R$, and $R^2$ elements behaves in exactly the same way as addition in $\mathbb{Z}_3$!

▷ **Quick Exercise.** Check that the pattern observed in the group table in Section 17.6 is the same as addition in $\mathbb{Z}_3$. ◁

Thus, we have a group homomorphism from a group where the operation is functional composition, to a group where the operation is addition (modulo 3). This example is again surjective but not injective.

**Example 22.8**

Consider the function $D : \mathbb{R}[x] \to \mathbb{R}[x]$ defined by $D(f) = f'$ (the derivative of the polynomial $f$). Of course, we know that the derivative of a polynomial is a polynomial. But we also know that the derivative of a sum is the sum of the derivatives. That is,

$$D(f + g) = D(f) + D(g).$$

This is exactly what is required to show that this function is a group homomorphism. Notice, however, that this function *does not* preserve multiplication (the product rule is not that simple), and so this function is *not* a ring homomorphism. (For more about the derivative function, see Exercise 4.7.) From what you know about derivatives from calculus it should be easy to infer that this is also surjective but not injective.

A group homomorphism certainly need not be surjective (see Example 22.3 above). However, if $\varphi : G \to H$ is not onto, the new function (which we still call $\varphi$) obtained by restricting the range to $\varphi(G)$ certainly is. Thus, $\varphi : G \to \varphi(G)$ is a surjective homomorphism, assuming that $\varphi(G)$ is in fact a subgroup of $H$. This is the case:

**Theorem 22.1** *Let $\varphi : G \to H$ be a homomorphism between groups $G$ and $H$. Then $\varphi(G)$ is a subgroup of $H$.*

**Proof:**   This is Exercise 6. Model your proof on the corresponding ring theory theorem (Theorem 11.2). □

## 22.3    Structure Preserved by Homomorphisms

We now record some of these further algebraic properties preserved by an arbitrary group homomorphism:

**Theorem 22.2** *Let $\varphi : G \to H$ be a homomorphism between the groups $G$ and $H$, and let $1_G$ and $1_H$ be the identities of $G$ and $H$, respectively.*

a. $\varphi(1_G) = 1_H$.

b. $\varphi(g^{-1}) = (\varphi(g))^{-1}$, *for any $g \in G$.*

c. *Suppose that $g \in G$, and $g$ has finite order. Then the order of $\varphi(g)$ divides the order of $g$. If $\varphi$ is an isomorphism, then $o(g) = o(\varphi(g))$.*

d. *Suppose that $G$ is abelian. Then $\varphi(G)$ is abelian.*

**Proof**: Parts (a) and (b) are proved just as in the ring case (Theorem 11.1, parts (a) and (b)); you have essentially proved these in Exercises 20.21 and 20.22. Your proofs there did not rely on the function being bijective.

(c): Let $g \in G$, and suppose that $o(g) = n$. Then

$$1 = \varphi(g^n) = (\varphi(g))^n.$$

But then by Theorem 21.2, $o(\varphi(g))$ divides $n = o(g)$, as claimed. If $\varphi$ is an isomorphism, it possesses an inverse function $\varphi^{-1}$ which is also an isomorphism. Applying the homomorphism result we just proved to $\varphi^{-1}$ tells us that $o(g)$ divides $o(\varphi(g))$ too, and because these are both positive integers, $o(g) = o(\varphi(g))$.

(d): This is Exercise 7. $\qquad\square$

**Example 22.9**

Consider the groups $S_3$ and $\mathbb{Z}_6$; they both have the same number of elements. But they cannot be isomorphic, because the second group has an element of order 6, while the first doesn't. But regardless of the order of the elements in each group, the second group is abelian, while the first is not. So we have two reasons why these groups are not isomorphic.

**Example 22.10**

The groups $\mathbb{Z}_4$ and $\mathbb{Z}_2 \times \mathbb{Z}_2$ are both abelian groups with four elements. But they are not isomorphic, because the first group has an element of order 4, while the second does not.

## 22.4 Direct Products

Let's recall a construction we made in rings: Given two groups $G$ and $H$, consider the set

$$G \times H = \{(g, h) : g \in G, h \in H\}$$

of all ordered pairs, with first entry an element of $G$ and second entry an element of $H$. Equip this set with the component-wise operation

$$(g_1, h_1)(g_2, h_2) = (g_1 g_2, h_1 h_2).$$

This makes $G \times H$ a group, called the **direct product** of $G$ and $H$. This turns out to be a handy way to build new groups from old.

▷ **Quick Exercise.**  Verify that this is a group. ◁

▷ **Quick Exercise.**  Review some examples of direct products of rings. By forgetting the multiplication, they become examples of direct products of groups. ◁

We can now use the direct product to construct two important examples of homomorphisms:

### Example 22.11

Let $G$ and $H$ be groups. Consider the function

$$\epsilon : G \to G \times H$$

defined by $\epsilon(g) = (g, 1)$. This is certainly a homomorphism:

$$\epsilon(g_1 g_2) = (g_1 g_2, 1) =$$
$$(g_1, 1)(g_2, 1) = \epsilon(g_1)\epsilon(g_2).$$

Note that this function is injective, but certainly not surjective (as long as $H$ has more than one element).

▷ **Quick Exercise.**  Check these claims. ◁

Of course, we could define a similar homomorphism on the second component. These homomorphisms are called **embeddings**.

Note that

$$\epsilon(G) = \{(g, 1) : g \in G\} = G \times \{1\}.$$

But now we know that $\epsilon$ is onto the group $G \times \{1\}$; since we have already observed that $\epsilon$ is injective, this means that the groups $G$ and $G \times \{1\}$ are isomorphic. By applying the injection to the other coordinate, we now have seen that $G \times H$ has two subgroups, one isomorphic to $G$ and the other isomorphic to $H$. In the next chapter we will pursue these ideas further.

### Example 22.12

Let $G$ and $H$ be groups. Consider the function

$$\pi : G \times H \to G$$

defined by $\pi(g, h) = g$. This is called the **projection** onto the first coordinate.

▷ **Quick Exercise.**  Check that this is surjective but not injective (as long as $H$ has more than one element). ◁

---

## Chapter Summary

In this chapter we introduced the idea of *group homomorphism*, a function between two groups that preserves the group operation, and we looked at numerous examples. We took note of the many other algebraic properties preserved by any group homomorphism. We also introduced the direct product of two groups along with the closely connected homomorphisms *embeddings* and *projections*.

# Warm-up Exercises

a. If $\varphi : G \to H$ is a group homomorphism and $G$ is an additive group, need $H$ be an additive group?

b. Is a ring homomorphism necessarily a group homomorphism?

c. Suppose we have a group homomorphism between the additive groups of two rings. Is this necessarily a ring homomorphism too?

d. Check that the determinant function preserves multiplication for two interesting matrices of your choice.

e. Does the determinant function preserve addition? Try it, for two interesting matrices of your choice.

f. We proved that the determinant function preserves multiplication for any pair of $2 \times 2$ matrices that are units. Show that this proof works for *any* pair of $2 \times 2$ matrices.

g. Does the observation in Exercise f mean that the determinant function defined from $M_2(\mathbb{R})$ onto $\mathbb{R}$ is a *ring* homomorphism? *Hint:* What does Exercise e say?

h. How many polynomials are sent to the polynomial 2 by the derivative homomorphism? Can you describe them all efficiently?

i. Find an injective group homomorphism $\varphi$ from $\mathbb{Z}_2$ to $\mathbb{Z}_4$. What is the subgroup $\varphi(\mathbb{Z}_2)$?

j. Let $G$ and $H$ be groups with operations $\circ$ and $*$, respectively, and $\varphi : G \to H$ a group homomorphism. Explain why $\varphi(g \circ h \circ k) = \varphi(g) * \varphi(h) * \varphi(k)$.

k. Generalize Exercise j to a product of $n$ terms. Does this mean that $\varphi(g^n) = (\varphi(g))^n$?

# Exercises

1. Show that the function $\varphi : \mathbb{Z}[i] \to \mathbb{Z}$ defined by $\varphi(a+bi) = a$ is a group homomorphism (although it is not a ring homomorphism).

2. Can you find a group homomorphism from $\mathbb{Z}_2 \times \mathbb{Z}_2$ onto the multiplicative group $\{1, -1, i, -i\}$?

3. Find three distinct group homomorphisms from $\mathbb{Z}_2 \times \mathbb{Z}_2$ onto $\mathbb{Z}_2$. How many more homomorphisms exist, if we remove the requirement that they be onto?

4. Let $G$ be an abelian group (written additively). Define

$$\psi : G \times G \to G$$

by $\psi(g, h) = g + h$. Make a two-column chart showing what this function does to each element of $G = \mathbb{Z}_3$. Prove that $\psi$ is a homomorphism, for any abelian group $G$.

5. Show by example that the corresponding homomorphism $\psi$ defined in Exercise 4 need *not* be a homomorphism, when the group is not abelian.

6. Prove Theorem 22.1: That is, let $\varphi : G \to H$ be a homomorphism between the groups $G$ and $H$. Prove that $\varphi(G)$ is a group.

7. Prove Theorem 22.2d: That is, let $\varphi : G \to H$ be a group homomorphism, and suppose $G$ is abelian. Prove that the group $\varphi(G)$ is abelian. Show by example that if $\varphi$ is not surjective, then the group $H$ need not be abelian.

8. Suppose that $G$ is a group for which every element has order a power $p^n$ of a fixed prime $p$. (We call such a group a $p$-**group**; we will discover much more about them in Section 28.1). Let $\varphi : G \to H$ be a surjective homomorphism. Prove that $H$ is a $p$-group too. Find an example where this is false if the homomorphism is not surjective.

9. Suppose that $R$ and $S$ are rings with unity, and $\varphi : R \to S$ is an onto ring homomorphism. Consider the function $\psi : U(R) \to S$ defined by $\psi(r) = \varphi(r)$. (That is, $\psi$ is just $\varphi$ restricted to the group of units $U(R)$.) Prove that $\psi$ is a group homomorphism from $U(R)$ into $U(S)$.

10. We know that $\log : \mathbb{R}^+ \to \mathbb{R}$ is a group homomorphism; see Example 22.4. Find a subgroup $G$ of $\mathbb{R}^+$, so that the restricted logarithm function maps $G$ onto the subgroup $\mathbb{Z}$ of $\mathbb{R}$. What about a subgroup of $\mathbb{R}^+$ that maps onto the subgroup $\mathbb{Q}$ of $\mathbb{R}$?

11. Consider the set $G$ of all differentiable real-valued functions. Why is this a group under addition? Is the derivative function $D$ a group homomorphism from $G$ to $G$? *Warning:* That the function preserves the operation is *not* the issue.

12. Consider the homomorphism in Example 22.6. Define the analogous homomorphism from $D_4 \to \{1, -1\}$, making use of your work in Exercise 17.1.

13. In this problem we consider group homomorphisms from a finite cyclic group to itself.

    (a) Find all homomorphisms from $\mathbb{Z}_5$ into $\mathbb{Z}_5$. How many of these are surjections? Now find all homomorphisms from $\mathbb{Z}_6$ into $\mathbb{Z}_6$. How many of these are surjections?

    (b) Suppose $G$ is a cyclic group of order $n$. How many homomorphisms are there from $G$ into $G$? Describe them. Describe those that are surjections.

14. In this exercise we generalize the ideas encountered in Exercise 13. A homomorphism from a group into itself is called an **endomorphism** of the group. For an abelian group $G$, let $\text{End}(G)$ be the set of all its endomorphisms $\varphi : G \to G$.

    (a) Let $\varphi, \psi \in \text{End}(G)$. Define the sum $\varphi + \psi$ of these two homomorphisms by

    $$(\varphi + \psi)(g) = \varphi(g) + \psi(g).$$

    Prove that $\varphi + \psi$ is an endomorphism of $G$.

    (b) Prove that $\text{End}(G)$, under the addition defined in part (a), is an abelian group.

    (c) Let $\varphi, \psi \in \text{End}(G)$. Define the product $\varphi\psi$ as the ordinary functional composition:

    $$\varphi\psi(g) = \varphi(\psi(g)).$$

    Show that $\text{End}(G)$ is a ring with unity when equipped with this multiplication. We call $\text{End}(G)$ the **endomorphism ring** for $G$.

    (d) Let $n$ be a positive integer. *Beware:* In this problem we will consider $\mathbb{Z}_n$ as an abelian group and as a commutative ring. Prove that the ring $\text{End}(\mathbb{Z}_n)$ is isomorphic as a ring to $\mathbb{Z}_n$.

15. In this problem we consider a particularly important example of a group endomorphism. Suppose $G$ is a group and $g \in G$. Define the function $\varphi : G \to G$ by setting $\varphi(h) = ghg^{-1}$.

    (a) Prove that $\varphi$ is a group isomorphism. That is, show that it preserves the group operation, and is also a bijection. We call such a function a **conjugation**, and say that $ghg^{-1}$ is a **conjugate** of the element $h$.

    (b) What can you say about $\varphi$ if the group is abelian? What if $g \in Z(G)$, the center of $G$ (see Exercise 20.16)?

    (c) Show that the order of $h$ is equal to the order of $\varphi(h) = ghg^{-1}$.

16. Given a group $G$ and $g \in G$, denote by $\varphi_g$ the function defined by $\varphi_g(h) = ghg^{-1}$ in the previous exercise. Note that in Exercise 20.27 we call a group isomorphism from a given group onto itself a group automorphism; we will call a function of the form $\varphi_g$ an **inner automorphism**. Let $\text{Inn}(G) = \{\varphi_g : g \in G\}$. Prove that $\text{Inn}(G)$ is a subgroup of $\text{Aut}(G)$. We call this the **group of inner automorphisms** of $G$. (See Exercise 20.27 for the group $\text{Aut}(G)$ of automorphisms of $G$.)

# Chapter 23

## Structure and Representation

When mathematicians study a structure like groups, they often take two different approaches. One is to find a *representation* for the group. What this typically means is to define an (often) injective homomorphism from the group into a larger structure that provides us with a more concrete notation and terminology, and with better computational techniques. In Chapter 17 we discovered that groups of symmetries of geometric figures could be better understood by representing them either as subgroups of a symmetric group of permutations, or else as a subgroup of a larger group of matrices. In linear algebra we learn a lot about computing with matrices, and in Chapter 18 we learned a good bit about computing with permutations. In this chapter we will discover that *all* (finite) groups can be represented as a subgroup of a symmetric group of permutations. (It is also actually true that any finite group can be represented as a group of matrices, but we will avoid the details in this book, because such representations often require matrices of large size, and more about linear algebra than we wish to use.)

An alternative to a representation is to look at the *internal structure* of the group. Can we better understand how a group works by looking at its subgroups and how they are related? The Chinese Remainder Theorem is a spectacular example of how this approach works for rings (and hence abelian groups) of the form $\mathbb{Z}_n$. In this chapter we will examine how groups can sometimes be thought of as a direct product of smaller pieces (or subgroups). This chapter will provide only a small start on this approach, and we will explore more sophisticated approaches in Chapters 28, 29, and 30.

We will discuss direct products first.

## 23.1 Characterizing Direct Products

Let's suppose that a group $G$ is isomorphic to a direct product of the two groups $H_1$ and $H_2$. In this section we would like to understand how we can tell this is true, by looking inside the group $G$. Let's suppose that

$$\varphi : H_1 \times H_2 \to G$$

is the isomorphism. Consider the two subgroups of the direct product

$$H_1 \times \{1\} = \{(a, 1) : a \in H_1\}$$

and

$$\{1\} \times H_2 = \{(1, b) : b \in H_2\}.$$

These subgroups are obviously isomorphic to $H_1$ and $H_2$, respectively.

▷ **Quick Exercise.** Give an isomorphism between $H_1 \times \{1\}$ and $H_1$ and an isomorphism between $\{1\} \times H_2$ and $H_2$. ◁

As we saw in Example 22.11, since the structure of the group $G$ is essentially the same as the structure of $H_1 \times H_2$, $G$ must have two corresponding subgroups: They are

$$G_1 = \varphi(H_1 \times \{1\})$$

and

$$G_2 = \varphi(\{1\} \times H_2).$$

Can we characterize abstractly the fact that $G$ is a direct product, in terms of the properties of $G_1$ and $G_2$? It turns out we can, and we do this in the next theorem. But first, let's look at an example.

**Example 23.1**

The group $\mathbb{Z}_6$ is isomorphic to the direct product $\mathbb{Z}_3 \times \mathbb{Z}_2$. (Of course, in this example the group operation is *addition*.) The isomorphism is

$$\varphi([a]_3, [b]_2) = [4a + 3b]_6.$$

(We have included the subscripts for emphasis, and will henceforth omit them.)

▷ **Quick Exercise.**  Check that this is a group isomorphism.  ◁

But then
$$G_1 = \varphi(\mathbb{Z}_3 \times \{[0]\}) = \{0, 2, 4\}, \text{ and}$$
$$G_2 = \varphi(\{[0]\} \times \mathbb{Z}_2) = \{0, 3\}.$$

We'll return to this example when we have looked at our theorem.

To state our theorem conveniently, we need a little notation. If $G_1$ and $G_2$ are subgroups of a group $G$, then
$$G_1 G_2 = \{g_1 g_2 : g_1 \in G_1, g_2 \in G_2\}.$$

That is, $G_1 G_2$ consists of all possible products, where the first factor comes from $G_1$ and the second factor comes from $G_2$. Notice that $G_1 G_2$ is a subset of the group $G$, but in general it need not be a subgroup. (See Warmup Exercises b and c.)

**Theorem 23.1     The Internal Characterization Theorem**
*The group $G$ is isomorphic to $H_1 \times H_2$ if and only if $G$ has two subgroups $G_1$ and $G_2$, so that*

a. $G_1$ *is isomorphic to* $H_1$ *and* $G_2$ *is isomorphic to* $H_2$.

b. $G = G_1 G_2$,

c. $G_1 \cap G_2 = \{1\}$, *and*

d. *every element of* $G_1$ *commutes with every element of* $G_2$.

Before we proceed, note that in Example 23.1 the properties (a), (b), (c), and (d) hold.

▷ **Quick Exercise.**  Check that properties (a), (b), (c), and (d) hold in this example. (Remember that our operation is addition.)  ◁

**Proof:**   We have two arguments to do. Suppose first that $G$ is isomorphic to the direct product of $H_1$ and $H_2$, via the isomorphism $\varphi$, as above. As before, we can define subgroups $G_1$ and $G_2$ and then part (a) is clearly satisfied.

To check part (b), let $g \in G$. Because the isomorphism $\varphi$ is surjective, there exist $h_1 \in H_1$ and $h_2 \in H_2$ so that $\varphi(h_1, h_2) = g$. But then

$$g = \varphi((h_1, 1)(1, h_2)) = \varphi(h_1, 1)\varphi(1, h_2) \in G_1 G_2,$$

as required.

To check part (c), suppose that $g \in G_1 \cap G_2$. Then $g = \varphi(h_1, 1)$, and $g = \varphi(1, h_2)$, for some elements $h_1 \in H_1$, and $h_2 \in H_2$. But then

$$1 = \varphi(h_1, 1)(\varphi(1, h_2))^{-1} = \varphi(h_1, h_2^{-1}).$$

Because $\varphi$ is injective, $h_1 = 1$ and $h_2 = 1$, and so $g = 1$ as required.

To check part (d), suppose that $g_1 \in G_1$ and $g_2 \in G_2$. Then $g_1 = \varphi(h_1, 1)$ and $g_2 = \varphi(1, h_2)$. But such elements clearly commute.

▷ **Quick Exercise.** Why do such elements commute? ◁

For the converse, suppose that $G$ has two subgroups $G_1$ and $G_2$ as specified in parts (a) to (d) above. We then define

$$\psi : G_1 \times G_2 \to G$$

by setting $\psi(g_1, g_2) = g_1 g_2$. We claim that this is an isomorphism, thus proving the theorem. But $\psi$ is a homomorphism, because of part (d); $\psi$ is onto, because of part (b); and $\psi$ is one-to-one, because of part (c). (You will check these three assertions in Exercise 3.) □

▷ **Quick Exercise.** Show that $\mathbb{Z}_{10}$ is isomorphic to $\mathbb{Z}_2 \times \mathbb{Z}_5$, by using the two subgroups $\langle 5 \rangle$ and $\langle 2 \rangle$, and the theorem. *Hint:* Examine Example 23.1. ◁

**Example 23.2**

> We work in the additive group of Gaussian integers $\mathbb{Z}[i]$. Consider the infinite cyclic subgroups $G_1 = \langle 1 \rangle$ and $G_2 = \langle i \rangle$. Given an arbitrary Gaussian integer $a + bi$, clearly $a \in G_1$ and $bi \in G_2$, and so part (b) above is satisfied. And because a non-zero number cannot be both real and imaginary, part (c) is satisfied as well. Part (d) holds because the group is abelian. Thus, $\mathbb{Z}[i]$ is isomorphic *as a group* to $G_1 \times G_2$, and hence to $\mathbb{Z} \times \mathbb{Z}$. (Of course, these are by no means isomorphic *as rings*.)

We call this the *Internal Characterization Theorem* because it tells us whether a given group is isomorphic to a direct product, by looking only *inside* the group.

We can generalize the Internal Characterization Theorem to a direct product of any finite number of groups. The proof of this is left as Exercise 8.

**Theorem 23.2** *The group $G$ is isomorphic to $H_1 \times H_2 \times \cdots \times H_n$ if and only if $G$ has $n$ subgroups $G_1, \ldots, G_n$, so that*

a. $G_i$ *is isomorphic to $H_i$ for $i = 1, \ldots, n$.*

b. $G = G_1 \cdots G_n$,

c. $G_i \cap G_j = \{1\}$ *for all $i \neq j$, and*

d. *every element of $G_i$ commutes with every element of $G_j$ for all $i \neq j$.*

**Example 23.3**

For an example of a non-abelian group that can be expressed as a direct product of two subgroups, first consider the group of matrices given in Example 19.15; let's call this group $M$. In Example 20.10, we discovered that $M$ is isomorphic to the symmetric group $S_3$. Now consider the set of all $3 \times 3$ matrices

$$G = \left\{ \begin{pmatrix} e & 0 \\ 0 & P \end{pmatrix} : e = \pm 1, P \in M \right\}.$$

▷ **Quick Exercise.**  Check that this is a subgroup of $U(M_3(\mathbb{R}))$.  ◁

Notice that this group has twelve elements. Now consider the two subgroups

$$G_1 = \left\{ \begin{pmatrix} 1 & 0 \\ 0 & I \end{pmatrix}, \begin{pmatrix} -1 & 0 \\ 0 & I \end{pmatrix} \right\},$$

where $I$ is the $2 \times 2$ identity matrix, and

$$G_2 = \left\{ \begin{pmatrix} 1 & 0 \\ 0 & P \end{pmatrix} : P \in M \right\}.$$

▷ **Quick Exercise.**  Check that these two sets are actually subgroups, with 2 and 6 elements, respectively.  ◁

We now claim that $G$ is isomorphic to the direct product $G_1 \times G_2$. To show this, we use the Internal Characterization Theorem 23.1.

▷ **Quick Exercise.**  Check each of four criteria we need.  ◁

Now the group $G_1$ is isomorphic to $\mathbb{Z}_2$ (it is cyclic of order two), and $G_2$ is isomorphic to $S_3$. We thus have that our group of matrices is isomorphic to $\mathbb{Z}_2 \times S_3$.

---

## 23.2   Cayley's Theorem

We have seen earlier that groups of symmetries of many geometric figures are isomorphic to some subgroup of a symmetric group $S_n$, for suitable choice of $n$. In this section we will prove the surprising result that all finite groups can be thought of as a subgroup of a symmetric group of permutations. In some sense, the symmetric groups $S_n$ include all of finite group theory. This result is Cayley's theorem, and we will devote the rest of this chapter to proving this.

We begin with a simple example, the cyclic group $\{1, a, a^2\}$ of order three. (Of course, this is isomorphic to the additive group $\mathbb{Z}_3$, but we will find it more convenient in this discussion to use multiplicative notation.) We wish to show that this group is a group of permutations of some set, that is, we claim it is a subgroup of $S_n$ for some $n$. In fact, we will represent the elements of this group as permutations of the set $\{1, a, a^2\}$ itself! Note that the set of all permutations of any finite set with $n$ elements is isomorphic to the corresponding symmetric group $S_n$.

This can be rather confusing at times because we will be thinking of the elements of $\{1, a, a^2\}$ as elements of the cyclic group, on the one hand, and as elements of the set to be permuted, on the other. Consider the function

$$\varphi_a : \{1, a, a^2\} \to \{1, a, a^2\}$$

defined by $\varphi_a(x) = ax$. That is, $\varphi_a$ merely multiplies on the left by $a$. This means that

$$\varphi_a(1) = a, \quad \varphi_a(a) = a^2, \quad \varphi_a(a^2) = a^3 = 1.$$

We could thus denote this function by

$$\begin{pmatrix} 1 & a & a^2 \\ a & a^2 & 1 \end{pmatrix}.$$

In a similar fashion we can define $\varphi_1$ and $\varphi_{a^2}$, which can be represented as

$$\begin{pmatrix} 1 & a & a^2 \\ 1 & a & a^2 \end{pmatrix} \quad \text{and} \quad \begin{pmatrix} 1 & a & a^2 \\ a^2 & 1 & a \end{pmatrix}.$$

Notice that the three functions $\varphi_1, \varphi_a, \varphi_{a^2}$ are all bijections defined on the three-element set $\{1, a, a^2\}$. Of course, if we re-label this set as $\{1, 2, 3\}$, we can then identify these three functions as permutations of $\{1, 2, 3\}$, that is, as elements of $S_3$. Explicitly, if we re-label 1 as 1, $a$ as 2, and $a^2$ as 3, then

$$\varphi_1 = \begin{pmatrix} 1 & 2 & 3 \\ 1 & 2 & 3 \end{pmatrix} = \iota, \ \varphi_a = \begin{pmatrix} 1 & 2 & 3 \\ 2 & 3 & 1 \end{pmatrix} = (123), \text{ and } \varphi_{a^2} = \begin{pmatrix} 1 & 2 & 3 \\ 3 & 1 & 2 \end{pmatrix} = (132).$$

We have demonstrated a means of associating with each element of the group $\{1, a, a^2\}$ an element of $S_3$. More generally, we wish to associate each element of a group that has $n$ elements with an element of $S_n$, in such a way that the group operation is preserved.

▷ **Quick Exercise.** Multiply two elements of the group

$$\{1, a, a^2\}.$$

Does the result correspond to the composition of the corresponding permutations of the three-element set $\{1, a, a^2\}$? ◁

With this example behind us, we are now ready to prove the general theorem:

**Theorem 23.3    Cayley's Theorem**    *Suppose that $G$ is a finite group with $n$ elements. Then $G$ is isomorphic to a subgroup of $S_n$.*

The idea of the proof that follows is actually very simple, but it is easy to get lost in the forest of details! This proof is merely a more general version of the argument just given for the cyclic group $\{1, a, a^2\}$.

**Proof:**    Our goal is to assign to each element of the group $G$ a permutation belonging to $S_n$. What we will actually do is assign to each group element a permutation of the set $G$ itself. We will call the group of such permutations $S_G$. But clearly $S_n$ and $S_G$ are isomorphic groups, and so this will be enough.

Because $G$ has finitely many elements, we can label its elements as

$$G = \{g_1, g_2, \cdots, g_n\}.$$

In other words, we have listed the elements of $G$ in some fixed order.

For each integer $i$, we define a function

$$\varphi_i : G \to G$$

by letting $\varphi_i(g_k) = g_i g_k$. That is, $\varphi_i$ merely multiplies each element of $G$ on the left by $g_i$, the $i$th element of $G$.

We will show that $\varphi_i$ is a permutation of the elements of $G$. That is, we will show that $\varphi_i$ is a bijection.

$\varphi_i$ *is one-to-one*: Suppose that $\varphi_i(g_k) = \varphi_i(g_j)$. This means that $g_i g_k = g_i g_j$. But if we multiply this equation on the left by $g_i^{-1}$, we see that $g_k = g_j$. Thus, $\varphi_i$ is injective.

$\varphi_i$ *is onto*: Actually, because the set in question (namely, $G$) is finite, and the function is injective, we could conclude immediately that the function is surjective too, by using the pigeonhole principle. However, for clarity we shall verify directly that $\varphi_i$ is onto. For that purpose, choose an arbitrary integer $j$, where $1 \leq j \leq n$. The equation $g_i x = g_j$ has a unique solution in $G$. Thus, there must be a $k$ so that $g_i g_k = g_j$. But this evidently means that $\varphi_i(g_k) = g_j$, and so $\varphi_i$ is surjective.

Let's denote by $\Phi$ the assignment $g_i \longmapsto \varphi_i$. That is, $\Phi(g_i) = \varphi_i$. Because $\varphi_i$ is a bijection (and hence, a permutation of $G$), we have that $\Phi$ is a function from $G$ into $S_G$. We claim now that this function translates the multiplication in $G$ into the functional composition in $S_G$; that is, we claim that $\Phi$ is a homomorphism.

$\Phi$ *is a homomorphism*: We must show that

$$\Phi(g_i g_k) = \Phi(g_i) \circ \Phi(g_k).$$

Note that both $\Phi(g_i g_k)$ and $\Phi(g_i) \circ \Phi(g_k)$ are functions defined on the set $G$. To show that two functions are equal, we should check that they have the same value at a generic element of their domain. So pick $g_m \in G$ and compute:

$$\Phi(g_i g_k)(g_m) = (g_i g_k)g_m = g_i(g_k g_m) = g_i(\varphi_k(g_m)) =$$

$$\varphi_i(\varphi_k(g_m)) = (\Phi(g_i) \circ \Phi(g_k))(g_m),$$

which is what we required.

Finally, we require that $\Phi$ *is an injection*: So suppose that $\Phi(g_i) = \Phi(g_k)$. Because $\Phi(g_i)$ and $\Phi(g_k)$ are functions, they are equal if they do the same thing to every element of their domain $G$. In particular, then, they must give the same element when applied to $1 \in G$. But then $g_i \cdot 1 = g_k \cdot 1$, or $g_i = g_k$. Thus, $\Phi$ is one-to-one.

We have thus proved that $G$ is isomorphic to the subgroup $\Phi(G)$ of the group of permutations $S_G$. Because $S_G$ is isomorphic to $S_n$, this completes the proof. $\square$

Cayley's Theorem is usually not very practical for gaining insight into a particular group. For example, if we were to apply it to a group with 8 elements, we would represent it as a subgroup of $S_8$, which has $8! = 40,320$ elements! Furthermore, if the original group is abelian, we have then represented it as a subgroup of a highly non-abelian group. Nonetheless, the theoretical importance of Cayley's Theorem should not be minimized.

We should perhaps remark that exactly the same proof as we gave above for Cayley's Theorem can be used for an infinite group. This shows that any group whatsoever can be represented as a subgroup of a group of permutations, only this time as permutations of an infinite set! In general, we will avoid groups of permutations of infinite sets in this book.

## Historical Remarks

Thus, in principle, to study finite groups we need only study groups of permutations. In the 19th century most group theorists did exactly that. Although the eminent British mathematician Arthur Cayley had enunciated an abstract definition of group in 1853, it was a long time before mathematicians felt comfortable working in an abstract and axiomatic context. The study of finite groups had grown out of work by Lagrange (in the late 18th century) and Galois (in the early 19th century) in studying the roots of polynomial equations. Their insight into such roots was enlarged by thinking about *permuting* them; we will pick up this theme in Chapter 46. Afterward, the French mathematician Cauchy studied permutation groups in their own right, and he introduced our $2 \times n$ matrix notation for permutations. Consequently, even when group theory had grown beyond the particular problems of Lagrange and Galois, the thought that it was still about permutations remained.

## Chapter Summary

In this chapter we introduced the idea of the *direct product of groups*, and characterized when this happens, looking inside the given group. We needed to use the idea of group isomorphism to do this. We also represented every finite group as a subgroup of a symmetric group $S_n$, the set of all permutations of a set with $n$ elements; this is called *Cayley's Theorem*.

## Warm-up Exercises

a. Suppose that $G$ is a group with 11 elements, and $H$ is a group with 9 elements. How many elements are there in $G \times H$?

b. Consider subgroups $G_1, G_2$ of a group $G$. If $G$ is abelian, why is $G_1 G_2$ a subgroup?

c. Consider $G_1 = \{\iota, (12)\}$ and $G_2 = \{\iota, (13)\}$, two subgroups of $S_3$. Show that $G_1 G_2$ is not a subgroup.

d. When someone says that every finite group is a group of permutations, what does that mean?

e. Every finite abelian group is isomorphic to a subgroup of a non-abelian group. Why?

## Exercises

1. Use the Internal Characterization Theorem 23.1 to show that $\mathbb{Z}_{341}$ is isomorphic to $\mathbb{Z}_{11} \times \mathbb{Z}_{31}$. Then specify an explicit isomorphism.

2. Suppose that $G, H$, and $K$ are groups. Prove that the direct products

$$(G \times H) \times K \quad \text{and} \quad G \times (H \times K)$$

are isomorphic. For this reason, we usually omit the parentheses when describing such groups.

3. Complete the proof the Internal Characterization Theorem 23.1; that is, show that the function $\psi$ defined from $G_1 \times G_2$ to $G$ is in fact an isomorphism.

4. Let $a, b, c \in \mathbb{R}$, not all zero.

   (a) Show that
   $$P = \{(x, y, z) \in \mathbb{R}^3 : ax + by + cz = 0\}$$
   is a subgroup of $\mathbb{R}^3$.

   (b) Show that
   $$L = \{(ak, bk, ck) \in \mathbb{R}^3 : k \in \mathbb{R}\}$$
   is a subgroup of $\mathbb{R}^3$.

(c) Use the Internal Characterization Theorem to show that $\mathbb{R}^3$ is isomorphic to $P \times L$. What does this mean geometrically?

5. By way of analogy with Example 23.2, show that the additive group $\mathbb{Z}[\sqrt{2}]$ is isomorphic to a direct product of two non-trivial groups.

6. Use the Internal Characterization Theorem to show that $U(\mathbb{Z}_{15})$ is isomorphic to a direct product of two non-trivial groups.

7. In the group $D_3$, let $H_1 = \{\iota, \varphi\}$ and $H_2 = \{\iota, \rho, \rho^2\}$. (Refer back to Chapter 17 for a description of $D_3$.) Determine which of the four criteria of the Internal Characterization Theorem are satisfied by these two subgroups.

8. Prove Theorem 23.2 that generalizes the Internal Characterization Theorem to a product of an arbitrary finite number of groups.

9. Prove that $\mathbb{Z}_{30}$ is isomorphic to $\mathbb{Z}_2 \times \mathbb{Z}_3 \times \mathbb{Z}_5$.

10. Suppose that $m$ and $n$ are positive integers, and $m < n$. Define $I : S_m \to S_n$ as follows: Given $\alpha \in S_m$, we let $I(\alpha)(k) = \alpha(k)$, if $k \leq m$, and $I(\alpha)(k) = k$, if $n \geq k > m$. Show that $I$ is an injective homomorphism, which is not surjective. *Note:* You first must check that $I(\alpha) \in S_n$. We can paraphrase the contents of this exercise by asserting that (up to isomorphism) $S_m$ is a subgroup of $S_n$.

11. Following the proof of Cayley's Theorem 23.3, determine explicitly which permutations of $S_4$ each of the elements of the group $\{1, -1, i, -i\}$ correspond to.

12. Repeat Exercise 11 for the group of quaternions.

13. Repeat Exercise 11 for $S_3$. Note that this gives us two representations of $S_3$ as a group of permutations: the original definition as a permutation of three elements, and the new one as a permutation of six elements!

14. Let $n$ be a positive integer and let $k$ be a fixed integer, $1 \leq k \leq n$. Let

$$\mathrm{Stab}(k) = \{\alpha \in S_n : \alpha(k) = k\}.$$

Prove that $\mathrm{Stab}(k)$ is a subgroup of $S_n$. It is called a **stabilizer subgroup** of $S_n$. How many elements belong to $\mathrm{Stab}(k)$?

15. Generalize Exercise 14: Let $K$ be any subset of $\{1, 2, \cdots, n\}$. Let

$$\mathrm{Stab}(K) = \{\alpha \in S_n : \alpha(k) = k, \text{for all } k \in K\}.$$

Prove that $\mathrm{Stab}(K)$ is a subgroup of $S_n$. How many elements belong to $\mathrm{Stab}(K)$?

16. Let $n$ be positive integer, and $k, j$ fixed integers with $1 \leq k, j \leq n$. Define the function $\Phi : \mathrm{Stab}(k) \to \mathrm{Stab}(j)$ on $\alpha \in \mathrm{Stab}(k)$ by setting $\Phi(\alpha) = (kj)\alpha(kj)$ (where $(kj)$ is the 2-cycle switching $k$ and $j$). Prove that $\Phi$ is a group isomorphism. Warning: An important verification is to check that $\Phi(\alpha) \in \mathrm{Stab}(j)$.

17. We generalize the previous exercise. Let $K$ and $J$ be subsets of $N = \{1, 2, 3, \cdots, n\}$, with the same number of elements. Since $K$ and $J$ have the same number of elements, there exists a bijection $\beta : K \to J$, and similarly there exists a bijection $\gamma : N \backslash K \to N \backslash J$.

(a) Define a function $\mu$ by setting $\mu(m) = \beta(m)$ if $m \in K$, and $\mu(m) = \gamma(m)$ if $m \in N \backslash K$. Argue that $\mu \in S_n$.

(b) Given $\alpha \in \text{Stab}(K)$, define $\Phi : \text{Stab}(K) \to \text{Stab}(J)$ by setting $\Phi(\alpha) = \mu \alpha \mu^{-1}$. Prove that this is a group isomorphism.

18. Suppose that $G$ and $H$ are groups, and that $\varphi : G \to \text{Aut}(H)$ is a group homomorphism (see Exercise 20.27 for the group of automorphisms $\text{Aut}(H)$). Let $G \times_\varphi H = \{(g, h) : g \in G, h \in H\}$. Define a binary operation on $G \times_\varphi H$ as follows:

$$(g_1, h_1)(g_2, h_2) = (g_1 g_2, h_1 \varphi(g_1)(h_2)).$$

(a) Prove that this operation makes $G \times_\varphi H$ a group, called the **semidirect product** of $G$ and $H$. Note: In the computation $h_1 \varphi(g_1)(h_2)$ we have that $\varphi(g_1)$ is an automorphism of the group $H$, and $\varphi(g_1)(h_2)$ is the element of $H$ that results when this automorphism is applied to $h_2 \in H$; we then multiply this element of $H$ on the left by $h_1$.

(b) This construction generalizes the idea of direct product. What homomorphism $\varphi : G \to \text{Aut}(H)$ gives the ordinary direct product?

(c) Consider the dihedral group $D_4 = \{I, R, R^2, R^3, F, FR, FR^2, FR^3\}$ (where we are using the notation of Exercise 20.29). Prove that $D_4$ is isomorphic to a semidirect product of its subgroups $\langle F \rangle$ and $\langle R \rangle$.

# Chapter 24

## Cosets and Lagrange's Theorem

In Chapter 22 we introduced the idea of *group homomorphism*. Let's recall the corresponding development that we followed after introducing the idea of *ring homomorphism*. We obtained the Fundamental Isomorphism Theorem for Rings 14.1, which asserts that knowing about homomorphisms is equivalent to knowing about ideals: Each homomorphism gives rise to an ideal (its *kernel*), and each ideal in turn gives rise to a homomorphism (of which it is the kernel) to a *ring of cosets*. We would like to emulate this powerful and useful theory in the theory of groups so that we can better understand group homomorphisms. This will be the goal of the next three chapters.

### 24.1   Cosets

We begin this development by considering the notion of *coset* in the group context. In rings, we started with an ideal and formed its cosets. In groups, we begin with a subgroup and form its cosets.

Let $G$ be a group (written multiplicatively), and suppose that $H$ is a subgroup. For each $g \in G$, we form the set

$$Hg = \{hg : h \in H\}.$$

We call such sets **right cosets** of $H$ in $G$. We use the term *right* coset, because we are multiplying by $g$ on the right. In a non-abelian group, it might make a difference whether we consider right cosets or left cosets. (We'll come back to this topic later.) Note that if $G$ is an additive group, we would write such a coset as

$$H + g = \{h + g : h \in H\}.$$

This is exactly the definition of coset we used in the case where $G$ is a ring and $H$ an ideal.

Let's look at some examples, before we go any further:

**Example 24.1**

Consider the additive group $\mathbb{Z}$ and its cyclic subgroup $\langle 4 \rangle$. Then the sets

$$\langle 4 \rangle + 0, \langle 4 \rangle + 1, \langle 4 \rangle + 2, \langle 4 \rangle + 3$$

are the distinct (right) cosets of $\langle 4 \rangle$ in $\mathbb{Z}$, just as in the ring context.

Of course, we can generate a long list of examples by referring back to ring theory by our customary means: Forget the multiplication, and look at the additive group. But let's consider a purely group-theoretic example.

## Example 24.2

Consider the subgroup $H = \{\iota, (12)\}$ of the symmetric group $S_3$. We then obtain three distinct right cosets as follows:

$$H\iota = H(12) = \{\iota, (12)\} = H,$$
$$H(123) = H(23) = \{(123), (23)\}, \text{ and}$$
$$H(132) = H(13) = \{(132), (13)\}.$$

The subgroup $H$ is obviously a coset of itself; we obtained this by choosing the two elements in $H$ itself. The other two cosets also have two elements each.

▷ **Quick Exercise.** Verify that the cosets given above do indeed consist of the elements listed. ◁

## Example 24.3

Let's compute the right cosets of the subgroup

$$K = \{\iota, (123), (132)\}$$

in the same group $S_3$. This time we obtain

$$K\iota = K(123) = K(132) = \{\iota, (123), (132)\}, \text{ and}$$
$$K(12) = K(13) = K(23) = \{(12), (13), (23)\}.$$

Once again $K$ is a coset of itself and there is one other coset, which also has three elements.

▷ **Quick Exercise.** Verify that the cosets listed are correct. ◁

## Example 24.4

Consider the group

$$U(\mathbb{Z}_{21}) = \{1, 2, 4, 5, 8, 10, 11, 13, 16, 17, 19, 20\}.$$

Let's compute the right cosets of the subgroup $H = \{1, 4, 16\}$: We obtain

$$H1 = H4 = H16 = \{1, 4, 16\},$$
$$H2 = H8 = H11 = \{2, 8, 11\},$$
$$H10 = H13 = H19 = \{10, 13, 19\}, \text{ and}$$
$$H5 = H20 = H17 = \{5, 20, 17\}.$$

▷ **Quick Exercise.** Verify that the cosets listed are correct. ◁

## 24.2 Lagrange's Theorem

The examples from finite group theory that we have looked at are very suggestive. In each case, all the cosets of a given subgroup are the same size, and this allows us to decompose the group into finitely many pairwise disjoint pieces of the same size. This observation, when made precise, is a very valuable tool for counting in finite groups; it is called *Lagrange's Theorem*. Before obtaining Lagrange's Theorem, we need to make precise the observations we've made about cosets. We here prove the Coset Theorem for groups, which is directly analogous to the Coset Theorem 13.1 for rings.

**Theorem 24.1    The Coset Theorem**    *Let $H$ be a subgroup of a group $G$, and $a, b \in G$. Then*

a. *If $Ha \subseteq Hb$, then $Ha = Hb$.*

b. *If $Ha \cap Hb \neq \emptyset$, then $Ha = Hb$.*

c. *$Ha = Hb$ if and only if $ab^{-1} \in H$.*

d. *There exists a bijection between any two right cosets $Ha$ and $Hb$. Thus, if $H$ has finitely many elements, every right coset has that same number of elements.*

**Proof:**    (a): Suppose that $H$ is a subgroup of the group $G$, and $a$ and $b$ are elements of the group for which $Ha \subseteq Hb$. Then

$$a = 1a \in Ha \subseteq Hb,$$

and so there exists $h \in H$ such that $a = hb$. But then $b = h^{-1}a \in Ha$. Now, if $k \in H$, $kb = kh^{-1}a \in Ha$, and so $Hb \subseteq Ha$. That is, $Ha = Hb$.

(b): Suppose that $Ha \cap Hb \neq \emptyset$. Choose $c$ in this intersection. Then $c \in Ha$, and so $Hc \subseteq Ha$. But then by part (a), $Hc = Ha$. But similarly, $Hc = Hb$, and so $Ha = Hb$.

(c): If $Ha = Hb$, then $a = 1a \in Ha = Hb$, and so there exists $h \in H$ such that $a = hb$. But then $ab^{-1} = h \in H$, as required. Conversely, if $ab^{-1} \in H$, then $a = ab^{-1}b \in Hb$. But then $a \in Ha \cap Hb$, and so by part (b) $Ha = Hb$.

(d): Define the function $\varphi : Ha \to Hb$ by $\varphi(x) = xa^{-1}b$. First, note that if $x \in Ha$, then $x = ha$, for $h \in H$. But then $\varphi(x) = \varphi(ha) = (ha)(a^{-1}b) = hb \in Hb$. Thus, our function is well defined. It is one-to-one, because if $\varphi(x) = \varphi(y)$, then $xa^{-1}b = ya^{-1}b$, and multiplying on the right by $b^{-1}a$ gives us that $x = y$. It is onto, because if we choose the arbitrary element $hb \in Hb$, then $\varphi(ha) = ha(a^{-1}b) = hb$.

For finite sets, a bijection establishes that two sets have the same number of elements. And because $H = H1$ is itself a right coset, all right cosets have the same size as $H$.    □

If we had defined **left cosets** instead as $gH = \{gh : h \in H\}$, we could have easily formulated an analogous theorem for left cosets; indeed all of the theory of this chapter would carry over with minor changes. You can pursue this approach in Exercises 3–5 below. We will return to left cosets in the next chapter.

We now introduce some notation, which allows us to state Lagrange's Theorem conveniently. If $X$ is a finite set, let $|X|$ be the number of elements in $X$. For a group $G$, we call $|G|$ its **order**. If $G$ is a finite group with subgroup $H$, we let $[G : H]$ denote the number of distinct right cosets of $H$ in $G$, called the **index of $H$ in $G$**.

**Example 24.5**

Let's consider the group $G = U(\mathbb{Z}_{21})$ and the subgroup $H = \{1, 4, 16\}$ as in Example 24.4. Then $|H| = 3$. Note that part (d) of the Coset Theorem tells us that $|Ha| = 3$ for any right coset $Ha$. Of course, we computed the right cosets explicitly above and saw that this is so. Finally, those computations show us that $[G : H] = 4$. Note that there are 12 elements in $G$ altogether (that is, $|G| = 12$), and we can arrive at this number by counting 4 cosets, each with 3 elements.

$\triangleright$ **Quick Exercise.**   Repeat this reasoning for Examples 24.2 and 24.3 above. $\triangleleft$

Let's now state the general theorem reflected in these examples:

**Theorem 24.2    Lagrange's Theorem**      *Let $G$ be a finite group with subgroup $H$.* *Then*
$$|G| = [G : H]|H|.$$
*In particular, this means that $|H|$ divides $|G|$.*

**Proof**:    Suppose that $[G : H] = m$. Every element of $G$ is in a coset of $H$, and part (b) of the Coset Theorem tells us we can decompose $G$ into a union of $m$ pairwise disjoint cosets:

$$G = H \cup Ha_1 \cup Ha_2 \cup \cdots \cup Ha_{m-1}.$$

But each of these cosets has $|H|$ elements. Thus, there must be $[G : H]|H|$ elements in $G$ altogether.                                                                              $\square$

This is illustrated in the following picture.

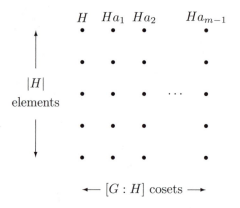

Lagrange did his work well before the notion of group had been defined. But he is honored for the theorem named after him because he applied a concrete version of this theorem in his arguments regarding permutations of the roots of a polynomial equation.

---

## 24.3   Applications of Lagrange's Theorem

Lagrange's Theorem is a very important one in the theory of finite groups. It places great restrictions on the sort of elements and the sort of subgroups a finite group can have. We make one such inference immediately:

**Corollary 24.3** *Let $G$ be a finite group, and $g \in G$. Then the order of $g$ divides the order of $G$.*

**Proof:** Consider the cyclic subgroup $\langle g \rangle$ of $G$ generated by $g$. If the order of $g$ is $n$, then this subgroup has $n$ elements. By Lagrange's Theorem, $n$ must divide $|G|$. □

Thus, no group of order 15 could possibly have elements of order 2 or 6.

▷ **Quick Exercise.** What are the orders of the elements in $S_4$? Do all these orders divide $4! = 24$? ◁

Let's apply this reasoning in the case where the number of elements in a group is a prime $p$. Now the only positive divisors of the prime $p$ are 1 and $p$ itself. Of course, every group has exactly one element of order 1: the identity. Thus, every other element of the group has order $p$. But this means that in a group with prime order, every non-identity element has that prime order. And so, the group must be cyclic! (In fact, it is cyclic where *every* non-identity element generates the group; see also Exercise 21.16.)

**Corollary 24.4** *Every group of prime order is cyclic.*

It is important to note that the converse of Lagrange's Theorem is false. That is, suppose that $G$ is a finite group with $n$ elements, and $m$ is a divisor of $n$. We *cannot* infer that $G$ has a subgroup with exactly $m$ elements. (This statement is true for abelian groups; we will meet this as Theorem 29.10.) Here is an example that shows this is true; we prove that the example works, as another application of Lagrange's Theorem:

**Example 24.6**

Consider the 12 element subgroup of $S_4$ we called $A_4$ in Example 20.11. Recall that the elements of this group are

$$\iota, \ (123), \ (132), \ (124), \ (142), \ (134), \ (143),$$
$$(234), \ (243), \ (12)(34), \ (13)(24), \ (14)(23).$$

If the converse of Lagrange's Theorem were true, then $A_4$ would have a subgroup $H$ with six elements. By Lagrange's Theorem, this subgroup would have only two distinct cosets. Now pick any element $\alpha$ of $A_4$ with order 3. Because $H$ has only two distinct cosets, at least two of the cosets $H\iota$, $H\alpha$, and $H\alpha^2$ must be the same. But then, by the Coset Theorem, $\alpha \in H$.

▷ **Quick Exercise.** Why? ◁

So all the elements of $A_4$ with order 3 belong to the subgroup $H$. But there are eight such elements, which must all belong to a subgroup with six elements. This contradiction implies that $A_4$ has no subgroup with six elements. (This elegant argument is due to Joseph Gallian.) For a more computational proof of this result, see Exercise 19.

We can however prove a very limited converse to Lagrange's Theorem, which asserts that *prime* divisors of the order of a finite group do lead to subgroups of that order; this is called Cauchy's Theorem. This result is a first important step towards several theorems that demonstrate the existence of certain subgroups in any finite group, and also count how many such subgroups are possible. These results are called the Sylow Theorems; we will pursue these ideas a good bit further in Chapters 28 and 29. The proof we offer here of Cauchy's Theorem is a clever one due to James McKay, based on counting a set in two different ways.

**Theorem 24.5     Cauchy's Theorem**     *Suppose that $p$ is a prime integer, $G$ is a group, and $p$ divides $|G|$. Then $G$ has an element of order $p$ (and so a subgroup of order $p$).*

**Proof:**     Consider the set $S$ of all lists $a_1, a_2, \cdots, a_p$, where each $a_i \in G$, and $a_1 a_2 \cdots a_p = 1$. Note that we are allowing for repetitions in the $a_i$'s, and in fact, we are hoping exactly to find an element (other than the identity) of $S$ for which all the entries are the same; this will give us the required element of $G$ with order $p$.

We are going to count the number of elements in $S$. Pick any element of $S$. For each entry in this list (until the last one), we can choose any element of the group; but the last entry in the list is required to be $(a_1 a_2 \cdots a_{p-1})^{-1}$. This means that $S$ has exactly $|G|^{p-1}$ elements.

We will now recount this set. Given a list $a_1, a_2, \cdots, a_p$, consider what happens when we cyclicly permute it. That is, we consider lists of the form $a_k, a_{k+1}, \cdots, a_p, a_1, \cdots, a_{k-1}$. First of all, note that these permuted lists remain in $S$.

▷ **Quick Exercise.**     Why must these permuted lists remain in $S$? ◁

Second, if at least two of the $a_i$'s are distinct, we obtain a total of $p$ lists from the original one (counting the original when we perform the "trivial" cyclic permutation). However, if all of the $a_i$'s are the same, we do not obtain any new lists. We can thus partition the elements of the set $S$ into the subsets consisting of a given element and the other lists resulting from its cyclic permutations. Some of these subsets have $p$ elements in them, while some have only 1 element in them, depending on whether any of the elements in one of the lists are distinct, or not. Those are the only two possibilities.

We now count the elements in $S$, by counting the number of elements in each of these subsets, and adding them up. So, let $m$ be the number of these subsets with 1 element in them, and let $n$ be the number of these subsets with $p$ elements. This means that a complete count of all elements in $S$ is given by $m \cdot 1 + n \cdot p$. Thus $|G|^{p-1} = m + np$.

Notice that $m > 0$, because there is at least one element (namely 1) that when multiplied by itself $p$ times gives 1. We will have completed the proof of the theorem if we can conclude that $m > 1$.

We now finally use the hypothesis of the theorem (that $p$ divides $|G|$) to conclude that $p$ divides $m$. Since $m > 0$, this means that $m \geq p > 1$. This proves the theorem.     □

**Example 24.7**

> Suppose $G$ is a group with 606 elements. Then Cayley's Theorem tells us that $G$ necessarily has elements of orders 2, 3, and 101.

We shall now use Lagrange's Theorem to determine the number of distinct groups of small size. Notice first that because all cyclic groups of a given order are isomorphic, there is essentially only one group of order 2, 3, 5, etc.: Namely, the cyclic groups

$$\mathbb{Z}_2, \mathbb{Z}_3, \mathbb{Z}_5, \cdots.$$

Let's take the smallest non-prime integer 4: What sort of groups of order 4 are there? Of course, there is $\mathbb{Z}_4$, the cyclic group of order 4. Are there any others?

If a group $G$ has order 4 and is not cyclic, then every non-identity element in the group must be of order 2 (because only 1, 2, and 4 divide 4). Choose two of these elements, and call them $a$ and $b$. Each is its own inverse (that's what being of order 2 means). So $ab$ has to be the third non-identity element; but so does $ba$, and so $ab = ba$. (Note that Exercise 19.8 also implies that $ab = ba$.) Thus, we conjecture that the multiplication table for $G$ must look like this:

|    | 1  | $a$ | $b$ | $ab$ |
|----|----|-----|-----|------|
| 1  | 1  | $a$ | $b$ | $ab$ |
| $a$ | $a$ | 1 | $ab$ | $b$ |
| $b$ | $b$ | $ab$ | 1 | $a$ |
| $ab$ | $ab$ | $b$ | $a$ | 1 |

To show this is a group, we need to verify that it satisfies all the group axioms. The most tedious of these is associativity. But, more easily, we see that this is just a multiplicative version of the (additive) group $\mathbb{Z}_2 \times \mathbb{Z}_2$. This group is often called the **Klein Four Group**.

▷ **Quick Exercise.** Establish an explicit isomorphism between the multiplicative and the additive representations we have given for the Klein Four Group. ◁

So we have concluded that there are essentially only two groups of order 4: the cyclic group and the Klein Four Group.

There's only one group of order 5 (because 5 is prime). But what about groups of order 6? If such a group had an element of order 6, it would be cyclic. So in a non-cyclic group of order 6, all non-identity elements must be of order 2 or 3. By Cauchy's Theorem we know that any such group must have at least one element $a$ with order 3, and at least one element $b$ of order 2. Then $1, a, a^2, b$ are all distinct elements of the group; note that $a^2 \neq b$, because $a^2$ also has order 3. We now claim that if we add the elements $ab$ and $a^2 b$ to this list, we will have a complete list of all elements in the group. To show all of these elements are distinct, let's examine a couple of typical cases. If we suppose that $ab = 1$, then $a^{-1} = a^2 = b$, which we have already concluded is impossible. And if $ab = a^2$, then $b = a$, which is clearly impossible.

▷ **Quick Exercise.** Finish checking that each of these elements is distinct from each of the others. ◁

We thus have that our group consists of the set $\{1, a, a^2, b, ab, a^2 b\}$. But obviously $ba$ must be an element too. It is easy to check that the only possibilities for this element are $ab$ and $a^2 b$. If $ba = ab$, then the group is abelian, and the element $ab$ is an element of order 6.

▷ **Quick Exercise.** Check that $ab$ is a generator in this case, either by direct calculation or else by using Exercise 21.6. ◁

So the only remaining case is if $ba = a^2 b$. It follows from this that $ba^2 = a^2 ba = a^4 b = ab$. Is there such a group? Of course! This is exactly $S_3$ (where $(123)$ could play the role of $a$, and $(12)$ could play the role of $b$).

We have thus concluded that there are essentially only two groups of order 6: $\mathbb{Z}_6$ and $S_3$. Note that this makes $S_3$ the non-abelian group with smallest order.

You will inquire in Exercises 11 and 12 into the groups of order 8. It turns out that there are five of them.

---

## Chapter Summary

In this chapter we defined the notion of *right coset* for a subgroup of a group. We determined that the set of right cosets of a subgroup divides the group into pairwise disjoint pieces, each of the same size. In the context of finite groups, this leads to the counting theorem known as *Lagrange's Theorem*. Lagrange's Theorem allows us to prove that the order of an element divides the order of the group, and that all groups of prime order are cyclic.

# Warm-up Exercises

a. Determine the set of right cosets of the subgroup $\langle 6 \rangle$ in $\mathbb{Z}_{20}$. Check that Lagrange's Theorem 24.2 holds in this case.

b. Determine the set of right cosets of the subgroup $\langle 7 \rangle$ in $U(\mathbb{Z}_{20})$. Check that Lagrange's Theorem holds in this case.

c. Determine the set of right cosets of the subgroup $\{1, -1\}$ in the group of quaternions $Q_8$. Check that Lagrange's Theorem holds in this case. (Note that we are here using the shorthand notation of Exercise 21.12 for the quaternions.)

d. If $H$ is a subgroup of $G$ and two cosets of $H$ share an element, then what can you say?

e. Is a coset ever a subgroup?

f. If $a \in Hb$, how are the two cosets $Ha$ and $Hb$ related?

g. Suppose that $G$ is an infinite group, and $H$ is a subgroup of $G$ with finitely many elements. How many distinct cosets does $H$ have?

h. Suppose that $G$ is an infinite group, and $H$ is a subgroup of $G$ with finitely many cosets. How many elements does $H$ have?

# Exercises

1. Determine explicitly the right cosets of the subgroup $\langle (124) \rangle$ in the group $A_4$ (see Example 20.11).

2. Determine explicitly the right cosets of the subgroup $\langle (124) \rangle$ in the whole group $S_4$.

3. Formulate and prove the left-handed version of the Coset Theorem 24.1.

4. Let $H$ be a subgroup of a group $G$, and pick $g \in G$. Prove that there is a bijection between the right coset $Hg$ and the left coset $gH$, given by the function $\alpha(hg) = gh$. (This is not a homomorphism!) This means that if $H$ is a finite set, all of its left cosets and all of its right cosets have $|H|$ elements.

5. Let $H$ be a subgroup of a group $G$. Show that there is a bijection from the set of all distinct left cosets of $H$ to the set of all distinct right cosets of $H$, given by the function $\beta(gH) = Hg^{-1}$. (Warning: You need to show that this function is well-defined, using the Coset Theorems.) This exercise says that if $G$ is finite, we can think of $[G : H]$ as counting either all distinct right cosets, or as counting all distinct left cosets.

6. Find the *left* cosets of the subgroup $H = \{\iota, (12)\}$ of $S_3$. How do these cosets compare with the right cosets of $H$ found in Example 24.2?

7. How do the left cosets of $K$ compare with the right cosets found in Example 24.3?

8. If $G$ is a group of order 8 and $G$ is not cyclic, why must $x^4 = 1$ for all $x$ in $G$? What similar statement can you make about a group of order 27? Generalize this further.

9. Let $G$ be a group of order $p^2$, where $p$ is prime. Show that every proper subgroup of $G$ is cyclic.

10. Replicate the argument about the number of groups of order 6 for groups of order 10. You should conclude that there are essentially only two such groups; namely, $\mathbb{Z}_{10}$ and the dihedral group $D_5$.

11. (a) What are the possible orders for a subgroup of a group of order 8?

    (b) Prove that any group of order 8 must have a subgroup of order 2. (This is easy, and does not require that you use either Theorem 24.5 or the results of Exercise 12.)

12. In this problem we analyze how many groups there are of order 8. Consider the following cases (with hints):

    *Case 1:* There is an element of order 8.

    *Case 2:* If there is no element of order 8, but there is an element $a$ of order 4, choose an element $b$ outside $\langle a \rangle$. Now, either you can find such a $b$ of order 2 or else all such $b$ have order 4. If you can find such a $b$ of order 2, then, depending on whether $a$ and $b$ commute, argue that we must obtain either $\mathbb{Z}_2 \times \mathbb{Z}_4$ or $D_4$. Now assume that all such $b$ have order 4. Argue that you must obtain the quaternions $Q_8$.

    *Case 3:* Now suppose that there are no elements of order 4. So, all non-identity elements have order 2. First, argue that the group must be abelian. Then, pick three elements of order 2 and argue that you must obtain $\mathbb{Z}_2 \times \mathbb{Z}_2 \times \mathbb{Z}_2$.

13. Show that every non-identity element of a group generates the group if and only if the group is cyclic of prime order. (This extends and strengthens Exercise 21.16.)

14. Let $n$ be a positive integer and $1 \le k \le n$. Consider the stabilizer subgroup $\text{Stab}(k)$ of $S_n$. (See Exercise 23.14.) What does Lagrange's Theorem say about $[S_n : \text{Stab}(k)]$? Demonstrate this explicitly, by placing the set of right cosets of $\text{Stab}(k)$ into a natural one-to-one correspondence with $\{1, 2, \cdots, n\}$.

15. (a) Suppose $G$ is a finite group and $a \in G$. Show that $a^{|G|} = 1$.

    (b) Now provide another proof for Fermat's Little Theorem 8.7: *If $p$ is prime and $0 < x < p$, then $x^{p-1} \equiv 1 \pmod{p}$.*

    (c) For any positive integer $n$, $\phi(n)$ is defined to be the number of positive integers less than $n$ that are also relatively prime to $n$. (This is called **Euler's phi function**.) So, $\phi(6) = 2$ because only 1 and 5 are less than 6 and also relatively prime to 6. Note that if $p$ is prime, $\phi(p) = p - 1$. Show that if $a$ is relatively prime to $n$, then

    $$a^{\phi(n)} = 1 \pmod{n}.$$

    This generalization of Fermat's Little Theorem 8.7 is called **Euler's Theorem**.

16. Refer to the definition in Exercise 15 of Euler's phi function.

    (a) Compute $\phi(7)$, $\phi(20)$, and $\phi(100)$.

    (b) Use Euler's Theorem to efficiently compute

    $$2^{38} \pmod 7, \qquad 17^{18} \pmod{20}, \qquad 7^{82} \pmod{100}.$$

17. Prove that the set of (right) cosets of the subgroup $\mathbb{Z}$ in $\mathbb{R}$ can be placed in a one-to-one correspondence with the set of real numbers $x$, with $0 \le x < 1$.

18. Consider the group $\mathbb{S}$, the unit circle, a multiplicative subgroup of $\mathbb{C}^*$. Provide a simple condition on complex numbers $\alpha, \beta$ characterizing when the cosets $\mathbb{S}\alpha$ and $\mathbb{S}\beta$ are equal, and prove that it works.

19. In this exercise you will obtain a more computational proof of the fact discussed in Example 24.6, that there is no subgroup of order 6 in the group $A_4$. We suppose by way of contradiction that there is such a subgroup called $H$.

    (a) Why does $H$ have an element of order 3 and an element of order 2?

    (b) Why can we assume without loss of generality that the element of order 3 has the form $(abc)$ and the element of order 2 either had form $(ab)(cd)$ or $(ac)(bd)$?

    (c) Now do computations with these elements to conclude that $H$ must have more than 6 elements.

20. Let $G$ be a finite group with $g \in G$. Recall from Exercise 20.15 the subgroup $C(g)$, the *centralizer* of $g$. Recall also that a conjugate of $g$ is an element of the form $hgh^{-1}$, for $h \in G$; see Exercise 22.15. Prove that the number of distinct conjugates of $g$ is equal to $[G : C(g)]$. Hint: Show that the mapping taking $hgh^{-1}$ to the left coset $hC(g)$ is well-defined and bijective.

# Chapter 25

## Groups of Cosets

Our goal now is to place a group structure on the set of right cosets of a subgroup of a given group, just as we did for the cosets of an ideal in a given ring. Given a group $G$ with a subgroup $H$, our experience with ring theory would suggest that the group operation on the set of cosets should be defined like this:

$$(Ha)(Hb) = Hab.$$

That is, to multiply two cosets, pick a representative of each, multiply them together, and then form its coset. It is clear that we must prove that this operation is *well defined*. (That is, it should be independent of the coset representative chosen.)

Before proceeding, let's look at an example of this; we return to Example 24.2. Let's try multiplying the cosets

$$H(123) = H(23) = \{(123), (23)\}, \text{ and}$$
$$H(132) = H(13) = \{(132), (13)\},$$

using our provisional definition. If our definition is to work, and we compute the product of *any* element from the first coset, times *any* element of the second coset, we should always land in the same coset. In this case there are four possible products:

$$(123)(132) = \iota, \quad (23)(132) = (12),$$
$$(123)(13) = (23), \text{ and} \quad (23)(13) = (123).$$

These elements *do not* all belong to the same coset! (Because, among other reasons, there are four distinct elements obtained and a coset of $H$ must have only two elements.) Consequently, it does not seem that here we can make a group out of the set of right cosets of $H$ in $G$.

---

## 25.1  Left Cosets

Let's beat a strategic retreat and return to a notion that we left behind in the previous chapter: There we defined *right* cosets. But as we pointed out in Section 24.2, we could have looked at *left* cosets instead. Given a subgroup $H$ of a group $G$, a **left coset** of $H$ in $G$ is a set of the form

$$gH = \{gh : h \in H\}.$$

Let's compute the left cosets for the subgroups $H$ and $K$ of $S_3$. We computed their right cosets in Examples 24.2 and 24.3, and you computed their left cosets if you did Exercises 24.6 and 24.7.

**Example 25.1**

(See Example 24.2 and Exercise 24.6.) We still obtain exactly 3 distinct cosets; this time they are

$$\iota H = (12)H = \{\iota, (12)\},$$
$$(123)H = (13)H = \{(123), (13)\}, \text{ and}$$
$$(132)H = (23)H = \{(132), (23)\}.$$

**Example 25.2**

(See Example 24.3 and Exercise 24.7.) We still obtain exactly 2 distinct cosets; this time they are

$$\iota K = (123)K = (132)K = \{\iota, (123), (132)\}, \text{ and}$$
$$(12)K = (13)K = (23)K = \{(12), (13), (23)\}.$$

A comparison of these two examples reveals a real difference! For $K$, each of its left cosets is equal to the corresponding right coset, while this is false for $H$. Of course, if the group had been abelian, then the left and right cosets would obviously be equal, because multiplying on the left, or on the right, amounts to the same thing in an abelian group. And we know that coset addition works for rings. This suggests that this property might be of importance. In fact, it turns out to be *exactly* what we need to build a group of cosets.

---

## 25.2 Normal Subgroups

Let $H$ be a subgroup of the group $G$. We say that $H$ is a **normal subgroup** if $gH = Hg$, for all $g \in G$. If $H$ is a normal subgroup, we can safely talk about its *cosets* (without an adjective). We denote the set of all these cosets by $G/H$.

**Example 25.3**

The subgroup $\langle (123) \rangle$ is a normal subgroup of $S_3$, while the subgroup $\langle (12) \rangle$ isn't.

**Example 25.4**

Every subgroup of an abelian group is normal.

In the next two chapters we will eventually obtain a host of examples of normal subgroups, but it's now time for the theorem we have been building up to. This is the group analogue of Theorem 13.2, where the idea of normal subgroup plays the role analogous to ideal in a ring.

**Theorem 25.1** *Let $H$ be a normal subgroup of $G$. Then the set $G/H$ of cosets of $H$ in $G$ is a group, under the operation*

$$(Ha)(Hb) = Hab.$$

**Proof:** We must first check that the operation specified in the statement of the theorem is *well defined*; that is, it should be independent of coset representatives. This verification is the crucial part of the proof and will depend in an essential way on the fact that $H$ is normal.

Suppose then that $Ha = Hc$ and $Hb = Hd$. We claim that $Hab = Hcd$. By Theorem 24.1c, this amounts to checking that $ab(cd)^{-1} \in H$. Because $Hb = Hd$, we know that $bd^{-1} \in H$. Now $H$ is a normal subgroup, and so $aH = Ha$. This means that

$$a(bd^{-1}) = ha$$

for some $h \in H$. Thus,

$$ab(cd)^{-1} = abd^{-1}c^{-1} = hac^{-1}.$$

But $Ha = Hc$ means exactly that $ac^{-1} \in H$, and so therefore $hac^{-1} \in H$ too. This completes the proof that the operation is well defined.

The operation is clearly associative:

$$(HaHb)(Hc) = HabHc = H(ab)c = Ha(bc) = HaHbc = Ha(HbHc).$$

The element $H1$ serves as the identity for $G/H$:

$$H1Ha = H1a = Ha = Ha1 = HaH1.$$

And the element $Ha$ has an inverse; namely, $Ha^{-1}$:

$$HaHa^{-1} = Haa^{-1} = H1 = Ha^{-1}a = Ha^{-1}Ha.$$

Thus, $G/H$ is a group, as we claim. □

The proof of this theorem is more complicated than that for the corresponding ring theorem, because the group need not be abelian. The crucial point in the proof is showing that the operation on $G/H$ is well defined. Loosely speaking, the normality of $H$ says that elements of $G$ commute with $H$ (but not necessarily with the individual elements of $H$), and this is just enough to make the argument work.

We call $G/H$ the **group of cosets of $G$ modulo $H$**; we also call $G/H$ a **quotient group**.

We have encountered many examples of this construction already: Merely take a ring of cosets as we discussed in Chapter 13 and thereafter, and forget the multiplicative structure. However, these examples do not call sufficient attention to the importance of the normality of the subgroup, because, as we've pointed out before, *all* subgroups of an abelian group are normal. Consequently, we must examine some purely group-theoretic examples, involving some non-abelian groups.

Before we do this, however, we will provide a characterization of normality that is slightly easier to work with. We will inquire further into this version of normality in the exercises, and in future chapters. For the moment, we view the following proposition only as a computational aid to checking normality.

**Theorem 25.2** *Let $H$ be a subgroup of the group $G$. Then $H$ is normal in $G$ if and only if $g^{-1}Hg = H$, for all $g \in G$. In fact, this is true if and only if $g^{-1}Hg \subseteq H$, for all $g \in G$.*

**Proof:** Suppose that $H$ is normal in $G$ and $g \in G$; then $Hg = gH$. So for any $h \in H$, there exists $h_1 \in H$ such that $hg = gh_1$. But then $g^{-1}hg = h_1 \in H$. Thus, $g^{-1}Hg \subseteq H$. A very similar argument, which we save for Exercise 1, shows that the reverse inclusion also holds.

Conversely, suppose that $g^{-1}Hg = H$ for all $g \in G$. Given any element $h \in H$, this means that $h = g^{-1}h_1g$, for some $h_1 \in H$. But then $gh = h_1g$; that is, $gH \subseteq Hg$. You will prove the reverse inclusion in Exercise 2.

If $g^{-1}Hg \subseteq H$ for *all* $g$, this applies to $g^{-1}$ as well: That is, $gHg^{-1} \subseteq H$. But this latter statement means that $H \subseteq g^{-1}Hg$, and so these two sets are actually equal.     □

## 25.3   Examples of Groups of Cosets

**Example 25.5**

Consider the group $A_4$ from Example 20.11:

$$A_4 = \{\iota, (234), (243), (143), (134), (124), (142),$$

$$(132), (123), (12)(34), (13)(24), (14)(23)\}.$$

Consider the subgroup

$$H = \{\iota, (12)(34), (13)(24), (14)(23)\}.$$

Notice that $H$ is a group with four elements that is not cyclic; that is, $H$ is isomorphic to the Klein Four Group (see the discussion in Section 24.3). We claim that $H$ is normal. Let's perform a few of the computations necessary to check this:

$$(132)^{-1}(12)(34)(132) = (123)(12)(34)(132) = (14)(23),$$
$$(243)^{-1}(12)(34)(243) = (234)(12)(34)(243) = (13)(24),$$
$$(124)^{-1}(13)(24)(124) = (142)(13)(24)(124) = (12)(34).$$

In each case, we see that an element of the form $\beta^{-1}\alpha\beta$, where $\beta \in A_4$ and $\alpha \in H$, remains in $H$. It would be tedious indeed to check all the other cases!

▷ **Quick Exercise.**   Perform two more of the required calculations.  ◁

In Exercise 25.16 you will provide a complete proof that $H$ is normal in $A_4$, cleverer than just verifying all the cases. We will also return to considering the normality of $H$ in Example 26.5. For the moment, we shall accept that this is true.

   We can thus construct the group $A_4/H$. Let's use Lagrange's Theorem 24.2 to determine the size of this group. The order of $A_4$ is 12, while the order of $H$ is 4. Thus, the order of $A_4/H$ must be $12/4 = 3$. But there is essentially only one group of order 3, the cyclic group. Let's write out the elements of this group of cosets, and see why it should be isomorphic to $\mathbb{Z}_3$:

$$A_4/H = \{\iota, H(123), H(132)\}.$$

▷ **Quick Exercise.**   Check that these three cosets are distinct, and so this list is really a complete list of the elements of $A_4/H$.  ◁

But now it is clear that $A_4/H$ is a cyclic group generated by $H(123)$.

**Example 25.6**

Consider now the group $\mathcal{G} = U(M_2(\mathbb{R}))$ of $2 \times 2$ invertible matrices. Recall that this consists of those matrices with non-zero determinant. (See Exercise 8.2.) Now consider the set

$$\mathcal{H} = \{A \in \mathcal{G} : \det(A) = 1\}.$$

First note that this is a subgroup. (You actually showed this in Exercise 20.18.) Given $A, B \in \mathcal{H}$, we need to check that $AB^{-1} \in \mathcal{H}$. But

$$\det(AB^{-1}) = \det(A)(\det(B))^{-1} = 1 \cdot 1 = 1.$$

This shows that $\mathcal{H}$ is a subgroup.

Furthermore, we claim that $\mathcal{H}$ is a normal subgroup. Because, given $A \in \mathcal{H}$ and $C \in \mathcal{G}$, we claim that $C^{-1}AC \in \mathcal{H}$. But

$$\det(C^{-1}AC) = (\det(C))^{-1} \det(A) \det(C) = \det(A) = 1.$$

This shows that $\mathcal{H}$ is normal.

We thus know that $\mathcal{G}/\mathcal{H}$ is a group. Let's see if we can determine the nature of this group. Let's look at a particular coset $\mathcal{H}A$. A typical element of this coset looks like $BA$, where $\det(B) = 1$. But

$$\det(BA) = \det(B) \det(A) = \det(A).$$

Thus, all elements of $\mathcal{H}A$ have the same determinant. In fact, we claim that $\mathcal{H}A$ consists exactly of all those matrices with determinant equal to the non-zero real number $\det A$. To see this, suppose that $C \in \mathcal{G}$, and $\det(C) = \det(A)$; we claim that $C \in \mathcal{H}A$. But $C = CA^{-1}A$, and $\det(CA^{-1}) = 1$, and so $C \in \mathcal{H}A$.

We thus have established a bijection between the elements of the group $\mathcal{G}/\mathcal{H}$ and the non-zero real numbers. This suggests the possibility that $\mathcal{G}/\mathcal{H}$ is isomorphic to $\mathbb{R}^*$, the multiplicative group of non-zero real numbers. We could prove this now but will instead wait until the next chapter, when we can prove this carefully, and easily, using the Fundamental Isomorphism Theorem for Groups 26.4; see Example 26.8.

We can much enlarge our fund of examples, by showing that in the special case when a subgroup has index 2 in the group, it is automatically normal:

**Theorem 25.3    Index 2 Theorem**    *Suppose that $G$ is a group with subgroup $H$, and $[G : H] = 2$. Then $H$ is a normal subgroup of $G$, and $G/H$ is isomorphic to $\mathbb{Z}_2$.*

**Proof**:    Suppose that $H$ is a subgroup of $G$ and $[G : H] = 2$. This means that there are only two right cosets of $H$ in $G$: We can write these two cosets as $H1$ and $Hg$, where $g$ is any element of $G$ that does not belong to $H$. But by Lagrange's Theorem 24.2 and Exercise 24.5, there can also be only two left cosets of $H$, and they must be of the form $1H$ and $gH$. But because $H1 = H = 1H$, this means that $Hg = gH$, for all $g \notin H$. In other words, $H$ is normal in $G$.

Because $[G : H] = 2$, we know that $G/H$ is a group with two elements and there is only one such group (up to isomorphism), namely, $\mathbb{Z}_2$.    □

**Example 25.7**

Consider again the group $A_4$, this time as a subgroup of $S_4$. This subgroup has 12 elements, and so its index in $S_4$ is 2. Theorem 25.3 then implies that $A_4$ is a normal subgroup of $S_4$.

**Example 25.8**

Consider the dihedral group $D_n$, where we will here use the notation introduced in Exercise 20.29. The cyclic subgroup $\langle R \rangle$ is clearly the subgroup of rotations in $D_n$ and has $n$ elements. But because $D_n$ has $2n$ elements, $\langle R \rangle$ is a normal subgroup of $D_n$.

---

## Chapter Summary

In this chapter we defined a *normal subgroup* and proved that if a subgroup is normal, then a group structure can be given to the set of cosets, which we then call a *quotient group*.

---

## Warm-up Exercises

a. Given a subgroup $H$ of a group $G$, at least one left coset of $H$ always equals its corresponding right coset. Why?

b. Why is every subgroup normal, if the group is abelian? (The converse is false; see Exercise 10 below.)

c. If $H$ is a normal subgroup of the group $G$, and $[G : H] = 7$, to what group is $G/H$ necessarily isomorphic?

d. The ring $\mathbb{Z}$ is a subring of $\mathbb{R}$ but is not an ideal. (Why?) We thus were unable to form the ring of cosets $\mathbb{R}/\mathbb{Z}$. But $\mathbb{Z}$ is clearly a normal subgroup, and so we are able to form the group of cosets $\mathbb{R}/\mathbb{Z}$. Explain this.

e. Show that $\{1\}$ is a normal subgroup of any multiplicative group $G$.

f. Consider the set
$$\mathcal{K} = \{A \in U(M_2(\mathbb{R})) : \det(A) = 3\}.$$

By using an argument just like that in Example 25.6, it is easy to see that if $A \in \mathcal{K}$, then $C^{-1}AC \in \mathcal{K}$, for all matrices $C \in U(M_2(\mathbb{R}))$. Does this make $\mathcal{K}$ a normal subgroup of $U(M_2(\mathbb{R}))$? (Be careful!)

---

## Exercises

1. Suppose that $H$ is a normal subgroup of the group $G$. Prove that $H \subseteq g^{-1}Hg$. *Note:* This is an omitted verification in the proof of Theorem 25.2.

2. Suppose that $H$ is a subgroup of the group $G$, and $g^{-1}Hg = H$, for all $g \in G$. Prove that $Hg \subseteq gH$. *Note:* This is another omitted verification in the proof of Theorem 25.2.

3. Are $\langle (124) \rangle$ and $\langle (12) \rangle$ normal subgroups of $S_4$? (Consider Exercise 24.2.) Is $\langle (124) \rangle$ a normal subgroup of $A_4$? (Consider Exercise 24.1.)

4. Is $\langle (123) \rangle$ a normal subgroup of $S_3$? (Consider Example 25.2.)

5. Suppose that $n > 2$ is a positive integer and $1 \leq k \leq n$. Show that the stabilizer subgroup $\text{Stab}(k)$ is not normal in $S_n$. (See Exercise 23.14.)

6. Let $H$ be a (not necessarily normal) subgroup of the group $G$, and $g \in G$.

   (a) Prove that $g^{-1}Hg$ is always a subgroup of $G$. What group is it, if $H$ is normal? The subgroup $g^{-1}Hg$ is called a **conjugate** subgroup of the subgroup $H$. (Note of course that $gHg^{-1} = \left(g^{-1}\right)^{-1} Hg^{-1}$ is a conjugate of $H$ too.)

   (b) See Exercise 22.15, where you showed that conjugation is a group isomorphism from $G$ onto $G$. Use this to argue that conjugate subgroups are isomorphic to one another.

   (c) Argue that if $H_1$ is a conjugate of the subgroup $H_2$, then $H_2$ is a conjugate of $H_1$ (that is, conjugacy is a *symmetric* relation).

7. Suppose that $n$ is a positive integer, and $d$ is a positive integer that (properly) divides $n$. Prove that

$$\mathbb{Z}_n/\langle d \rangle$$

   is isomorphic to $\mathbb{Z}_d$. *Note:* This exercise foreshadows developments in the next chapter and will be easy to do with the main theorem of that chapter; your experience with rings should suggest the appropriate approach. See Exercise 26.4.

8. Suppose that $H$ and $K$ are normal subgroups of the group $G$. Prove that $H \cap K$ is a normal subgroup. *Note:* In Exercise 20.5 you have already shown that $H \cap K$ is a subgroup.

9. Consider the direct product $G \times H$ of the groups $G$ and $H$. Prove that the subgroups $G \times \{1\}$ and $\{1\} \times H$ are normal subgroups of $G \times H$.

10. Show that every subgroup of the group $Q_8$ of quaternions is normal, even though the group is not abelian. (See Warm-up Exercise b. For the standard notation for the quaternions, see Exercise 21.12.)

11. Consider the subgroup $H = \{1, -1\}$ of $Q_8$. In the previous exercise, you showed that $H$ is a normal subgroup of $Q_8$. Give the group table for $Q_8/H$. To what group is this group of cosets isomorphic?

12. Consider the group of cosets $\mathbb{R}^*/\mathbb{Q}^*$. Exhibit an element of this group with order 2. What about an element of order $n$, for any positive integer $n$? Do you believe that there are elements of infinite order in this group? (You should be able to conjecture an element that works, but you will not be able to prove this rigorously.)

13. Suppose that $H$ and $K$ are subgroups of the group $G$, and $K$ is a normal subgroup. Let $HK = \{hk : h \in H, k \in K\}$. (Note that we defined the product of two subgroups in Section 23.1; we observed in Warmup Exercise 23.b and 23.c that this need not be a subgroup, but is a subgroup if $G$ is abelian.)

   (a) Prove that in this case $HK$ is a subgroup of $G$.

   (b) Suppose in addition that $H$ is normal. Prove that $HK$ is a normal subgroup of $G$.

14. Suppose that $G$ is a group with subgroups $H, K$. Suppose further that $H$ is a normal subgroup and $H \subseteq K \subseteq G$. Prove that $H$ is a normal subgroup of the group $K$.

15. Suppose that $G$ is a group with subgroups $H, K$. Suppose further that $K$ is a normal subgroup and $H \subseteq K \subseteq G$. Prove that $H$ need not be a normal subgroup of $K$. (Compare to the previous exercise.)

16. Consider the groups $A_4$ and $H$ described in Example 25.5. Use Exercise 22.15 to provide a complete proof that $H$ is normal in $A_4$.

17. Let $G$ be an abelian group, and let

$$t(G) = \{g \in G : o(g) \text{ is finite}\}.$$

We call $t(G)$ the **torsion subgroup** of $G$.

   (a) Prove that $t(G)$ is a subgroup of $G$.

   (b) Prove that $G/t(G)$ is torsion-free. (Recall that this means that $G$ has no non-identity elements of finite order.)

18. Suppose that $H$ is a normal subgroup of the group $G$. Prove that $G/H$ is an abelian group if and only if $g^{-1}h^{-1}gh \in H$, for all $g, h \in G$. *Note:* Elements of the form $g^{-1}h^{-1}gh$ are called **commutators** in the group $G$.

19. We generalize Exercise 18 slightly. Given a group $G$, define $H$ to be the smallest subgroup of $G$ that contains all the commutators. The subgroup $H$ is called the **commutator subgroup**. Argue that $H$ consists of all finite products of commutators. For a normal subgroup $K$ of $G$, prove that $G/K$ is abelian if and only if $K \supseteq H$.

20. Let $G$ be a group, and consider its center $Z(G)$, as described in Exercise 20.16. Suppose that $H$ is a subgroup of $Z(G)$.

   (a) Prove that $H$ is a normal subgroup of $G$.

   (b) Now suppose that $G/Z(G)$ is a cyclic group. Prove that $G$ is abelian.

# Chapter 26

## The Isomorphism Theorem for Groups

In this chapter we prove the *Isomorphism Theorem for Groups*, the important theorem analogous to the ring theory result we obtained in Chapter 14. The ring theory theorem asserts that knowing about ideals is essentially the same as knowing about ring homomorphisms. We will establish a similar connection between normal subgroups and group homomorphisms.

## 26.1 The Kernel

Let's begin with a homomorphism $\varphi : G \to H$ between the groups $G$ and $H$. By way of analogy with ring theory, we define the **kernel** of $\varphi$ to be

$$\ker(\varphi) = \{g \in G : \varphi(g) = 1\}.$$

We can express this colloquially: The kernel of $\varphi$ is the set of elements in $G$ that are sent to the identity in $H$ by $\varphi$. We can use a slightly different notation to emphasize the definition of the kernel. It is the **pre-image** of the identity; that is,

$$\ker(\varphi) = \varphi^{-1}(1).$$

The following theorem should not be a surprise:

**Theorem 26.1** *Let* $\varphi : G \to H$ *be a homomorphism between the groups* $G$ *and* $H$. *Then* $\ker(\varphi)$ *is a normal subgroup of* $G$.

**Proof:** To show that $\ker(\varphi)$ is a subgroup, choose two elements $a, b \in \ker(\varphi)$; we must show that $ab^{-1}$ is in the kernel (by the Subgroup Theorem 20.2). To show this, we compute:

$$\varphi\left(ab^{-1}\right) = \varphi(a)(\varphi(b))^{-1} = 1.$$

That is, $ab^{-1} \in \ker(\varphi)$.

To show that $\ker(\varphi)$ is normal, we must choose an arbitrary element $g \in G$, and check that

$$g^{-1} \ker(\varphi) g \subseteq \ker(\varphi)$$

(by Theorem 25.2). For that purpose, choose $a \in \ker(\varphi)$, and compute again:

$$\varphi\left(g^{-1}ag\right) = (\varphi(g))^{-1}\varphi(a)\varphi(g) = (\varphi(g))^{-1}1\varphi(g) = 1.$$

That is, $g^{-1}ag \in \ker(\varphi)$, as required. $\qquad\square$

Let's look at some examples of kernels of homomorphisms.

▷ **Quick Exercise.** Review several of the examples of kernels of ring homomorphisms, discussed in Chapter 12. ◁

Next, let's compute the kernels of a number of the homomorphisms in the examples in Chapter 22.

## Example 26.1

(Example 22.3) Consider the function $\rho : \mathbb{Z} \to \mathbb{Z}$ that multiplies elements by 3. The equation $\rho(n) = 3n = 0$ has only one solution, and so the kernel is $\{0\}$, which is evidently a normal subgroup.

## Example 26.2

(Example 22.4) Consider the logarithm function $\log : \mathbb{R}^+ \to \mathbb{R}$. The equation $\log(r) = 0$ has a unique solution, and so again the kernel here is the trivial subgroup, this time written multiplicatively: $\{1\}$.

## Example 26.3

(Example 22.5) Consider the determinant function

$$\det : U(M_2(\mathbb{R})) \to \mathbb{R}^*.$$

Here, the kernel is the subgroup

$$\mathcal{H} = \{A \in U(M_2(\mathbb{R})) : \det(A) = 1\}.$$

It is difficult to describe the subgroup $\mathcal{H}$ more succinctly. But note that it has infinitely many elements.

▷ **Quick Exercise.** Provide 4 distinct matrices belonging to $\mathcal{H}$. ◁

By Theorem 26.1, $\mathcal{H}$ is normal; in Example 25.6 we showed directly that this is the case. We will return to this example yet again, later in this chapter.

## Example 26.4

(Example 22.6) Consider the homomorphism $\varphi : D_3 \to \{1, -1\}$ that takes the rotations $\{1, \rho, \rho^2\}$ to 1, and the remaining elements (the flips) to $-1$. Obviously, the kernel is the subgroup $\langle \rho \rangle$. According to Theorem 26.1, this is a normal subgroup.

## Example 26.5

(Example 22.7) Consider the homomorphism $\Phi : G \to \mathbb{Z}_3$, where $G$ is the group of symmetries of the tetrahedron. (We've seen in Example 20.11 that $G$ is (isomorphic to) the group $A_4$.) If you check the discussion in Example 22.7, it is obvious that the kernel is

$$\{\iota, \varphi_1, \varphi_2, \varphi_3\}.$$

In Example 25.5 we started to verify that this subgroup (or rather, the corresponding subgroup of the corresponding group of permutations) is normal, by brute force. It follows easily here, now that we note that it is the kernel of a homomorphism. (If you did Exercise 25.16, you obtained another proof that this subgroup is normal.)

**Example 26.6**

(Example 22.8) Consider the differentiation function $D : \mathbb{R}[x] \to \mathbb{R}[x]$. The kernel consists of exactly those polynomials whose derivative is the zero polynomial. That is, the kernel consists of the constant polynomials:

$$\ker(D) = \mathbb{R} \subseteq \mathbb{R}[x].$$

**Example 26.7**

(Examples 22.11 and 22.12) The kernel of the embedding homomorphism $\epsilon : G \to G \times H$ defined by $\epsilon(g) = (g, 1)$ is the trivial subgroup of $G$. The kernel of the projection homomorphism $\pi : G \times H \to G$ defined by $\pi(g, h) = g$ is the subgroup

$$\{1\} \times H = \{(1, h) : h \in H\}.$$

▷ **Quick Exercise.** Verify that the kernels of this example are what is claimed. Check directly that these are normal subgroups (or see Exercise 25.9). ◁

## 26.2   Cosets of the Kernel

The kernel contains indirectly more information than just the pre-image of the identity. Just as for rings, we have:

**Theorem 26.2** *Let $\varphi : G \to H$ be a homomorphism between the groups $G$ and $H$. Then $\varphi^{-1}(h)$ equals the coset $\ker(\varphi)g$, where $g$ is any given element of $\varphi^{-1}(h)$.*

**Proof:**   This is left for you to prove, using the proof of the corresponding ring result (Theorem 12.2) as a model; see Exercise 6. □

An important special consequence of this theorem follows in the case when the kernel is the trivial subgroup. In that case, *every* inverse image of the homomorphism consists of a single element. In other words, the homomorphism in question is injective. This is directly analogous to the results for rings (Corollary 12.4).

**Corollary 26.3** *A group homomorphism is injective if and only if its kernel is the trivial subgroup.*

▷ **Quick Exercise.**   Check this corollary for each of the examples above. ◁

We now know that each group homomorphism leads to a normal subgroup, namely, its kernel. But each normal subgroup likewise leads to a homomorphism (of which it is the kernel). This requires another definition, again analogous to the ring case:

Let $H$ be a normal subgroup of the group $G$. Then form the group of cosets $G/H$. Consider the function

$$\nu : G \to G/H$$

defined by $\nu(g) = Hg$. It is quite evident (from the definition of the group operation on $G/H$) that this is a homomorphism, that its kernel is exactly $H$, and that it is surjective. (See Exercise 3.)

We call $\nu$ the **natural homomorphism from $G$ onto $G/H$**.

## 26.3    The Fundamental Theorem

It remains to show that any onto group homomorphism is 'essentially the same' as the natural homomorphism from the domain group onto the group of cosets formed from the kernel. This is the next theorem.

**Theorem 26.4    Fundamental Isomorphism Theorem for Groups**      *Let $\varphi : G \to H$ be a surjective homomorphism between groups, and let $\nu : G \to G/\ker(\varphi)$ be the usual natural homomorphism. Then there exists an isomorphism $\mu : G/\ker(\varphi) \to H$ such that $\mu \circ \nu = \varphi$.*

We exhibit this situation in the following diagram:

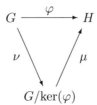

**Proof:**    Suppose that $G$ and $H$ are groups, $\varphi : G \to H$ is a surjective group homomorphism, and $\nu : G \to G/\ker(\varphi)$ is the natural homomorphism. Clearly, what we need to do is define a function $\mu$, and prove that it has the desired properties. In this proof (unlike the ring case), you will be doing the required verifications.

Choose an arbitrary element of $G/\ker(\varphi)$. Such an element is a coset of the form $\ker(\varphi)g$, where $g \in G$. What element of $H$ should it correspond to? If the composition of functions required in the theorem is to work as we wish, we must have that

$$\mu(\ker(\varphi)g) = \varphi(g).$$

And so, this is how we define the map $\mu : G/\ker(\varphi) \to H$.

There is an immediate problem we must solve: This definition apparently *depends on the particular representative g*. That is, our function does not appear to be unambiguously defined. But it is well defined; you will check this in Exercise 5.

After showing that $\mu$ is well defined, we must next show that $\mu$ is a group isomorphism. This requires showing that:

$\mu$ preserves the group operation,

$\mu$ is surjective,

and that

$\mu$ is injective.

You check these things in Exercise 5.                                          □

We now look at a couple of examples of this theorem.

▷ **Quick Exercise.**    Review the ring theory examples in Chapter 14. ◁

## Example 26.8

Consider again the determinant homomorphism

$$\det : U(M_2(\mathbb{R})) \to \mathbb{R}^*.$$

This homomorphism is surjective.

▷ **Quick Exercise.** Why? ◁

We have calculated its kernel in Example 26.3: It is $\mathcal{H}$, the set of all matrices with determinant 1. The Fundamental Isomorphism Theorem for Groups 26.4 now allows us to conclude that the groups $U(M_2(\mathbb{R}))/\mathcal{H}$ and $\mathbb{R}^*$ are isomorphic, as we conjectured in Example 25.6. In fact, from the proof of the Fundamental Isomorphism Theorem we can extract the function that establishes the isomorphism; namely, $\mu(\mathcal{H}A) = \det(A)$.

The inverse function, which is also necessarily an isomorphism, should assign to each non-zero number $r$ the coset of a matrix whose determinant is $r$. In the previous chapter we laboriously determined that each coset of $\mathcal{H}$ consists of those matrices with a certain determinant. We proved this again in this chapter, in a more general context, because we proved that the cosets of $\mathcal{H} = \ker(\varphi)$ are exactly the inverse images of real numbers under the function det. Because of this, to define the inverse function of $\mu$, *any* choice of such a matrix with the appropriate determinant will work as a representative for the coset. Here is a particularly understandable version:

$$\mu^{-1}(r) = \mathcal{H} \begin{pmatrix} r & 0 \\ 0 & 1 \end{pmatrix}.$$

## Example 26.9

Let's return to the derivative function of Example 26.6. It is evidently onto $\mathbb{R}[x]$.

▷ **Quick Exercise.** Prove that the derivative function is surjective. (Use a little calculus!) ◁

Thus, the Fundamental Isomorphism Theorem 26.4 says that the groups $\mathbb{R}[x]/\mathbb{R}$ and $\mathbb{R}[x]$ are isomorphic. (Isn't this strange?)

## Chapter Summary

In this chapter we introduced the notion of the *kernel* of a group homomorphism. It is a normal subgroup, and every normal subgroup is the kernel of some homomorphism. Furthermore, we proved the *Fundamental Isomorphism Theorem for Groups*, which asserts that every onto homomorphism can be viewed as a natural homomorphism onto the group modulo its kernel.

## Warm-up Exercises

a. Explain why every normal subgroup is the kernel of some homomorphism.

b. Explain why knowing the kernel of a homomorphism tells us essentially everything about the homomorphism.

c. Can you tell if a homomorphism is surjective, by just looking at its kernel?

d. Can you tell if a homomorphism is injective, by just looking at its kernel?

## Exercises

1. Consider the function $\varphi : \mathbb{R} \to \mathbb{S}$ defined by

$$\varphi(r) = e^{2\pi r i} = \cos(2\pi r) + i\sin(2\pi r),$$

   where $\mathbb{S}$ is the unit circle of Example 19.14. Show that $\varphi$ is a surjective homomorphism. What is the kernel of $\varphi$? What two groups are isomorphic, according to the Fundamental Isomorphism Theorem 26.4?

2. Consider the subgroup $G$ of $U(M_2(\mathbb{R}))$ given by

$$G = \left\{ \begin{pmatrix} a & 0 \\ b & 1 \end{pmatrix} : a \neq 0 \right\}.$$

   (See Exercise 20.19.) Define $\varphi : G \to \mathbb{R}^*$ by letting

$$\varphi \begin{pmatrix} a & 0 \\ b & 1 \end{pmatrix} = a.$$

   (a) Prove that this is a surjective group homomorphism.

   (b) What is the kernel of this homomorphism?

   (c) What two groups does the Fundamental Isomorphism Theorem assert are isomorphic?

   (d) What is the relationship between this exercise and Example 26.8?

3. Suppose $H$ is a normal subgroup of $G$. Show that the natural homomorphism $\nu : G \to G/H$ is indeed a homomorphism, is surjective, and has kernel equal to $H$, as claimed in Section 26.2.

4. Let's redo Exercise 25.7, in light of the Fundamental Isomorphism Theorem. Suppose that $n$ is a positive integer and $d$ is a positive integer which (properly) divides $n$. Prove that

$$\mathbb{Z}_n / \langle d \rangle$$

   is isomorphic to $\mathbb{Z}_d$.

5. Complete the proof of the Fundamental Isomorphism Theorem 26.4, relying if necessary on the corresponding proofs for the Fundamental Isomorphism Theorem for Commutative Rings 14.1:

   (a) Show that the function $\mu$ is well defined (that is, it does not depend on the coset representative chosen).

   (b) Show that the function $\mu$ is a group isomorphism (that is, that it preserves the group operation and is a bijection).

6. Prove Theorem 26.2, using Theorem 12.2 as a model, if necessary: Suppose that $\varphi : G \to H$ is a surjective group homomorphism, and $h \in H$. Prove that $\varphi^{-1}(h)$ equals the coset of $\ker(\varphi)g$, where $g$ is any element in $\varphi^{-1}(h)$.

7. Let $G$ be a group and fix some $g \in G$. Consider the map $\varphi : \mathbb{Z} \to G$ defined by $\varphi(n) = g^n$. Show that $\varphi$ is a group homomorphism. What are the possible kernels for $\varphi$? Describe the image of $\varphi$ in $G$.

8. Consider the function $\varphi : \mathbb{C}^* \to \mathbb{R}^+$ defined by $\varphi(\alpha) = |\alpha|$. Prove that $\varphi$ is a surjective homomorphism. What two groups are isomorphic, according to the Fundamental Isomorphism Theorem?

9. Suppose that $G$ is a group with normal subgroups $H, K$. Consider the normal subgroups $H \cap K$ and $HK$. (See Exercises 25.8, 25.13, 25.14.) Prove that the groups $H/(H \cap K)$ and $HK/K$ are isomorphic. (You should compare this to the corresponding result in ring theory, in Exercise 14.23.)

10. Consider the group $\mathbb{R}^3$, and suppose that $a, b, c \in \mathbb{R}$, not all zero. Define $\varphi : \mathbb{R}^3 \to \mathbb{R}$ by $\varphi(x, y, z) = ax + by + cz$. Show that $\varphi$ is a group homomorphism. What is the kernel of $\varphi$ (considered *geometrically*)? What does the Fundamental Isomorphism Theorem say in this context? What is the geometric meaning of this assertion? (Compare this exercise to Exercise 23.4.)

11. Consider the dihedral group $D_4$. Every element of $D_4$ can be written in the form $F^i R^j$, where $i = 0, 1$, and $j = 0, 1, 2, 3$, where $F$ is the 'flip' about the vertical axis $(12)(34)$, and $R$ is the counterclockwise rotation $(1432)$. (See Exercise 20.29.) Define

$$\psi : D_4 \to \mathbb{Z}_2 \times \mathbb{Z}_2$$

by setting $\psi(F^i R^j) = ([i]_2, [j]_2)$.

   (a) Prove that this is a surjective homomorphism.

   (b) What is the kernel of this homomorphism?

   (c) The subgroup you obtained in part (b) is necessarily normal. (Why?) Prove this directly.

   (d) What two groups does the Fundamental Isomorphism Theorem assert are isomorphic?

12. Suppose that $G$ is a group with normal subgroup $H$.

   (a) Define a correspondence between the subgroups of $G$ containing $H$, and the subgroups of $G/H$, by assigning $K \to K/H$. Why is Exercise 25.14 relevant to this definition?

   (b) Show that the correspondence from part (a) is a bijection.

   (c) Furthermore, show that this correspondence also establishes a bijection between *normal* subgroups of $G$ containing $H$ and *normal* subgroups of $G/H$.

13. Suppose that $H$ and $K$ are normal subgroups of the group $G$, and that $K \subseteq H$. Prove that the group $G/H$ is isomorphic to

$$(G/K)/(H/K).$$

(You should compare this to the corresponding result in ring theory, in Exercise 14.21.)

14. Suppose that $H$ is a finite normal subgroup of the group $G$, and that $K$ is a normal subgroup of $G$ for which $G/K$ is finite. Suppose further that the integers $|H|$ and $|G/K|$ are relatively prime. Prove that $H \subseteq K$. Hint: Use Exercise 9.

15. Let

$$S = \{(\ldots, x_{-1}, x_0, x_1, \ldots) : x_k \in \mathbb{Z}\}.$$

That is, $S$ is the set of integer-valued sequences indexed by $\mathbb{Z}$ (rather than $\mathbb{N}$). Then define

$$W = \{(a, \vec{x}) : a \in \mathbb{Z}, \vec{x} \in S\}.$$

We define the following binary operation $+$ on $W$:

$$(a, \vec{x}) + (b, \vec{y}) = (a + b, \vec{z})\},$$

where $z_k = x_k + y_{k+a}$.

(a) Show that the operation $+$ is *not abelian* (even though we are using additive notation).

(b) Show that $+$ is associative.

(c) Show that $+$ makes $W$ a group.

(d) Consider $B = \{(0, \vec{x}) : \vec{x} \in S\}$. Show that $B$ is an abelian subgroup which is normal in $W$.

(e) Find a group homomorphism with kernel $B$. What two groups are isomorphic, according to the Fundamental Isomorphism Theorem?

(f) If you have done Exercise 23.18, interpret the group $W$ as a semidirect product of $\mathbb{Z}$ and $S$.

16. For a group $G$, recall the group of inner automorphisms $\text{Inn}(G)$, considered in Exercise 22.16. Define $\Phi : G \to \text{Inn}(G)$ by setting $\Phi(g) = \varphi_g$, where $\varphi$ is the inner automorphism of $G$ defined by $\varphi_g(h) = ghg^{-1}$.

(a) Prove that $\Phi$ is a surjective homomorphism.

(b) Prove that its kernel is the center $Z(G)$ (see Exercise 20.16).

(c) What two groups does the Fundamental Isomorphism Theorem assert are isomorphic?

# Section V in a Nutshell

This section starts by considering group homomorphisms, by way of analogy with ring homomorphisms: A function between groups $\varphi : G \to S$ is a *group homomorphism* if $\varphi(gh) = \varphi(g)\varphi(h)$ for every $g, h \in G$. A group homomorphism preserves the group identity and inverses, the image of $G$ is a subgroup of $S$, and if $G$ is abelian then so is $\varphi(G)$ (Theorems 22.1 and 22.2). If $\varphi$ is a bijection, then we say $\varphi$ is an *isomorphism*, in which case $\ker(\varphi) = 1_G$.

A subgroup $H$ of $G$ is *normal* if $gH = Hg$ for all $g \in G$; that is, if the left and right cosets of $H$ are the same for each element of $G$. This is equivalent to saying that $g^{-1}Hg \subseteq H$ for all $g \in G$. If $H$ is normal in $G$, then the collection of cosets of $H$ in $G$, denoted $G/H$, forms a group under the operation $(Ha)(Hb) = Hab$ (Theorem 25.1). $G/H$ is then called the *group of cosets of G mod H* or the *quotient group of G mod H*. Normal subgroups are to groups what ideals are to rings. The kernel $\ker(\varphi)$ of any group homomorphism is a normal subgroup of $G$.

Paralleling rings, there is the *Fundamental Isomorphism Theorem for Groups* (Theorem 26.4) which says that if $\varphi : G \to S$ is an onto homomorphism, then $G/\ker(\varphi)$ is isomorphic to $S$.

Whether the subgroup $H$ is normal in $G$ or not, the number of left cosets of $H$ is the same as the number of right cosets. Assuming $G$ is finite, this number is called the *index of H in G* and denoted by $[G : H]$. *Lagrange's Theorem* (Theorem 24.2) says that $|G| = [G : H]|H|$. Thus the order of a subgroup must divide the order of the group and so the order of any element must divide the order of the group. Thus, any group of prime order is cyclic.

A group $G$ may not have a subgroup for every order dividing $|G|$; however $G$ does have a subgroup of order $p$ for each prime $p$ that divides $|G|$ (Cauchy's Theorem 24.5).

Any group can be thought of as a subgroup of a group of permutations; indeed, *Cayley's Theorem* (Theorem 23.3) says that every group of order $n$ is isomorphic to a subgroup of $S_n$.

We can characterize when a group is a direct product of smaller groups. The *Internal Characterization Theorem* for two subgroups (Theorem 23.1) says:
The group $G$ is isomorphic to $H_1 \times H_2$ if and only if $G$ has two subgroups $G_1$ and $G_2$, so that

a. $G_1$ is isomorphic to $H_1$, and $G_2$ is isomorphic to $H_2$.

b. $G = G_1 G_2$,

c. $G_1 \cap G_2 = \{1\}$, and

d. every element of $G_1$ commutes with every element of $G_2$.

Theorem 23.2 generalizes this theorem to more than two subgroups.

# Part VI

# Topics from Group Theory

# Chapter 27

## The Alternating Groups

In this chapter we inquire further into the symmetry groups $S_n$. In the process we gain further insight into the importance and significance of normal subgroups.

## 27.1 Transpositions

For obvious reasons, we call a cycle of length 2 a **transposition**.

We now show that any permutation can be factored as a product of transpositions.

**Theorem 27.1** *Any permutation can be factored as a product of transpositions.*

**Proof**: Because every permutation is a product of cycles (Theorem 18.3), it clearly suffices to show that each cycle can be factored as a product of transpositions. But this is easy: Consider the cycle $(a_1 a_2 \cdots a_n)$ of length $n$. Clearly,

$$(a_1 a_2 \cdots a_n) = (a_1 a_n)(a_1 a_{n-1}) \cdots (a_1 a_3)(a_1 a_2).$$

$\square$

Notice that the method of factorization suggested by the proof of the theorem provides $n - 1$ transpositions for each cycle of length $n$. So the 5-cycle $(12345)$ can be factored into a product of 4 transpositions:

$$(12345) = (15)(14)(13)(12).$$

The product of cycles can then also be factored into a product of transpositions:

$$(1345)(267) = (15)(14)(13)(27)(26).$$

However, this method of factoring into transpositions does *not* give us a factorization which is unique (unlike the factorization into disjoint cycles in Theorem 18.3).

### Example 27.1

Here's a different factorization of the last permutation into transpositions:

$$(1345)(267) = (25)(67)(57)(25)(45)(35)(15).$$

(Don't worry about where this factorization came from.)

▷ **Quick Exercise.** Check that this factorization works. Can you come up with another essentially different factorization of this permutation into transpositions? ◁

**Example 27.2**

A surprising place where permutations and transpositions occur is in the theory of *bell ringing*. A bell ringer rings $n$ bells. *Ringing the changes* means ringing each of the $n!$ orders of the $n$ bells exactly once. The $n$ bells are typically arranged in a row on a table and the ringer starts on the left and rings each bell in sequence. According to bell ringing practice, after a certain permutation is rung, the next permutation to be rung must be the same, except for one transposition. This is true because the bell ringer has only two hands with which to interchange two bells! Our theorem in essence asserts that we can get from any one order of the bells to any other, by repeated transpositions.

▷ **Quick Exercise.** Suppose that a bell ringer has six bells, in order $1, 2, 3, 4, 5, 6$. Give a list of transpositions, which when done one after another, results in the bells in order $6, 1, 4, 3, 5, 2$. ◁

---

## 27.2 The Parity of a Permutation

Our earlier example revealed that two distinct factorizations of a given permutation into transpositions need not even have the same number of factors. For example, our first factorization of $(1345)(267)$ had five factors, while our second had seven factors. In fact, any given permutation can be expressed as a product of transpositions in an infinite number of ways. To see this, we need only notice that $\iota = (12)(12)$ and that $(ab) = (ca)(cb)(ca)$.

▷ **Quick Exercise.** Why does this mean that any permutation can be expressed as a product of transpositions in infinitely many ways? ◁

However, it does turn out that *any* factorization for the permutation $(1345)(267)$ will have an odd number of factors. In fact, for any permutation, all the factorizations of that permutation into transpositions involve an even number of transpositions, or they all involve an odd number of transpositions. To see that the number of factors is always even or always odd, consider the polynomial

$$g_n = (x_1 - x_2)(x_1 - x_3)\cdots(x_1 - x_n)(x_2 - x_3)\cdots(x_{n-1} - x_n)$$

$$= \prod_{i<j}(x_i - x_j).$$

Notice that this polynomial consists of the product of all terms of the form $x_i - x_j$, where $i < j$, and so each pair $x_i, x_j$ appears together in a factor of this product exactly once. Now consider a permutation of the $n$ terms $x_1, \ldots, x_n$. We can think of such a permutation as acting on the polynomial $g_n$, and it will clearly map $g_n$ either to $g_n$ or to $-g_n$, as the order of the subscripts $i$ and $j$ are scrambled by the permutation.

Now clearly a single transposition maps $g_n$ to $-g_n$. So, the product of an even number of transpositions will map $g_n$ to $g_n$ and the product of an odd number of transpositions will map $g_n$ to $-g_n$. Thus, if a particular permutation of $n$ elements leaves $g_n$ unchanged, then clearly *any* representation of this permutation as a product of transpositions must do the same, and so such a product must have an even number of tranpositions. Similarly, if the permutation changes $g_n$ to $-g_n$, any representation of this permutation as a product of transpositions must contain an odd number of transpositions.

So, we can now call a permutation **even** if it can be expressed as a product of an even number of transpositions, and **odd** if it can be expressed as a product of an odd number of tranpositions. Saying whether a permutation is even or odd is to specify its **parity.**

▷ **Quick Exercise.** Determine whether the following permutations are even or odd: $\iota$, $(123)(4789)$, $(123)(342)$. ◁

---

## 27.3 The Alternating Groups

Let $A_n$ be the set of even permutations. Take two such permutations, and multiply them together. Any representation of this product as a product of transpositions will have an even number of transpositions. Thus, the product of two even permutations is even. That is, $A_n$ is closed under multiplication. But what about inverses? Given a permutation represented as a product of transpositions, its inverse can be computed as the product of the same transpositions, in the opposite order!

▷ **Quick Exercise.** Why is the inverse of a product of transpositions the same product in the opposite order? ◁

Thus, the inverse of an even transposition is still even. This all means that $A_n$ is a subgroup of $S_n$. We call $A_n$ the **alternating group**.

Note that the argument we have just given does *not* mean that the set of odd permutations forms a subgroup: The sum of two odd integers is even, and so the product of two odd permutations is an even permutation. And there's an even simpler reason why the odd permutations cannot form a subgroup: Any subgroup must contain the identity, and the identity is an even permutation (because 0 is even).

Note that cycles of *odd* length are the *even* permutations!

▷ **Quick Exercise.** Why are cycles of odd length even permutations? ◁

**Example 27.3**

> We have looked at the group $A_4$ before (see Example 20.11). It is isomorphic to the group of symmetries of the tetrahedron and consists of the twelve permutations
>
> $$\iota, \ (123), \ (132), \ (124), \ (142), \ (234), \ (243),$$
> $$(134), \ (143), \ (12)(34), \ (13)(24), \ (14)(23).$$

▷ **Quick Exercise.** List the elements in the group $A_3$. ◁

In the examples we've looked at, it certainly appears that exactly half the permutations are even, which would mean that the group $A_n$ has exactly $n!/2$ elements (where $n > 1$, to avoid the trivial case $A_1$). This is quite easy to prove. Consider the function

$$\Lambda : A_n \to S_n \backslash A_n$$

defined by $\Lambda(\alpha) = (12)\alpha$. That is, $\Lambda$ merely multiplies each element by the transposition $(12)$. If $\alpha$ is even, quite evidently $(12)\alpha$ is odd, and so this function is well defined. But $\Lambda$ is also a bijection (see Exercise 27.2.) Thus, there are as many even permutations as odd. Notice of course that $\Lambda$ establishes that the sets $A_n$ and $S_n \backslash A_n$ have the same number of elements, but it is certainly *not* a homomorphism.

## 27.4 The Alternating Subgroup Is Normal

It is now easy to see that the alternating group $A_n$ is a normal subgroup of $S_n$; the argument we've just done shows that $[S_n : A_n] = 2$, and so the result follows from the Index 2 Theorem 25.3. We will however provide here another proof, which will give us added insight into the arithmetic of permutations.

By Theorem 25.2, to show that $A_n$ is normal, we must check that for any permutation $\alpha$,

$$\alpha^{-1} A_n \alpha \subseteq A_n.$$

In other words, we must show that if $\beta$ is even, then $\alpha^{-1}\beta\alpha$ is too.

We introduce some general terminology and notation to deal with this. Let $G$ be a group and $g, h \in G$. We call the element $g^{-1}hg$ the **conjugate** of $h$ by $g$ (we introduced this terminology in Exercise 22.15). Our goal then can be rephrased as this: We must show that *conjugation preserves the parity of permutations*. (Note that in Exercise 22.15 you proved that conjugates have the same order.) But in fact, much more is true:

**Theorem 27.2  Conjugation Theorem**  *Conjugation in $S_n$ preserves disjoint cycle structure. That is, if we express an element $\alpha$ as a product of $m$ disjoint cycles of lengths $k_1, k_2, \cdots, k_m$, then any conjugate of $\alpha$ can be expressed as a product of $m$ disjoint cycles, of lengths $k_1, k_2, \cdots, k_m$.*

Before proceeding to the proof of this theorem, let's look at a couple of examples.

**Example 27.4**

Consider the element $\alpha = (12)(3456)(789)$, the product of a 2-cycle, a 3-cycle, and a 4-cycle. Consider some other permutation, like $\beta = (14567)(239)$. Then $\beta^{-1} = (76541)(932)$, and

$$\beta^{-1}\alpha\beta = \beta^{-1}(153724689) = (1452)(368)(79).$$

This new element has exactly the same disjoint cycle structure as $\alpha$.

▷ **Quick Exercise.**  Now pick some other permutation $\beta$ at random, and use it to conjugate $\alpha$. Is the disjoint cycle structure preserved? ◁

**Example 27.5**

In Chapter 17 we looked at $S_3$ as the set of symmetries of the equilateral triangle. Let's consider what conjugation might mean geometrically in this context, and thus understand why we should expect it to preserve disjoint cycle structure.

In particular, consider the symmetry $\varphi$, which is represented by the permutation $(12)$. Recall that $\varphi$ is the reflection of the triangle through the vertical line $\ell$. Let's see what the conjugate $\rho^{-1}\varphi\rho$ of $\varphi$ by $\rho$ means, where $\rho$ rotates the triangle $120°$ counterclockwise. The rotation $\rho$ re-orients the triangle, so that the flip $\varphi$ will permute the vertices at original positions 1 and 2. We then rotate back via $\rho^{-1}$. We have in effect accomplished the flip $\rho\varphi$ through the line $m$. (See the diagram below.) In other words, $\rho^{-1}\varphi\rho = (13)$.

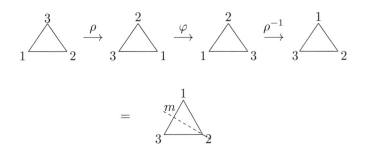

A metaphor for the interpretation of conjugation in this example might be illustrative. Suppose you, an English-speaking mathematics student, are asked to solve a mathematical problem written in French! Your thought process would probably be this: Translate the problem into English (operation $\beta$), solve the problem (operation $\alpha$), and then translate back into French ($\beta^{-1}$). The general message is this: If a group element can be thought of as an operation or process, then any conjugate of it will be an operation or process of a similar type. Thus the conjugate of a 3-cycle is a 3-cycle; the conjugate of a flip is a flip.

This discussion suggests the general method of proof for our theorem:

**Proof of the Conjugation Theorem:** We prove the theorem first in the case where $\alpha$ is a cycle of length $k$. So suppose that $\alpha = (a_1 a_2 \cdots a_k)$, and $\beta$ is an arbitrary permutation. Then we claim that $\beta^{-1}\alpha\beta$ is the $k$-cycle

$$\chi = (\beta^{-1}a_1\ \beta^{-1}a_2\ \cdots\ \beta^{-1}a_k).$$

By direct computation it is easy to see that the conjugate $\beta^{-1}\alpha\beta$ and the $k$-cycle $\chi$ behave in the same way on the support of $\chi$

$$\mathrm{Supp}(\chi) = \{\beta^{-1}a_1, \beta^{-1}a_2, \cdots, \beta^{-1}a_k\}.$$

▷ **Quick Exercise.** Check this. ◁

Now suppose $m$ is some integer not belonging to this set; that is, $\chi(m) = m$. Then $\beta(m)$ is not in the set $\{a_1, \cdots, a_k\}$. That is, $\beta(m)$ is not in the support of $\alpha$. Thus,

$$\beta^{-1}\alpha\beta(m) = \beta^{-1}\beta(m) = m = \chi(m).$$

We have just shown that $\chi = \beta^{-1}\alpha\beta$, and so $\beta^{-1}\alpha\beta$ is also a cycle of length $k$.

Now suppose that $\alpha$ has been factored as a product

$$\chi_1\chi_2\cdots\chi_n$$

of disjoint cycles. Note that

$$\beta^{-1}\alpha\beta =$$
$$\beta^{-1}\chi_1\beta\beta^{-1}\chi_2\beta\cdots\beta^{-1}\chi_n\beta,$$

and so the conjugate of $\alpha$ is evidently a product of cycles of the same length as the corresponding cycles making up $\alpha$. It remains only to show that these cycles are disjoint. But if $\beta^{-1}\chi_i\beta$ and $\beta^{-1}\chi_j\beta$ both move $r$, then $\chi_i$ and $\chi_j$ would both move $\beta^{-1}(r)$, which is impossible, since $\chi_i$ and $\chi_j$ are disjoint. □

▷ **Quick Exercise.** Check this theorem for three distinct conjugates of the permutation $(13)(54789)$. ◁

Now it is easy to see (again) that $A_n$ is a normal subgroup of $S_n$:

**Corollary 27.3** *The alternating group is a normal subgroup of the symmetric group.*

**Proof**:    A conjugate of an even permutation has the same disjoint cycle structure. But this means that the conjugate is even, too.                                            □

**Example 27.6**

As another example of the conjugation theorem, consider again the subgroup

$$\{\iota, (12)(34), (13)(24), (14)(23)\}$$

of the alternating group $A_4$. Any conjugate of a non-identity element of this group has the same disjoint cycle structure (a product of two disjoint transpositions). But the subgroup consists of the complete collection of elements of this form (together with $\iota$). We first looked at this in Example 25.5 where we found it tedious to prove completely the normality of this subgroup. In Example 26.5 we argued that it is normal by exhibiting a homomorphism, of which it is the kernel.

---

## 27.5   Simple Groups

As we've seen already, $A_n$ is always normal in $S_n$, for all positive integers $n$. Does $A_n$ have any normal subgroups? The answer for $n = 4$ is certainly yes (as we've seen above in Example 27.6). But it is possible to prove that for $n \geq 5$, $A_n$ itself has no proper normal subgroups (except the trivial subgroup). This proof is quite a bit more technical, but similar in flavor, to the proof of the Conjugation Theorem 27.2. Note that these groups $A_n$ have many non-trivial proper subgroups (just consider any cyclic subgroup). However, no such subgroup is normal.

This situation merits a definition: We call a group with no proper normal subgroups (except the trivial subgroup) a **simple** group. The following theorem gives a large collection of simple groups.

**Theorem 27.4** *The groups $A_n$ are simple, for $n \neq 4$.*

**Proof**:    For $n < 4$, showing that $A_n$ is simple is left as Exercise 27.4. We will now prove that $A_n$ is simple, for $n \geq 5$. For that purpose, suppose that $N$ is a non-trivial normal subgroup of $A_n$. Choose a non-identity element $\alpha \in N$ whose support is of minimal size. We wish to show that $\alpha$ is a 3-cycle.

Note first that in its disjoint cycle representation, all factors of $\alpha$ must be cycles of the same size. For if $\alpha = (a_1 \cdots a_k)(b_1 \cdots b_m) \cdots$, where $k < m$, then $\alpha^k \in N$, but $\alpha^k$ leaves the integers $a_i$ fixed, and so has a smaller support than $\alpha$.

Suppose next that $\alpha$ is a product of disjoint 2-cycles (that is, transpositions). So, we may suppose (without loss of generality) that $\alpha = (12)(34) \cdots$, where the remainder of $\alpha$ consists of further disjoint transpositions. Because $N$ is normal, we then have that $\beta = (123)\alpha(123)^{-1}\alpha^{-1} \in N$. But $\beta = (13)(24)$, and so because $\alpha$ has a support of minimal size in $N$, we must have that $\alpha = (12)(34)$. Now $n > 4$, and so $(125) \in A_n$. But then $(125)(12)(34)(125)^{-1}((12)(34))^{-1} \in N$. But this latter element is $(152)$, and this contradicts the minimality of the support of $\alpha$.

We now may suppose that $\alpha$ is a product of disjoint $k$-cycles, where $k \geq 3$. Without loss of generality we then have that

$$\alpha = (123\cdots)(456\cdots)\cdots.$$

But then $(12534) = (124)\alpha(124)^{-1}\alpha^{-1} \in N$, which again contradicts the minimality of the support of $\alpha$.

At this point we have concluded that $\alpha$ must be a single odd $k$ cycle, with $k \geq 3$. Now if $k > 3$, we may suppose that $\alpha = (12345\cdots)$. But then $(124) = (123)\alpha(123)^{-1}\alpha^{-1} \in N$. This yet again contradicts the minimality of the support of $\alpha$, and so we conclude that $\alpha$ itself must be a 3-cycle.

We may as well suppose that $\alpha = (123)$. But for any other 3-cycle $(abc)$, we can construct a permutation $\tau$ by first setting $\tau(a) = 1, \tau(b) = 2, \tau(c) = 3$, and then filling in the other values of $\tau$ in such a way that we have a permutation. It is then clear that $\tau^{-1}(123)\tau = (abc)$. By normality, this would say that $(abc) \in N$, if $\tau \in A_n$. However, as we have constructed it $\tau$ may not be an even permutation (that is, an element of $A_n$). However, if the permutation $\tau$ we have first constructed is odd, the permutation $\tau(45)$ will certainly belong to $A_n$, and still take $(123)$ to $(abc)$ by conjugation. Of course, this last step is permissible only because $n \geq 5$.

We have thus shown that if $N$ is a non-trivial normal subgroup of $A_n$, it contains all 3-cycles. But by Exercise 27.12, the set of 3-cycles generates $A_n$, and so $N = A_n$, as we claimed. $\square$

It turns out that this theorem will be of considerable importance to us in Chapter 48, when we prove that polynomial equations with degree greater than 4 cannot always be solved by ordinary arithmetic and root extractions.

There are other simple groups that are very familiar to us:

**Example 27.7**

> Consider the cyclic group $\mathbb{Z}_5$. This group has no non-trivial proper subgroups. If $H$ is any non-trivial subgroup, choose a non-identity element $h \in H$. By Lagrange's Theorem 24.2, this non-identity element has order 5, and so $\langle h \rangle = \mathbb{Z}_5$. Thus, $\mathbb{Z}_5$ is simple.

The argument of this example can clearly be extended, and so we have:

**Theorem 27.5** *The groups $\mathbb{Z}_p$ are simple, for all primes $p$.*

▷ **Quick Exercise.** Prove this by repeating the argument above for $\mathbb{Z}_p$, $p$ prime. ◁

# Historical Remarks

In the theory of finite groups that has been developed in the 20th century, simple groups can be viewed as the building blocks out of which more complicated groups can be constructed. Thus, to understand all finite groups, we need to understand simple groups, and how they can be put together to form more complicated groups. Both questions are very difficult.

One of the triumphs of 20th-century mathematics has been the complete classification of all finite simple groups. Dozens of mathematicians contributed to the solution of this problem, and a full proof would require many hundreds of pages of technical mathematics. We have met two families of finite simple groups—namely, the cyclic groups of prime order, and the alternating groups—and there are other such families. But part of the difficulty in the classification theorem lies in the fact that there are some finite simple groups that do not belong to such naturally defined families. These simple groups are called *sporadic*, and the last of these 26 groups was finally constructed in the 1980s.

The problem of putting together simple groups (called the *extension problem*) is far from solved. The easiest way to put together smaller groups to build more complicated ones is the idea of direct product, which we understand well. But there are much more complicated ways (including the idea of semidirect product we explored in Exercise 23.18). We will turn our attention to how abelian groups can be put together in certain ways to form a much larger class called *solvable* groups, in Chapter 30.

The general program for understanding finite groups is typical in much of algebra. We wish to prove a *structure theorem*: All objects of a given type can be built in a specified way, from well-understood pieces. In Section 29.3 we will discuss such a program for finite abelian groups. In this case, the well-understood pieces turn out to be cyclic groups, and the method of putting them together is the idea of direct product. In Chapter 45, we will provide a structure theorem for finite fields.

---

## Chapter Summary

In this chapter we proved that the set of even permutations forms a subgroup of the symmetric group, called the *alternating group*. We proved the *Conjugation Theorem*, which says that conjugation preserves disjoint cycle structure. It follows that $A_n$ is normal in $S_n$. If $n \neq 4$, then $A_n$ has no proper normal subgroups; such a group is *simple*.

---

## Warm-up Exercises

a. Compute explicitly three conjugates of the permutation

$$(145)(96)(237),$$

   checking that its disjoint cycle structure is preserved.

b. How many elements are there in $A_5$? $A_6$?

c. Because conjugation preserves parity, the conjugate of an odd permutation is odd. Why doesn't that make the set of odd permutations a normal subgroup?

d. Why can't a simple group have a non-trivial subgroup of index 2?

e. Give an example of a non-normal subgroup, whose index is 3. (Thus, Theorem 25.3 is false if 2 is replaced by 3.)

f. Is the identity permutation even or odd?

g. Suppose the operation $\beta$ means "paint the north wall of the room," and the operation $\alpha$ means "move to your sister's room." What is the meaning of the conjugate $\alpha^{-1}\beta\alpha$? Remember that functional composition is read from right to left.

# Exercises

1. Is the product of an even permutation and an odd permutation always an odd permutation? Prove this, or give a counterexample.

2. Show that the map $\Lambda : A_n \rightarrow S_n \backslash A_n$ defined by $\Lambda(\alpha) = (12)\alpha$ is a bijection, completing the argument of Section 27.3 that there are as many even as odd permutations. (The notation $S_n \backslash A_n$ means the elements of $S_n$ that are not in $A_n$.)

3. (a) List the elements of $A_5$. (See Exercise b.)

   (b) List all cyclic subgroups of $A_5$.

   (c) Let $\alpha \in A_5$. Show $\langle \alpha \rangle$ is not a normal subgroup of $A_5$.

4. Show that $A_n$ is simple for $n < 4$, proving part of Theorem 27.4.

5. By following the proof of the Conjugation Theorem 27.2, demonstrate explicitly that

$$(1456)(29) \quad \text{and} \quad (7895)(13)$$

are conjugates in $S_9$.

6. Let $n \geq 5$. Prove that $A_n$ is the only non-trivial normal subgroup of $S_n$.

7. Prove that any permutation can be expressed as a product of the transpositions $(12), (13), (14), \cdots, (1n)$.

8. Prove that any permutation can be expressed as a product of the transpositions $(12), (23), (34), \cdots, (n-1 \; n)$.

9. Prove that $S_n$ is generated by the two elements $(1234 \cdots n)$ and $(12)$.

10. In Example 25.6 we showed that $A_4$ has no subgroup with 6 elements. Provide a different proof for this fact: Assume that $H$ is such a subgroup; then $H$ must be normal. (Why?) Now count the three cycles. Show that at least one of the three cycles must be in $H$. Then argue that $H$ must contain all the three cycles, which is absurd.

11. Let $n \geq 2$ and let $H$ be a subgroup of $S_n$. Prove either that all elements of $H$ are even, or else exactly half of the elements of $H$ are even.

12. Let $n$ be a positive integer, $n \geq 3$. Prove that $A_n$ is generated by its 3-cycles. That is, prove that every element of $A_n$ can be expressed as a product of 3-cycles.

13. Suppose that $H$ is a subgroup of the permutation group $S_n$. We say that $H$ is a **transitive subgroup** if for any integers $k, j$ with $1 \leq k, j \leq n$, we can find $\alpha \in H$ so that $\alpha(k) = j$.

    (a) Determine all transitive subgroups of $S_4$.

    (b) Argue that $A_n$ is a transitive subgroup of $S_n$, for any $n > 2$.

    (c) Suppose that $H$ is a transitive subgroup of $S_n$, and $\alpha \in S_n$. Prove that the conjugate subgroup $\alpha^{-1}H\alpha$ is also a transitive subgroup of $S_n$. (See Exercise 25.6 for more about conjugate subgroups.)

# Chapter 28

## Sylow Theory: The Preliminaries

Lagrange's Theorem 24.2 tells us that the order of every subgroup of a finite group must divide the order of the group. It is natural to ask if the converse of this theorem holds. We get a partial answer from Cauchy's Theorem 24.5 that says that if the order of the group is divisible by prime $p$, then the group must have an element of order $p$. Thus the group must at least have a subgroup of order $p$. However, the converse of Lagrange's Theorem does not hold in general, as we encountered in Example 24.6, where we discovered that the twelve element group $A_4$ does not have a subgroup with 6 elements. (For alternative proofs for this result, see Exercise 24.19 or Exercise 27.10.)

Therefore, we cannot guarantee that in general there is a subgroup for every possible divisor of the order of the group. However, we can guarantee the existence of subgroups of some orders, find out the relationship between such subgroups, and even count how many such subgroups there are. These results are called the Sylow theorems (first proved by the Norwegian mathematician Peter Sylow, 1832–1918). These results focus attention on one prime divisor at a time.

So suppose that $G$ is a finite group, and $p$ is a prime integer that divides the order of $G$. We may then suppose that $|G| = p^n m$, where the exponent $n$ is positive and $m$ is relatively prime to $p$. The first Sylow theorem will show that $G$ *does* have subgroups of all orders $p^k$, for $k = 0, 1, \ldots n$. To prove this theorem (and the theorems that follow) we need some preliminary material, and those preliminaries will be the focus of this chapter.

---

## 28.1 $p$-groups

We first turn our attention to finite groups whose order is divisible by only one prime. Given a prime integer $p$, a $p$-**group** is a group where the order of every element is some power of $p$. (There are according to this definition $p$-groups with infinitely many elements, but in these chapters our emphasis is entirely on finite groups; see Exercise 1.) We record first an easy consequence:

**Theorem 28.1** *Let $G$ be a finite group. Then $G$ is p-group if and only if $|G| = p^n$.*

**Proof:** If $G$ is a $p$-group, and some other prime $q$ divided the order of $G$, then by Cauchy's Theorem 24.5 $G$ would have an element of order $q$, which is impossible. Conversely, if $|G| = p^n$, Corollary 24.3 to Lagrange's Theorem says that every element has order $p^k$, $k \leq n$.

### Example 28.1

The abelian groups $\mathbb{Z}_2$, $\mathbb{Z}_{16}$, $\mathbb{Z}_2 \times \mathbb{Z}_4$, and $\mathbb{Z}_{1024}$ are all 2-groups. The non-abelian groups $D_4, Q_8$ are also 2-groups. We could directly check that every element of these groups has order equal to a power of 2, or by Theorem 28.1 just compute the order of the group, in each case.

Now one might suppose that in any finite group $G$, if we consider the set of elements of order $p^k$, for a fixed prime $p$ and varying exponents $k$, that this might form a $p$-group which is a subgroup of $G$. However, this is definitely false! Note for example, Theorem 27.1, which asserts that *every* element of the group $S_n$ is a product of transpositions, which are, of course, elements of order two. Similarly, in Exercise 27.12, you prove that every even permutation in $S_n$ can be expressed as a product of 3-cycles, which are of course of order three. In short, in an arbitrary group the elements of order $p^k$ for a fixed prime $p$ are not close to being a subgroup. However, this statement is true for abelian groups. The following theorem shows this, and more.

**Theorem 28.2** *Let $G$ be a finite abelian group with order $p^n m$, where $p$ is a prime, and $p$ and $m$ are relatively prime. Let $P$ be the set of all elements in $G$ whose orders divide $p^n$, and let $M$ be the set of all elements in $G$ whose orders divide $m$. Then $G$ is isomorphic to the direct product of $P$ and $M$.*

**Proof**: We first show that $P$ is in fact a subgroup. But if $g, h \in P$, then $(gh)^{p^n} = g^{p^n} h^{p^n} = 1$, and so the order of $gh$ divides $p^n$. Thus $gh \in P$. Since the order of $g$ and $g^{-1}$ are the same, $P$ is obviously closed under taking inverses too, and so $P$ is a subgroup. The argument that $M$ is also a subgroup is essentially the same.

Now we use the Internal Characterization Theorem 23.1. Note that any element in $P \cap M$ must have an order that divides both $p^n$ and $m$. But these integers are relatively prime, and so such an element must be the identity. Since $G$ is abelian, it is clear that elements from $P$ commute with elements from $M$.

To apply the Internal Characterization Theorem it thus remains to prove that $G = PM$. Now since $p^n$ and $m$ are relatively prime, by the GCD identity there exist integers $r, s$ with

$$rp^n + sm = 1.$$

But then we have that

$$g = g^1 = g^{rp^n + sm} = g^{rp^n} g^{sm}.$$

But consider the element $b = g^{sm}$. Clearly $b^{p^n} = 1$ and so $b \in P$. Similarly, if $a = g^{rp^n}$, then $a^m = 1$ and so $a \in M$. This means that $g = ab \in PM$, as required. The theorem now follows from Theorem 23.1. $\square$

An easy induction on the number of primes that divide the order the abelian group yields the following corollary (see Exercise 3):

**Corollary 28.3** *Suppose that the order of the finite abelian group $G$ is divisible by the distinct prime integers $p_1, \ldots, p_n$. Let $P_i$ be the subgroup of $G$ consisting of those elements whose orders are powers of $p_i$. Then $G$ is isomorphic to the direct product of the $n$ groups $P_i$.*

To rephrase: Every finite abelian group can be represented as a direct product of $p_i$-groups, where $p_1, \ldots, p_n$ is the list of the prime factors of $|G|$. There is actually more to say about finite abelian groups than this, and we will return to this topic in Section 29.3.

Sylow theory then asks how much of this structure can we recover, should the group not be abelian.

## 28.2 Groups Acting on Sets

We have another preliminary topic to consider, before we can begin the proof of the Sylow theorems. To aid us in this task, we need to consider a group as a group of permutations of a set. We have seen instances when it is useful to consider groups whose elements permute elements of a set. The most obvious examples are the full symmetric groups $S_n$, which consist of *all* the permutations of $\{1, 2, \ldots, n\}$ (which we first considered in Chapter 18), and also the groups of symmetries of geometric figures, which we considered in Chapter 17 and thereafter. An important general instance of this approach is Cayley's Theorem 23.3, which says that *every* finite group $G$ is isomorphic to a subgroup of $S_G$, the permutations of the set of elements of the group itself. In the proof of Cayley's Theorem we found an isomorphism that mapped each element of $G$ to a permutation of the elements of $G$ in a natural way. Note that for even relatively modest sized $G$, $S_G$ can be enormous. We shall see below that this mapping from the proof of Cayley's Theorem is by no means the only way a group can be realized as a group of permutations of some set.

For yet another example of this approach, we can turn to a multiplicative group of matrices, like $U(M_2(\mathbb{R}))$; we recall from Example 19.11 that this group consists of the $2 \times 2$ matrices whose determinants are non-zero. You may recall from linear algebra that such matrices act bijectively on the set of vectors in the plane, by matrix multiplication (see Exercise 4).

We now make our general definition: We say the group $G$ **acts on a set** $X$ if there is a homomorphism $\sigma : G \to S_X$. Thus $\sigma(G)$ is a subgroup of $S_X$, the group of all permutations of $X$. Equivalently we say $G$ is a **group of permutations** of $X$. The notation can become a bit cumbersome. We would write $\sigma(g)$ as the permutation of $X$ that $g$ is mapped to and so if $x \in X$, $\sigma(g)(x)$ would be the element of $X$ that $x$ is mapped to under the permutation $\sigma(g)$. To make the notation less cumbersome, we will simply eliminate the $\sigma$ and *identify* the group element with the permutation to which it is associated, and then use functional notation. That is, we will understand $g(x)$ as the element of $X$ that the permutation associated with $g$ maps $x$ to.

Using this functional notation for the permutation associated with an element of the group, the fact that $\sigma$ is a homomorphism from the group operation in $G$ to the functional composition in the group of permutations amounts to asserting that $(gh)(x) = g(h(x))$, for all $g, h \in G$ and $x \in X$. Below we will describe this by saying that the action is **homomorphic**. Note we do not insist that $\sigma$ be an injection here. Indeed, sometimes two different elements of $G$ may be associated with the same permutation on $X$, as we will see in some of the examples below.

We now illustrate the idea of a group acting on a set with some examples, with special emphasis on those actions we will be using to prove the Sylow theorems in the next chapter.

**Example 28.2**

Recall in the proof of Cayley's Theorem 23.3 that we have the finite group $G$ acting on $G$ (so $X = G$) by *left multiplication*; that is, $g(x) = gx$. The notation is confusing the first time you see it. With the term $g(x)$ we think of the element $g$ as a permutation of the set $G$; that is, an element of $S_G$. But in the term $gx$ we use $g$ as an element of the group $G$ and the operation is the group operation of $G$.

First note that the action described really is a permutation on the elements of $G$. For clearly $g(x) = gx \in G$ and $x \neq y$ if and only if $gx \neq gy$. Note furthermore that this action is homomorphic: We must show that if $g, h, x \in G$, then $g(h(x)) = (gh)(x)$. But

$$g(h(x)) = g(hx) = ghx = (gh)(x).$$

▷ **Quick Exercise.** Show furthermore that in this case the map taking the group element $g$ to a permutation is injective (in the proof of Cayley's theorem we called this map $\Phi$). That is, if $g \neq h$, then we claim that $g(x) \neq h(x)$ for some $x \in X = G$. (This would make $g$ and $h$ two different permutations of $X$ since $g$ and $h$ would map at least one element of $X$ to different places. In fact here $g$ and $h$ map *every* element of $X$ to different places. Right?) ◁

## Example 28.3

Let $H$ be a subgroup of $G$. Let $G$ act on the set of *left cosets* of $H$ by *left multiplication*. That is, the action is $g(kH) = gkH$. We claim that this gives a permutation of the left cosets. But recall from the left-handed version of the Coset Theorem (Exercise 24.3) that $kH = hH$ if and only if $k^{-1}h \in H$. But $(gk)^{-1}(gh) = k^{-1}g^{-1}gh = k^{-1}h$ and so different left cosets are taken to different left cosets. Furthermore, this action is homomorphic: $g(h(kH)) = ghkH = (gh)(kH)$.

We will consider this action for any subgroup $H$ of our group $G$, whether or not $H$ is a normal subgroup. In this chapter we will use the notation $G/H$ for the set of left cosets of $H$, whether or not $G/H$ is actually a group. In this example we have thus considered our group $G$ as acting on the set $X = G/H$.

Let's look at this action for a particular group $G = S_3$, and a particular subgroup $H = \langle(123)\rangle$. Here $X = G/H = \{\iota H, (12)H\}$ (see Example 25.2). Every element of $G = S_3$ then corresponds to a permutation on the two element set $X = G/H$. Then the three elements $\iota, (123), (132)$ all perform the action which interchanges the two elements of $X$, while the elements $(12), (23), (13)$ all correspond to the identity permutation on $X$. Note that while this action is homomorphic, it is not injective: More than one element of the group $G = S_3$ corresponds to the same element of $S_X$.

▷ **Quick Exercise.** Check the claims of the previous paragraph by explicit calculuation. ◁

In Exercise 2 you will consider the action of $G = S_3$ on the set $G/\langle(12)\rangle$. (See Example 25.1 for this set of left cosets.)

Notice that we could also restrict this action by considering just $H$ as acting on the set of its left cosets by left multiplication. This action is actually only interesting if $H$ is not a normal subgroup, because if $H$ is normal, then

$$h(kH) = hkH = kk^{-1}hkH = kH,$$

where the latter equality holds because $k^{-1}hk \in H$, when $H$ is normal. Thus, in this case the action is the identity on $X$, for all $h \in H$. But this restricted action in the non-normal case will be useful for us in the proof of the First Sylow Theorem.

## Example 28.4

Let $G$ act on $G$ by *conjugation* given by $g(x) = gxg^{-1}$; see Exercise 22.15. In that exercise you show that the function $g(x)$ is a bijection (that is, a permutation of $G$). We claim also that this action is homomorphic: That is, we claim that $g(h(x)) = (gh)(x)$.

▷ **Quick Exercise.** Show this. (This is essentially Exercise 26.16a.) ◁

## Example 28.5

Let $X$ be the set of all subgroups of $G$. Let $G$ act on $X$ by conjugation; that is, $g(H) = gHg^{-1}$. We have shown in Exercise 25.6 that if $H$ is a subgroup, then so is $gHg^{-1}$; recall that we call $gHg^{-1}$ a *conjugate* of the subgroup $H$. Thus this action really does map subgroups to subgroups. We claim that for any given $g \in G$, this establishes a bijection from the set $X$ onto itself. This is most easily seen by observing that the function $g(x) : X \to X$ has an inverse function, namely conjugation by $g^{-1}$.

To see that $G$ acts on $X$ it remains to show that this action is homomorphic. But,

$$g(h(H)) = g(hHh^{-1}) = ghHh^{-1}g^{-1} = ghH(gh)^{-1} = (gh)(H).$$

Notice of course that if a subgroup $H$ is a normal subgroup, then $gHg^{-1} = H$ for all $g \in G$. That is, normal subgroups are left fixed by this action.

Now if $G$ is a group of permutations of the set $X$ then two elements $x, y$ of $X$ are called **$G$-equivalent** if and only if there exists $g \in G$ such that $g(x) = y$. You will show in Exercise 5 that $G$-equivalence is indeed an equivalence relation.

The $G$-equivalence classes are called the **orbits** of $G$. So the orbit of $x \in X$ is

$$\mathrm{Orb}(x) = \{y \in X : g(x) = y, \text{ for some } g \in G\}.$$

We actually used this terminology in Section 18.4, when we first discussed permutation groups. For a general permutation $\alpha \in S_n$ acting on the set $X = \{1, 2, \ldots, n\}$, its orbits are thus precisely the elements of $X$ making up the disjoint cycles in the factorization in Theorem 18.3.

So for any group of permutations on a set $X$, we have that the orbits *partition* $X$ into pairwise disjoint sets, since the orbits of the elements are the equivalence classes. So, we have the following counting theorem:

**Corollary 28.4** *Suppose $G$ is a group of permutations of the finite set $X$. If $x_1, \ldots, x_n$ are representatives chosen from each of the disjoint orbits of $G$, then*

$$|X| = \sum_{i=1}^{n} |Orb(x_i)|.$$

## Example 28.6

Consider Example 28.2, where $G$ acts on itself by left multiplication. Here, $\mathrm{Orb}(x) = \{y : y = gx, \text{ for some } g \in G\}$. But for every $y$ there is always such a $g$, namely $yx^{-1}$. So $\mathrm{Orb}(x) = G$ for every $x \in G$. That is, there is but one orbit.

## Example 28.7

Consider Example 28.3, where $G$ acts on the left cosets $G/H$ by multiplication on the left. Here $\mathrm{Orb}(gH) = \{kH : kH = hgH, \text{ for some } h \in G\}$. But here too there is only a single orbit, because for any $k \in G$, $kH = kg^{-1}gH$.

If $G$ is a group of permutations of the set $X$ and there is only a single orbit, then we say that $G$ acts **transitively** on $X$. Speaking more concretely, this means that for any $x, y \in X$, there exists $g \in G$ for which $g(x) = y$. The previous two examples illustrate this case. (We introduced this terminology earlier in Exercise 27.13.)

**Example 28.8**

Consider Example 28.4, where $G$ acts on itself by conjugation. If $x \in G$, then $\text{Orb}(x) = \{y : y = gxg^{-1}, \text{ for some } g \in G\}$. That is, $\text{Orb}(x)$ is the set of conjugates of $x$. For an interesting example of this, suppose that $G = S_n$. Then by the Conjugation Theorem 27.2, the orbit of a given permutation is precisely the set of all permutations with the same disjoint cycle structure.

**Example 28.9**

Consider Example 28.5, where $G$ acts on its set $X$ of subgroups by conjugation. For a given subgroup $H$ of $G$, we then have that $\text{Orb}(H) = \{gHg^{-1} : g \in G\}$, which is precisely its set of conjugate subgroups. Now since conjugation is a group isomorphism (Exercise 22.15), this means that if $H$ is finite, then the subgroups in its orbit all have the same number of elements.

If $G$ acts on $X$ and $x \in X$, then the **stabilizer** of $x$ is

$$\text{Stab}(x) = \{g \in G : g(x) = x\}.$$

That is, $\text{Stab}(x)$ is the set of those elements of $G$ that do not move $x$. Note that we introduced this concept in the context of permutation groups in Exercises 23.14–23.17, and in particular, verified that $\text{Stab}(x)$ is a subgroup of $G$.

▷ **Quick Exercise.**   Check that $\text{Stab}(x)$ is a subgroup of $G$ (referring to Exercise 23.14 if necessary). ◁

**Example 28.10**

In Example 28.4, $h \in \text{Stab}(g)$ if and only if $hgh^{-1} = g$. But this means that $hg = gh$. Thus $\text{Stab}(g) = \{h : hg = gh\}$, the subgroup of all elements that commute with $g$ . In Exercise 20.15 we called this the *centralizer* of $g$, and denoted it by $C(g)$.

**Example 28.11**

In Example 28.5, $\text{Stab}(H) = \{g \in G : gHg^{-1} = H\}$. We call this set the **normalizer** of $H$ and denote it by $N(H)$. We then have that $N(H)$ is a subgroup of $G$ and that $H$ is a normal subgroup of $N(H)$. Furthermore, $N(H)$ is the largest subgroup of $G$ in which $H$ is normal. You will prove this (and more) in Exercise 9.

It is important to remember that if $G$ is a group acting on the set $X$, then $\text{Orb}(x)$ is a subset of $X$ and $\text{Stab}(x)$ is a subgroup of $G$. Surprisingly, perhaps, there is a strong connection between these two sets, as the next theorem shows.

**Theorem 28.5** *If $G$ is a finite group of permutations of the set $X$, then the size of $\text{Orb}(x)$ is the index of the subgroup $\text{Stab}(x)$ in $G$; that is*

$$|\text{Orb}(x)| = [G : \text{Stab}(x)].$$

**Proof:**    Suppose that $g, h \in G$, $x \in X$, and $g(x) = h(x)$. Then $g^{-1}h(x) = x$ and this happens if and only if $g^{-1}h \in \text{Stab}(x)$. But by the left-handed Coset Theorem 24.3, this is true if and only if $h\,\text{Stab}(x) = g\,\text{Stab}(x)$. That is, $g$ and $h$ perform the same action on $x$ if and only if $g$ and $h$ are in the same left coset of $\text{Stab}(x)$.

The orbit of $x$ consists of the points $\{g(x) : g \in G\}$. So, there is a bijection between distinct points in the orbit of $x$ and the left cosets of $\text{Stab}(x)$. Thus, $|\text{Orb}(x)| = [G : \text{Stab}(x)]$. □

**Corollary 28.6** *If $G$ is a finite group of permutations of the finite set $X$, then the size of every orbit of $G$ divides $|G|$.*

**Proof:** $|G| = |\text{Stab}(x)|[G : \text{Stab}(x)] = |\text{Stab}(x)| \, |\text{Orb}(x)|.$ □

**Example 28.12**

For a simple example of this theorem, consider the subgroup of $S_5$ given by

$$G = \{\alpha \in S_5 : \alpha(\{1, 2\}) = \{1, 2\}\}.$$

It of course still acts on the set $X = \{1, 2, 3, 4, 5\}$. This group has $12 = 3!2!$ elements, because every element is determined by a permutation of $\{3, 4, 5\}$ and a permutation of $\{1, 2\}$.

▷ **Quick Exercise.** List these 12 elements. ◁

Now consider the orbit $\text{Orb}(3)$. Clearly, this is $\{3, 4, 5\}$, and we have that 3 divides 12. Similarly, $\text{Orb}(1)$ has two elements, which also divides 12.

Now if $G$ is a group of permutations of the set $X$, we say that $x$ is a **fixed point** for the element $g$ if $g(x) = x$. As in Section 18.4, we denote by $\text{Fix}(g)$ the set of all such points; this is of course a subset of the set $X$, called the **fixed set** of $g$, and we can then consider the set of points in $X$ left fixed by *all* the elements of the group $G$, and we denote this by $\text{Fix}(G)$, called the **fixed set** of $G$. Clearly $\text{Fix}(G) = \cap\{\text{Fix}(g) : g \in G\}$.

Our final theorem in this chapter brings together these ideas for a $p$-group, connecting the number of elements in $X$ with the number of fixed points. This result will be very useful in proving the Sylow theorems in the next chapter.

**Theorem 28.7    Fixed Point Theorem**    *Let $G$ be a finite $p$-group acting on a finite set $X$. Then*

$$|X| \equiv |Fix(G)| \bmod p.$$

**Proof:** Suppose that $G$ acts on $X$, and has $n$ distinct orbits; we choose $x_1, \ldots, x_n$ to be elements of each of these disjoint orbits. Then, by Corollary 28.4

$$|X| = \sum_{i=1}^{n} |\text{Orb}(x_i)|.$$

But by Theorem 28.5 $|\text{Orb}(x_i)| = [G : \text{Stab}(x_i)]$. Because $G$ is a $p$-group we know by Theorem 28.1 that $|G|$ is a power of $p$, so $[G : \text{Stab}(x_i)]$ is also a power of $p$, including possibly $p^0 = 1$. Thus $|\text{Orb}(x_i)| \equiv 0 \bmod p$ unless $\text{Stab}(x_i) = G$, in which case $x_i$ is a fixed point, and so $|\text{Orb}(x_i)| = 1$. So, the terms of $\sum_{i=1}^{n} |\text{Orb}(x_i)|$ are all 0 modulo $p$ except for the fixed points. The result then follows. □

This now concludes the preliminaries we will need to prove the Sylow theorems themselves in the next chapter.

---

# Chapter Summary

In this chapter we introduced the idea of $p$-group, and proved that a finite abelian group is isomorphic to a direct product of $p$-groups. We then considered the general notion of a group acting on a set, and proved some important counting theorems relating to the orbits, stabilizers and fixed sets defined in this situation.

# Warm-up Exercises

a. Give two non-isomorphic 3-groups with 9 elements.

b. Suppose that $G$ is an abelian group with 45 elements. Why does $G$ have an element with order 15?

c. Find two elements in $S_5$ with order 3, whose product has order 5.

d. Every finite group can be thought of as a group acting on a set. Why?

e. Let $G$ be a group acting on the set $X$, and suppose that $H$ is a subgroup of $G$. Why can we consider $H$ a group acting on $X$?

f. Consider the subgroup $G = \langle (123) \rangle$ of the group $S_4$, which is acting on the set $\{1, 2, 3, 4\}$. Note that $G$ is a 3-group. Interpret Theorem 28.7 for this example.

# Exercises

1. Consider the set $G$ of all infinite sequences, with entries drawn from the additive group $\mathbb{Z}_p$, where $p$ is a prime integer. That is

$$G = \{(x_1, x_2, \ldots) : x_n \in \mathbb{Z}_p, n \in \mathbb{N}\}.$$

   Show that $G$ is a group under addition, and all of its non-zero elements have order $p$. That is, this is an infinite $p$-group.

2. Consider the group $G = S_3$, and consider its action on the set $X = G/\langle (12) \rangle$ of left cosets of the cyclic subgroup in $G$ generated by the permutation $(12)$. Explicitly write down the correspondence between the elements of the group, and the permutations of the set $X$. Is the action homomorphic? Is the map from $G$ to $S_X$ injective? (See Example 28.3.)

3. Prove Corollary 28.3, using an induction argument on the number of prime factors of the order of the group.

4. Consider the group $G = U(M_2(\mathbb{Z}))$, of all $2 \times 2$ matrices with real entries, whose determinants are non-zero (we discussed this example in Section 28.2). Argue that $G$ acts on the set of all 2-vectors

$$X = \left\{ \begin{pmatrix} a \\ b \end{pmatrix} : a, b \in \mathbb{R} \right\}$$

   by multiplication on the left. Be sure to argue that this action does in fact establish a bijection on $X$, and that the action is homomorphic. Argue also that the map from $G$ to $S_X$ is injective.

5. Suppose that $G$ is a group of permutations of the set $X$. Recall from Section 28.2 that $x, y \in X$ are $G$-equivalent if there exists $g \in G$ with $g(x) = y$. Argue that this is an equivalence relation.

6. Let $S$ be a subgroup of the group $G$. We wish to consider $G$ acting on the right cosets of $S$ by right multiplication. That is, $g(Sh) = Shg$. Show that this does *not* work in general. That is, $G$ is not a group of permutations of the right cosets of $S$. What goes wrong here? (The problem in this example, and the ones given in the next two exercises, can be rectified if we consider "right actions" instead of "left actions" which is our approach in the text. This takes us too far afield from our discourse, and we leave the interested readers to research this on their own.)

7. We wish to let $G$ act on $G$ by the conjugation $g(x) = g^{-1}xg$. Note this is different from the conjugation we used in Example 28.4. Show that this does not work in general. What goes wrong here?

8. We ask the same question as the previous exercise, where we try to have $G$ act on its set of subgroups by the conjugation, $g(S) = g^{-1}Sg$. What goes wrong here?

9. Given a group $G$ and a subgroup $H$, consider the normalizer $N(H)$ as defined in Example 28.11.

   (a) Since the normalizer is the stabilizer of an element under a group action, we know that it is a subgroup. Give a direct proof of this fact, from the definition of the normalizer.

   (b) Show that $N(H)$ is the largest subgroup of $G$ which contains $H$, in which $H$ is a normal subgroup. What is $N(H)$ if $H$ is a normal subgroup of $G$?

   (c) Recall the subgroup $C(H)$, the centralizer of $H$, as defined in Exercise 20.15. Why is $C(H)$ a normal subgroup of $N(H)$?

   (d) Prove that $N(H)/C(H)$ is isomorphic to a subgroup of $\operatorname{Aut}(H)$, the group of automorphisms of $H$ discussed in Exercise 20.27.

10. Suppose that $\varphi : G_1 \to G_2$ is a surjective group homomorphism, and that $G_2$ is a group acting on the set $X$. How can we consider $G_1$ as a group acting on $X$?

# Chapter 29

## Sylow Theory: The Theorems

We will now apply the theory from the last chapter to obtain the Sylow theorems. These theorems will enable us to understand the $p$-subgroups of any finite group, taken one prime $p$ at a time. Throughout this chapter we will assume that $G$ is a finite group, and that it has $p^n m$ elements, where $p$ is a prime, and $m$ is relatively prime to $p$. We will be able to show that $G$ has subgroups of order $p^n$ (called *Sylow $p$-subgroups*), that all such subgroups are isomorphic, and we will often be able to count how many of these subgroups there are.

### 29.1   The Sylow Theorems

Given our finite group $G$ with $p^n m$ elements, suppose that $H$ is a subgroup of $G$ and a $p$-group. We then say that $H$ is a $p$-**subgroup** of $G$. We call $H$ a **Sylow $p$-subgroup** of $G$ if it is a maximal $p$-subgroup of $G$. That is, if $H$ is a Sylow $p$-subgroup of $G$ and $K$ is a $p$-subgroup of $G$ that contains $H$, then $K = H$. Because $G$ is a finite group, it is then obvious that every $p$-subgroup is contained in a Sylow $p$-subgroup. What is not yet obvious is how many elements a Sylow $p$-subgroup has.

We claim first that every conjugate of a Sylow $p$-subgroup is also a Sylow $p$-subgroup. To see this, first recall that conjugation of subgroups of $G$ gives an isomorphism between subgroups (see Exercise 25.6). Now suppose that $P$ is a Sylow $p$-subgroup of $G$ and $g \in G$, but suppose by way of contradiction that $gPg^{-1}$ is not a Sylow $p$-subgroup of $G$. Then there exists a $p$-subgroup $Q$ of $G$ with proper subgroup $gPg^{-1}$. But then $g^{-1}Qg$ properly contains $P$ and is a $p$-subgroup of $G$, contradicting that $P$ is a maximal $p$-subgroup. Notice that if $G$ has only one Sylow $p$-subgroup, then it must be a normal subgroup of $G$.

**Example 29.1**

> Consider the alternating group $A_4$. This group has $12 = 2^2 \cdot 3$ elements. We have considered in both Example 25.5 and 27.6 the 4 element subgroup $\{\iota, (12)(34), (13)(24), (14)(23)\}$. Because this subgroup includes all of the elements of $A_4$ of order two, this is the only Sylow 2-subgroup of $A_4$ and so (as we have seen before) it is normal in $A_4$. In contrast, a Sylow 3-subgroup of $A_4$ can have only 3 elements and hence must be a cyclic subgroup of order 3; we know that all of these subgroups are conjugate to one another. There are exactly 4 such subgroups, namely
>
> $$\langle (123) \rangle, \quad \langle (124) \rangle, \quad \langle (234) \rangle, \quad \langle (134) \rangle.$$
>
> This is the complete list, because we have included all eight elements of order 3.

We are now ready to prove the *First Sylow Theorem*, which generalizes Cauchy's Theorem 24.5 that says if $p$ divides the order of a finite group $G$, then $G$ contains an element (and hence a subgroup) of order $p$.

**Theorem 29.1    First Sylow Theorem**    *Let $G$ be a finite group, $p$ a prime, and $|G| = p^n m$ where $p$ does not divide $m$. Then $G$ has a subgroup of order $p^k$ for $k = 0, 1, \ldots, n$. Indeed, for $k = 0, \ldots, n-1$, each subgroup of order $p^k$ is a normal subgroup of a subgroup of order $p^{k+1}$.*

**Proof:**    Let $|G| = p^n m$, with $p$ prime and $p$ not dividing $m$. We need to show that $G$ has a subgroup of order $k$ for $k = 0, 1, \ldots, n$.

If $k = 0$, then $p^0 = 1$ and so the trivial subgroup (the subgroup consisting of only the identity) is the desired $p$-subgroup. We now proceed by induction on $k$, and so will assume the existence of a $p$-subgroup $H$ of order $p^k$, with $k < n$, and show $H$ is contained in a subgroup of order $p^{k+1}$.

So consider a subgroup $H$ of $G$ with order $p^k$, where $k < n$. We now consider $H$ acting on the left cosets $G/H$ of $H$ in $G$ by left multiplication, as in Example 28.3. Remember of course that probably $H$ is not normal in $G$, and so $G/H$ need not be a group. To find our subgroup of order $p^{k+1}$ we will actually need a group of cosets; we will find this below.

Now by the Fixed Point Theorem 28.7 we have that

$$|G/H| = [G : H] \equiv |\text{Fix}(G/H)| \bmod p.$$

Let's now consider these fixed points. If $gH$ is a point fixed by $H$, then $hgH = gH$ for all $h \in H$. But by the left-handed Coset Theorem Exercise 24.3 this is true if and only if $g^{-1}hg \in H$ for all $h \in H$, that is, exactly if $g^{-1}Hg \subseteq H$. But since $|g^{-1}Hg| = |H|$, $g^{-1}Hg \subseteq H$ if and only if $gHg^{-1} = H$. This means exactly that $g \in N(H)$, the normalizer of $H$ in $G$ (see Example 28.11). Thus the set of fixed points of $G/H$ under left multiplication by $H$ is $\{gH : g \in N(H)\} = N(H)/H$, which is a group. Therefore the above fixed point equation becomes

$$[G : H] \equiv [N(H) : H] \bmod p.$$

Since $|H| = p^k$ with $k < n$, then $p$ divides $[G : H] = p^n m / p^k$. The fixed point equation tells us then that $p$ divides $|N(H)/H|$. Cauchy's Theorem 24.5 then says that the group $N(H)/H$ has a subgroup of order $p$ and so there is a subgroup $K$ of $N(H)$ that contains $H$ where $K/H$ is isomorphic to this subgroup in $N(H)/H$ (see Exercise 26.12). Thus $|K| = p \cdot p^k = p^{k+1}$, as we desired.

Note also that since $K$ is a subgroup of $N(H)$, $H$ is normal in $K$ (see Exercise 25.14). And so each subgroup of order $p^k$, for $k = 0, \ldots, n-1$ is a normal subgroup of some subgroup of order $p^{k+1}$.                                                                    □

The proof of this theorem reveals even more about $p$-subgroups. *Every* $p$-subgroup of order $p^k$ is a normal subgroup of a $p$-subgroup of order $p^{k+1}$, and in fact can be found in an increasing tower of $p$-subgroups, which culminate in a $p$-subgroup of size $p^n$, which is then necessarily a maximal $p$-subgroup, that is, a Sylow $p$-subgroup. The following two corollaries record some of this formally:

**Corollary 29.2** *If $G$ is a finite group and $p$ is prime, then every $p$-subgroup of $G$ is contained in a Sylow $p$-subgroup.*

**Corollary 29.3** *If $|G| = p^n m$, where $p$ is prime and $p$ does not divide $m$, then all Sylow $p$-subgroups of $G$ are of order $p^n$.*

From this last corollary it is easy to see that every conjugate of a Sylow $p$-subgroup is also a Sylow $p$-subgroup, because conjugation is a group isomorphism and therefore preserves the order of the subgroup. The Second Sylow Theorem says that the converse is also true.

**Theorem 29.4     Second Sylow Theorem**     *For every prime p, the Sylow p-subgroups of G are conjugate.*

**Proof**:     Let $P$ and $Q$ be two Sylow $p$-subgroups of $G$. Let $Q$ act on $G/P$ by left multiplication. Again, we are not assuming $P$ is normal in $G$ and so not assuming that the left cosets $G/P$ form a group. By the Fixed Point Theorem 28.7,

$$|G/P| \equiv |\text{Fix}(G/P)| \bmod p.$$

But $|G/P| = [G : P]$ which is not divisible by $p$, since $P$ is a Sylow $p$-subgroup. So the number of fixed points of $G/P$ cannot be zero. Let $gP$ be one such fixed point. So, $qgP = gP$ for all $q \in Q$. By the left-handed Coset Theorem Exercise 24.3, this means that $g^{-1}Qg \subseteq P$. But then $g^{-1}Qg = P$ since $Q$ and $P$ are both finite and the same size. So $Q$ and $P$ are conjugate.     $\square$

**Corollary 29.5** *There is exactly one Sylow p-subgroup of a group G if and only if it is normal.*

We conclude this section with the third Sylow theorem, which provides us some valuable information for counting the number of Sylow $p$-subgroups a finite group has:

**Theorem 29.6     Third Sylow Theorem**     *Let G be a finite group where $|G| = p^n m$ with p prime and p does not divide m. The number of Sylow p-subgroups divides m. Furthermore, this number is congruent to 1 modulo p.*

**Proof**:     Let $n_p$ be the number of Sylow $p$-subgroups of $G$ and let $P$ be one of these subgroups. Let $P$ act on the set $X$ of all Sylow $p$-subgroups of $G$ by conjugation. By the Second Sylow Theorem, this action does indeed permute the Sylow $p$-subgroups. By the Fixed-Point Theorem 28.7,

$$n_p \equiv |\text{Fix}(P)| \bmod p.$$

Fixed points are Sylow $p$-subgroups $Q$ such that $gQg^{-1} = Q$ for all $g \in P$. Note $P$ is one such fixed point. In general, if $Q$ is fixed by conjugation by the elements in $P$, then $P \subseteq N(Q)$. Obviously $Q \subseteq N(Q)$, so $P$ and $Q$ are both Sylow $p$-subgroups of $N(Q)$ and so by the Second Sylow Theorem they are conjugate in $N(Q)$. But $Q$ is a normal subgroup of $N(Q)$. That is, the only conjugate in $N(Q)$ of $Q$ is $Q$ itself. Thus, $P = Q$ and so $P$ is the only fixed point. Therefore,

$$n_p \equiv 1 \bmod p.$$

Now we show that $n_p$ divides $m$. Consider $G$ acting on the Sylow $p$-subgroups of $G$ by conjugation. Since all Sylow $p$-subgroups are conjugates, this action has one orbit and so its size is $n_p$. The size of the orbit must divide $|G|$, by Corollary 28.6, and so $n_p$ divides $|G|$. But $n_p \equiv 1 \bmod p$, and thus $n_p$ is relatively prime to $p$. This means that $n_p$ must divide $m$.     $\square$

It is worth stating a weaker version of the Third Sylow Theorem as a corollary:

**Corollary 29.7** *The number of Sylow p-subgroups of a finite group G divides $|G|$.*

## 29.2     Applications of the Sylow Theorems

We conclude this chapter with a few applications of the Sylow Theorems.

## Example 29.2

Consider $S_4$, the group of all permutations of $\{1, 2, 3, 4\}$. Now $|S_4| = 4! = 24 = 2^3 \cdot 3$. We thus know that the Sylow 2-subgroups of $S_4$ must all have order 8. Furthermore, the Second Sylow Theorem says they all must be conjugates and therefore isomorphic. We know however that there are five non-isomorphic groups of order eight (see Exercise 24.12). We claim that the Sylow 2-subgroups are in fact all isomorphic to $D_4$, the group of symmetries of the square. But recall from Chapter 17 that by labeling the vertices of the square, we see $D_4$ is group of permutations of $\{1, 2, 3, 4\}$ and so is isomorphic to a subgroup of $S_4$. Thus *all* the Sylow 2-subgoups must be isomorphic to $D_4$. This means that $S_4$ has no subgroup isomorphic to $\mathbb{Z}_8$, $\mathbb{Z}_2 \times \mathbb{Z}_4$, $\mathbb{Z}_2 \times \mathbb{Z}_2 \times \mathbb{Z}_2$, and the quaternions, $Q_8$, the other groups of order eight.

Now, how many of these Sylow 2-subgroups do we have? The Third Sylow Theorem says that the number of these subgroups must be an odd divisor of 24. If there were only one such, it would be normal; the other alternative is that there are three of them. But we can specify three distinct 2-subgroups of order 4, namely

$$\langle (1234) \rangle, \quad \langle (1324) \rangle, \quad \langle (1243) \rangle.$$

▷ **Quick Exercise.** Verify that these three cyclic groups of order 4 are all distinct, and that there are no other such subgroups of $S_4$. To do this, you need only consider all possible elements of order 4 – they are all 4-cycles. ◁

Since $D_4$ has but one subgroup of order 4 these 3 subgroups must belong to 3 distinct Sylow 2-subgroups (each isomorphic to $D_4$).

Let's now consider the Sylow 3-subgroups of $S_4$. Clearly they are all cyclic groups of order 3. But we have already analyzed such subgroups as Sylow 3-subgroups of $A_4$ in Example 29.1 above. Notice that according to the Third Sylow Theorem, the number of these subgroups must be a divisor of 24 which is congruent to 1 modulo 3. That is, there must be 4 of them, as we have discovered.

The following is a nice application of the Third Sylow Theorem.

**Theorem 29.8** *If the order of a group $G$ is $pq$ where $p$ and $q$ are prime with $p < q$ and $p$ does not divide $q - 1$, then $G$ is cyclic.*

**Proof:**    Let $H$ be a Sylow $p$-subgroup and K be a Sylow $q$-subgroup. Let $n_p$ and $n_q$ be number of Sylow $p$-subgroups and Sylow $q$-subgroups, respectively. By the Third Sylow Theorem $n_p \equiv 1 \mod p$ and $n_p$ divides $q$. Thus $n_p = 1 + kp$. Therefore $n_p = 1$ or $q$. But if $n_p = q$, then $p | q - 1$, a contradiction. Therefore $n_p = 1$.

Similarly, $n_q = 1$.

▷ **Quick Exercise.**    Show that $n_q = 1$. ◁

Since there is only one of each Sylow subgroup, they are both normal in $G$. Take non-identity elements $x \in H$ and $y \in K$ and consider $x^{-1}y^{-1}xy$. By the normality of $H, K$ we have that

$$x^{-1}y^{-1}xy = x^{-1}(y^{-1}xy) \in H \quad \text{and} \quad x^{-1}y^{-1}xy = (x^{-1}y^{-1}x)y \in K.$$

But Lagrange's Theorem 24.2 implies that $H \cap K = \{1\}$. Therefore $x^{-1}y^{-1}xy = 1$ and so $xy = yx$. Thus the order of $xy$ is $pq$ and so $G$ is cyclic. ◻

From this theorem we see immediately that the only group of order 15 is $\mathbb{Z}_{15}$. A similar statement is true for groups of order 35, 65, 33, and infinitely many others.

▷ **Quick Exercise.**    Find three more groups of order less than 100 that satisfy the conditions for Theorem 29.8. ◁

We use the First Sylow Theorem to prove the following:

**Theorem 29.9** *Suppose the order of group $G$ is $p^2$ for prime $p$. Then $G$ is abelian.*

**Proof**: If $G$ had an element of order $p^2$ it would be cyclic and so abelian. So we assume that all non-identity elements have order $p$. Take non-identity $a \in G$ and $b \notin \langle a \rangle$. Then we claim that $\langle a \rangle \cap \langle b \rangle = \{1\}$. For if we had non-identity $c \in \langle a \rangle \cap \langle b \rangle$, then $c = a^m$ and $c = b^n$. It follows that $\langle c \rangle = \langle a \rangle$ and $\langle c \rangle = \langle b \rangle$, since the order of $a$ and $b$ are prime. But then $\langle a \rangle = \langle b \rangle$, a contradiction.

By the First Sylow Theorem, $H = \langle a \rangle$ is a normal subgroup of a subgroup of $G$ that has order $p^2$. But then this subgroup must be $G$ itself and so $H$ is a normal subgroup of $G$. Similarly $K = \langle b \rangle$ is a normal subgroup of $G$. But then by the same argument as in the proof of 29.8 we have that $ab = ba$. Finally, $HK = G$ because there are evidently $p^2$ distinct elements of this set. We therefore have all conditions required by the Internal Characterization Theorem 23.1 to see that $G$ is isomorphic to $H \times K$. But this group must be abelian, since $H$ and $K$ are cyclic, and hence abelian. □

Note that this theorem says that if $|G| = p^2$, for prime $p$, then either $G$ is isomorphic either to $\mathbb{Z}_{p^2}$ or $\mathbb{Z}_p \times \mathbb{Z}_p$.

Thus from Theorem 29.9 we know all the structure of all groups of orders

$$4, 9, 25, 49, 121, \ldots.$$

As another application of the Sylow theorems, let us prove that the converse of Lagrange's Theorem is true for abelian groups.

**Theorem 29.10** *Let $m$ divide the order of a finite abelian group $G$. Then $G$ has a subgroup of order $m$.*

**Proof**: We may assume that $|G| = p_1^{e_1} p_2^{e_2} \cdots p_k^{e_k}$, where the $p_i$ are distinct primes. From Corollary 28.3 we know that $G$ is isomorphic to a direct product of the Sylow $p_i$-groups, each of order $p_i^{e_i}$. Because $m$ is a divisor of the order of $G$, we know that $m = p_1^{d_1} p_2^{d_2} \cdots p_k^{d_k}$, where $d_i \leq e_i$. By the First Sylow Theorem we know that $G$ has a subgroup of order $p_i^{d_i}$, for each $i$. But we can consider in the direct product the subgroup which consists of the direct product of these subgroups, and this subgroup clearly has exactly $m$ elements. □

We now generalize Corollary 28.3 to nonabelian groups.

**Theorem 29.11** *All the Sylow $p$-subgroups of a finite group $G$ are normal if and only if $G$ is isomorphic to the direct product of its Sylow $p$-subgroups.*

**Proof**: In one direction, if $G$ is isomorphic to a direct product of its Sylow $p$-subgroups, then each factor of the direct product is normal in $G$, as is the case for all direct products. Conversely, suppose all the Sylow $p$-subgroups of $G$ are normal. We use the Internal Characterization Theorem 23.1 to show $G$ is isomorphic to the direct product of its Sylow $p$-subgroups. Let $H_{p_1}, \ldots, H_{p_k}$ be the distinct nontrivial Sylow $p$-subgroups of $G$.

Suppose $x \in G$ is of order $n$. By the Fundamental Theorem of Arithmetic,

$$n = p_1^{e_1} p_2^{e_2} \cdots p_k^{e_k}.$$

Let $n_i = n/p_i^{e_i}$. Then the $\gcd(n_1, n_2, \ldots, n_k) = 1$ (for there is no prime dividing all the $n_i$) and so there exist $m_i$ such that $\sum m_i n_i = 1$. So,

$$x^{\sum m_i n_i} = x^{m_1 n_1} \cdots x^{m_k n_k} = x.$$

But $(x^{m_i n_i})^{p_i^{e_i}} = x^{m_i n} = 1$, and thus $x^{m_i n_i} \in H_{p_i}$, as the subgroup $H_{p_i}$ contains all the elements of order some power of $p_i$. Thus every element of $G$ is the product of elements from the $H_{p_i}$.

Now suppose $x \in H_{p_i} \cap H_{p_j}$ for distinct primes $p_i$ and $p_j$. Then $x = 1$, by Lagrange's Theorem. Then, if $x_i \in H_{p_i}$ and $x_j \in H_{p_j}$, then

$$x_i^{-1} x_j^{-1} x_i x_j = (x_i^{-1} x_j^{-1} x_i) x_j = x_i^{-1} (x_j^{-1} x_i x_j) \in H_{p_i} \cap H_{p_j} = \{1\}.$$

Thus $x_i x_j = x_j x_i$.

So, by Theorem 23.1, $G$ is isomorphic to a direct product of the $H_{p_i}$.      □

**Example 29.3**

We show that all groups of order 45 are abelian. If $|G| = 45$, its Sylow 3-subgroups are of order 9 and its Sylow-5 subgroups are of order 5. By the Third Sylow Theorem, $n_3 \equiv 1 \bmod 3$ and $n_3 | 5$. This implies $n_3 = 1$. Also $n_5 \equiv 1 \bmod 5$ and $n_5 | 3$ which implies $n_5 = 1$. Thus there is only one Sylow 3-subgroup and one Sylow 5-subgroup and so they are each normal, by Corollary 29.5. Let $H_3$ be the unique Sylow 3-subgroup and $H_5$ be unique Sylow 5-subgroup. Then $G$ is isomorphic to $H_3 \times H_5$, by Theorem 29.11. Now $H_3$ is abelian (being of order $p^2$) and $H_5$ is cyclic and so abelian. Thus $G$ must also be abelian.

**Example 29.4**

Suppose that $|G| = 100$; we claim that $G$ has a normal Sylow 5-subgroup. This is true because by the Third Sylow Theorem $n_5 \equiv 1 \bmod 5$ and $n_5 | 4$. Thus $n_5 = 1$. Therefore any group of order 100 has at least one normal subgroup. Using the terminology from Section 27.5, this means that $G$ is not simple.

▷ **Quick Exercise.** Show that if $|G| = 20$, then $G$ is not simple. ◁

## 29.3    The Fundamental Theorem for Finite Abelian Groups

In this section we state the *Fundamental Theorem for Finite Abelian Groups*, which completely describes the structure of such groups. This powerful theorem provides an easy-to-understand recipe by which all such groups can be constructed, using the two familiar notions of *cyclic group* and *direct product*. The theorem is relatively difficult to prove, and so we will not prove it here. The interested reader can refer to any introductory text in group theory. However, we can see how part of the proof goes. Thanks to Corollary 28.3, if $G$ is a finite abelian group, it is the direct product of its Sylow $p$-subgroups. Now if $G$ is (isomorphic to) a direct product of $H$ and $K$ and $H$ is a direct product of $H_1$ and $H_2$, then $G$ is a direct product of the three groups $H_1$, $H_2$, and $K$. (See Exercise 29.18.) This, of course, generalizes to longer direct products. What remains to prove (and the part we omit) is that an abelian $p$-group is a direct product of cyclic $p$-groups.

**Example 29.5**

What finite abelian groups of order 8 are possible? Clearly, the cyclic groups used to build such groups can only have order 2, 4, or 8, and the product of the orders of the cyclic groups must be 8. So the only possible abelian groups of order 8 are

$$\mathbb{Z}_8, \quad \mathbb{Z}_4 \times \mathbb{Z}_2, \quad \mathbb{Z}_2 \times \mathbb{Z}_2 \times \mathbb{Z}_2.$$

▷ **Quick Exercise.** We did not include $\mathbb{Z}_2 \times \mathbb{Z}_4$, because it is isomorphic to $\mathbb{Z}_4 \times \mathbb{Z}_2$. Give an isomorphism. ◁

Are these three groups really non-isomorphic? We can answer this affirmatively by looking at the orders of elements in these groups. Of them, only $\mathbb{Z}_8$ has any elements of order 8; $\mathbb{Z}_4 \times \mathbb{Z}_2$ has elements of order 4, while all non-identity elements of $\mathbb{Z}_2 \times \mathbb{Z}_2 \times \mathbb{Z}_2$ are of order 2.

▷ **Quick Exercise.** Verify our assertions in the last sentence, thus checking that these groups are not isomorphic. ◁

**Theorem 29.12   The Fundamental Theorem for Finite Abelian Groups** *Every finite abelian group is isomorphic to a direct product of cyclic groups; each cyclic group in this decomposition is of order $p^n$, where $p$ is prime. That is, each finite abelian group is isomorphic to a group of the form*

$$\mathbb{Z}_{p_1^{k_1}} \times \mathbb{Z}_{p_2^{k_2}} \times \cdots \times \mathbb{Z}_{p_n^{k_n}}$$

*where the $p_i$'s are primes (not necessarily distinct), and the $k_i$'s are positive integers (not necessarily distinct). This representation is unique, up to order and isomorphism.*

Note that the group

$$\mathbb{Z}_{p_1^{k_1}} \times \mathbb{Z}_{p_2^{k_2}} \times \cdots \times \mathbb{Z}_{p_n^{k_n}}$$

has order $p_1^{k_1} p_2^{k_2} \cdots p_n^{k_n}$.

**Example 29.6**

What finite abelian groups of order 10 are possible? By the Fundamental Theorem for Finite Abelian Groups 29.12, since $10 = 2 \cdot 5$ (a product of primes), there is only one possibility: $\mathbb{Z}_2 \times \mathbb{Z}_5$. However, you might notice that $\mathbb{Z}_{10}$ is also abelian and of order 10. But these two groups are isomorphic. This is easy to see, because the element $(1, 1) \in \mathbb{Z}_2 \times \mathbb{Z}_5$ has order 10, and so this group is cyclic of order 10.

More generally, these two groups are necessarily isomorphic, on account of the Chinese Remainder Theorem 16.1. Of course, that theorem is phrased as a theorem about rings, but any cyclic group of the form $\mathbb{Z}_n$ can be equipped with a ring structure, and so we can use it here.

▷ **Quick Exercise.** Review Theorem 16.1. ◁

Thus, there is (up to isomorphism) only one abelian group of order 10.

## Example 29.7

The non-isomorphic abelian groups of order 12 are $\mathbb{Z}_3 \times \mathbb{Z}_4$ and $\mathbb{Z}_3 \times \mathbb{Z}_2 \times \mathbb{Z}_2$.

▷ **Quick Exercise.** Which of these two groups is isomorphic to $\mathbb{Z}_{12}$? ◁

▷ **Quick Exercise.** How many distinct abelian groups of order 9 are there? What are they? How about order 40? ◁

---

# Chapter Summary

For a finite group $G$ with $p^n m$ elements, where $p$ is prime and $m$ is relatively prime to $p$, we proved the three Sylow Theorems. They assert that $G$ always has subgroups of order $p^n$, and these subgroups are called the Sylow $p$-subgroups. Each such subgroup is conjugate to every other one. If a group has only one such subgroup it is necessarily normal. The number of such subgroups is always congruent to 1 modulo $p$, and that number must divide the order of the group. By way of illustrating the power of these theorems, we applied them to numerous special cases.

We gave the structure of a finite abelian group as a direct product of cyclic $p$-groups in the Fundamental Theorem for Finite Abelian Groups; we thus know the structure of all such groups.

---

# Warm-up Exercises

a. Suppose 9 divides the order of a finite group $G$. Why does $G$ have a subgroup of order 9?

b. Replace 9 by 6 in the previous question. Why does your answer now fail?

c. Suppose that $G$ is a finite group and $|G| = 16m$, where $m$ is odd. Explain why $G$ cannot have two subgroups, one isomorphic to $\mathbb{Z}_{16}$, and the other isomorphic to $\mathbb{Z}_4 \times \mathbb{Z}_4$.

d. Describe all groups of order 25. Of order 49.

e. Suppose $G$ is a non-abelian group with order $3m$, where $m$ is odd. Argue that the order of $G$ is at least 21.

f. How many distinct abelian groups are there of the following orders: 14, 18, 25, 29?

g. Describe the following finite abelian groups as direct products as specified by the Fundamental Theorem for Finite Abelian Groups 29.12:

$$\mathbb{Z}_4 \times \mathbb{Z}_6, \quad \mathbb{Z}_{150}, \quad U(\mathbb{Z}_{20}), \quad U(\mathbb{Z}_{19}), \quad U(\mathbb{Z}_{21}).$$

h. Up to isomorphism, how many abelian groups are there of order $pq$ where $p$ and $q$ are distinct primes?

i. Up to isomorphism, how many abelian groups are there of order $p^2$, for prime $p$?

j. What is the largest order of an element in the finite abelian group $\mathbb{Z}_2 \times \mathbb{Z}_4 \times \mathbb{Z}_9 \times \mathbb{Z}_5$? Is this group cyclic?

k. What is the largest order of an element in the finite abelian group $\mathbb{Z}_4 \times \mathbb{Z}_9 \times \mathbb{Z}_5$? Is this group cyclic?

# Exercises

1. In Example 29.1 we described all Sylow 3-subgroups of $A_4$. Show explicitly that any one of these subgroups is conjugate to any other.

2. Find all groups of order less than 100 that satisfy the conditions of Theorem 29.8.

3. Show that if $|G| = p^2 q$ with primes $p, q$ where $p$ does not divide $q - 1$ and $p^2 < q$, then $G$ is abelian.

4. Show that if $H$ is a normal subgroup of $G$ and $H$ is a $p$-group, then $H$ is contained in every Sylow $p$-subgroup of $G$.

5. If $|G| = 21$ and not cyclic, how many Sylow 3-subgroups can $G$ have?

6. If $|G| = pq$ for distinct primes $p$ and $q$, show that $G$ is not simple. (Note the slight difference from Theorem 29.8.)

7. If $|G| = p^k m$ for prime $p$ with $m < p$, show that $G$ is not simple.

8. Show that if $|G| = 12$ then $G$ has a normal Sylow 2-subgroup or a normal Sylow 3-subgroup.

9. Find the Sylow 2-subgroups and Sylow 3-subgroups of $S_3$.

10. $S_4$ has 3 Sylow 2-subgroups, all isomorphic to $D_4$, as we have seen in Example 29.2. Find them.

11. Use the Third Sylow Theorem to determine how many Sylow $p$-subgroups are possible in $S_5$, for $p = 2, 3, 5$. Then obtain a list of all such subgroups, to verify your results. Determine first what group the Sylow $p$-subgroups must be isomorphic to, and then describe them in terms of their cycle structure and generators.

12. Show that if $p$ and $q$ are distinct primes, that $\mathbb{Z}_{p^k} \times \mathbb{Z}_{q^j}$ is isomorphic to $\mathbb{Z}_{p^k q^j}$ by finding an element of $\mathbb{Z}_{p^k} \times \mathbb{Z}_{q^j}$ of order $p^k q^j$. (Of course, this result follows from Theorem 16.1 or Theorem 29.12, but we want you to actually find a generator.)

13. Show that $\mathbb{Z}_{12} \times \mathbb{Z}_{10}$ has no element of order 8 and, hence, no subgroup isomorphic to $\mathbb{Z}_8$.

14. Suppose $G$ is a finite abelian group of order $m$, where $m$ is square-free. (That is, if $p$ divides $m$, then $p^2$ does not.) Show that $G$ is cyclic.

15. If there are $k$ abelian groups of order $m$ and $j$ abelian groups of order $n$, how many abelian groups are there of order $mn$ if $m$ and $n$ are relatively prime?

16. Suppose that $G$, $H$, and $K$ are finite abelian groups. Suppose that the direct products $G \times K$ and $H \times K$ are isomorphic as groups. Prove that $G$ and $H$ are isomorphic.

17. The conclusion of the previous exercise seems so plausible that you might conjecture that it is true for all groups (or at least for all abelian groups). Construct a counterexample to this, using abelian groups.

18. Suppose $G$ is isomorphic to $H \times K$ and $H$ is isomorphic to $H_1 \times H_2$. Show that $G$ is isomorphic to $H_1 \times H_2 \times K$. Now write a theorem that generalizes this result.

19. Suppose that $G$ is a finite group with $pm$ elements, where $p$ is prime and $m$ is relatively prime to $p$, and $p > m$. How many elements of order $p$ does $G$ have? Prove your answer.

20. Suppose that $G$ is a group with 21 elements.

   (a) Explain why Theorem 29.8 does not apply to this group.

   (b) What are the possible values for $n_3$ and $n_7$, the number of Sylow 3-subgroups, and the number of Sylow 7-subgroups, respectively?

   (c) Describe and justify the case which leads to the group $\mathbb{Z}_{21}$.

   (d) There remains one other case. Determine exactly how many elements of order 3 and 7 we would have for such a group. It turns out there does exist a non-abelian group with elements distributed in this way, and we can find such a group as a subgroup of $S_7$. We will not pursue the explicit construction of this group here, however.

# Chapter 30

## Solvable Groups

In this chapter we will introduce solvable groups, as a natural generalization of abelian groups. Solvable groups can be viewed as groups built out of finitely many abelian pieces. This gives us a chance to see a bit of how the extension problem in groups is approached; we discussed these ideas briefly in the historical note at the end of Chapter 27. We will use the notion of solvability in Chapter 48, when we discuss whether a polynomial equation can be solved using ordinary arithmetic and the extraction of roots.

## 30.1 Solvability

The symmetry group $S_3$ is not abelian, but we can think of it as consisting of two abelian pieces, put together in the appropriate way. Namely, $S_3$ has an abelian subgroup $A_3$, and the homomorphic image $S_3/A_3$ is also abelian. We generalize this idea to finitely many steps in our definition. We say that a group $G$ is **solvable** if it has a finite collection of subgroups $G_0, G_1, \cdots G_n$, so that

$$G_n = \{1\} \subseteq G_{n-1} \subseteq G_{n-2} \subseteq \cdots \subseteq G_1 \subseteq G_0 = G,$$

and furthermore each $G_{i+1}$ is normal in $G_i$, and the group of cosets $G_i/G_{i+1}$ is abelian. We shall call such a finite sequence of subgroups satisfying the definition of solvability a **subnormal series with abelian quotients** for the group $G$.

The terminology we have chosen for the series required for a solvable group is cumbersome, but it is consistent with the mathematical literature. In general, a subnormal series requires only that each $G_{i+1}$ is normal in $G_i$, without any requirement on the group $G_i/G_{i+1}$. The word 'subnormal' places emphasis on the fact that we require only that $G_{i+1}$ is normal in the next larger subgroup $G_i$; it is not necessarily normal in the entire group $G$.

### Example 30.1

We note first that any abelian group $G$ is of course solvable because $\{1\} \subseteq G$ is a subnormal series with abelian quotients.

### Example 30.2

The group $S_3$ is solvable because $\{\iota\} \subseteq A_3 \subseteq S_3$, and $S_3/A_3$ and $A_3/\{\iota\} = A_3$ are abelian. We can also think of $S_3$ as the symmetries of an equilateral triangle, that is, as the dihedral group $D_3$. We can now generalize this example to see that any of the dihedral groups $D_n$ are solvable, because the subgroup of rotations is abelian, and of index two in $D_n$.

▷ **Quick Exercise.** Check this. ◁

**Example 30.3**

The group $S_4$ is solvable, although it will require more than two subgroups to build a subnormal series with abelian quotients. We will use the notation from Example 25.5, where we denote by $H$ a subgroup of $S_4$ isomorphic to the Klein Four Group:

$$\{\iota\} \subseteq H \subseteq A_4 \subseteq S_4.$$

Each of these subgroups is normal in the next bigger subgroup. The quotient groups are abelian because they are isomorphic to $H$, $\mathbb{Z}_3$ and $\mathbb{Z}_2$, respectively.

▷ **Quick Exercise.**   Check these assertions. ◁

**Example 30.4**

The group $S_5$ is not solvable. By Exercise 27.6 we know that $A_5$ is the only non-trivial normal subgroup of $S_5$. But by Theorem 27.4, $A_5$ is a non-abelian group with no non-trivial normal subgroups, and so $S_5$ cannot have a solvable series. The fact that $S_5$ is not solvable will be important for us in Chapter 48.

The next two examples make use of Sylow theory, from Chapter 29.

**Example 30.5**

Let $G$ be a finite $p$-group, where $p$ is a prime integer. Then by Theorem 28.1 we know that $G$ has $p^n$ elements. By the First Sylow Theorem 29.1, we know that $G$ has a sequence of subgroups

$$\{1\} = G_0 \subset G_1 \subset G_2 \subset \cdots G_{n-1} \subset G_n = G,$$

where $G_k$ has $p^k$ elements, and $G_k$ is a normal subgroup of $G_{k+1}$, for $k = 0, \ldots, n-1$. But then $[G_{k+1} : G_k] = p$ and so $G_{k+1}/G_k$ is a cyclic group of order $p$, and is consequently abelian. Thus every finite $p$-group is solvable.

**Example 30.6**

Let $G$ be a group with 100 elements. In Example 29.4 we discovered that $G$ has exactly one Sylow 5-subgroup $P$, which has 25 elements. By Theorem 29.9 $P$ is abelian. By Corollary 29.5 $P$ is normal in $G$. Now $|G/P| = 4$, and so $G/P$ is abelian. Thus any group with 100 elements is solvable.

## 30.2   New Solvable Groups from Old

In this section we will present some natural ways to get more solvable groups from a given solvable group.

**Theorem 30.1** *Every homomorphic image of a solvable group is solvable.*

**Proof:** Suppose that $G$ is a solvable group with normal subgroup $H$. We shall assume that $G$ has the following subnormal series with abelian quotients:

$$G_n = \{1\} \subseteq G_{n-1} \subseteq G_{n-2} \subseteq \cdots \subseteq G_1 \subseteq G_0 = G.$$

Consider each of the sets $HG_i = \{hg : h \in H, g \in G_i\}$. By Exercise 25.13a we know that $HG_i$ is a subgroup of $HG_{i-1}$. We claim that $HG_i$ is normal in $HG_{i-1}$. To prove this, choose $h \in H$ and $a \in G_{i-1}$. The right coset $HG_iha = HG_ia$, because $H \subseteq HG_i$. But then

$$HG_iha = HG_ia = H(G_ia) = H(aG_i) = (Ha)G_i = (aH)G_i = aHG_i,$$

because $G_i$ is normal in $G_{i-1}$ and $H$ is normal in $G$. But $aHG_i = aa^{-1}haHG_i = haHG_i$, because $a^{-1}ha \in H$. This equality of right and left cosets means that $HG_i$ is normal in $HG_{i-1}$.

By Exercise 26.12 we have for each $i$ that $HG_i/H$ is a normal subgroup of $HG_{i-1}/H$, and so we have a subnormal series for $G/H$:

$$H/H \subseteq HG_{n-1}/H \subseteq HG_{n-2}/H \subseteq \cdots \subseteq HG_1/H \subseteq G/H.$$

Now by Exercise 26.13

$$(HG_{i-1}/H)/(HG_i/H)$$

is isomorphic to $HG_{i-1}/HG_i$.

It remains only to show that the latter group is abelian to conclude that $G/H$ is solvable. For that purpose, suppose that $h, k \in H$ and $a, b \in G_{i-1}$. Then we can use normality and the fact that $G_{i-1}/G_i$ is abelian to conclude the following:

$$
\begin{aligned}
HG_i(ha)(kb) &= HG_iakb = HG_i(aka^{-1})ab = HG_iab \\
&= H(G_iab) = H(G_iba) = HG_iba \\
&= HG_i(bhb^{-1})ba = HG_ibha = HG_i(kb)(ha).
\end{aligned}
$$

$\square$

A similar proof shows the following:

**Theorem 30.2** *Every subgroup of a solvable group is solvable.*

**Proof:** We leave this as Exercise 2. $\square$

We can also build larger solvable groups, by putting together a normal subgroup and the corresponding homomorphic image:

**Theorem 30.3** *Suppose that $G$ is a group with normal subgroup $H$. Then $G$ is solvable if and only if $G/H$ is solvable and $H$ is solvable.*

**Proof:** If $G$ is solvable, the conclusions about $H$ and $G/H$ are just the previous two theorems.

Now suppose that $H$ and $G/H$ are solvable. Then we have subnormal series with abelian quotients for both these groups. By Exercise 26.12, we can write the subgroups of $G/H$ in the form $G_i/H$:

$$H_m = \{1\} \subseteq H_{m-1} \subseteq \cdots \subseteq H_1 \subseteq H_0 = H$$

and

$$G_n/H = H/H \subseteq G_{n-1}/H \subseteq \cdots \subseteq G_1/H \subseteq G_0/H = G/H.$$

But then we can put together these subgroups to obtain a subnormal series with abelian quotients for $G$:

$$
\begin{aligned}
H_m \;=\;& \{1\} \subseteq H_{m-1} \subseteq \cdots \subseteq H_1 \subseteq H_0 = H \\
\subseteq\;& G_{n-1} \subseteq \cdots \subseteq G_1 \subseteq G_0 = G.
\end{aligned}
$$

▷ **Quick Exercise.**  Check that the above series is subnormal with abelian quotients. Why is Exercise 26.13 relevant?  ◁                                □

We can informally paraphrase this theorem by asserting that a solvable extension of a solvable group is solvable. It is easy to extend this inductively to obtain the following:

**Theorem 30.4** *A group $G$ is solvable if and only if it has a subnormal series*

$$
G_n = \{1\} \subseteq G_{n-1} \subseteq G_{n-2} \subseteq \cdots \subseteq G_1 \subseteq G_0 = G,
$$

*where each quotient group $G_i/G_{i+1}$ is solvable.*

**Proof**:   If $G$ is solvable, it has a subnormal series with abelian quotients, which are of course solvable.

For the converse, suppose that $G$ has such a subnormal series. We will proceed by induction on $n$. If $n = 1$, then $G = G_0$ is obviously solvable. We now suppose that if $G$ has such a subnormal series with length $n - 1$, it is in fact solvable. Given the series above of length $n$, we can now conclude by induction that the group $G_1$ is in fact solvable. But then $G_1$ is a solvable normal subgroup of $G$, and by assumption $G/G_1$ is solvable. By Theorem 30.3 $G$ is then solvable.                                □

## Historical Remarks

We have barely scratched the surface here regarding the study of group extensions. The general project of understanding groups in terms of extensions built of simpler pieces is a large one that we cannot develop fully here. In the general theory, the 'simpler pieces' to use are precisely the simple groups, whose classification was such a major part of 20th century group theory — for more information you should consult the Historical Remarks following Chapter 27. This idea leads to subnormal series whose quotients are simple; these are called **composition series**. (In Exercise 3 you look at this in the specific context of finite solvable groups.) Any finite group has such a composition series (see Exercise 7). That the factor groups in such a series are unique (up to order) is an important theorem proved by Camille Jordan in the context of permutation groups, and in the general case by Otto Hölder.

One of the most important steps in the classification problem for finite simple groups was the difficult theorem of John Thompson and Walter Feit that all finite groups of odd order are in fact solvable! Consequently, we need look only to groups with even order in our search for finite simple groups. The Feit-Thompson Theorem took up an entire issue of the Pacific Journal of Mathematics, when the theorem was first published in 1963.

---

## Chapter Summary

In this chapter we explored the notion of *solvable group*. Such groups can be built as finitely many abelian extensions of an abelian group, and as such are a natural generalization of the abelian groups. We then saw how the class of solvable groups is closed under taking subgroups, homomorphic images, and building group extensions.

# Warm-up Exercises

a. Is every abelian group solvable? If not, give an example of a non-solvable abelian group.

b. Is every solvable group abelian? If not, give an example of a non-abelian solvable group.

c. Show that the group of quaternions $Q_8$ is solvable. That is, give an appropriate subnormal series for this group.

d. Show that all groups of order 12 are solvable.

e. Show that all groups of order 18 are solvable.

# Exercises

1. Suppose that $G$ and $H$ are solvable groups. Prove that $G \times H$ is solvable.

2. Prove Theorem 30.2.

3. Suppose $G$ is a finite abelian group. Prove that $G$ has a composition series; that is, show that $G$ has a collection of subgroups $G_i$ so that $G_n = \{1\} \subseteq G_{n-1} \subseteq \cdots \subseteq G_1 \subseteq G_0 = G$, and $G_{i-1}/G_i$ is a simple group.

4. Suppose that $G$ is a finite solvable group. Prove that $G$ has a composition series.

5. Prove that $\mathbb{Z}$ does not have a composition series.

6. Let $G$ be a group and $H$ a proper normal subgroup. Prove that $G/H$ is a simple group if and only if $H$ is a maximal normal subgroup.

7. Generalize Exercise 4. That is, prove that every finite group has a composition series.

8. Show that all groups of order $294 = 2 * 3 * 7^2$ are solvable.

9. Use Sylow Theory to prove that all groups of order 105 are solvable. You might find Exercise 29.20 useful; you should of course not use the difficult Feit-Thompson Theorem that asserts that all groups of odd order are solvable.

# Section VI in a Nutshell

We examine more deeply the permutations of the symmetric groups $S_n$. A *transposition* is a cycle of length two. Any permutation can be factored as a product of transpositions (Theorem 27.1), and while this factorization is not unique, all factorizations of a given permutation have the same parity; thus we can classify a permutation as either even or odd. The set $A_n$ of even permutations is a group called the $n$th *alternating group*. The alternating group $A_n$ is a normal subgroup of the symmetric group $S_n$ and $[S_n : A_n] = 2$. Furthermore, $A_n$ is simple for $n \neq 4$.

A group $G$ *acts on a set* $X$ if there is a homomorphism $\sigma : G \to S_X$. Notationally we simplify things by letting $g(x) = y$ if $g \in G$ acts of the element $x$ by sending it to $y$. (Obviously, $x, y \in X$.) If $G$ acts on $X$, then the *orbit* of $x \in X$ is

$$\mathrm{Orb}(x) = \{y \in X : g(x) = y, \text{ for some } g \in G\}.$$

The orbits partition $X$. So, if $x_1, \ldots, x_k$ are representatives of the disjoint orbits, $|X| = \sum |\mathrm{Orb}(x_i)|$.

For $x \in X$, the *stabilizer* of $x$ is

$$\mathrm{Stab}(x) = \{g \in G : g(x) = x\}.$$

$\mathrm{Stab}(x)$ is a subgroup of $G$ and

$$|\mathrm{Orb}(x)| = [G : \mathrm{Stab}(x)]. \text{ (Theorem 28.5.)}$$

Thus the size of every orbit divides the order of $G$ (Corollary 28.6). Groups acting on sets is the main technique used in proving the three Sylow Theorems of Chapter 29.

Given prime $p$, a group is a $p$-group if the order of every element is a power of $p$. A finite group $G$ is a $p$-group if and only if $|G| = p^n$ for some $n$ (Theorem 28.1). A Sylow $p$-subgroup of a group $G$ is a maximal $p$-subgroup of $G$. The three Sylow Theorems are:

- (First Sylow Theorem 29.1)  Let $G$ be a finite group, $p$ a prime, and $|G| = p^n m$ where $p$ does not divide $m$. Then $G$ has a subgroup of order $p^k$ for $k = 0, 1, \ldots, n$. Indeed, for $k = 0, \ldots, n-1$, each subgroup of order $p^k$ is a normal subgroup of a subgroup of order $p^{k+1}$.

- (Second Sylow Theorem 29.4)  For every prime $p$, the Sylow $p$-subgroups of $G$ are conjugate.

- (Third Sylow Theorem 29.6)  Let $G$ be a finite group where $|G| = p^n m$ with $p$ prime and $p$ does not divide $m$. The number of Sylow $p$-subgroups divides $m$. Furthermore, this number if congruent to 1 modulo $p$.

Thus Sylow $p$-subgroups exist (First Sylow Theorem), all Sylow $p$-subgroups are conjugate for each prime $p$ (Second Sylow Theorem), and the number of Sylow $p$-subgroups satisfies some restrictions (Third Sylow Theorem). The First Sylow Theorem gives us a partial converse to Lagrange's Theorem guaranteeing the existence of $p$-subgroups of $G$, for all possible orders $p^k$.

One important consequence of the Sylow Theorems is that if $G$ is a finite abelian group of order $r$ and $m$ divides $r$, then $G$ has a subgroup of order $m$ (Corollary 29.10). That is, $G$ has subgroups of every possible order. Thus, in the case of finite abelian groups, the converse of Lagrange's Theorem is true.

Another consequence is that there is exactly one Sylow $p$-subgroup of a group $G$ if and only if it is normal (Corollary 29.5). All the Sylow $p$-subgroups of a finite group $G$ are normal if and only if $G$ is isomorphic to the direct product of its Sylow $p$-subgroups (Theorem 29.11).

All finite abelian groups are direct products of $p$-groups (Corollary 28.3). Indeed, if $p_1, p_2, \ldots, p_n$ are the distinct primes dividing $|G|$, then $G$ is a direct product of $p_i$-groups.

All finite abelian groups can be completely described by the *Fundamental Theorem for Finite Abelian Groups* (Theorem 29.12): Every finite abelian group is isomorphic to a direct product of cyclic groups; each cyclic group in this decomposition is of order $p^n$, where $p$ is prime. That is, each finite abelian group is isomorphic to a group of the form

$$\mathbb{Z}_{p_1^{k_1}} \times \mathbb{Z}_{p_2^{k_2}} \times \cdots \times \mathbb{Z}_{p_n^{k_n}}$$

where the $p_i$'s are primes (not necessarily distinct), and the $k_i$'s are positive integers (not necessarily distinct).

Finally, the idea of abelian group is generalized by the idea of *solvable* groups. A group is solvable if it has a finite collection of subgroups $G_0, G_1, \cdots G_n$, so that

$$G_n = \{1\} \subseteq G_{n-1} \subseteq G_{n-2} \subseteq \cdots \subseteq G_1 \subseteq G_0 = G,$$

and furthermore each $G_{i+1}$ is normal in $G_i$, and the group of cosets $G_i/G_{i+1}$ is abelian. Such a sequence of subgroups is called a *subnormal series with abelian quotients* for $G$. An important group that is *not* solvable is $S_5$ (Example 30.4). We will use this fact in Section X.

Every homomorphic image of a solvable group is also solvable (Theorem 30.1) as is every subgroup of a solvable group (Theorem 30.2). In fact, if $G$ has a normal subgroup $H$, then $G$ is solvable if and only if both $H$ and $G/H$ are solvable (Theorem 30.3).

# Part VII

# Unique Factorization

# Chapter 31

## Quadratic Extensions of the Integers

Our goal in this section of the text is to discover which integral domains have a unique factorization theorem like the Fundamental Theorem of Arithmetic. That is, we wish to obtain a common generalization of $\mathbb{Z}$ and $\mathbb{Q}[x]$. As we shall see, there are two stages to obtain such a unique factorization theorem. First, we need to see whether elements in an arbitrary integral domain can even be factored into irreducibles at all, and then focus our attention on whether such a factorization is unique (up to order and unit factors).

To better understand these questions, we need a richer fund of examples. For that purpose we will in this chapter explore rings of the form $\mathbb{Z}[\sqrt{n}]$; we first met these rings in Exercise 7.2. We will conclude this chapter by proving that in all such rings, we can at least factor every non-unit into a product of irreducible elements.

---

### 31.1 Quadratic Extensions of the Integers

Let $n$ be an integer (positive or negative), other than 1. Suppose further that $n$ has no non-trivial factors that are perfect squares. We call such an integer **square-free**. Thus, $-21$ is square-free, while 18 is not. As we did in Exercise 7.2, we define a subset of $\mathbb{C}$ as follows:

$$\mathbb{Z}[\sqrt{n}] = \{a + b\sqrt{n} : a, b \in \mathbb{Z}\},$$

which we call a **quadratic extension** of $\mathbb{Z}$. Our restriction to $n$ being square-free is merely to ensure that no rational simplification is necessary under the radical sign. In particular, if $n$ is square-free, we know that $\sqrt{n}$ is an irrational number (see Exercise 15). Notice that if $n$ is positive, the quadratic extension consists of a set of real numbers. The case where $n = -1$ is of particular importance; this is called the ring of **Gaussian integers** and is usually written $\mathbb{Z}[i]$ (as we noted in Exercise 6.12).

It is an easy matter to use the Subring Theorem 7.1 to check that these sets are subrings of $\mathbb{C}$ (and you explored this in Exercises 6.12, 7.1, and 7.2.) We need merely to see that $\mathbb{Z}[\sqrt{n}]$ is non-empty, and closed under subtraction and multiplication:

▷ **Quick Exercise.** Carry out these verifications (or consult your solution to Exercise 7.2). ◁

What are the units and irreducibles of these rings? This general question actually turns out to be a fairly sophisticated inquiry from number theory, and we cannot give a general answer here. We will make some progress on this question, particularly in special cases.

## 31.2   Units in Quadratic Extensions

Let's see first what the question regarding units amounts to. If $a + b\sqrt{n}$ is a unit, then we have an equation of the form

$$(a + b\sqrt{n})(c + d\sqrt{n}) = (ac + nbd) + (ad + bc)\sqrt{n} = 1.$$

This means that we must find all simultaneous solutions $a, b, c, d$ to the equations $ac + nbd = 1$ and $ad + bc = 0$, where $a, b, c, d$ are *integers* (recall that here $n$ is fixed). Equations where we require integer solutions are called **Diophantine equations**. Study of such equations makes up an important branch of number theory. In Exercise 1, you will try to solve this pair of equations directly in the particular case of the Gaussian integers (that is, where $n = -1$).

However, in the long run we will be better off by recalling our geometric representation for complex numbers introduced in Chapter 8. Let us view the Gaussian integers as points in the *complex plane*.

Recall that for a complex number $\rho = r + si$, we computed its modulus (or length) as $|\rho| = \sqrt{r^2 + s^2}$. (Actually, in what follows we will often use the square of the modulus instead, because in our present context that will be an integer.) What is important for us is that the modulus function *preserve multiplication*. By this we mean that $|\rho\tau| = |\rho|\,|\tau|$, for any pair of complex numbers $\rho, \tau$. If you don't recall this fact, it would be well worth reviewing the proof of Theorem 8.3 now.

The fact that this function preserves multiplication means that we can translate certain algebraic questions about $\mathbb{Z}[i]$ into questions in the integers $\mathbb{Z}$, where they are presumably easier to answer.

For example, suppose that $\alpha \in \mathbb{Z}[i]$ is a unit. Then there exists another Gaussian integer $\beta$ for which $\alpha\beta = 1$. But then we have that

$$|\alpha|^2\,|\beta|^2 = |\alpha\beta|^2 = |1|^2 = 1.$$

Since $|\alpha|^2$ and $|\beta|^2$ are positive integers, this means that $|\alpha|^2 = |\beta|^2 = 1$. Hence, if a Gaussian integer $\alpha = a + bi$ is a unit, then $a^2 + b^2 = 1$. It is then easy to check in $\mathbb{Z}$ that the only solutions to this equation are $a = \pm 1$ and $b = 0$, or vice versa. That is, the units of $\mathbb{Z}[i]$ consist precisely of $\{1, -1, i, -i\}$, the four points on the unit circle in the complex plane at angles $0, \pi/2, \pi, 3\pi/2$. Notice that we have arrived at this conclusion much more easily than by directly solving the appropriate Diophantine equations, as in Exercise 1. (Notice also that this discussion is precisely a solution to Exercise 8.5.)

What we would really like to do is to equip $\mathbb{Z}[\sqrt{n}]$ with a function playing the same role as the square of the modulus does for the Gaussian integers, where $n$ is any square-free integer. The function we propose to use is the following. Let $N : \mathbb{Z}[\sqrt{n}] \mapsto \mathbb{N} \cup \{0\}$ be defined by setting

$$N(a + b\sqrt{n}) = |(a + b\sqrt{n})(a - b\sqrt{n})| = |a^2 - nb^2|.$$

We call $N(\alpha)$ the **norm** of $\alpha$.

Notice that if $n$ is negative (as in the case of the Gaussian integers), this is merely the square of the ordinary complex modulus, and is automatically non-negative; consequently, the absolute value sign in the definition is superfluous. However, if $n$ is positive, we need the absolute value, or else we might not get non-negative integer values. And note that as long as $n$ is square-free, $N(\alpha) = 0$ only if $\alpha = 0$ (otherwise, $\sqrt{n}$ would be rational, which contradicts Exercise 15).

In order to use the function $N$ as we did $|\cdot|^2$ for the Gaussian integers, we need to show that it preserves multiplication:

**Theorem 31.1** *Let $n$ be a square-free integer, and let $\alpha, \beta \in \mathbb{Z}[\sqrt{n}]$. Then $N(\alpha\beta) = N(\alpha)N(\beta)$.*

**Proof:** This proof is similar in flavor to that given for Theorem 8.3 and is left to the reader; see Exercise 2. □

We now show that the norm is useful for identifying units, in any of the rings $\mathbb{Z}[\sqrt{n}]$:

**Theorem 31.2** *Let $n$ be a square-free integer, and let $\alpha \in \mathbb{Z}[\sqrt{n}]$. Then $\alpha$ is a unit if and only if $N(\alpha) = 1$.*

**Proof:** Suppose that $n$ is a square-free integer, and $\alpha \in \mathbb{Z}[\sqrt{n}]$ is a unit. Then there exists $\beta$ with $\alpha\beta = 1$. But then application of the function $N$ gives us $N(\alpha)N(\beta) = 1$. Because all values of the norm function are non-negative integers, we must have that $N(\alpha) = 1$, as required.

Conversely, suppose that $N(\alpha) = 1$. But if $\alpha = a + b\sqrt{n}$, this means that $(a + b\sqrt{n})(a - b\sqrt{n}) = 1$ or $-1$, which means that $a - b\sqrt{n}$ (or $-a + b\sqrt{n}$) is the required multiplicative inverse. □

▷ **Quick Exercise.** Notice that $5^2 - 6(2)^2 = 1$. What units does this provide for the ring $\mathbb{Z}[\sqrt{6}]$? ◁

Let us for the moment restrict ourselves to the question of explicitly determining the units in quadratic extensions of the form $\mathbb{Z}[\sqrt{n}]$, where $n$ is negative. If in this case $\alpha = a + b\sqrt{n}$ is a unit, then $1 = N(\alpha) = a^2 - nb^2$. If $n = -1$, this is the case of the Gaussian integers, and we get $\{1, -1, i, -i\}$ as the units, exactly as before. But if $n < -1$, then there are no solutions to this equation if $b > 0$. And so the only units in this case are $\pm 1$.

▷ **Quick Exercise.** Check for yourself the solutions to the equation $1 = a^2 - nb^2$, when $n < -1$. ◁

Once we know the units for a ring, we know which elements are associates of which. For example, what are the associates of $3 + i$ in $\mathbb{Z}[i]$? They are $(1)(3 + i) = 3 + i, (-1)(3 + i) = -3 - i, (i)(3 + i) = -1 + 3i$, and $(-i)(3 + i) = 1 - 3i$.

▷ **Quick Exercise.** What are the associates of $a + bi$ in $\mathbb{Z}[i]$? ◁

**Example 31.1**

What are the associates of a given element in $\mathbb{Z}[\sqrt{-5}]$? The only units in this ring are $\pm 1 + 0\sqrt{-5} = \pm 1$. Thus, the only associates of an element in $\mathbb{Z}[\sqrt{-5}]$ are the element itself and its negative.

Let's now turn to the case $\mathbb{Z}[\sqrt{n}]$, for $n > 0$. If we wish to find the units of $\mathbb{Z}[\sqrt{n}]$, when $n$ is a positive square-free integer, we must solve the Diophantine equation $|a^2 - nb^2| = 1$. In the Quick Exercise after Theorem 31.2 we actually gave a solution to this equation for $n = 6$, which then gave us units in $\mathbb{Z}[\sqrt{6}]$.

Unfortunately however, this is in general a rather more subtle question than the corresponding question when $n < 0$. We are looking for integers $a, b$ satisfying either $a^2 = nb^2 + 1$, or else $nb^2 = a^2 + 1$, where $n$ is a constant positive square-free integer. Such Diophantine equations have a long history and are known as *Pell's equations*. The description of how such equations are solved in general is a fascinating piece of mathematics that is unfortunately beyond the scope of our text.

**Example 31.2**

For $n = 2$, observe that $a = 1$ and $b = 1$ does in fact give a solution, because $|1^2 - 2(1)^2| = 1$. Thus, $1 + 1\sqrt{2}$ is a unit for $\mathbb{Z}[\sqrt{2}]$.

We have already observed in Section 8.2 that in any ring the set of units is closed under multiplication, and so this means that $(1 + \sqrt{2})^2$ must also be a unit in $\mathbb{Z}[\sqrt{2}]$, and indeed so is $(1 + \sqrt{2})^n$, for any positive integer $n$. Furthermore, each of these units is distinct: Since $1 + \sqrt{2} > 1$, these numbers form a list of strictly increasing real numbers.

Notice that we could also have concluded that the positive powers of $\alpha$ are units by observing that if $N(\alpha) = 1$, then $N(\alpha^n) = (N(\alpha))^n = 1^n = 1$. But this even makes sense for negative exponents. After all, because $1 + \sqrt{2}$ is a unit, $(1 + \sqrt{2})^{-1} = -1 + \sqrt{2}$ is a unit of $\mathbb{Z}[\sqrt{2}]$, and so all elements of the form $(1 + \sqrt{2})^{-n} = ((1 + \sqrt{2})^{-1})^n = (-1 + \sqrt{2})^n$ are units too.

▷ **Quick Exercise.** Why is $(1 + \sqrt{2})^{-1} = -1 + \sqrt{2}$? ◁

But $-1$ is a unit too (it is its own multiplicative inverse), and so we obtain the following lists of units of $\mathbb{Z}[\sqrt{2}]$:

$$1, \quad 1 + \sqrt{2}, \quad (1 + \sqrt{2})^2 = 3 + 2\sqrt{2}, \quad (1 + \sqrt{2})^3 = 7 + 5\sqrt{2}, \cdots,$$

$$-1, \quad -1 - \sqrt{2}, \quad -3 - 2\sqrt{2}, \quad -7 - 5\sqrt{2}, \cdots,$$

$$(1 + \sqrt{2})^{-1} = -1 + \sqrt{2}, \quad (1 + \sqrt{2})^{-2} = -3 + 2\sqrt{2},$$

$$(1 + \sqrt{2})^{-3} = -7 + 5\sqrt{2}, \cdots,$$

and

$$1 - \sqrt{2}, \quad 3 - 2\sqrt{2}, \quad 7 - 5\sqrt{2}, \cdots.$$

It turns out (though we will not prove it here) that this is a complete list.

This infinite list of units means that detecting whether or not two elements in $\mathbb{Z}[\sqrt{2}]$ are associates is not the trivial matter it is in $\mathbb{Z}$ (or even in $\mathbb{Z}[i]$). For example, $4 + \sqrt{2}$ and $8 - 5\sqrt{2}$ are associates, because $(4 + \sqrt{2})(3 - 2\sqrt{2}) = 8 - 5\sqrt{2}$, and $3 - 2\sqrt{2}$ is a unit.

▷ **Quick Exercise.** Calculate several distinct associates of $3 + 5\sqrt{2}$ in $\mathbb{Z}[\sqrt{2}]$. ◁

You encountered some of these computations before, in Exercise 8.6 and Example 9.8.

▷ **Quick Exercise.** Find several units of $\mathbb{Z}[\sqrt{3}]$ (other than $\pm 1$), and then compute several associates of $\sqrt{3}$. ◁

Notice that in any of these quadratic extension rings, it is obvious that if two elements are associates, then they have the same norm.

▷ **Quick Exercise.** Why do two elements that are associates have the same norm? ◁

It is tempting to conjecture that the converse of this statement is true, but you will demonstrate by explicit example in Exercise 12 two elements of such a ring, with the same norm, which are not associates.

## 31.3 Irreducibles in Quadratic Extensions

In order to pursue a Fundamental Theorem of Arithmetic for quadratic extensions of $\mathbb{Z}$, we must now explore the irreducible elements of these rings; as we discussed in Section 15.1, elements are irreducible if they admit no 'non-trivial' factorizations.

So what about irreducibles for $\mathbb{Z}[\sqrt{n}]$ (for any square-free $n$)? First observe that if $N(\alpha)$ is an irreducible (integer), then certainly $\alpha$ itself is an irreducible. For if $\alpha = \beta\gamma$, then the norm of exactly one of $\beta$ and $\gamma$ must be 1 (because $N(\alpha)$ is a (positive) irreducible in $\mathbb{Z}$) and units are exactly those elements of $\mathbb{Z}[\sqrt{n}]$ which have norm one. We state this result for emphasis:

**Theorem 31.3** *Let $n$ be a square-free integer, and $\alpha \in \mathbb{Z}[\sqrt{n}]$. If $N(\alpha)$ is a prime integer, then $\alpha$ is irreducible in $\mathbb{Z}[\sqrt{n}]$.*

### Example 31.3

Thus, $2 + 5\sqrt{-5}$ is irreducible in $\mathbb{Z}[\sqrt{-5}]$, because $N(2+5\sqrt{-5}) = 129$, which is prime (in $\mathbb{Z}$). And $1 + 2\sqrt{2}$ is irreducible in $\mathbb{Z}[\sqrt{2}]$, because its norm is 7.

### Example 31.4

Notice that $1 + i$ is irreducible in the Gaussian integers. But because $2 = (1+i)(1-i)$, 2 (an irreducible in $\mathbb{Z}$) is *not* an irreducible in $\mathbb{Z}[i]$.

▷ **Quick Exercise.** Use Theorem 31.3 to find some irreducible elements in $\mathbb{Z}[\sqrt{6}]$ and in $\mathbb{Z}[\sqrt{-3}]$. ◁

Unfortunately, the converse of Theorem 31.3 is false. That is, there exist irreducibles that do not have prime norm.

### Example 31.5

For example, we claim that $1 + \sqrt{-5}$ (which has norm 6) is irreducible in $\mathbb{Z}[\sqrt{-5}]$. For if it had a non-trivial factorization, the factors would have to have norms of 2 and 3. But this would require integer solutions of the Diophantine equations $a^2 + 5b^2 = 2$ and $a^2 + 5b^2 = 3$. Obviously, no such solutions are possible.

### Example 31.6

Similarly, we claim that 3 is irreducible in $\mathbb{Z}[\sqrt{2}]$, even though the norm of 3 is 9 (which is not prime in $\mathbb{Z}$). In this case we would need solutions of at least one of the Diophantine equations $a^2 = 2b^2 + 3$ or $2b^2 = a^2 + 3$. Suppose by way of contradiction that the first of these equations did in fact have a solution. Then $a$ would have to be an odd integer, and so $a = 2k + 1$. But then $(2k + 1)^2 = 2b^2 + 3$, or after a little algebra, $2k^2 + 2k = b^2 + 1$. This means that $b$ must be odd, and so $b = 2m + 1$. Then $2k^2 + 2k = (2m + 1)^2 + 1$, or after simplification, $k^2 + k = 2m^2 + 2m + 1$. But $k^2 + k$, as the product of consecutive integers, is necessarily even, while $2m^2 + 2m + 1$ is necessarily odd. This contradiction shows that no integer solution to the equation $a^2 = 2b^2 + 3$ is possible. We leave it to you to check (see Exercise 4) that there is no solution to the other equation either. This means there are no members of $\mathbb{Z}[\sqrt{2}]$ which have norm 3, and so 3 (with norm 9) has no non-trivial factorizations.

The fact that it required this excursion into number theory to prove that 3 is irreducible in $\mathbb{Z}[\sqrt{2}]$ might convince you that the general question of determining all irreducibles for quadratic extensions of $\mathbb{Z}$ is difficult. This is in fact the case, and we will not pursue the matter here.

## 31.4    Factorization for Quadratic Extensions

You should now recall the proof that every integer can be factored into irreducibles (and the analogous proof for $\mathbb{Q}[x]$). Both of these proofs depend heavily on the fact that $\mathbb{N}$ is well ordered: By continuing to extract factors from a positive integer, we decrease its size, and we cannot continue this indefinitely. And for $\mathbb{Q}[x]$, by continuing to extract non-trivial factors from a polynomial of degree greater than 1, we decrease the degree, and we cannot continue this indefinitely. Can we apply this technique to $\mathbb{Z}[\sqrt{n}]$?

We have exactly the appropriate tool at hand: The norm function $N$ provides a measure of size. In fact, it shouldn't be too surprising that this might work, because the norm function takes values precisely in the set of non-negative integers. We can now prove the following:

**Theorem 31.4    Factorization Theorem for Quadratic Extensions of $\mathbb{Z}$**    *Let $n$ be a square-free integer. Then every non-zero non-unit of $\mathbb{Z}[\sqrt{n}]$ is either irreducible or a product of irreducibles.*

**Proof**:    Let $\alpha \neq 0$ be a non-unit of $\mathbb{Z}[\sqrt{n}]$. We proceed by induction on $N(\alpha)$. By Theorem 31.2, $N(\alpha) \neq 1$. Now if $N(\alpha) = 2$, then $\alpha$ is itself irreducible, by Theorem 31.3.

Now suppose the theorem holds true for all $\beta$ with $N(\beta) < m$. If $\alpha$ is irreducible already, we are done. If not, then $\alpha = \beta\gamma$, where both factors are non-units. But because $N(\alpha) = N(\beta)N(\gamma)$ and $N(\beta) > 1$ and $N(\gamma) > 1$, we have that $N(\beta) < m$ and $N(\gamma) < m$. By the induction hypothesis both $\beta$ and $\gamma$ can be factored as a product of irreducibles and, thus, so can their product $\alpha$.    □

▷ **Quick Exercise.**    Find irreducible elements in $\mathbb{Z}[i]$ and $\mathbb{Z}[\sqrt{-7}]$ that have norm 2. ◁

It is of great importance to note that we have neither claimed nor proved that the factorization into irreducibles provided by this theorem is unique. There is good reason for this: For suitable choice of $n$, such factorizations are *not* unique.

To see this, consider the following two factorizations of 6 in $\mathbb{Z}[\sqrt{-5}]$:

$$6 = (1 + \sqrt{-5}) \cdot (1 - \sqrt{-5}) = 2 \cdot 3.$$

We argued in Example 31.5 that $1 + \sqrt{-5}$ is irreducible, and a similar argument applies for the other factors in the two given factorizations.

▷ **Quick Exercise.**    Verify that $1 - \sqrt{-5}$, 2, and 3 are irreducible in $\mathbb{Z}[\sqrt{-5}]$, by considering their norms. ◁

Now, if a unique factorization theorem applied for $\mathbb{Z}[\sqrt{-5}]$, this would mean that these two factorizations would be the same, up to order and unit factors. But it is quite easy to see that 2 is not an associate of either $1 + \sqrt{-5}$ or $1 - \sqrt{-5}$, because associates must have the same norm. In Exercise 7, you will provide two essentially distinct factorizations of 8 in $\mathbb{Z}[\sqrt{-7}]$.

In contrast, we shall see later that we will have a unique factorization into irreducibles, for *some* quadratic extensions of the integers. We have thus seen that these rings provide us with some crucial examples which will delineate the theory.

---

## Chapter Summary

We introduced the *quadratic extensions of the integers* and looked at units and irreducibles in such rings, making heavy use of the *norm function*. We were able to prove that in any such ring, non-units can be factored into irreducible elements. However, we discovered by example that such factorizations are not always unique.

---

## Warm-up Exercises

a. Determine four distinct associates of $3 + 2\sqrt{2}$ in $\mathbb{Z}[\sqrt{2}]$.

b. Determine all associates of $5 + i$ in $\mathbb{Z}[i]$.

c. Let $n$ be any square-free integer. Why is $n$ not irreducible in $\mathbb{Z}[\sqrt{n}]$?

d. Do the two factorizations
$$5 = \sqrt{-5} \cdot -\sqrt{-5} = 1 \cdot 5$$
provide another example that factorization into irreducibles is not unique in $\mathbb{Z}[\sqrt{-5}]$? Why or why not?

e. Do the two factorizations
$$6 = 3 \cdot 2 = (-2 + 2\sqrt{2})(3 + 3\sqrt{2})$$
provide an example to show that factorization into irreducibles is not unique in $\mathbb{Z}[\sqrt{2}]$? Why or why not?

f. Give a nice description of the elements of the ideal $\langle \sqrt{7} \rangle$ in the ring $\mathbb{Z}[\sqrt{7}]$.

g. Determine four distinct irreducibles in $\mathbb{Z}[\sqrt{2}]$. *Hint:* Look for elements with prime norm.

h. Do irreducible elements of $\mathbb{Z}[\sqrt{n}]$ necessarily have prime norm?

i. Determine which of the following elements are irreducible:

    (a) $9 + \sqrt{10}$ in $\mathbb{Z}[\sqrt{10}]$.

    (b) $5 + \sqrt{5}$ in $\mathbb{Z}[\sqrt{5}]$.

## Exercises

1. Find all simultaneous integer solutions to the Diophantine equations $ac - bd = 1, ad + bc = 0$ directly, by eliminating variables; interpret your solutions as determining all units in the Gaussian integers.

2. Prove Theorem 31.1. That is, let $n$ be a square-free integer. As in the text, define $N(a + b\sqrt{n}) = |a^2 - nb^2|$. Prove that $N$ preserves multiplication, that is, $N(\alpha\beta) = N(\alpha)N(\beta)$.

3. Suppose that $n, m$ are distinct square-free integers. Prove that

$$\mathbb{Z}[\sqrt{n}] \cap \mathbb{Z}[\sqrt{m}] = \mathbb{Z}.$$

   This is not true if at least one of the integers $n$ and $m$ is not square-free. Give an example to show this.

4. Prove that the Diophantine equation $2b^2 = a^2 + 3$ has no integer solutions, proceeding similarly as the problem $a^2 = 2b^2 + 3$ is handled in the text in Example 31.6.

5. Find infinitely many distinct units in $\mathbb{Z}[\sqrt{7}]$. Then list infinitely many associates of $\sqrt{7}$ in $\mathbb{Z}[\sqrt{7}]$.

6. Suppose that $n$ is a square-free integer and $n > 0$. Prove that $\mathbb{Z}[\sqrt{-n}]$ has only finitely many units.

7. Find two factorizations of 8 into irreducibles in $\mathbb{Z}[\sqrt{-7}]$ that are essentially distinct.

8. Prove that if $p$ is a prime in $\mathbb{Z}$ and $p$ is congruent to 3 mod 4, then $p$ is irreducible in $\mathbb{Z}[i]$.

9. Prove 2 is irreducible in $\mathbb{Z}[\sqrt{n}]$ for all square-free $n < -2$.

10. Find two distinct square-free integers $n$ (with $n > 1$) for which 2 is not irreducible in $\mathbb{Z}[\sqrt{n}]$.

11. Suppose that $p$ is a positive prime integer. Prove that $\sqrt{p}$ is irreducible in $\mathbb{Z}[\sqrt{p}]$. Show by example that this is false if $p$ is not prime; in particular, consider $p = 6$.

12. Show that $11 + 6\sqrt{-5}$ and $16 + 3\sqrt{-5}$ are irreducible elements in $\mathbb{Z}[\sqrt{-5}]$, with the same norm. Show that these elements are not associates.

13. Using arguments similar to those used in Examples 14.6 and 15.11 for $\mathbb{Z}[\sqrt{-5}]$, show that

$$\langle 2, 1 + \sqrt{-7} \rangle$$

   is a maximal ideal in $\mathbb{Z}[\sqrt{-7}]$ that is not principal.

14. Consider the ideal $\langle 7 + \sqrt{-5}, 9 \rangle$ in $\mathbb{Z}[\sqrt{-5}]$. Show that this ideal is neither maximal nor principal.

15. Suppose that $n$ is a square-free integer. Prove that $\sqrt{n}$ is irrational.

16. Consider the ring $\mathbb{Z}[\sqrt{2}]$.

(a) Show that
$$I = \{a + b\sqrt{2} \in \mathbb{Z}[\sqrt{2}] : a \text{ is even}\}$$
is an ideal.

(b) Show that $I$ is principal. *Hint:* Think of 'small' elements of $I$.

(c) Show that $I$ is a maximal ideal.

(d) Show that
$$J = \{a + b\sqrt{2} \in \mathbb{Z}[\sqrt{2}] : b \text{ is even}\}$$
is closed under subtraction but is not an ideal.

17. Let $I$ be an ideal of $\mathbb{Z}[\sqrt{n}]$, where $n$ is a square-free integer. Define
$$\bar{I} = \{a + b\sqrt{n} : a - b\sqrt{n} \in I\}.$$

(a) Prove that $\bar{I}$ is an ideal of $\mathbb{Z}[\sqrt{n}]$.

(b) Provide particular examples of such ideals, where $I \neq \bar{I}$, and where $I = \bar{I}$.

(c) Prove that $I$ is a principal ideal if and only if $\bar{I}$ is a principal ideal.

# Chapter 32

## Factorization

In this chapter we first discuss what is required to be able to factor each non-unit as a product of irreducibles. We will then inquire into the condition necessary to force that such factorizations be unique.

## 32.1  How Might Factorization Fail?

To make a more general attack on the factorization problem, let us think about how factorization into irreducibles in a domain $R$ could fail. Suppose then that $0 \neq a_1 \in R$ is not irreducible and is not the product of irreducibles. Then there exists a factorization $a_1 = a_2 b_2$, where neither $a_2$ nor $b_2$ is a unit. Furthermore, because $a_1$ cannot be factored into a product of irreducibles, this must be true of at least one of $a_2$ or $b_2$. To be specific, let's suppose that $a_2$ can't be so factored. But then $a_2$ can be factored as $a_2 = a_3 b_3$, where neither $a_3$ nor $b_3$ is a unit, and where $a_3$ cannot be factored into a product of irreducibles. If we continue in this fashion, we obtain an infinite sequence of non-trivial factorizations:

$$a_1 = a_2 b_2, \quad a_2 = a_3 b_3, \quad a_3 = a_4 b_4, \quad \cdots .$$

where every element in sight is a non-unit.

But we can rephrase this infinite sequence, using the language of principal ideals $\langle a \rangle$, which consist precisely of the multiples of $a$ in $R$. Our sequence of factorizations clearly in part asserts that

$$\langle a_1 \rangle \subseteq \langle a_2 \rangle \subseteq \langle a_3 \rangle \subseteq \cdots .$$

But we claim further that these containments are proper. This follows immediately from Theorem 9.3, which asserts that two principal ideals in an integral domain are equal exactly if the elements are associates. Since in our sequence the factorizations are all non-trivial, no element $a_n$ in the sequence can be an associate of the next element $a_{n+1}$.

We thus have that our infinite sequence of proper factorizations leads to an infinite *ascending* sequence of principal ideals, each being strictly larger than its predecessor.

### Example 32.1

Let's consider an example of these ideas and this notation in the ring $\mathbb{Z}$. Consider the integer 360. By successively factoring 360, we might obtain the following chain of inclusions:

$$\langle 360 \rangle \subset \langle 180 \rangle \subset \langle 60 \rangle \subset \langle 20 \rangle \subset \langle 10 \rangle \subset \langle 5 \rangle .$$

We cannot continue any further in this example, precisely because 5 is irreducible and so admits no further non-trivial factorization. Of course, we know already that $\mathbb{Z}$ has a factorization theorem, and so we should have expected this chain of inclusions to halt.

The argument above about the potential lack of factorization is important enough for us that we will record its conclusion as a lemma:

**Lemma 32.1** *Let $R$ be a domain and $0 \neq a_1$ an element of $R$ that is neither irreducible nor the product of irreducibles. Then there exist non-units $a_2, a_3, a_4, \cdots$, such that*

$$\langle a_1 \rangle \subset \langle a_2 \rangle \subset \langle a_3 \rangle \subset \cdots .$$

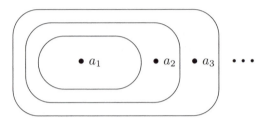

For a specific example of a domain which has a non-zero non-unit which is neither irreducible, nor can it be factored as a product of irreducibles, see Exercise 10.

## 32.2   PIDs Have Unique Factorization

We now obtain the following elegant theorem that provides a sufficient condition for factorization into irreducibles.

**Theorem 32.2    Factorization Theorem for PIDs**    *In a principal ideal domain, every non-zero non-unit is either irreducible or a product of irreducibles.*

**Proof:**    Suppose by way of contradiction that $R$ is a PID with at least one non-zero non-unit that is neither irreducible nor factorable as a product of irreducibles. By Lemma 32.1, we then have a properly ascending chain of principal ideals, where each $a_i$ is a non-unit:

$$\langle a_1 \rangle \subset \langle a_2 \rangle \subset \langle a_3 \rangle \subset \cdots .$$

Consider now the set $I = \bigcup \{ \langle a_i \rangle : i \in \mathbb{N} \}$. We claim that this is an ideal. First, we check that $I$ is closed under subtraction. Given $x, y \in I$, there exist $i$ and $j$ such that $x \in \langle a_i \rangle$ and $y \in \langle a_j \rangle$. Suppose (without loss of generality) that $j$ is the larger of $i$ and $j$. Then $x, y \in \langle a_j \rangle$, and so $x - y \in \langle a_j \rangle \subseteq I$. Now $I$ also satisfies the multiplicative absorption property: Suppose that $x \in I$ and $r \in R$. There exists $i$ so that $x \in \langle a_i \rangle$, an ideal. So, $rx \in \langle a_i \rangle \subseteq I$. Thus, $I$ is an ideal.

Because $R$ is a PID, $I = \langle a \rangle$, for some $a \in R$. But $I$ is the union of the $\langle a_j \rangle$'s, and so $a \in \langle a_j \rangle$, for some $j$; thus, $\langle a \rangle = \langle a_j \rangle$. But then

$$\langle a \rangle = \langle a_j \rangle = \langle a_{j+1} \rangle = \langle a_{j+2} \rangle = \cdots ,$$

which is contrary to our assumption. This contradiction means that every non-unit can be factored into irreducibles.    $\square$

This abstract proof now gives an alternate approach to seeing that factorization holds for $\mathbb{Z}$ and for $\mathbb{Q}[x]$. The additional abstraction of this proof makes it more powerful. As we shall see, it applies to many more domains than our two familiar examples.

An important generalization of one of the ideas in the proof above is due to the great German mathematician Emmy Noether, who in the 1920s laid much of the important groundwork for axiomatic ring theory. She isolated as particularly important those commutative rings where every ascending sequence of ideals is finite. That is, if a set of ideals $I_n$ is totally ordered under inclusion

$$I_1 \subseteq I_2 \quad \subseteq I_3 \subseteq \cdots \subseteq I_n \subseteq \cdots$$

then there must exist an integer $j$ for which $I_j = I_{j+1} = \cdots$. We have seen in the proof above that PIDs satisfy this property. Such commutative rings are said to have the **ascending chain condition** (ACC for short) on ideals, and are called **Noetherian**.

## 32.3   Primes

You should now recall the proof of the *uniqueness* of factorization into irreducibles for $\mathbb{Z}$ (or $\mathbb{Q}[x]$). Our intent in this section is to construct a more general context in which this proof is true. This proof relies heavily on the fact that irreducible elements in $\mathbb{Z}$ (or $\mathbb{Q}[x]$) are in fact prime. It should not then be surprising that a general unique factorization theorem should rely on the same considerations. We now define prime elements in an arbitrary domain: A non-unit $p \neq 0$ of a domain $R$ is a **prime** if, whenever $p$ divides $ab$, then $p$ divides $a$ or $b$.

**Example 32.2**

> Under this new definition, the prime integers remain prime in $\mathbb{Z}$. And the irreducible polynomials in $\mathbb{Q}[x]$ are prime also.

**Example 32.3**

> Let's show directly that the element $\sqrt{3}$ is prime in the domain $\mathbb{Z}[\sqrt{3}]$. For that purpose, we must suppose that $\sqrt{3}$ divides a product $\alpha\beta$, where $\alpha, \beta \in \mathbb{Z}[\sqrt{3}]$. Now $\alpha = a + b\sqrt{3}$ and $\beta = c + d\sqrt{3}$, and
> $$\alpha\beta = (ac + 3bd) + (ad + bc)\sqrt{3}.$$
> If $\sqrt{3}$ divides this product, it must clearly divide the rational part $ac + 3bd$. But $\sqrt{3}$ obviously divides 3, and so $\sqrt{3}$ must divide $ac$ in $\mathbb{Z}[\sqrt{3}]$. But $ac$ is an integer, and so the only way this can occur is if 3 actually divides $ac$. Now 3 is a prime *integer*, and so (without loss of generality) 3 divides $a$. But then $\sqrt{3}$ divides $a + b\sqrt{3}$, as required.

Note that an easy inductive proof (identical to that for $\mathbb{Z}$) shows that if $p$ is prime and it divides a product of $n$ terms, then $p$ divides at least one of the factors. (See Exercise 2.5 and Exercise 5.3.)

Now examine the proof that for $\mathbb{Z}$, irreducible and prime elements are the same (Theorem 2.7). Notice that the proof that primeness implies irreducibility is general, while the converse depends on the GCD identity (a theorem which need not hold for arbitrary domains). We thus have the following:

**Theorem 32.3** *In any domain, prime elements are irreducible.*

**Proof:**   Check that the proof for $\mathbb{Z}$ (Theorem 2.7) holds in general. (This is Exercise 1.)□

## Example 32.4

The converse of this theorem is in general false, and the factorization

$$6 = 2 \cdot 3 = (1 + \sqrt{-5})(1 - \sqrt{-5})$$

in $\mathbb{Z}[\sqrt{-5}]$ that we considered in Section 31.4 provides the counterexample, for 2 is irreducible, yet it divides $(1 + \sqrt{-5})(1 - \sqrt{-5})$ without dividing either factor and hence *is not* prime.

The previous example finally justifies our long-standing careful distinction between the concepts of irreducibility and primeness.

The following theorem now shows that uniqueness of factorization fails in a domain exactly when the concepts of irreducibility and primeness fail to coincide:

**Theorem 32.4** *Consider a domain in which every non-zero non-unit is either an irreducible or a product of irreducibles. Then all irreducible elements are prime if and only if the factorization of non-units into irreducibles is unique, up to order and unit factors.*

**Proof**:    Suppose that $D$ is a domain with unique factorization, and $p$ is an irreducible element. We wish to show that $p$ is prime and toward that end assume that $p|ab$. Then $pc = ab$, for some $c \in D$. Now factor $c$, $a$, and $b$ into irreducibles, thus obtaining the following equation:

$$pc_1c_2 \ldots c_k = a_1a_2 \ldots a_m b_1 b_2 \ldots b_n.$$

By the uniqueness of the factorization of $pc$, $p$ must be an associate of some $a_i$ or $b_j$; that is, $p|a$ or $p|b$, as required.

For the other direction, check that the proof for $\mathbb{Q}[x]$ works in general. (This is Exercise 2; see Theorem 5.4.)    □

## Example 32.5

In Exercise 31.7 you obtained two essentially distinct factorizations of 8 in $\mathbb{Z}[\sqrt{-7}]$; namely, as

$$2 \cdot 2 \cdot 2 = (1 + \sqrt{-7})(1 - \sqrt{-7}).$$

Theorem 32.4 means that of necessity $\mathbb{Z}[\sqrt{-7}]$ must possess some element that is irreducible but not prime. It is in fact easy in this case to identify a particular element that plays this role.

▷ **Quick Exercise.**   What element is that? ◁

If you did Exercise 15.2, you have encountered the idea of a prime element before. While not directly relevant to our study of factorization here, it remains interesting that the notion of prime element can be rephrased in terms of the corresponding principal ideal: The non-unit $p \neq 0$ is a prime element if and only if the principal ideal $\langle p \rangle$ is a prime ideal. This can and should be placed alongside the characterization of irreducible elements in Theorem 15.1 as those elements $p$ for which the $\langle p \rangle$ is maximal among principal ideals. In order to obtain uniqueness of factorization in a domain with factorization, we need that these two conditions are the same.

# Warm-up Exercises

a. Give examples of the following (or explain why no example exists). You should specify both a domain and an element of it:

   (a) A prime element that isn't irreducible.

   (b) An irreducible element that isn't prime.

   (c) A non-unit that is neither prime nor irreducible.

b. Argue that $\sqrt{2}$ is prime in $\mathbb{Z}[\sqrt{2}]$.

c. Suppose $u$ is a unit in the commutative ring $R$. Why is it true that if $u$ divides the product $ab$, then $u$ divides $a$ or $b$? Does this make $u$ prime?

d. Why do we care whether all irreducible elements of a domain are prime?

e. Give examples in the integral domain $\mathbb{Q}[x]$ of principal ideals $\langle f \rangle, \langle g \rangle$, where $\langle f \rangle \subset \langle g \rangle$ and where $\langle f \rangle = \langle g \rangle$. Explain why there cannot be an infinite ascending sequence of ideals in this ring which are getting strictly larger.

# Exercises

1. Prove Theorem 32.3: In any domain, prime elements are irreducible.

2. Complete the proof of Theorem 32.4. That is, suppose we have a domain in which all non-zero non-units are irreducible, or a product of irreducibles. Furthermore, suppose all irreducible elements are prime. Prove that the factorization of any non-unit into irreducibles is unique (up to order and unit factors).

3. Exhibit in the ring $\mathbb{Z}_6$ a non-unit $a$ for which $a^n = a$, for all positive integers $n$. Why does this mean that there is no unique factorization into irreducibles for this ring? Now repeat this exercise for $\mathbb{Z} \times \mathbb{Z}$.

4. Prove that $x$ is a prime element of $\mathbb{Z}[x]$.

5. Consider $1 + i \in \mathbb{Z}[i]$.

   (a) Show that

$$\begin{aligned} \langle 1 + i \rangle &= \{a + bi : a + b \text{ is even}\} \\ &= \{\alpha \in \mathbb{Z}[i] : N(\alpha) \text{ is even}\}. \end{aligned}$$

   Note that the first equation is a repeat of Exercise 13.4a.

   (b) Use the second equation in part (a) to prove that $1 + i$ is a prime element of $\mathbb{Z}[i]$.

6. Find two integers $n$ with $n \leq -5$ where 2 is not prime in $\mathbb{Z}[\sqrt{n}]$. (Note by Exercise 31.9 that 2 is irreducible in these rings.)

7. In this exercise we generalize an argument used in the proof of Theorem 32.2. Let $R$ be a commutative ring with unity, and suppose that $I_n$ is a proper ideal for all positive integers $n$. Suppose further that

$$I_1 \subseteq I_2 \cdots \subseteq I_n \subseteq I_{n+1} \subseteq \cdots.$$

Let $I = \bigcup I_n$. Prove that $I$ is a proper ideal.

8. Let $n$ be a positive integer and consider the ideals

$$\langle x^n \rangle$$

in $\mathbb{Q}[x]$. Describe succinctly the elements of $\langle x^n \rangle$. What containment relations hold among these ideals? Explain why $\mathbb{Q}[x]$ is Noetherian. Explain why the ideals $\langle x^n \rangle$ do not contradict the assertion that $\mathbb{Q}[x]$ is Noetherian.

9. Recall from Exercise 9.12 what it means for an ideal in a commutative ring with unity to be *finitely generated*. Prove that a commutative ring with unity is Noetherian if and only if all of its ideals are finitely generated. Note that Exercise 9.29 then gives us a non-Noetherian ring.

10. Let

$$D = \{f \in \mathbb{Q}[x] : \text{the constant term of } f \text{ is an integer}\}.$$

Find a non-unit $d \neq 0$ which is neither irreducible, nor a product of irreducibles. Exhibit explicitly the infinite ascending chain of principal ideals guaranteed by Lemma 32.1.

# Chapter 33

## Unique Factorization

We know that in a principal ideal domain, every non-unit can be factored into irreducibles, and we know what we must show to guarantee the uniqueness of such a factorization: We must show that every irreducible element is prime. In this chapter we are able to show that this is true for any PID.

### 33.1  UFDs

We now introduce some convenient terminology: A **unique factorization domain** (or **UFD**) is a domain in which all non-zero non-units can be factored uniquely (up to units and order) into irreducibles. More concretely, uniqueness of factorization means that if $a$ is a non-unit, and

$$a = a_1 a_2 \cdots a_n = b_1 b_2 \cdots b_m,$$

where the $a_i$ and $b_i$ are irreducibles, then $n = m$, and under some rearrangement of the $b_i$ we find $a_i$ and $b_i$ are associates, for $i = 1, \cdots n$.

Let's now rephrase Theorem 32.4 in light of our new terminology: We showed that if factorization of non-units into irreducibles is always possible, then the domain is a UFD if and only if each irreducible element is prime. Following the model of our PIDs $\mathbb{Z}$ and $\mathbb{Q}[x]$, we'd now like to prove that *all* PIDs are UFDs. Because we already know that factorization occurs in PIDs (Theorem 32.2), we can complete this proof by showing that in a PID all irreducible elements are prime.

### 33.2  A Comparison between $\mathbb{Z}$ and $\mathbb{Z}[\sqrt{-5}]$

Before we prove that every PID is a UFD, we should look more closely into our example where unique factorization fails, namely in the domain $\mathbb{Z}[\sqrt{-5}]$. In Section 31.4 we showed that the element 6 can be factored into irreducibles in two ways, namely

$$6 = 2 \cdot 3 = (1 + \sqrt{-5}) \cdot (1 - \sqrt{-5}).$$

We would like to look at this example in terms of the ideals in this domain. Consider for example the irreducible 3 in $\mathbb{Z}[\sqrt{-5}]$. Because 3 is irreducible, we know by Theorem 15.1 that $\langle 3 \rangle$ is maximal among proper principal ideals; but it need not be maximal among *all* proper ideals (because $\mathbb{Z}[\sqrt{-5}]$ may have non-principal ideals). And indeed, we showed in Example 14.6 that $\langle 3, 1 + \sqrt{-5} \rangle$ is an ideal by providing a ring homomorphism of which it is the kernel. Now clearly $1 + \sqrt{-5} \notin \langle 3 \rangle$, and so $\langle 3 \rangle \subset \langle 3, 1 + \sqrt{-5} \rangle$.

▷ **Quick Exercise.** Verify that $1 + \sqrt{-5} \notin \langle 3 \rangle$. ◁

Now since $\langle 3 \rangle$ is maximal among all principal ideals, this must mean that $\langle 3, 1 + \sqrt{-5} \rangle$ is not principal, as long as it is a proper ideal! But it is clearly proper, because in Example 14.6 we showed that

$$\langle 3, 1 + \sqrt{-5} \rangle = \{a + b\sqrt{-5} : 3|(a - b)\}.$$

▷ **Quick Exercise.** Use this characterization to give elements of $\mathbb{Z}[\sqrt{-5}]$ that do not belong to the ideal. ◁

Our argument that $\langle 3, 1 + \sqrt{-5} \rangle$ is not a principal ideal, was indirect. Let's provide an explicit computational argument also. So suppose by way of contradiction that this ideal is generated by $\alpha = a + b\sqrt{-5}$. But then 3 and $1 + \sqrt{-5}$ are multiples of $\alpha$, and so the norm $N(\alpha)$ divides both $N(3) = 9$ and $N(1 + \sqrt{-5}) = 6$. This means that $N(\alpha)$ is either 1 or 3. It can't be 1 (because then it would be a unit, which can't be an element of a proper ideal), and it can't be 3 because (as we observed in Example 31.5) the Diophantine equation $a^2 + 5b^2 = 3$ has no solutions. Thus, we have concluded again that the ideal $\langle 3, 1 + \sqrt{-5} \rangle$ is a non-principal ideal in $\mathbb{Z}[\sqrt{-5}]$. (If you have done Exercise 14.10, you have already made this argument.)

Now we claim that the ideal $\langle 3, 1 + \sqrt{-5} \rangle$ *is* maximal in $\mathbb{Z}[\sqrt{-5}]$. But we concluded this in Example 15.11, simply because the ideal is the kernel of a ring homomorphism onto a field.

But perhaps a direct computational argument would be illuminating. We need to show that if $a + b\sqrt{-5} \notin \langle 3, 1 + \sqrt{-5} \rangle$, then $\langle 3, 1 + \sqrt{-5}, a + b\sqrt{-5} \rangle$ is the whole domain. We show this by giving a linear combination of these three elements equal to 1. But recall that $\langle 3, 1 + \sqrt{-5} \rangle = \{x + y\sqrt{-5} : 3|(x - y)\}$. Because $a + b\sqrt{-5}$ is not in this ideal, this means that $a - b$ is not divisible by 3 (and hence, relatively prime to 3). Thus, by the GCD identity for $\mathbb{Z}$, there are integers $z$ and $w$ with $3z + (a - b)w = 1$. But then

$$1 = z(3) + (-bw)(1 + \sqrt{-5}) + (w)(a + b\sqrt{-5}),$$

the linear combination which we required.

Notice the distinction between $\mathbb{Z}$ and $\mathbb{Z}[\sqrt{-5}]$: In $\mathbb{Z}$ the ideal $\langle a, b \rangle$ is always principal, and a generator is $\gcd(a, b)$, which can be expressed as $ax + by$ for some $x$ and $y$; see Exercise 15.7. In particular, if $a$ and $b$ have no common divisors (other than $\pm 1$), then 1 can be so expressed. That is, $\langle 1 \rangle = \langle a, b \rangle$ if and only if $a$ and $b$ have no non-trivial common divisors.

On the other hand, in $\mathbb{Z}[\sqrt{-5}]$, 3 and $1 + \sqrt{-5}$ have no common divisors (except units), but 'ought to', because

$$1 \notin \langle 3, 1 + \sqrt{-5} \rangle \subset \mathbb{Z}[\sqrt{-5}].$$

This is precisely why the 19th-century German mathematician Ernst Kummer first used the term 'ideal'. He viewed an ideal like $\langle 3, 1 + \sqrt{-5} \rangle$ as an 'ideal number' playing the role of a gcd for 3 and $1 + \sqrt{-5}$, and thus filling in a gap in the multiplicative structure of $\mathbb{Z}[\sqrt{-5}]$. This, as we shall see, is one of the advantages of considering ideals, instead of elements.

Actually, Kummer's formulation of ideal was different than ours; our definition is due to Richard Dedekind. The crux of the Dedekind version of the definition is to identify the 'ideal number' with the set of all those numbers that ought to be its multiples. The set-theoretic nature of this definition is typical of a modern mathematical definition; Dedekind was a pioneer of this set-theoretic approach, and not only to the definition of ideal, but also in his axiomatic construction of the real numbers.

## 33.3 All PIDs Are UFDs

We can now use what we have learned about ideals in PIDs to prove the following crucial theorem, which allows us to then conclude that *all* PIDs are UFDs:

**Theorem 33.1** *In a PID, all irreducibles are prime.*

**Proof**: Suppose $p$ is irreducible in the PID $D$. To show that $p$ is prime, we suppose further that $p|ab$. We claim that $p|a$ or $p|b$. Suppose that $p$ does *not* divide $a$. Then $a \notin \langle p \rangle$, and so $\langle p \rangle \subset \langle a, p \rangle$. Because $\langle p \rangle$ is maximal this means that $\langle a, p \rangle = D$, and so $1 \in \langle a, p \rangle$. That is, $1 = ax + py$, for some $x, y \in D$. But then $b = abx + pby$. Now $p|pby$ and $p|abx$, and so $p|b$. Thus, $p$ is prime, as claimed. $\square$

We have actually encountered special cases of the previous argument twice before: Namely, for $\mathbb{Z}$ (in the proof of Theorem 2.7) and for $\mathbb{Q}[x]$ (in the proof of Theorem 5.2, where you did the proving). What we used in those arguments was the GCD identity. In the argument above we arrive at the conclusion $1 = ax + py$ by using the fact that $\langle p \rangle$ is a maximal ideal.

From Chapter 32 we know that any non-unit in a PID is irreducible or factorable into irreducibles (Theorem 32.2). From Theorem 32.4 we know that uniqueness of factorization is equivalent to irreducible elements being prime. Because we've just proved that this is true in a PID, we have the following theorem:

**Theorem 33.2** *Any PID is a UFD.*

This theorem thus encompasses both the Fundamental Theorem of Arithmetic for $\mathbb{Z}$, and the Unique Factorization Theorem for $\mathbb{Q}[x]$. In Chapter 35 we will discover other PIDs as well, and hence other UFDs too.

The natural question to ask at this point is: Is the converse of this theorem true? That is, are all UFDs PIDs? The answer is no, and we have already encountered a domain which will serve as a counterexample. That domain is $\mathbb{Z}[x]$. In Example 9.12 we argued that $\langle 2, x \rangle$ is a non-principal ideal, and so $\mathbb{Z}[x]$ is not a PID. It thus remains to show that this domain is in fact a UFD. In order to accomplish this, we will need to inquire more carefully into the relationship between $\mathbb{Z}[x]$ and $\mathbb{Q}[x]$. Note that the relationship cannot be too simple, because the former ring is not a PID, while the latter ring is. This inquiry is the subject of Chapter 34; it requires Gauss's Lemma 5.5.

## Chapter Summary

In this chapter we defined the notion of a *unique factorization domain*. We proved that in a PID every irreducible element is prime, and consequently concluded that every PID is a UFD.

## Warm-up Exercises

a. Explain which implies which: UFD and PID.

b. Have we given an example yet of a UFD which is not a PID?

c. Give examples of the following, or else explain why such an example does not exist:

   (a) a domain where every prime element is irreducible;

   (b) a domain where every irreducible element is prime;

   (c) a domain where every non-zero non-unit can be factored into irreducibles, but not all such factorizations are unique;

   (d) a domain without any irreducible elements.

d. Why is a finite integral domain always a PID?

e. What does the GCD identity have to do with the fact that in $\mathbb{Z}$, $\langle 9, 50 \rangle = \mathbb{Z}$?

---

## Exercises

1. Find another element of $\mathbb{Z}[\sqrt{-5}]$ other than 6 which has two distinct factorizations into irreducibles (it is easiest to do this for another integer). Prove that your factors are irreducible, but that the factors in one factorization are not associates of the other factors. Then use your factorization to provide another example of a principal ideal in this domain which is maximal among all principal ideals, but not a maximal ideal; give explicitly a proper ideal strictly larger than the principal ideal.

2. Let $n > 2$ be a square-free integer. Prove that $\mathbb{Z}[\sqrt{-n}]$ is not a PID.

3. In this exercise we generalize the notion of *greatest common divisor* to any PID. If $R$ is a PID, and $a, b \in R$ are non-zero non-units, then $\langle a, b \rangle = \langle d \rangle$, for some $d \in R$. We call $d$ a **greatest common divisor** for $a$ and $b$. You will now prove that $d$ satisfies the appropriate properties:

   (a) *d is a common divisor*: Prove that $d$ divides both $a$ and $b$; that is, $a = dx$ and $b = dy$ for some $x, y \in R$.

   (b) *d is a greatest common divisor*: Suppose that $c$ is an element that divides both $a$ and $b$. Prove that $c$ divides $d$.

   (c) *d is unique, up to trivial factors*: If $d$ and $e$ are greatest common divisors of $a, b$, according to this definition, then $d$ and $e$ are associates.

   (d) *d satisfies the GCD identity*: There exist $x, y \in R$ for which $d = ax + by$.

# Chapter 34

## Polynomials with Integer Coefficients

Our aim in this chapter is to prove that $\mathbb{Z}[x]$, the ring of polynomials with integer coefficients, is a UFD. It probably seems plausible that every polynomial with integer coefficients can be factored uniquely into irreducibles, but the proof of the analogous statement for $\mathbb{Q}[x]$ will not work for $\mathbb{Z}[x]$.

## 34.1  The Proof That $\mathbb{Q}[x]$ Is a UFD

Let's recall how we prove that $\mathbb{Q}[x]$ is a UFD, to see where the argument fails for $\mathbb{Z}[x]$. We proved this result for $\mathbb{Q}[x]$ originally in Theorems 5.1 and 5.4 and then provided another proof in Theorem 33.2 which depends on the fact that $\mathbb{Q}[x]$ is a PID. In either proof we showed two things: (1) non-units factor into irreducibles, and (2) such factorizations are unique. We proved in Chapter 5 that non-units factor by using induction on degree, and in Chapter 32 by using the fact that $\mathbb{Q}[x]$ is a PID. We then proved that such factorizations are unique by showing that in $\mathbb{Q}[x]$ irreducible elements are prime. In the first version of our proof, this latter depended on the GCD identity; in the proof in Chapter 32, this depended on the fact that in a PID an element $a$ is irreducible if and only if $\langle a \rangle$ is a maximal ideal.

Because the idea of degree still makes good sense, we will find that proving the existence of factorizations is not difficult. However, there is no GCD identity for $\mathbb{Z}[x]$. Consider 2 and $x$ in $\mathbb{Z}[x]$. A gcd in $\mathbb{Z}[x]$ for these two elements is 1, but 1 cannot be written as a linear combination of 2 and $x$. Furthermore, $\mathbb{Z}[x]$ is not a PID: We saw already in Example 9.12 that $\langle 2, x \rangle$ is not a principal ideal. Note that the lack of a GCD identity (for the elements 2 and $x$) and the fact that $\langle 2, x \rangle$ is not principal, amount to the same thing. So, we cannot use either of the proofs for $\mathbb{Q}[x]$ to show that all irreducible elements are prime in $\mathbb{Z}[x]$.

## 34.2  Factoring Integers out of Polynomials

We shall now concentrate on showing that every non-zero non-unit in $\mathbb{Z}[x]$ can be factored into irreducibles (recall that 1 and $-1$ are the only units in $\mathbb{Z}[x]$). Let's restrict our attention first to those factorizations that involve elements of $\mathbb{Z}$. For example, the factorization

$$6x^2 - 12x + 24 = 2 \cdot 3(x^2 - 2x + 4),$$

while considered trivial in $\mathbb{Q}[x]$, is of interest in $\mathbb{Z}[x]$ because 2 and 3 are *not* units in $\mathbb{Z}[x]$.

Consider now an irreducible $p$ from $\mathbb{Z}$. We can also consider $p$ as an element of $\mathbb{Z}[x]$ (of degree 0), and so we can ask whether $p$ is an irreducible element of $\mathbb{Z}[x]$. But if $p = fg$, where $f, g \in \mathbb{Z}[x]$, then $f$ and $g$ would both have to have degree 0 (because $\deg(fg) =$

$\deg(f) + \deg(g)$). This means that $p = fg$ can be considered a factorization in $\mathbb{Z}$, and so one of $f$ and $g$ must be a unit in $\mathbb{Z}$ (and hence in $\mathbb{Z}[x]$). Thus, irreducible elements in $\mathbb{Z}$ are also irreducible elements in $\mathbb{Z}[x]$. For example, 2 (considered as a degree 0 polynomial) is an irreducible element of $\mathbb{Z}[x]$.

**Example 34.1**

> It is *not* always the case that an irreducible element in a smaller ring stays irreducible in a larger one: 2 is irreducible in $\mathbb{Z}$ but is *not* irreducible in $\mathbb{Z}[i]$, because $2 = (1+i)(1-i)$. (See Example 31.4.)

We know that an element $\mathbb{Z}$ is prime if and only if it is irreducible. On the other hand, we have not (yet) shown that $\mathbb{Z}[x]$ is a UFD, and, consequently, we cannot yet infer that irreducible elements in $\mathbb{Z}[x]$ are prime. We will show this eventually, but you should exercise care until then to preserve the (potential) distinction between irreducible and prime elements.

---

## 34.3   The Content of a Polynomial

So, a first step toward factoring elements in $\mathbb{Z}[x]$ is to factor out any non-trivial constant (that is, degree 0) elements. To describe this efficiently, we introduce the following terminology:

Given a polynomial

$$f = a_n x^n + a_{n-1} x^{n-1} + \cdots + a_0 \in \mathbb{Z}[x],$$

consider a gcd in $\mathbb{Z}$ of the elements $a_n, a_{n-1}, \cdots a_0$. We shall call such a gcd the **content** of $f$ and denote it by cont $f$. Of course, the content is only well defined up to plus or minus (that is, unit multiples), but the notation cont $f$ is convenient enough that we will live with this harmless ambiguity.

**Example 34.2**

> The content of the polynomial $6x^2 - 12x + 24$ in $\mathbb{Z}[x]$ is 6 (or $-6$).

Recall from Example 15.5 that we called a polynomial in $\mathbb{Z}[x]$ **primitive** if its coefficients have no non-trivial common factor; that is, it is primitive if its content is 1. Our first important theorem asserts that the set of primitive polynomials is closed under multiplication.

**Example 34.3**

> Both $2x + 3$ and $3x^2 + 4x + 1$ are primitive polynomials; note that their product $6x^3 + 17x^2 + 24x + 18$ is then primitive as well.

▷ **Quick Exercise.**   Choose two other primitive polynomials and check that their product remains primitive. ◁

**Theorem 34.1** *Suppose that $f$ and $g$ are primitive polynomials in $\mathbb{Z}[x]$. Then $fg$ is primitive.*

**Proof**: This is just Gauss's Lemma 5.5 in disguise. Suppose that $f$ and $g$ are primitive polynomials, and let $d = \text{cont } fg$. Consider the polynomial $h = (1/d)fg$; this is an element of $\mathbb{Z}[x]$. Now $h = ((1/d)f)(g)$ is a factorization in $\mathbb{Q}[x]$; by Gauss's Lemma, this leads to a factorization in $\mathbb{Z}[x]$. That is, there are rational numbers $A, B$ so that $h = (A(1/d)f)(Bg)$, with $A(1/d)f$ and $Bg$ being elements of $\mathbb{Z}[x]$; note that $AB = 1$. Because $g$ is primitive and $Bg \in \mathbb{Z}[x]$, we must have that $B$ is an integer, because if $B$ were not an integer, then the denominator of $B$ would be cancelled by a non-trivial integer factor of all the coefficients of $g$; that is, cont $g$ would not be 1. But because $f$ is primitive, this means that $A/d$ must be an integer too. Because $AB = 1$, this means that $A$, $B$, and $d$ must each be $\pm 1$. That is, cont $fg = \pm 1$, and so $fg$ is primitive. $\qquad\square$

An important consequence of this theorem is that the content function preserves multiplication:

**Corollary 34.2** *Given $f, g \in \mathbb{Z}[x]$, we have that*

$$\text{cont } fg = \text{cont } f \cdot \text{cont } g.$$

**Proof**: Given $f, g \in \mathbb{Z}[x]$, we have that $f = (\text{cont } f)f_1$ and $g = (\text{cont } g)g_1$, where $f_1, g_1$ are primitive. But then

$$fg = (\text{cont } f \cdot \text{cont } g)f_1 g_1$$

and by Theorem 14.1, $f_1 g_1$ is primitive. Hence, cont $fg = \text{cont } f \cdot \text{cont } g$. $\qquad\square$

▷ **Quick Exercise.** Check this corollary by multiplying together two polynomials from $\mathbb{Z}[x]$ of your choice. ◁

To achieve our goal of proving that $\mathbb{Z}[x]$ is a UFD we must first prove that every element of $\mathbb{Z}[x]$ either is irreducible or can be factored into a product of irreducibles. As promised, this is easy to prove, using induction on degree.

**Theorem 34.3** *Every non-zero non-unit of $\mathbb{Z}[x]$ is either irreducible or a product of irreducibles.*

**Proof**: Suppose that $0 \neq f \in \mathbb{Z}[x]$ is a non-unit. We proceed by induction on $\deg(f)$. If $\deg(f) = 0$, then $f \in \mathbb{Z}$. Because $\mathbb{Z}$ is a UFD, $f$ is either irreducible in $\mathbb{Z}$, or a product of irreducibles in $\mathbb{Z}$; but irreducibles in $\mathbb{Z}$ are irreducible in $\mathbb{Z}[x]$, and so we have the required result.

Now suppose that $\deg(f) = n > 0$. If $f$ itself is irreducible, we are done. Otherwise, we first factor out of $f$ the element cont $f$; this is an element of $\mathbb{Z}$ that we can factor into irreducibles of $\mathbb{Z}$ (and hence of $\mathbb{Z}[x]$). So we may suppose that $f$ is primitive. If $f$ isn't irreducible, we can then factor it as $f = gh$. Because $f$ is primitive, we must have that $\deg(g) > 0$ and $\deg(h) > 0$. So by the induction hypothesis, $g$ and $h$ are irreducible or the product of irreducibles, and therefore so is $f$. By the Principle of Mathematical Induction, this proves the theorem. $\qquad\square$

## 34.4 Irreducibles in $\mathbb{Z}[x]$ Are Prime

So, what remains to be done to prove that $\mathbb{Z}[x]$ is a UFD? Theorem 32.4 asserts that a domain in which factorization into irreducibles is possible has unique factorization if and only if all irreducible elements are prime (remember, of course, that prime elements are *always* irreducible). We thus must show that all irreducible elements of $\mathbb{Z}[x]$ are actually prime. We will do this by relating factorizations of polynomials in $\mathbb{Z}[x]$ to factorizations in $\mathbb{Q}[x]$, using Gauss's Lemma 5.5.

**Theorem 34.4** *In $\mathbb{Z}[x]$, all irreducible elements are prime; consequently $\mathbb{Z}[x]$ is a UFD.*

**Proof:** Suppose that $f$ is an irreducible polynomial in $\mathbb{Z}[x]$. Note first that because $f$ has no non-trivial factors from $\mathbb{Z}$, this means that $f$ is primitive. By Gauss's Lemma, $f$ must be irreducible in $\mathbb{Q}[x]$ as well. Because $\mathbb{Q}[x]$ is a PID, $f$ is then a prime element of $\mathbb{Q}[x]$; we must show that $f$ is a prime element of $\mathbb{Z}[x]$.

To show this, suppose that $f$ divides $gh$, where $g, h \in \mathbb{Z}[x]$. Now consider $g$ and $h$ as elements of $\mathbb{Q}[x]$. Because $f$ *is* prime in $\mathbb{Q}[x]$, this means that $f$ must divide one of them (say, $g$) in $\mathbb{Q}[x]$. That is, $g = fg_1$, where $g_1 \in \mathbb{Q}[x]$. But by Gauss's Lemma 5.5 there exist rational numbers $A$ and $B$, so that $AB = 1$ and $Af, Bg_1 \in \mathbb{Z}[x]$. But because $f$ is primitive and $Af \in \mathbb{Z}[x]$, $A$ must be an integer; otherwise, the denominator of $A$ would be cancelled by a non-trivial factor of cont $f$. Thus, $g = f(ABg_1)$ is a factorization in $\mathbb{Z}[x]$, and so $f$ divides $g$ in $\mathbb{Z}[x]$. Hence, $f$ is prime in $\mathbb{Z}[x]$, as required. It then follows immediately that $\mathbb{Z}[x]$ is a UFD. $\square$

This means that $\mathbb{Z}[x]$ is an example of a UFD that is not a PID, thus showing that the converse of Theorem 33.2 is false.

The method of this chapter can actually be generalized to prove that any time $D$ is a PID, then $D[x]$ is a UFD that is not a PID. The general proof, while quite similar, requires a bit more machinery than we presently have available to us, and so we will not pursue it here.

---

# Chapter Summary

In this chapter, we used Gauss's Lemma to prove that in $\mathbb{Z}[x]$, irreducible elements are prime, and so $\mathbb{Z}[x]$ is a UFD.

---

# Warm-up Exercises

a. What properties does $\mathbb{Z}[x]$ lack that prevent us from proving that it is a UFD just as we did for $\mathbb{Q}[x]$?

b. Determine the content of

$$30x^4 - 12x^2 + 42x - 54 \text{ and } 49x^3 + 70x^2 - 14.$$

What is the content of the product of these two polynomials?

c. Why would it be silly to try to talk about the content of polynomials from $\mathbb{Q}[x]$?

d. Factor the following polynomials completely into irreducibles in $\mathbb{Z}[x]$; do they have the same irreducibles as factors in $\mathbb{Q}[x]$?

  (a) $6x^3 - 6$.
  (b) $3x^4 - 6x$.

(c) $5x^4 - x^3 - 15x^2 - 7x + 2$.

e. Give examples of the following (or say why they don't exist):

    (a) A PID that isn't a UFD.

    (b) A UFD that isn't a PID.

f. Give examples of the following polynomials from $\mathbb{Z}[x]$ (or say why they don't exist):

    (a) An irreducible polynomial of degree 4.

    (b) An irreducible polynomial that isn't prime.

    (c) An irreducible polynomial of degree 0.

    (d) A unit.

    (e) A polynomial that can't be non-trivially factored in $\mathbb{Z}[x]$, but can be non-trivially factored in $\mathbb{Q}[x]$.

    (f) A polynomial that can't be non-trivially factored in $\mathbb{Z}[x]$, but can be non-trivially factored in $\mathbb{R}[x]$.

    (g) A polynomial that can't be non-trivially factored in $\mathbb{Q}[x]$, but can be non-trivially factored in $\mathbb{Z}[x]$.

---

## Exercises

1. Suppose that $f \in \mathbb{Z}[x]$ and the coefficient on its highest power of $x$ is 1 (we say that such a polynomial is called **monic**).

    (a) Why is a monic polynomial necessarily primitive?

    (b) Give an example of a primitive polynomial in $\mathbb{Z}[x]$ that isn't monic.

2. Suppose that $f \in \mathbb{Z}[x]$ is monic (see Exercise 1).

    (a) Prove that all rational roots of $f$ are integers.

    (b) Prove that all integer roots of $f$ divide its constant term.

    (c) Give examples of primitive polynomials for which parts (a) and (b) fail.

3. Use Exercise 2 and the Root Theorem 4.3 to show that $x^3 + 2x + 7$ is prime in $\mathbb{Z}[x]$.

4. Consider the ring $\mathbb{Z}[i][x]$ of polynomials with coefficients from the Gaussian integers $\mathbb{Z}[i]$. Argue that this ring is not a PID, by an argument analogous to that for $\mathbb{Z}$. (The argument in the text that $\mathbb{Z}[x]$ is a UFD can actually be generalized to this case too, but we will not do this here.)

5. Prove that in $\mathbb{Z}[x]$, $\langle 3x + 1, x + 1 \rangle = \langle 2, x - 1 \rangle$. Show that this ideal is not principal.

# Chapter 35

## Euclidean Domains

The difficulty we encountered in the last chapter in proving that $\mathbb{Z}[x]$ is a UFD might convince you of the advantage of having a Division Theorem available: The Division Theorem makes proving that $\mathbb{Z}$ and $\mathbb{Q}[x]$ are UFDs relatively easy. It seems natural then to define a more general class of domains (including both $\mathbb{Z}$ and $\mathbb{Q}[x]$) that *have* a Division Theorem. We name this class of domains in honor of Euclid, in whose *Elements* we find the first reference to that corollary of the Division Theorem, Euclid's Algorithm. This is a typical gambit of mathematicians. We have identified an important tool we'd like to study in general (in this case, a Division Theorem for domains), and so we isolate those domains having this tool by means of a definition.

## 35.1   Euclidean Domains

A **Euclidean domain** is a domain $D$ that can be equipped with a function $v : D \backslash \{0\} \to \mathbb{N}$ that satisfies the following two criteria:

a. For $a, b \in D$ with $a \neq 0$, there exist $q, r \in D$ (called the **quotient** and **remainder**, respectively) such that $b = aq + r$, with $r = 0$ or $v(r) < v(a)$.

b. For all $a, b \neq 0$, $v(a)v(b) = v(ab)$.

The function $v$ is called a **Euclidean valuation** for $D$.

You should think of the function $v$ as a measure of 'size' of the elements. Thus, the first condition says that we can always divide an element of $D$ by a non-zero element; either the division is exact, or we have a remainder 'smaller' than the divisor. To be consistent with our earlier terminology, we will call this the **Division Theorem** for Euclidean domains (even though we have built it right into the definition). The second condition allows us to relate divisibility in $D$ to divisibility in $\mathbb{Z}$ (which we presumably know more about). Note that because the function $v$ takes on its values in $\mathbb{N}$, we have available to us all we know about this set, including such tools as the Well-ordering Principle and Mathematical Induction.

Before making any abstract inferences about Euclidean domains, we shall first list a number of examples, showing that this concept is in fact a common generalization of a number of the domains we have already considered.

**Example 35.1**

$\mathbb{Z}$ is a Euclidean domain. Here, the valuation (or 'size' function) is obvious; just let $v(n) = |n|$. The first condition is then the Division Theorem of $\mathbb{Z}$ (Theorem 2.1), and the second condition holds because $|nm| = |n||m|$.

**Example 35.2**

$\mathbb{Q}[x]$ is a Euclidean domain. How did we measure 'size' of elements for polynomials? We just used their degree; thus, $v(f) = \deg(f)$ seems a natural definition; condition (1) is now just the Division Theorem for $\mathbb{Q}[x]$ (Theorem 4.2). However, the second condition fails, because after all, $\deg(fg) = \deg(f) + \deg(g)$ (rather than $\deg(f)\deg(g)$). We can escape this problem by means of a trick: Just let $v(f) = 2^{\deg(f)}$. Because exponentiation turns addition into multiplication, condition (2) is now satisfied. But what about condition (1)? This still works because $\deg(f) < \deg(g)$ exactly when $v(f) < v(g)$.

**Example 35.3**

Any field is a Euclidean domain. In a field we can divide any element by any non-zero element, with no remainder. Thus, all non-zero elements should be 'small'. A function that works easily is just $v(a) = 1$, for all $a \neq 0$. This seems almost too easy, but, after all, we are only really interested in a Division Theorem where we have at least some elements that do *not* divide one another exactly.

---

## 35.2   The Gaussian Integers

We will now show that the ring of Gaussian integers $\mathbb{Z}[i]$ is a Euclidean domain. In Chapter 31 we had already measured the 'size' of Gaussian integers, by means of the norm function $N(a + bi) = a^2 + b^2$. We wish to show that the norm function serves as a Euclidean valuation. Note that the second condition holds because $N(\alpha\beta) = N(\alpha)N(\beta)$ (Theorem 31.1). Unfortunately, the Division Theorem is not really obvious. What should the quotient and remainder be if we divide one Gaussian integer by another?

**Example 35.4**

Let's look at a particular example: Consider $11 + 6i$ divided by $2 + 3i$. Now we can certainly perform this division in the field $\mathbb{C}$. We obtain the following:

$$\frac{11 + 6i}{2 + 3i} = \frac{(11 + 6i)(2 - 3i)}{(2 + 3i)(2 - 3i)} = \frac{40 - 21i}{13} = \frac{40}{13} - \frac{21}{13}i.$$

What is the Gaussian integer closest to this quotient? Because $\frac{40}{13} = 3\frac{1}{13}$ and $\frac{21}{13} = 1\frac{8}{13}$, the answer to this question seems to be $3 - 2i$. If this is to be our quotient $q$, then the remainder $r$ must be given by

$$(11 + 6i) - (2 + 3i)(3 - 2i) = -1 + i.$$

Note that $v(-1 + i) = 2 < 13 = v(2 + 3i)$.

This example makes plausible the assertion that the norm function is in fact a Euclidean valuation for the Gaussian integers. Let us now prove this carefully:

**Theorem 35.1** $\mathbb{Z}[i]$ *is a Euclidean domain.*

**Proof:** As observed above, we clearly need check only that the Division Theorem (that is, condition (1) above) really works. So suppose that $\beta = m + ni$ and $\alpha = r + si$ are Gaussian integers, and we wish to verify the Division Theorem for them, where $\alpha \neq 0$ will serve as the divisor. Using the example above as our model, consider the following computation in $\mathbb{C}$:

$$\frac{\beta}{\alpha} = \frac{m + ni}{r + si} = \frac{mr + ns}{r^2 + s^2} + \frac{nr - ms}{r^2 + s^2}i.$$

The real and imaginary parts of this complex number certainly need not be integers, but we shall now choose integers $q_1$ and $q_2$ as close as possible to them. Any real number is within $\frac{1}{2}$ of an integer, and so this amounts to saying that

$$\left| \frac{mr + ns}{r^2 + s^2} - q_1 \right| \leq \frac{1}{2}$$

and

$$\left| \frac{nr - ms}{r^2 + s^2} - q_2 \right| \leq \frac{1}{2}.$$

We will use $\gamma = q_1 + q_2i$ as the proposed quotient in the Division Theorem. The remainder $\rho$ will then necessarily be given by

$$\rho = \beta - \alpha\gamma = (m - rq_1 + sq_2) + (n - rq_2 - sq_1)i.$$

It remains to check that the valuation of $\rho$ is less than the valuation of $\alpha$ (or else $\rho = 0$). To verify this we use the fact that the function $N$ preserves multiplication even in $\mathbb{C}$ (Theorem 8.3). We can thus do the following computation:

$$
\begin{aligned}
v(\rho) &= N(\rho) = N(\beta - \alpha\gamma) = N\left( \alpha \cdot \left( \frac{\beta}{\alpha} - \gamma \right) \right) \\
&= N(\alpha)N\left( \frac{\beta}{\alpha} - \gamma \right) \\
&= N(\alpha)\left( \left( \frac{mr + ns}{r^2 + s^2} - q_1 \right)^2 + \left( \frac{nr - ms}{r^2 + s^2} - q_2 \right)^2 \right) \\
&\leq N(\alpha)\left( \frac{1}{4} + \frac{1}{4} \right) < N(\alpha) = v(\alpha).
\end{aligned}
$$

This shows that the remainder $\rho$ is indeed suitably 'small'. $\qquad\square$

You will use similar means to prove that $\mathbb{Z}[\sqrt{2}]$ is a Euclidean domain in Exercise 1. Not all such domains are; it is a deep and not completely solved problem of number theory to distinguish which are Euclidean domains, and which aren't.

The proof of Theorem 35.1 can be interpreted geometrically. The crucial step is the choice of the quotient $\gamma$; we do this by finding a point in the complex plane with integer coordinates, whose distance to $\beta/\alpha$ (the quotient in $\mathbb{C}$) is less than 1. If the complex number $\beta/\alpha$ happened to fall between points with consecutive integer coordinates, as illustrated in the diagram below, we would actually have two possible choices for the quotient:

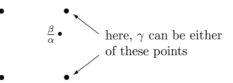

here, $\gamma$ can be either of these points

Thus, the quotient and remainder *need not be unique* in a Euclidean domain; the careful reader might have noted earlier the absence of the word 'unique' in the original definition. We will illustrate this point algebraically, in the following example, where we actually have four choices for the quotient available:

## Example 35.5

Let's divide $3 + 5i$ by 2. Because the quotient in $\mathbb{C}$ is $3/2 + (5/2)i$, which element of $\mathbb{Z}[i]$ should we choose as the quotient? In this case, each of the four nearby points with integer coordinates are at distance less than 1 from $3/2 + (5/2)i$, and so we have four different quotient/remainder pairs:

$$
\begin{aligned}
3 + 5i &= (1 + 2i)(2) + (1 + i) \\
&= (1 + 3i)(2) + (1 - i) \\
&= (2 + 2i)(2) + (-1 + i) \\
&= (2 + 3i)(2) + (-1 - i).
\end{aligned}
$$

## 35.3 Euclidean Domains Are PIDs

We are now ready to make some abstract observations about Euclidean domains. We first prove that the units of a Euclidean domain are identifiable by means of its valuation:

**Theorem 35.2** *Let $D$ be a Euclidean domain with valuation $v$. Then $u$ is a unit of $D$ if and only if $v(u) = 1$.*

**Proof:** Notice first that because $1^2 = 1$, $v(1)^2 = v(1)$; the only positive integer with this property is 1, and so $v(1) = 1$. But then if $u$ is a unit, $v(u)v(u^{-1}) = v(uu^{-1}) = v(1) = 1$, and so $v(u) = 1$. Conversely, if $v(u) = 1$, apply the Division Theorem to 1 and $u$ to obtain $1 = qu + r$. But $v(r) < v(u) = 1$, which is impossible, and so $r = 0$. That is, $u$ is a unit with inverse $q$. □

We next show that all Euclidean domains are PIDs, using very much the same sort of proof as we used for $\mathbb{Z}$ and $\mathbb{Q}[x]$.

**Theorem 35.3** *Each Euclidean domain is a PID.*

**Proof:** Suppose that $I$ is a proper ideal of the Euclidean domain $D$. What element of $I$ might serve as its generator? We answer this question (as we did for $\mathbb{Z}$) by using the Well-ordering Principle. Choose an element $d$ of $I$ with smallest valuation; there may be many choices for $d$. Note that this valuation is necessarily larger than 1 because $I$ contains no units. We claim that $\langle d \rangle = I$. Because $I$ is an ideal, it is clear that $\langle d \rangle \subseteq I$. Suppose now that $b \in I$. We claim $b$ is a multiple of $d$. To check this, we should clearly use the Division Theorem to obtain a quotient $q$ and remainder $r$: $b = qd + r$. Because $r = b - qd$, this means that $r \in I$. But $v(r) < v(d)$, which is impossible unless $r = 0$. Thus, $b = qd \in \langle d \rangle$. □

Because every PID is a UFD, we have:

**Corollary 35.4** *Each Euclidean domain is a UFD.*

**Example 35.6**

We thus know that the ring $\mathbb{Z}[i]$ of Gaussian integers is a PID. The crucial idea in the proof above is that a generator for an ideal is an element with smallest valuation. As an example of this procedure in $\mathbb{Z}[i]$, consider the set

$$I = \{a + bi \in \mathbb{Z}[i] : 5 \mid (2a - b)\}.$$

We claim that $I$ is an ideal. It is quite easy to show that $I$ is closed under subtraction.

▷ **Quick Exercise.** Check this. ◁

Suppose now that $a + bi \in I$ and $c + di \in \mathbb{Z}[i]$. Then

$$(a + bi)(c + di) = (ac - bd) + i(bc + ad)$$

and so we require that $2ac - 2bd - bc - ad$ is an integer divisible by 5. But

$$2ac - 2bd - bc - ad =$$
$$c(2a - b) - d(2b + a) = c(2a - b) - d(5a - 2(2a - b)),$$

and this is divisible by 5 because $2a - b$ is.

The theorem asserts that $I$ is a principal ideal, and the proof tells that we can find a generator by picking an element of $I$ with smallest valuation. To find out what this is, suppose that $a + bi \in I$; then $2a - b = 5k$, for $k \in \mathbb{Z}$. So

$$v(a + bi) = a^2 + b^2 = a^2 + (2a - 5k)^2 = 5\left(a^2 - 4ak + 5k^2\right),$$

and so this smallest valuation is divisible by 5. But note that $2 - i$ has valuation 5 and $2 - i \in I$; consequently, we must have that $\langle 2 - i \rangle = I$. Let's do some arithmetic in $\mathbb{C}$, to see explicitly that every element of $I$ is a multiple of $2 - i$. For that purpose, let $a + bi \in I$; then

$$\frac{a + bi}{2 - i} = \frac{(a + bi)(2 + i)}{(2 - i)(2 + i)} = \frac{1}{5}((2a - b) + (a + 2b)i)$$
$$= \frac{1}{5}((2a - b) + (3(2a - b) - 5(a - b))i)$$

and this latter element is in $\mathbb{Z}[i]$, because the real and imaginary parts of the numerator are divisible by 5. Thus, any element of $I$ is a multiple of $2 - i$, as claimed. Note that in practice it might be quite difficult to find the generator for an ideal by this method.

---

## 35.4  Some PIDs Are Not Euclidean

A natural question to ask at this point is: Are there PIDs that are not Euclidean? The answer is yes, but it is difficult to prove this. It is usually quite easy to show that a domain is not a PID: Just exhibit a particular ideal and prove it is not principal. However, to show that a domain $D$ is not Euclidean, one must prove that *all* functions $v : D\backslash\{0\} \to \mathbb{N}$ fail to satisfy at least one of the two defining conditions. The difference between these two tasks is that the definition of PID is a *universal* statement—*all* ideals are principal—while the definition of a Euclidean domain is an *existential* statement—*there exists* a function satisfying particular properties. And the negation of a universal statement is existential, while the negation of an existential statement is universal.

The picture below provides a nice summary of the relationships between the various sorts of rings we have studied in this book. We have not included an example of a PID which is not Euclidean, although such an example does exist; this is beyond the scope of this book.

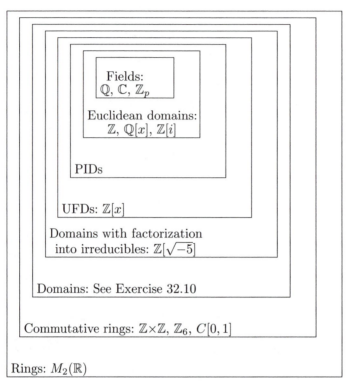

---

## Chapter Summary

In this chapter we defined the concept of *Euclidean domain* and proved that all Euclidean domains are PIDs. We showed that $\mathbb{Z}[i]$ is a Euclidean domain, in addition to $\mathbb{Z}$ and $\mathbb{Q}[x]$.

---

## Warm-up Exercises

a. Compute a quotient and remainder for the following pairs of elements, in the given Euclidean domain.

   (a) 116 divided by 7 in $\mathbb{Z}$.

   (b) $x^4 - 2x^3 + 5x - 3$ divided by $2x^2 - 1$ in $\mathbb{Q}[x]$.

   (c) $13 + 5i$ divided by $3 + 2i$ in $\mathbb{Z}[i]$.

   (d) 7 divided by $\pi$ in $\mathbb{R}$.

b. Given an ideal $I$ in a Euclidean domain, we know that $I$ is principal. Describe conceptually how to determine the generator of $I$. (This description is very simple; it might not be so simple to carry out in practice.)

## Exercises

1. Using a similar argument to that in the text for $\mathbb{Z}[i]$, prove that $\mathbb{Z}[\sqrt{2}]$ is a Euclidean domain, using the function $N$ as the valuation.

2. Compute a quotient and remainder for $17+32\sqrt{2}$ divided by $3-4\sqrt{2}$ in the Euclidean domain $\mathbb{Z}[\sqrt{2}]$. Check that your remainder has a smaller valuation than $3-4\sqrt{2}$.

3. Show by example that quotients and remainders are not unique in the Euclidean domain $\mathbb{Z}[\sqrt{2}]$.

4. Make a definition for *greatest common divisor* of two elements in a Euclidean domain.

5. Suppose that $a, b \in D$, and $D$ is a Euclidean domain. Consider the ideal $\langle a, b \rangle = \{ax + by : x, y \in D\}$; see Exercise 9.11. Let $d$ be an element of $\langle a, b \rangle$ with least valuation.

   (a) Why does the element $d$ exist?

   (b) Prove that $d$ is a gcd of $a, b$. (See Exercise 4.)

   (c) Why does this mean that Euclidean domains possess a GCD identity?

   (d) We now have a GCD identity for Euclidean domains in terms of ideals: $\langle a, b \rangle = \langle d \rangle$ if and only if $d$ is a gcd of $a$ and $b$. Prove this.

6. (a) Explain why Euclid's Algorithm for finding a gcd makes sense in a Euclidean domain.

   (b) Apply Euclid's Algorithm to $5 + 133i$ and $17 + 34i$ in $\mathbb{Z}[i]$.

   (c) Backtrack through Euclid's Algorithm to show explicitly that $\langle 5+133i, 17+34i \rangle$ is a principal ideal.

7. Suppose that $a$ and $b$ are integers. Prove that their gcd in $\mathbb{Z}[i]$ is the same as in $\mathbb{Z}$, up to unit multiples in $\mathbb{Z}[i]$.

8. Prove that if $D$ is a Euclidean domain and $a, b \in D$ are associates, then $v(a) = v(b)$. (This is easy!) Show by example that the converse is false in $\mathbb{Z}[\sqrt{2}]$.

9. Provide a geometric interpretation of the proof that $\mathbb{Z}[\sqrt{2}]$ is a Euclidean domain, analogous to the interpretation we discussed for $\mathbb{Z}[i]$.

# Section VII in a Nutshell

This section examines conditions that $\mathbb{Z}$ and $\mathbb{Q}[x]$ share, which provide them with a unique factorization theorem into irreducibles, like the Fundamental Theorem of Arithmetic.

For $\mathbb{Z}$ and $\mathbb{Q}[x]$, all ideals are principal. An integral domain where this holds is called a *principal ideal domain* (or PID). For a PID, we have factorization into irreducibles (Theorem 32.2). To prove uniqueness of such factorization requires precisely that the concepts of irreducibility and primeness coincide (Theorems 32.4), as they do for the integers. An integral domain that has unique factorization into irreducibles is called a *unique factorization domain* (or UFD).

Integral domains of the form $\mathbb{Z}[\sqrt{n}] = \{a + b\sqrt{n} : a, b \in \mathbb{Z}\}$ are called *quadratic extensions* of the integers. These are important examples; some such rings are PIDs, but not all. The ring $\mathbb{Z}[\sqrt{-5}]$ is not a PID and is not even a UFD.

The ring $\mathbb{Z}[x]$ of polynomials with integer coefficients is not a PID, but it is a UFD; this follows essentially from Gauss's Lemma (Theorem 34.4).

Some quadratic extensions (such as the *Gaussian integers* $\mathbb{Z}[i]$) share even more properties in common with $\mathbb{Z}$ and $\mathbb{Q}[x]$ and are called *Euclidean domains*: Elements in such domains have a notion of 'size', which equips them with a Division Theorem. This makes it easy to prove (Theorem 35.3) that they are PIDs, and hence UFDs.

# Part VIII

# Constructibility Problems

# Chapter 36

## Constructions with Compass and Straightedge

You probably recall doing various constructions with a compass and a straightedge in high school geometry. We will imagine *idealized* tools. Thus, the straightedge is an unmarked ruler, which in principle can be as long as necessary. With it we can draw line segments of arbitrary length, perhaps passing through a particular point or connecting two given points. Likewise, the compass is as large as necessary; with it we can draw arcs and circles and duplicate distances. For instance, if $A$ and $B$ are marked on one line and point $C$ is marked on another, the compass allows us to mark a point $D$ on the second line so that the distance between $C$ and $D$ is the same as the distance between $A$ and $B$. Of course, in actually carrying out these constructions with real rulers and compasses, there is always error involved. However, we are concerned with *idealized* constructions—perfect constructions with no error.

## 36.1 Construction Problems

From around the fifth century B.C. Greek mathematicians wondered what constructions could be carried out using only a compass and a straightedge. In the axioms that begin Euclid's *Elements* we discover the postulation of an idealized compass and straightedge. The theorems about plane geometry that follow show how successful the Greeks were at doing plane geometry with compass and straightedge. You should recall some of the constructions possible:

- find the midpoint of a line segment;

- construct a line perpendicular to a given line through a given point on the line;

- construct a line perpendicular to a given line through a point off the line;

- construct a line parallel to a given line through a point off the given line;

- given an angle, bisect it.

▷ **Quick Exercise.** How are these constructions done? Better yet, get out your compass and straightedge and do them. ◁

There were, however, three famous constructions that ancient Greek mathematicians could *not* accomplish with compass and straightedge:

- **Doubling the cube:** Given a line segment, which represents the edge of a cube, construct another line segment representing the edge of another cube, whose volume is twice that of the original cube.

- **Trisecting an angle:** Given an arbitrary angle, divide it into three equal parts.

- **Squaring the circle:** Construct a square that has the same area as a given circle.

The question of whether such constructions are possible bedeviled mathematicians for over 2000 years.

It is important to note that in each case we desire a general method that works for *all* instances of the given problem. For example, the angle trisection problem is to find a method of trisection that works for all angles. *Certain* angles can be easily trisected, but this does not mean that the general problem has been solved. For example, a 90° angle can be trisected, because this is equivalent to constructing a 30° angle, which is easy to do. In fact, there are a number of ways to do this. For instance, you could construct an equilateral triangle and then bisect one of the angles. Or, you could directly construct a right triangle with angles of 60° and 30°.

▷ **Quick Exercise.**   Construct an equilateral triangle. Also, directly construct a 30°–60°–90° triangle, by starting with a shorter leg of length 1.  ◁

In the 19th century all three of the famous constructions above were shown to be impossible; the proofs were largely algebraic, rather than geometric. Think for a moment about what a proof of impossibility means. It is not at all clear how it is possible to show that a construction is *impossible*—we certainly can't try all possibilities! Proving the impossibility of these three classical construction problems was one of the great triumphs of mathematics in general and algebra in particular.

---

## 36.2   Constructible Lengths and Numbers

We start the attack on these problems by first deciding which lengths can be constructed. Specifically, we want to answer the following question: *Given a line segment in the plane, which we say is of length 1, for what values $\alpha$ can we construct a line segment of length $\alpha$?*

Note that if we talk about lengths, we must start with some unit of measure; hence, we designate some particular line segment as the *unit line segment*. Our goal is to give a complete algebraic description of which numbers we can construct, when starting with a unit line segment. Our modest beginning is the observation that all the natural numbers can be constructed:

**Lemma 36.1** *Given a line segment of length 1 and a natural number $n$, it is possible to construct a line segment of length $n$.*

**Proof:**   Merely use the compass to lay off $n$ copies of the unit line segment, next to one another on a line (which can be made as long as necessary, using the straightedge).     □

Hence, we say that the natural numbers are constructible. In general, we say that the real number $\alpha$ is **constructible** if, given a line segment of length 1, it is possible to construct a line segment of length $|\alpha|$. Thus, the integers are constructible because the natural numbers are.

**Theorem 36.2** *Given line segments of length 1, $a$, and $b$, it is possible to construct segments of lengths $a + b$, $a - b$ (if $a \geq b$), $ab$, and $a/b$.*

**Proof**: The constructibility of $a + b$ and $a - b$ are obvious. To construct $ab$, consider the figure below. Start by constructing two rays with vertex $V$. On one ray mark $A$ so that the length of the line segment from $V$ to $A$ is $a$. (We will write $|\overline{VA}|$ for the length of the line segment from $V$ to $A$.) Then on the other ray mark points $P$ and $B$ so that $|\overline{VP}| = 1$ and $|\overline{VB}| = b$. Now draw the line segment from $P$ to $A$. Finally, construct a line parallel to this line segment and through the point $B$, as shown in the figure. Label the point where this line intersects the other ray as $Q$.

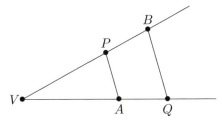

Now, $\triangle VAP$ is similar to $\triangle VQB$; thus,

$$\frac{|\overline{VA}|}{|\overline{VP}|} = \frac{|\overline{VQ}|}{|\overline{VB}|}, \quad \text{or}$$

$$\frac{a}{1} = \frac{|\overline{VQ}|}{b}, \quad \text{and so}$$

$$ab = |\overline{VQ}|.$$

A similar construction can be made for $a/b$. This is left as Exercise 1. $\square$

The previous theorem asserts that the sum, difference, product, and quotient of two constructible numbers is constructible. This makes the next corollary obvious:

**Corollary 36.3** *Given a line segment of length 1, the set of all constructible numbers is a field.*

We will therefore refer to this set as the **field of constructible numbers** and denote it by $\mathbb{K}$ (the German word for 'construct' is 'konstruieren', and for 'field' is 'Körper'). Our lemma tells us that $\mathbb{Z}$ is a subring of $\mathbb{K}$. But $\mathbb{K}$ is closed under division, and so $\mathbb{K}$ contains all quotients of integers; that is, the field $\mathbb{Q}$ of rational numbers is a subfield of $\mathbb{K}$. Furthermore, by definition all constructible numbers must be real numbers. That is, $\mathbb{K}$ is a subfield of $\mathbb{R}$.

Are there constructible numbers other than the rationals? To answer this, construct a square with side one. Then draw the diagonal; this means that $\sqrt{2}$ is a constructible number! And we've seen before (Exercise 2.14) that $\sqrt{2}$ is an irrational number, and so not all constructible numbers are rational. We shall see, however, that not all real numbers are constructible. Indeed, the highlight of the next chapter is to exactly describe the field of constructible numbers. There, we shall see that constructing the square root is of utmost importance.

Accordingly, we generalize the fact that $\sqrt{2}$ is constructible, in the next theorem:

**Theorem 36.4** *If $\alpha$ is constructible, then so is $\sqrt{|\alpha|}$.*

**Proof**: We assume that $\alpha$ is positive. Here is one way of constructing $\sqrt{\alpha}$. Refer to the figure below. First mark points $P$, $O$, and $Q$ on a line so that $|\overline{PO}| = \alpha$ and $|\overline{OQ}| = 1$. Now find the midpoint of the line segment from $P$ to $Q$ and using that as the center, draw a semicircle of radius $(\alpha + 1)/2$, as shown. Draw a perpendicular to the line through the point $O$. Label the point where this perpendicular intersects the semicircle as $X$.

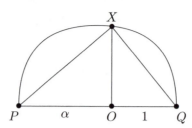

Note that $\triangle XOQ$ is similar to $\triangle POX$. Hence,

$$\frac{|\overline{PO}|}{|\overline{OX}|} \;=\; \frac{|\overline{OX}|}{|\overline{OQ}|}, \quad \text{or}$$

$$\frac{\alpha}{|\overline{OX}|} \;=\; \frac{|\overline{OX}|}{1}, \quad \text{and so}$$

$$\alpha \;=\; |\overline{OX}|^2.$$

That is, $|\overline{OX}| = \sqrt{\alpha}$. $\qquad\qquad\square$

By repeatedly applying the last theorem, we can construct $\sqrt[k]{\alpha}$ if $\alpha$ is a positive constructible number and $k$ is a power of two. So, for instance, $\sqrt{5}$, $\sqrt[4]{5}$, and $\sqrt[8]{5}$ are all constructible. Thus, there are infinitely many constructible numbers in addition to the rationals. Our goal in the next chapter is to characterize algebraically all elements of the field $\mathbb{K}$.

Although the constructions we have discussed in this chapter would have been familiar and understandable to an ancient Greek geometer, our emphasis on constructing *numbers*, rather than *line segments*, would have seemed strange. But we must change our point of view to bring to bear our modern algebraic tools.

## Chapter Summary

We introduced the three famous construction problems of the ancient Greeks: Is it possible, using only a compass and a straightedge, to

- double the cube,

- trisect an angle, or

- square the circle?

We defined a *constructible number* and showed that the rationals are constructible. Indeed, we showed that the set $\mathbb{K}$ of constructible numbers is a field. We also showed that $\sqrt{|\alpha|}$ is constructible if $\alpha$ is.

## Warm-up Exercises

a. Can you trisect the angle $45°$?

b. Suppose you could square a particular circle. Could you then square any circle?

c. Did we discuss a real number in this chapter that is *not* constructible?

d. Is

$$\sqrt{2 - \sqrt{5}}$$

constructible? (Be careful!)

e. Suppose that $\sqrt{\pi}$ were a constructible number. Why would this mean that we could square a circle of radius 1? (We'll see in Chapter 38 that $\sqrt{\pi}$ is not constructible.)

f. Suppose that $\sqrt[3]{2}$ were a constructible number. Why would this mean that we could double a cube with edge length 1? (We'll see in Chapter 38 that $\sqrt[3]{2}$ is not constructible.)

g. Euclid's first postulate states: "Let the following be postulated: to draw a straight line from any point to any point." His second postulate says: "To produce a finite straight line continuously in a straight line." With what tool do these postulates equip us? Why is the tool 'idealized'?

h. Euclid's third postulate states: "To describe a circle with any center and any distance." With what tool does this postulate equip us? Why is the tool 'idealized'?

## Exercises

1. Give the construction for $a/b$ required in the Theorem 36.2.

2. (a) Construct $\sqrt{5}$, using the method described in the proof of Theorem 36.4.

   (b) Now provide an easier construction of $\sqrt{5}$, by constructing the diagonal of an appropriately chosen rectangle.

3. Explain the steps necessary to construct $\sqrt[4]{3} + 1$ and $\sqrt{\frac{5}{2} + \sqrt{7}}$, when starting with a line segment of length 1.

4. Explain the steps necessary to construct $\sqrt{5 + 2\sqrt{6}}$ and $\sqrt{2} + \sqrt{3}$. Then show that these two numbers are the same.

5. Show that $\sqrt{3 + 2\sqrt{2}}$ is of the form $a + b\sqrt{2}$, for some integers $a$ and $b$. This exercise reinforces the lesson of Exercise 4: A given element of $\mathbb{K}$ may be constructible by quite a distinct list of steps!

6. (a) Perform the following construction with compass and straightedge, justifying each step: Take a line segment $\overline{AB}$, and construct on it a square $ABDC$. Then find the midpoint $E$ of $\overline{AC}$. Extend line segment $\overline{AC}$ so that $A$ is between $E$ and $H$, and $\overline{BE} = \overline{EH}$. (See diagram below.)

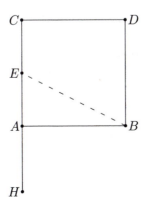

(b) Now apply the Pythagorean Theorem to the right triangle $EAB$ to show that

$$\overline{AB}(\overline{AB} - \overline{AH}) = \overline{AH}^2.$$

(c) If $|\overline{AB}| = 1$, what quadratic equation in $\rho = |\overline{AH}|$ do we obtain from part (b)?

(d) Show that $1/\rho = \rho + 1$.

(e) What is the value of the constructible number $\rho$? This number (or its reciprocal) is called the **Golden Section**, and the construction we have given for it appears as Proposition 11 in Book 2 of Euclid's *Elements*.

7. In this exercise you will show that the regular pentagon is constructible.

   (a) In the diagram below, show that $\angle BAC = 36°$ and $\angle ABC = \angle ACB = 72°$.

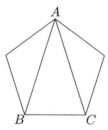

   (b) Let $E$ be the point where the bisector of $\angle ABC$ intersects the opposite side.

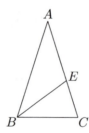

   Note that $\triangle ABC$ is similar to $\triangle BCE$. If $\overline{BC} = 1$ and $\overline{EC} = x$, show that $x^2 + x - 1 = 0$.

   (c) From Exercise 6c, you should have noted that the positive solution to this quadratic is the golden section, $\rho = \frac{-1+\sqrt{5}}{2}$. Now argue that the regular pentagon is constructible.

8. Technically speaking, the compass given by Euclid's third postulate (see Exercise h) is *collapsible*: It cannot be used directly to transfer distances from one line segment to another. Prove that distances can be transferred, with a collapsible compass and a straightedge, and so the modern compass is mathematically equivalent to the Greek collapsible compass.

# Chapter 37

## Constructibility and Quadratic Field Extensions

In the previous chapter we saw that constructible numbers can be obtained from the unit segment by addition, subtraction, multiplication, division (the field operations), and by taking square roots of positive numbers. We may, of course, perform any number of these operations in any order we please. Note that being closed under taking square roots of positive elements is a property that $\mathbb{K}$ does not share with all fields. For example, $2 \in \mathbb{Q}$ (the smallest field of constructible numbers, as we have seen) but $\sqrt{2} \notin \mathbb{Q}$.

## 37.1   Quadratic Field Extensions

Let's build a bigger field than $\mathbb{Q}$ that does contain $\sqrt{2}$. Consider the set

$$\mathbb{Q}(\sqrt{2}) = \{a + b\sqrt{2} : a, b \in \mathbb{Q}\}.$$

It is straightforward to show that $\mathbb{Q}(\sqrt{2})$ satisfies all the field axioms.

▷ **Quick Exercise.**   Check that $\mathbb{Q}(\sqrt{2})$ is a subring of $\mathbb{R}$, and then verify that it is in fact a field. ◁

In fact, $\mathbb{Q}(\sqrt{2})$ is the *smallest* field containing both $\mathbb{Q}$ and $\sqrt{2}$; that is, $\mathbb{Q}(\sqrt{2})$ is contained in every subfield of $\mathbb{C}$ containing both $\mathbb{Q}$ and $\sqrt{2}$. For if $\sqrt{2}$ is an element of a field $F \subseteq \mathbb{C}$ because $\mathbb{Q}$ necessarily is contained in $F$, then so is $a + b\sqrt{2}$ for all $a$ in $\mathbb{Q}$. Hence, $\mathbb{Q}(\sqrt{2}) \subseteq F$. Because $F$ was arbitrary, $\mathbb{Q}(\sqrt{2})$ is contained in all subfields of $\mathbb{C}$ containing both $\sqrt{2}$ and $\mathbb{Q}$.

We say that $\mathbb{Q}(\sqrt{2})$ is a **field extension** of $\mathbb{Q}$, because $\mathbb{Q}$ is a subfield of $\mathbb{Q}(\sqrt{2})$; since $\mathbb{Q}(\sqrt{2})$ is a strictly larger field, we call this a **proper** field extension. More specifically, $\mathbb{Q}(\sqrt{2})$ is a **quadratic field extension** of $\mathbb{Q}$. In general, a field $H \subseteq \mathbb{C}$ is a quadratic field extension of a field $F$ if

$$H = \{a + b\sqrt{k} : a, b, \in F\}$$

for some $k \in F$ such that $\sqrt{k} \notin F$. (If $\sqrt{k} \in F$, then $H = F$, and so $H$ would not be a proper extension.) We will return to this definition of quadratic field extension in Section 42.2, when we will put it in a more general context. Note also that when talking about field extensions, we will often omit the word 'field' if the context is clear.

▷ **Quick Exercise.**   Show that if $F \subseteq \mathbb{C}$ is a field and $k \in F$, then

$$H = \{a + b\sqrt{k} : a, b \in F\}$$

is a field. If $\sqrt{k} \notin F$, then $H$ properly contains $F$. That is, $H$ is a quadratic field extension of $F$. (This is simply a generalization of $\mathbb{Q}(\sqrt{2})$ being a quadratic field extension of $\mathbb{Q}$.) ◁

**Example 37.1**

> Consider the field $\mathbb{Q}(i)$; it is a quadratic extension of $\mathbb{Q}$ because $i = \sqrt{-1}$. This is the field of all complex numbers whose real and imaginary parts are rational. Of course, $\mathbb{Q}(i)$ does not consist entirely of constructible numbers, because $\mathbb{K} \subseteq \mathbb{R}$.

**Example 37.2**

> Let's build a quadratic extension of the field $\mathbb{Q}(\sqrt{2})$. To do so, we need an element belonging to this field whose square root does not. Does $\sqrt{3} \in \mathbb{Q}(\sqrt{2})$? If so, then for some $a$ and $b$ in $\mathbb{Q}$,
>
> $$\sqrt{3} = a + b\sqrt{2},$$
>
> and thus
>
> $$3 = (a + b\sqrt{2})^2 = (a^2 + 2b^2) + (2ab)\sqrt{2}.$$
>
> But then $2ab = 0$, and so at least one of $a$ and $b$ must be zero. Hence, there is no solution to the equation $3 = a^2 + 2b^2$. Thus, we can consider the quadratic extension field $\mathbb{Q}(\sqrt{2})(\sqrt{3})$, which we usually write as $\mathbb{Q}(\sqrt{2}, \sqrt{3})$.
>
> But what are the elements of this field? A typical element should look like
>
> $$(a + b\sqrt{2}) + (c + d\sqrt{2})\sqrt{3} = a + b\sqrt{2} + c\sqrt{3} + d\sqrt{6}.$$
>
> ▷ **Quick Exercise.** Check that the quadratic extension $\mathbb{Q}(\sqrt{3})(\sqrt{2})$ consists of the same elements as $\mathbb{Q}(\sqrt{2})(\sqrt{3})$. Thus, the notation $\mathbb{Q}(\sqrt{2}, \sqrt{3})$ is not ambiguous about the order in which the elements $\sqrt{2}$ and $\sqrt{3}$ are adjoined to $\mathbb{Q}$. ◁

In this second example (unlike the first) we obtain a field of constructible numbers. This is more generally true: If $F$ is a field of constructible numbers — that is, $F$ is a subfield of $\mathbb{K}$ — and $k$ is an element of $F$, then $F(\sqrt{|k|})$ is also a field of constructible numbers. We state this in the following lemma.

**Lemma 37.1** *Suppose that $F$ is a field of constructible numbers and $k \in F$ with $k > 0$. Then $F(\sqrt{k}) \subseteq \mathbb{K}$.*

**Proof:**  If $\sqrt{k} \in F$, then $F(\sqrt{k}) = F$, and we are done. So assume that $\sqrt{k} \notin F$. By Theorem 36.4, $\sqrt{k}$ is constructible. If $\alpha \in F(\sqrt{k})$, then $\alpha = a + b\sqrt{k}$ for some $a$ and $b$ in $F$, and so $\alpha$ is also constructible, because $\mathbb{K}$ is a field.  □

---

## 37.2    Sequences of Quadratic Field Extensions

We can thus start with the rational numbers and repeatedly take quadratic extensions by square roots of positive elements, in the process building larger and larger fields, always contained within the field of constructible numbers. We formally state this in the following theorem.

**Theorem 37.2** *Suppose that*

$$\mathbb{Q} = F_0 \subset F_1 \subset \cdots \subset F_n$$

*is a sequence of fields such that $F_{i+1} = F_i(\sqrt{k_i})$ for some $k_i \in F_i$, with $k_i > 0$ for $i = 0, 1, \ldots, n - 1$. Then $F_n \subseteq \mathbb{K}$.*

**Proof:** Use induction and the previous lemma. □

Are there any constructible numbers that cannot be obtained by this process of repeatedly extending our field by taking square roots of positive elements? The surprising answer is 'No'! Proving this is our next task.

Before we do, let's consider an example of a constructible number, say,

$$\sqrt{6 + \frac{4}{3}\sqrt{2 + 2\sqrt{7}}}.$$

This number could be constructed by successively constructing the following sequence of numbers:

$$
\begin{aligned}
X_1 &= 7, \\
X_2 &= \sqrt{7}, \\
X_3 &= \sqrt{7} + \sqrt{7} = 2\sqrt{7}, \\
X_4 &= 2, \\
X_5 &= 2 + 2\sqrt{7}, \\
X_6 &= \sqrt{2 + 2\sqrt{7}}, \\
X_7 &= \frac{4}{3}, \\
X_8 &= \frac{4}{3}\sqrt{2 + 2\sqrt{7}}, \\
X_9 &= 6, \\
X_{10} &= 6 + \frac{4}{3}\sqrt{2 + 2\sqrt{7}}, \\
X_{11} &= \sqrt{6 + \frac{4}{3}\sqrt{2 + 2\sqrt{7}}}.
\end{aligned}
$$

The fields corresponding to the above sequence of numbers would be $F_1 = \mathbb{Q}$, $F_2 = F_1(\sqrt{7})$, $F_5 = F_4 = F_3 = F_2$, $F_6 = F_5\left(\sqrt{2 + 2\sqrt{7}}\right)$, $F_{10} = F_9 = F_8 = F_7 = F_6$, and

$$F_{11} = F_{10}\left(\sqrt{6 + \frac{4}{3}\sqrt{2 + 2\sqrt{7}}}\right).$$

So, compressing this sequence, we see that the sequence of quadratic field extensions given in the above theorem would be

$$\mathbb{Q} \subset F_1 \subset F_2 \subset F_3,$$

$$\text{where} \quad F_1 = \mathbb{Q}(\sqrt{7}), \ F_2 = F_1(\sqrt{2 + 2\sqrt{7}}),$$

$$\text{and} \quad F_3 = F_2(\sqrt{6 + \frac{4}{3}\sqrt{2 + 2\sqrt{7}}}).$$

(We've abbreviated the original sequence of points somewhat. For instance, to construct 7 we might have constructed $2 \ (= 1 + 1)$, $3 \ (= 2 + 1)$, $4 \ (= 2 + 2)$, and finally $7 \ (= 3 + 4)$. Likewise, 6 and $\frac{4}{3}$ would take some intermediate steps. All those 'missing' numbers are in $\mathbb{Q}$, however.)

Note that there may be different paths to reach a given number and hence a different sequence of fields reflecting the order in which the numbers are constructed. (For examples illustrating this, look at Exercises 36.4 and 36.5.) The important point is that each field extension is a quadratic extension and only a finite sequence of extensions is needed.

## 37.3    The Rational Plane

We now return to using the compass and straightedge. This discussion will clarify what is actually meant by constructing with these tools. In constructing numbers with a compass and a straightedge, we start with a given unit segment somewhere in the plane and start constructing lengths. We can impose a Cartesian coordinate system on the plane so that the left-hand endpoint of the given unit segment is at the origin, and the right-hand endpoint is on the $x$-axis at location $(1, 0)$. All rational numbers on the $x$-axis can be located by applying only the constructions necessary to carry out field operations (addition, subtraction, multiplication, division). Note that a rational number being constructible means that a line segment of the appropriate length can be constructed somewhere in the plane. But we can easily transfer this length to the $x$-axis so that one end of the line segment is at the origin. Thus, on the $x$-axis we can locate $\pm q$ for any $q \in \mathbb{Q}$. We can easily transfer these points to the $y$-axis with the compass, and so we can locate any point $(p, q)$ in the plane where $p$ and $q$ are in $\mathbb{Q}$.

▷ **Quick Exercise.**   How do you locate $(p, q)$ in the plane once $p$ has been located on the $x$-axis and $q$ has been located on the $y$-axis? ◁

We have thus located all points in the plane that we can construct by means of only the field operations: We call this the **rational plane**, or the **plane of** $\mathbb{Q}$.

## 37.4    Planes of Constructible Numbers

Suppose now that somewhere in the plane we construct the length $\sqrt{k}$, where $k$ is some positive rational number, and $\sqrt{k} \notin \mathbb{Q}$. By applying only the field operations to $\sqrt{k}$ and elements from $\mathbb{Q}$, we can locate points on the $x$ and $y$ axes that are in the quadratic extension $\mathbb{Q}(\sqrt{k})$. Thus, we can locate all points in the plane with coordinates $(p, q)$, where $p$ and $q$ are in $\mathbb{Q}(\sqrt{k})$; we call this set of points the **plane of** $\mathbb{Q}(\sqrt{k})$.

More generally, suppose that $F$ is any subfield of $\mathbb{K}$ — that is, a field of constructible numbers. Then the **plane of** $F$ consists of the set of points $(p, q)$ for which $p$ and $q$ are in $F$.

We now wish to consider what further points can be reached using compass and straightedge alone, when working in the plane of $F$, where $F$ is some field of constructible numbers. Let's consider the compass first. To use the compass we need to know two points at which to place our compass: one for the center, and one for some point on the circumference. In the case we're describing, these two points must belong to the plane of $F$. We will call such a circle **a circle in the plane of** $F$. Likewise, with a straightedge we can draw a line that passes through two points in the plane of $F$. We will call such a line **a line in the plane of** $F$. Other constructions are a combination of these two simple constructions.

At this point you might object and say that a constructed circle could have center at *any* point in the entire plane with *any* radius — simply close your eyes and put the compass down. Or, a line drawn with the straightedge need not pass through two points in the plane of $F$ — again, lay the straightedge down arbitrarily. But steps like these are not permitted by the axioms of Greek geometry. Instead, we require step-by-step procedures that can be *replicated*. For example, a successful solution to the angle trisection problem should be an unambiguous list of constructions that when carried out by anyone leads to the same solution. This means that when we draw lines or circles with the straightedge or compass, we must have unambiguous information: A circle is determined by its center and a point on its circumference, a line by two points.

Let's be explicit algebraically about what equations for lines and circles in the plane of a field $F$ look like.

The equation for a line in the plane of $F$ is of the form

$$ax + by + c = 0$$

where $a$, $b$, and $c$ are in $F$, and $a$ and $b$ are not both zero. For if the line passes through the points $(x_1, y_1)$ and $(x_2, y_2)$, then any point $(x, y)$ on this line must satisfy

$$\frac{y - y_1}{x - x_1} = \frac{y_2 - y_1}{x_2 - x_1},$$

provided $x_1 \neq x_2$. If $(x_1, y_1)$ and $(x_2, y_2)$ are both in the plane of $F$, then putting the above equation into the form $ax + by + c = 0$ will give us $a$, $b$, and $c$ in the field $F$. The case where $x_1 = x_2$ is an easily handled special case.

▷ **Quick Exercise.** Show that we can obtain an equation of the form $ax + by + c = 0$ with $a, b, c \in F$ for the line through two points in the plane of $F$ with equal $x$-coordinates. ◁

▷ **Quick Exercise.** Determine the equation of the line passing through $(2 + \sqrt{5}, -\sqrt{5})$ and $(4 + 3\sqrt{5}, 2 + 7\sqrt{5})$, and check that the coefficients $a$, $b$, and $c$ do belong to $\mathbb{Q}(\sqrt{5})$. ◁

The equation for a circle in the plane of $F$ is of the form

$$x^2 + y^2 + dx + ey + f = 0,$$

where $d$, $e$, and $f$ are in $F$. For if the circle has center at $(x_1, y_1)$ and a point on the circumference is $(x_2, y_2)$, then the radius is

$$\sqrt{(x_2 - x_1)^2 + (y_2 - y_1)^2}$$

and so if $(x, y)$ is any point on the circle, it must satisfy

$$(x - x_1)^2 + (y - y_1)^2 = (x_2 - x_1)^2 + (y_2 - y_1)^2.$$

This can be put into the desired form where $d$, $e$, and $f$ are in the field $F$ if $(x_1, y_1)$ and $(x_2, y_2)$ are in the plane of $F$.

▷ **Quick Exercise.** Put this equation into the form

$$x^2 + y^2 + dx + ey + f = 0$$

and note that $d$, $e$, and $f$ are elements of $F$. ◁

▷ **Quick Exercise.** Determine the equation of the circle with center $(2 + \sqrt{5}, -\sqrt{5})$ and $(4 + 3\sqrt{5}, 2 + 7\sqrt{5})$ on the circumference, and check that the coefficients $d$, $e$, and $f$ are in $\mathbb{Q}(\sqrt{5})$. ◁

So, given the circle and line constructions we can make in the plane of $F$, what new points can we locate? New points can be located in one of three ways:

1. The intersection of two lines in the plane of $F$.

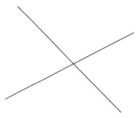

2. The intersection of a circle in the plane of $F$ with a line in the plane of $F$.

3. The intersection of two circles in the plane of $F$.

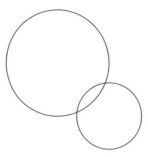

We can easily locate points found by method (1) by solving the system of two linear equations. Notice that the solution, if the lines are not parallel, is a point $(p_1, p_2)$ with $p_1$ and $p_2$ in the field $F$, because our method of solving two equations in two unknowns involves *only field operations*. In other words, method (1) can locate no points outside of the plane of $F$.

▷ **Quick Exercise.**   Find the simultaneous solution to the equations $a_1 x + b_1 y + c_1 = 0$ and $a_2 x + b_2 y + c_2 = 0$. Under what conditions will these two equations have no simultaneous solution and thus represent parallel lines?  ◁

Method (2) involves the simultaneous solution of an equation for a circle and an equation for a line. Method (3) involves the simultaneous solution of two equations for circles. Notice that method (3) reduces to method (2), for if

$$x^2 + y^2 + d_1 x + e_1 y + f_1 = 0$$

is subtracted from

$$x^2 + y^2 + d_2 x + e_2 y + f_2 = 0,$$

we get

$$(d_2 - d_1)x + (e_2 - e_1)y + (f_2 - f_1) = 0,$$

and so the simultaneous solution of this linear equation (which gives the equation for the common chord; see Exercise 8) with either of the two circle equations is the desired solution. Thus, what remains is to find which points can be obtained from method (2).

So, suppose we wish to solve simultaneously

$$x^2 + y^2 + dx + ey + f = 0, \text{ and}$$
$$ax + by + c = 0,$$

where all the coefficients belong to the field $F$. Because $a$ and $b$ can't both be zero we will assume that $b \neq 0$ (the case where $a \neq 0$ is similar). We solve for $y$ in terms of $x$:

$$y = -\frac{a}{b}x - \frac{c}{b}.$$

Substituting this into the circle equation yields:

$$x^2 + \left(-\frac{a}{b}x - \frac{c}{b}\right)^2 + dx + e\left(-\frac{a}{b}x - \frac{c}{b}\right) + f = 0$$

which is a quadratic equation in $x$. We could use the quadratic formula to solve it; we will not carry out the calculations explicitly.

▷ **Quick Exercise.** Apply the quadratic formula to this quadratic equation, and obtain the solutions in terms of the coefficients. ◁

The solution you just obtained has the form $A \pm B\sqrt{k}$, with $A, B, k \in F$. If $k < 0$, then there are no *real* solutions, and this means geometrically that there is no intersection. If $k = 0$, then there is a unique solution; this means geometrically that the line is *tangent* to the circle. Finally, if $k > 0$, we have two distinct intersections.

In case we have solutions, we can determine $y$ by substitution: This yields an expression of the form $A' \pm B'\sqrt{k}$, with $A', B' \in F$.

▷ **Quick Exercise.** Show that if there are solutions for $x$ of the form $A \pm B\sqrt{k}$, with $A, B, k \in F$, then $y$ has the form $A' \pm B'\sqrt{k}$, with $A', B' \in F$. ◁

So, notice that both $x$ and $y$ are in the field $F(\sqrt{k})$; if $\sqrt{k} \in F$, then of course $F(\sqrt{k}) = F$, and so $x$ and $y$ are in $F$. That is, the point $(x, y)$ is in the plane of $F(\sqrt{k})$.

---

## 37.5  The Constructible Number Theorem

We are now ready to prove the main result of this chapter.

**Theorem 37.3    Constructible Number Theorem**    *The following two statements are equivalent:*

 a. *The number $\alpha$ is constructible.*

 b. *There exists a finite sequence of fields*

$$\mathbb{Q} = F_0 \subset F_1 \subset \cdots \subset F_N$$

*with $\alpha \in F_N$ and $F_{i+1} = F_i(\sqrt{k_i})$ for some $k_i \in F_i$, with $k_i > 0$ for $i = 0, \ldots, N-1$.*

**Proof:** That (b) implies (a) is the previous theorem.

To show that (a) implies (b), suppose $\alpha$ is constructible. Then we can construct the point $(\alpha, 0)$, which we will label $P$, starting only with the segment of length 1 along the positive $x$-axis. To construct $P$, we must construct a finite sequence of points, $P_0, P_1, \ldots, P_M = P$. Because the points $(0,0)$ and $(1,0)$ are the first two points constructed, we set $P_0 = (0,0)$ and $P_1 = (1,0)$. Of course, $P_0$ and $P_1$ are both elements of the plane of $\mathbb{Q}$. Now $P_2$ is constructed using only $P_0$, $P_1$ and one of the three constructions for locating new points in the plane. Hence, by the above discussion, $P_2 \in \mathbb{Q}(\sqrt{k})$ for some positive $k \in \mathbb{Q}$. (This field may be equal to $\mathbb{Q}$ if $\sqrt{k} \in \mathbb{Q}$.) Then $F_0 = F_1 = \mathbb{Q}$ and $F_2 = \mathbb{Q}(\sqrt{k})$.

We now proceed inductively. Let $F_i$ be the smallest field containing the points $P_0, \ldots, P_i$. By the induction hypothesis, for each $i = 3, 4, \ldots, M$, $P_i$ was constructed using only the points constructed before it and one of the three constructions for locating points in the plane. Hence, $P_i \in F_{i-1}(\sqrt{k_{i-1}})$ for some positive $k_{i-1} \in F_{i-1}$.

Noting only those times the field $F_i$ is a proper extension of $F_{i-1}$, we thus have that $\alpha$ is in the field $F_N$ where

$$\mathbb{Q} = F_0 \subset F_1 \subset \cdots \subset F_N$$

and $F_{i+1} = F_i(\sqrt{k_i})$ where $k_i \in F_i$ with $k_i \geq 0$, for $i = 0, 1, \ldots, N-1$. (Notice that $N$, the number of different fields needed in the construction of $\alpha$, is probably smaller than $M$, the number of points actually constructed.) $\square$

Thus, for any constructible number, there is a *finite* sequence of quadratic extension fields, the last of which will contain the given number. Of course, this was exactly what we discovered for the particular constructible number

$$\sqrt{6 + \frac{4}{3}\sqrt{2 + 2\sqrt{7}}}.$$

In Chapters 41–43 we will provide a more general theory of field extensions (including degrees other than 2); but this will require some new concepts, which we introduce in Chapters 39 and 40. See Exercise 5 for an example of a non-quadratic extension.

---

## Chapter Summary

We defined *quadratic field extension*. We examined what new points could be constructed from points previously constructed and found that a new point was always in a quadratic field extension of the field determined by the old points.

Finally, we proved the main result of this section, the *Constructible Number Theorem*, which completely describes in *algebraic* terms which numbers are constructible.

---

## Warm-up Exercises

   a. Explain why the Constructible Number Theorem 37.3 guarantees that the following numbers are constructible:

$$\sqrt{6}, \quad \sqrt[8]{6}, \quad \sqrt{2 + \sqrt[4]{5}}.$$

   b. What is the quadratic field extension $\mathbb{R}(i)$ usually known as?

   c. Does $\mathbb{C}$ admit any proper quadratic field extensions?

d. Does $\mathbb{K}$ admit any proper quadratic field extensions? Does it admit any proper quadratic field extensions that are subfields of $\mathbb{R}$?

e. Did we discuss a real number in this chapter that is *not* constructible?

f. Suppose that $\ell_1$ and $\ell_2$ are lines in the plane of a field $F$ of constructible numbers. What can you say about the intersection of $\ell_1$ and $\ell_2$?

g. Suppose that $a$, $b$, and $c$ are constructible numbers, and $ax^2 + bx + c$ has real roots. Are these roots constructible numbers?

---

## Exercises

1. Show that $\sqrt{5} \notin \mathbb{Q}(\sqrt{3})$ by showing that 5 cannot be the square of a number of the form $a + b\sqrt{3}$ where $a$ and $b$ are in $\mathbb{Q}$.

2. If $F = \mathbb{Q}(\sqrt{3})$, describe the set of elements of the field $F(\sqrt{5})$. Show that this field is the same as the quadratic extension

$$\mathbb{Q}(\sqrt{5})(\sqrt{3}).$$

3. Generalize Exercise 2: Suppose that $p$ and $q$ are distinct positive prime integers; prove that

$$\mathbb{Q}(\sqrt{p})(\sqrt{q}) = \mathbb{Q}(\sqrt{q})(\sqrt{p}),$$

and describe the elements of the field.

4. Give a sequence of numbers necessary to construct the number

$$\sqrt[4]{2 + 4\sqrt{3}}.$$

Give the corresponding sequence of fields.

5. We've been able to give a nice description of the smallest field containing both $\sqrt{2}$ and $\mathbb{Q}$; namely, $\{a + b\sqrt{2} : a, b \in \mathbb{Q}\}$. In this problem we try to give a description of the smallest field containing $\sqrt[3]{2}$ and $\mathbb{Q}$. Why is $\{a + b\sqrt[3]{2} : a, b \in \mathbb{Q}\}$ not the answer? Now consider $\{a + b\sqrt[3]{2} + c\sqrt[3]{4} : a, b, c \in \mathbb{Q}\}$. When proving this is a field, the difficult part is showing that a typical element $a + b\sqrt[3]{2} + c\sqrt[3]{4}$, with $a, b, c \in \mathbb{Q}$ and not all zero, has a multiplicative inverse. You might not be successful in this part, and we will give an explicit hint below. But do at least verify that this set is indeed a commutative ring with unity.

Here is a hint for how to show that a typical element indeed has a multiplicative inverse. In fact, the multiplicative inverse of $a + b\sqrt[3]{2} + c\sqrt[3]{4}$ is $d + e\sqrt[3]{2} + f\sqrt[3]{4}$ where

$$
\begin{aligned}
d &= \frac{-2bc + a^2}{2b^3 + 4c^3 + a^3 - 6bca} \\
e &= \frac{2c^2 - ba}{2b^3 + 4c^3 + a^3 - 6bca}, \quad \text{and} \\
f &= \frac{b^2 - ca}{2b^3 + 4c^3 + a^3 - 6bca}.
\end{aligned}
$$

You can verify this by computing $(a + b\sqrt[3]{2} + c\sqrt[3]{4})(d + e\sqrt[3]{2} + f\sqrt[3]{4})$ and seeing that it simplifies to 1. (This is obviously a non-trivial calculation!)

We will show (in a less grubby manner) that this set is a field in Theorem 42.3. Note also that we have by no means claimed that $\sqrt[3]{2}$ is a constructible number (see Section 38.1).

6. Describe the elements of the quadratic extension field $F(\sqrt[4]{2})$, where $F = \mathbb{Q}(\sqrt{2})$.

7. Suppose that $a$, $b$, and $c$ are constructible numbers. Consider the polynomial $ax^4 + bx^2 + c$: Are its real roots constructible? (See Warm-up Exercise g above.)

8. Argue that the linear equation that results when two circles are intersected algebraically gives the equation of the common chord of the two circles. (See the figure below.)

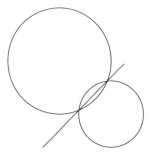

9. Find the equation of tangent line to the circle

$$(x-1)^2 + y^2 = 5$$

at the point $(3, 1)$ by algebra: This is the only line that intersects this circle in exactly this point. *Note:* This was Rene Descartes' method for determining a tangent line. Doing this problem should make you appreciate the *geometric* approach to finding the tangent (it is perpendicular to a radial line) and the *calculus* approach (calculate $dy/dx$).

# Chapter 38

## The Impossibility of Certain Constructions

The Constructible Number Theorem 37.3 is the most important piece of machinery necessary to show that the three construction problems of the Greeks are indeed impossible with a compass and a straightedge.

### 38.1  Doubling the Cube

We first tackle the problem of doubling the cube. Recall the problem:

*Given a line segment representing the edge of a cube, construct another line segment representing the edge of a cube with twice the volume of the original cube.*

We start with a line segment of length 1, and consider the cube with this segment as one edge. A cube of twice the volume would have edges of length $\sqrt[3]{2}$. So, doubling the cube then amounts to constructing the number $\sqrt[3]{2}$. We will show that $\sqrt[3]{2}$ is not constructible by using the Constructible Number Theorem. We must prove that $\sqrt[3]{2}$ cannot be an element of a field at the end of a finite sequence of quadratic field extensions that starts with the rational numbers. The next lemma is the key to showing this.

**Lemma 38.1** *Let $F(\sqrt{k})$ be a real quadratic field extension of a field $F$. If $\sqrt[3]{2} \in F(\sqrt{k})$, then $\sqrt[3]{2} \in F$.*

**Proof:**  Suppose that $\sqrt[3]{2} \in F(\sqrt{k})$, a proper quadratic extension of the field $F$. Then $\sqrt[3]{2} = a + b\sqrt{k}$, with $a, b \in F$ and $\sqrt{k} \notin F$. We want to show that $b = 0$. But,

$$
\begin{aligned}
2 &= (a + b\sqrt{k})^3 \\
&= a^3 + 3a^2 b\sqrt{k} + 3ab^2 k + b^3 k\sqrt{k} \\
&= (a^3 + 3ab^2 k) + (3a^2 b + b^3 k)\sqrt{k}.
\end{aligned}
$$

If $3a^2 b + b^3 k \neq 0$, then we could solve the above equation for $\sqrt{k}$; the resulting equation would show that $\sqrt{k} \in F$. Hence, $3a^2 b + b^3 k = 0$. But then,

$$
\begin{aligned}
(a - b\sqrt{k})^3 &= (a^3 + 3ab^2 k) - (3a^2 b + b^3 k)\sqrt{k} \\
&= a^3 + 3ab^2 k \\
&= 2,
\end{aligned}
$$

and so $a - b\sqrt{k}$ is also a cube root of 2. Thus, $a + b\sqrt{k}$ and $a - b\sqrt{k}$ are both real roots of $x^3 - 2$. But

$$
x^3 - 2 = (x - \sqrt[3]{2})(x^2 + \sqrt[3]{2}x + \sqrt[3]{4}),
$$

and the quadratic factor is irreducible in $\mathbb{R}[x]$.

▷ **Quick Exercise.** Use the quadratic formula to check that this quadratic factor is irreducible in $\mathbb{R}[x]$. ◁

Hence, $a + b\sqrt{k} = a - b\sqrt{k}$ which implies that $b = 0$, as we wished. □

It now follows from Lemma 38.1 that:

**Theorem 38.2** *It is impossible to double the cube.*

**Proof:** As noted, doubling the cube is equivalent to constructing $\sqrt[3]{2}$. But if $\sqrt[3]{2}$ were constructible, by the Constructible Number Theorem 37.3, there would exist a finite sequence of quadratic field extensions, $\mathbb{Q} = F_0 \subset F_1 \subset \cdots \subset F_N$, with $\sqrt[3]{2} \in F_N$.

But Lemma 38.1 says that if $\sqrt[3]{2} \in F_N = F_{N-1}(\sqrt{k})$, then $\sqrt[3]{2} \in F_{N-1}$. Repeating this argument inductively implies that $\sqrt[3]{2} \in \mathbb{Q}$, which is false. (See Exercise 5.13.) □

---

## 38.2 Trisecting the Angle

A similar approach is used to show that trisecting an angle is impossible. Again, recall the problem:

*Given an arbitrary angle, divide it into three equal parts.*

The problem calls for a method to trisect all angles. If we can exhibit just one angle that can't be trisected, then such a method does not exist. The angle we will use is the 60° angle. As we've noted before, the 60° angle is constructible—as an angle of an equilateral triangle, for instance. Trisecting a 60° angle implies the construction of a 20° angle. But if we can construct an angle $\alpha$, we can construct the cosine of $\alpha$, as the figure below shows.

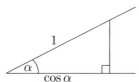

So being able to trisect a 60° angle implies that $\cos 20°$ is a constructible number; this is what we will show is impossible, using the Constructible Number Theorem 37.3.

We use the formulas for the sine and cosine of the sum of angles to perform the following derivation:

$$
\begin{aligned}
\cos 3\theta &= \cos(2\theta + \theta) \\
&= \cos 2\theta \cos \theta - \sin 2\theta \sin \theta \\
&= (\cos^2 \theta - \sin^2 \theta) \cos \theta - (2 \sin \theta \cos \theta) \sin \theta \\
&= \cos^3 \theta - 3 \sin^2 \theta \cos \theta \\
&= \cos^3 \theta - 3(1 - \cos^2 \theta) \cos \theta \\
&= 4 \cos^3 \theta - 3 \cos \theta
\end{aligned}
$$

Setting $\theta = 20°$, $\cos 3\theta = \cos 60° = 1/2$, and so $\cos 20°$ is a solution to $4x^3 - 3x - 1/2 = 0$ or $8x^3 - 6x - 1 = 0$; thus, $2\cos 20°$ is a solution to $x^3 - 3x - 1 = 0$. So, if $\cos 20°$ were constructible then so would $2\cos 20°$, which is a root of $x^3 - 3x - 1$. Thus, if $\cos 20°$ were constructible, it would be possible to construct a real root of $x^3 - 3x - 1$. We will show that this is impossible, using the next lemma, which has the same flavor as Lemma 38.1 above.

**Lemma 38.3** *Let $F(\sqrt{k})$ be a quadratic field extension of a field $F$. If the equation $x^3 - 3x - 1 = 0$ has a solution in $F(\sqrt{k})$, then it has a solution in $F$.*

**Proof:** Let $a + b\sqrt{k}$ be a root of $x^3 - 3x - 1$ in $F(\sqrt{k})$, a proper quadratic extension of the field $F$. If $b = 0$, then the root is $a$, which is in $F$. If $b \neq 0$, we will show that $-2a$ is a root; this will still mean that $x^3 - 3x - 1$ has a root in $F$. So, if $b \neq 0$,

$$
\begin{aligned}
0 &= (a + b\sqrt{k})^3 - 3(a + b\sqrt{k}) - 1 \\
&= a^3 + 3a^2 b\sqrt{k} + 3ab^2 k + b^3 k\sqrt{k} - 3a - 3b\sqrt{k} - 1 \\
&= (a^3 + 3ab^2 k - 3a - 1) + (3a^2 b + b^3 k - 3b)\sqrt{k}.
\end{aligned}
$$

But $3a^2 b + b^3 k - 3b = 0$, for otherwise, $\sqrt{k} \in F$. But then $a^3 + 3ab^2 k - 3a - 1 = 0$. After dividing the first equation by $b$ (we know $b \neq 0$), we have $3a^2 + b^2 k - 3 = 0$, and so $b^2 k = 3 - 3a^2$. Substituting this into the second equation we have,

$$
\begin{aligned}
0 &= a^3 + 3a(3 - 3a^2) - 3a - 1 \\
&= a^3 + 9a - 9a^3 - 3a - 1 \\
&= -8a^3 + 6a - 1 \\
&= (-2a)^3 - 3(-2a) - 1.
\end{aligned}
$$

In other words, $-2a$ is a root of $x^3 - 3x - 1$. □

**Theorem 38.4** *It is not possible to trisect an arbitrary angle.*

**Proof:** As noted before, if we could trisect a $60°$ angle, we could construct a $20°$ angle. This means we could construct the number $\cos 20°$, and this implies that we can construct a root of $x^3 - 3x - 1$. Using Lemma 38.3 and the Constructible Number Theorem 37.3 and arguing as in the previous theorem, we see that this implies that there is a rational root of $x^3 - 3x - 1$. But, by the Root Theorem 4.3, this implies that $x^3 - 3x - 1$ factors in $\mathbb{Q}[x]$. However, we can see that this polynomial is irreducible in $\mathbb{Z}[x]$ and so, by Gauss's Lemma 5.5, is also irreducible in $\mathbb{Q}[x]$.

▷ **Quick Exercise.** Why is $x^3 - 3x - 1$ irreducible in $\mathbb{Z}[x]$? ◁

Hence, $x^3 - 3x - 1$ has no rational root, and so we cannot trisect a $60°$ angle. □

---

## 38.3 Squaring the Circle

Finally, we turn our attention to squaring the circle. Let's state the problem carefully:

*Given a circle, construct a square with the same area.*

We are unable to give the full proof here of the impossibility of squaring the circle, because one step is quite difficult. Given a circle with radius 1, its area is $\pi$, and so to square the circle we must be able to construct the number $\sqrt{\pi}$ (to be the side of the required square). Obviously, this is possible, if we could construct the number $\pi$ itself. (This is the argument for Exercise 36.e.)

In the 18th century the German mathematician Johann Heinrich Lambert managed to prove that $\pi$ is not a rational number, by showing that if $x$ is rational, then $\tan x$ cannot be;

because $\tan(\pi/4) = 1$, Lambert's theorem means that $\pi/4$ (and hence $\pi$) cannot be rational. Lambert conjectured that $\pi$ is a **transcendental** number; that is, a number that is not the root of any polynomial in $\mathbb{Q}[x]$. This was finally proved by another German mathematician, Ferdinand Lindemann, in 1882; he made heavy use of the work of the Frenchman Charles Hermite, who had proved the transcendence of $e$ a decade earlier. As we shall see shortly, Lindemann's theorem finally laid to rest the last of the great constructibility problems of the ancient Greeks.

**Theorem 38.5    Lindemann's Theorem**    *$\pi$ is transcendental.*

**Proof:**    The proof is long, difficult, and analytic, rather than algebraic. For an accessible version, see *Field Theory and its Classical Problems*, by Charles Hadlock.    □

To use Lindemann's Theorem to prove the impossibility of squaring the circle, we first need a little more machinery.

**Lemma 38.6** *Suppose $F(\sqrt{k})$ is a quadratic field extension of $F$. If $\alpha$ is a root of a polynomial in $F(\sqrt{k})[x]$ of degree $n$, then it is the root of a polynomial in $F[x]$ of degree $2n$.*

**Proof:**    Assume that $\alpha$ is a solution to an equation of the form

$$(a_n + b_n\sqrt{k})x^n + \cdots + (a_0 + b_0\sqrt{k}) = 0$$

where all the $a_i$'s and $b_i$'s are in the field $F$ and $a_n$ and $b_n$ are not both zero. Moving the terms with $\sqrt{k}$ to the right side of the equation, we get

$$a_n x^n + \cdots + a_0 = -\sqrt{k}(b_n x^n + \cdots + b_0).$$

Squaring both sides and moving all terms to the left gives a polynomial in $F[x]$ that also has $\alpha$ as a root. Now the leading term of this polynomial is $(a_n^2 - kb_n^2)x^{2n}$. This coefficient is not zero, because otherwise $k = a_n^2/b_n^2$, contradicting the fact that $\sqrt{k} \notin F$.    □

**Theorem 38.7** *If $\alpha$ is constructible, then $\alpha$ is the root of a polynomial in $\mathbb{Q}[x]$ of degree $2^r$, for some $r \in \mathbb{N}$.*

**Proof:**    By the Constructible Number Theorem 37.3, $\alpha \in F_N$, where $\mathbb{Q} = F_0 \subset F_1 \subset \cdots \subset F_N$ is a sequence of quadratic field extensions. But $\alpha$ is the root of a linear equation in $F_N$, namely, $x - \alpha$. By repeated application of Lemma 38.6, $\alpha$ is the root of a polynomial in $\mathbb{Q}[x]$ of degree $2^N$.    □

So, if $\pi$ were constructible, it would have to be the root of some polynomial in $\mathbb{Q}[x]$; Lindemann's Theorem says this is not true, and so we have:

**Theorem 38.8** *It is not possible to square the circle.*

## Historical Remarks

Our proofs that it is impossible to duplicate the cube and to trisect an arbitrary angle are similar in flavor to the first such proofs, by Pierre Wantzel, which appeared in 1837. His version of the Constructible Number Theorem asserted that any constructible number is the root of an irreducible polynomial with degree a power of two, and hence numbers like $\sqrt[3]{2}$ and $\cos(20°)$ are not constructible.

Another important impossibility result had been obtained a decade earlier by the Norwegian mathematician Niels Abel, who showed that it is impossible to solve an arbitrary fifth-degree equation, using only elementary algebra and the extraction of roots. We made reference to this result in the Historical Remarks following Chapter 10.

It turns out that both these achievements can be viewed most elegantly as part of a general theory of field extensions of the rational numbers, and it was the French mathematician Evariste Galois who laid the important groundwork for this theory. In the remainder of this book we will look into this important area of algebra.

## Chapter Summary

We proved that it is impossible to double the cube, trisect an angle, or square the circle using only a compass and a straightedge. The outline of the proofs are the same: If the construction were possible, then we could construct a certain number. (For us, the numbers are $\sqrt[3]{2}$ for doubling the cube, $\cos 20°$ for trisecting an angle, and $\pi$ for squaring the circle.) But the Constructible Number Theorem or its corollaries show that each number is not constructible; therefore, the three constructions are impossible.

## Warm-up Exercises

a. We have finally discussed some real numbers that are *not* constructible! Give some examples.

b. During our discussion of trisecting the angle, we proved that if an angle $\theta$ is constructible, then $\cos \theta$ is constructible. Explain why the converse of this statement is true too.

c. Lambert proved in the 18th century that $\pi$ is not rational; why is this *not* sufficient to show that the circle cannot be squared?

d. Explain why it is possible to double the square.

e. Explain why it is possible to 'octuple' the cube.

## Exercises

1. (a) The number $\sqrt{2} + \sqrt{3}$ is constructible. It is an element of $F(\sqrt{3})$ where $F = \mathbb{Q}(\sqrt{2})$. In fact, $\sqrt{2} + \sqrt{3}$ is the root of the polynomial $x - (\sqrt{2} + \sqrt{3})$ in $F(\sqrt{3})$. Find a polynomial in $\mathbb{Q}[x]$ for which $\sqrt{2} + \sqrt{3}$ is a root.

    (b) Find a polynomial in $\mathbb{Q}[x]$ for which $\sqrt{4 + \sqrt{7}}$ is a root.

2. Show that $\sqrt[3]{2} + \sqrt{2}$ is a root of the polynomial

$$x^3 - 3\sqrt{2}x^2 + 6x - (2 + 2\sqrt{2})$$

in $\mathbb{Q}(\sqrt{2})[x]$. Then obtain a sixth-degree polynomial in $\mathbb{Q}[x]$ that has this number as a root.

3. While it is impossible to square the circle, it is possible to square the rectangle. That is, given a rectangle, it is possible to construct a square of the same area. Do this.

*Hint:* Consider the diagram below. Show that $c^2 = ab$. Thus, if one constructed a square with sides of length $c$, it would have the same area as the given rectangle with sides $a$ and $b$.

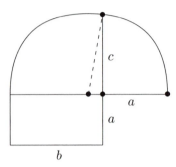

4.  (a) Show how to square an arbitrary triangle.

    (b) Consider a **polygon** in the plane (by this, we mean just a bounded figure with edges that are line segments). How could you use part (a) to square such a figure?

5. Suppose that $p$ and $q$ are elements of $F$, a subfield of the field of real numbers. Let $F(\sqrt{k})$ be a proper quadratic field extension of the field $F$. Prove that if the equation

$$x^3 + px + q = 0$$

has a solution in $F(\sqrt{k})$, then it has a solution in $F$.

6. Use the trig identity

$$\cos 3\theta = 4\cos^3 \theta - 3\cos \theta,$$

and the previous exercise to prove that the angle $2\pi/9$ is not constructible.

7. Suppose that $n$ is a positive integer, with no integer cube root. Let $F(\sqrt{k})$ be a quadratic field extension of a field $F$. Prove that if $\sqrt[3]{n} \in F(\sqrt{k})$, then $\sqrt[3]{n} \in F$.

8. Let $n$ be a positive integer, with no integer cube root. Use the previous exercise to prove that we cannot construct a cube with volume equal to $n$ times that of a given cube. (Thus, we cannot triple or quadruple the cube with compass and straightedge.)

9. Consider a parabola (by analytic geometry, we may consider a curve of the form $f(x) = ax^2 + bx + c$). Pick two points

$$P_1(x_1, f(x_1)) \text{ and } P_2(x_2, f(x_2))$$

on the parabola, and consider the region bounded by the parabola and the line segment between the two points. This is called a **segment of the parabola**.

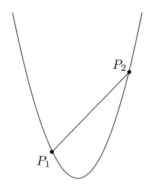

(a) Prove (using calculus) that the area of the segment is equal to $\frac{4}{3}$ times the area of the triangle $P_1P_2P_3$, where $P_3$ is the point on the parabola with $x$ coordinate $(x_1 + x_2)/2$.

(b) Explain why this means that we can square the segment of the parabola. This result was known to Archimedes (ca. 250 B.C.); he proved it using the geometry of the parabola, rather than calculus.

10. In the proof of Lemma 38.1 we needed to show that $x^3 - 2$ has only one real root; we did this using algebra. Prove this result using calculus.

11. Suppose that $\theta$ is any fixed angle with positive radian measure. In this problem you will show that it is possible to construct an angle arbitrarily close in size to $\theta$.

   (a) Suppose that $\epsilon$ is an arbitrary positive number. Describe how to construct an angle $\psi$ whose angle measure is smaller than $\epsilon$.

   (b) Explain why an integer multiple $n\psi$ of the angle constructed in part (a) must be larger than $\theta$.

   (c) Consider the *smallest* positive integer $n$ so that $n\psi$ is larger than $\theta$ (why must $n$ exist?). Use $n$ to find an angle within $\epsilon$ radians of $\theta$.

   (d) Why does part (c) mean that we can come arbitrarily close to constructing the trisection of any given angle?

   (e) What is the philosophical difference between the construction in part (d) and the sought-for (impossible) trisection construction?

# Section VIII in a Nutshell

This section presents three famous compass and straightedge construction problems of the ancient Greeks: doubling the cube, trisecting an angle, and squaring the circle. These problems were unsolved by the Greeks. We show that it is not in principle possible to make these constructions, using modern algebraic (and not geometric) techniques.

When starting with a line segment of length 1, we call the length of any line segment we can construct after a finite number of compass and straightedge construction steps to be a *constructible number*. First, we show that the set of constructible numbers is a field (Corollary 36.3). We can construct all rational numbers (Lemma 36.1 and Theorem 36.2) and all square roots of constructible numbers (Theorem 36.4). This development leads to the *Constructible Number Theorem* (Theorem 37.3), which asserts that a number $\alpha$ is constructible exactly if the following condition holds:

There exists a finite sequence of fields

$$\mathbb{Q} = F_0 \subset F_1 \subset \cdots \subset F_N$$

with $\alpha \in F_N$ and $F_{i+1} = F_i(\sqrt{k_i})$ for some $k_i \in F_i$, with $k_i > 0$ for $i = 0, \ldots, N-1$.

We show that it is impossible to double the cube by showing that $\sqrt[3]{2}$ is not constructible (Lemmas 38.1 and Theorem 38.2). We show that it is impossible to trisect a $60°$ angle (that is, construct a $20°$ angle) by showing that to do so would imply being able to construct a solution to $x^3 - 3x - 1 = 0$, which we show is impossible (Lemma 38.3 and Theorem 38.4). Finally, we consider the problem of squaring the circle. If this were possible, then $\pi$ would be a constructible number. *Lindemann's Theorem* (Theorem 38.5 — which we do not prove) says that $\pi$ is transcendental: That is, $\pi$ is not the root of any polynomial with rational coefficients. But we show that any constructible number is the root of a polynomial in $\mathbb{Q}[x]$ of degree $2^n$ (Theorem 38.7) and so is not transcendental. Thus it is impossible to square the circle.

# Part IX

# Vector Spaces and Field Extensions

# Chapter 39

## Vector Spaces I

The three previous chapters showed the impossibility of the three famous Greek constructibility problems. Some of the grubbier parts of the proofs in the previous chapters can be replaced by more elegant arguments—provided we know more about the algebraic structures involved. The next five chapters will develop the machinery needed for these more sophisticated arguments. The proofs we presented in Chapters 36–38 are correct and have the advantage of not needing a great deal of sophistication in order to understand them. However, the arguments to be presented next have the advantage of being much more elegant and concise (at the expense of being less accessible). This is not very surprising as the more we know, the easier it is to express ourselves. As an added bonus, our additional machinery will enable us to prove another impossibility result, regarding the solution of polynomial equations using arithmetic and root extraction.

Putting our applications aside, the topics covered in these next few chapters are important in their own right. The first such topic we need to discuss is the notion of *vector space*; this will allow us to better understand the ideas of field extensions. As we shall see, a vector space (like a ring, group, or field) is just a set, equipped with operations satisfying certain nice rules.

The study of vector spaces is a subject in its own right, called *linear algebra*, and you may have the opportunity to take an entire course about this topic, if you haven't already. In the next two chapters we will develop only enough of the theory from this subject in order to better understand fields. There is much more (both computational and theoretical) to linear algebra than we will be able to present here.

---

## 39.1 Vectors

In calculus you studied vectors in two and three dimensions. Such vectors provide a very useful way of looking at the geometry of the plane and three-dimensional space. In this chapter, we wish to generalize the properties of such vectors, to cases that are not quite so easily visualized. We will emphasize a more abstract and algebraic approach to vectors, but you should keep in mind that the familiar vectors from calculus are important motivating examples.

Let's start by examining some of the algebraic properties of the familiar three-dimensional vectors. We denote this set of vectors by $\mathbb{R}^3$. Recall that $\mathbb{R}^3 = \{(r_1, r_2, r_3) : r_i \in \mathbb{R}\}$. Two of these vectors may be added coordinate-wise to get another vector. Addition of two vectors has a nice geometric interpretation, as illustrated below.

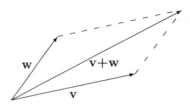

You may not have given much thought to it at the time, but this vector addition has the following properties: For $\mathbf{v}$, $\mathbf{w}$, and $\mathbf{u} \in \mathbb{R}^3$,

i. $\mathbf{v} + \mathbf{w} = \mathbf{w} + \mathbf{v}$,

ii. $\mathbf{v} + (\mathbf{w} + \mathbf{u}) = (\mathbf{v} + \mathbf{w}) + \mathbf{u}$,

iii. there exists a **zero vector 0**, with the property that $\mathbf{v} + \mathbf{0} = \mathbf{v}$, and

iv. every vector $\mathbf{v}$ has an **additive inverse** $-\mathbf{v}$, with the property that $\mathbf{v} + (-\mathbf{v}) = \mathbf{0}$.

You should recognize these as the same defining properties possessed by addition in a ring (see Chapter 6) or the properties of the operation in an abelian group (see Chapter 19). Here, the additive identity is the zero vector $\mathbf{0}$. In $\mathbb{R}^3$, $\mathbf{0} = (0, 0, 0)$ and if $\mathbf{v} = (v_1, v_2, v_3)$, then $-\mathbf{v} = (-v_1, -v_2, -v_3)$.

The other arithmetic operation in the algebra of vectors is scalar multiplication, in which vectors are multiplied by scalars to give other vectors. Here, our scalars come from the field $\mathbb{R}$. Scalar multiplication has the following properties: For $r, s \in \mathbb{R}$ and $\mathbf{v}, \mathbf{w} \in \mathbb{R}^3$,

v. $(r + s)\mathbf{v} = r\mathbf{v} + s\mathbf{v}$,

vi. $(rs)\mathbf{v} = r(s\mathbf{v})$,

vii. $r(\mathbf{v} + \mathbf{w}) = r\mathbf{v} + r\mathbf{w}$, and

viii. $1\mathbf{v} = \mathbf{v}$.

Note that the 1 in property (viii) is the scalar 1. You can easily show that these eight properties hold in $\mathbb{R}^2$ as well as $\mathbb{R}^3$. Note that when we considered $\mathbb{R}^3$ as a ring we also defined coordinate-wise multiplication. However, when thinking of elements of $\mathbb{R}^3$ as vectors, we will consider no such operation, because coordinate-wise multiplication of two vectors has no simple geometric interpretation.

## 39.2   Vector Spaces

There are other algebraic objects for which the above eight properties hold. A **vector space** $V$ **over a field** $F$ is a set with a binary operation called *addition* that satisfies properties (i) through (iv) above, together with a *scalar multiplication* of vectors from $V$ by scalars from $F$ so that if $r, s \in F$ and $\mathbf{v}, \mathbf{w} \in V$, then properties (v) through (viii) listed above hold.

Note that scalar multiplication is not a binary operation in the sense we've discussed before. Instead, scalar multiplication by the field element $r$ maps each vector in $V$ to another vector in $V$. We write $r\mathbf{v}$ to denote the vector that $\mathbf{v}$ gets mapped to and speak of 'multiplying $\mathbf{v}$ by $r$'.

It is important to think carefully about the meaning of the axioms (v) through (viii) and in particular where the operations are taking place. For example, the $+$ on the left side of the equation in axiom (v) denotes the addition in the field $F$, while the $+$ on the right side denotes the addition of vectors in $V$. The juxtaposition of $r$ with $s$ in (vi) denotes multiplication in the field, while the juxtaposition of $(rs)$ with $\mathbf{v}$ denotes the scalar multiplication. You should be careful to keep track of which operation is which.

▷ **Quick Exercise.** Look at the other axioms, and make sure you understand which of the operations are meant in each case. ◁

We now examine some examples of vector spaces:

## Example 39.1

The field $\mathbb{C}$ of complex numbers is a vector space over $\mathbb{R}$, the field of real numbers. Here, the vectors are complex numbers and the scalars are real numbers.

To show that this is a vector space, we first note that the set of vectors is closed with respect to addition (because $\mathbb{C}$ is a field, which is closed under addition) and that properties (i) through (iv) are properties of addition in all fields. Now, let's look at property (v). Let $r$ and $s$ be real numbers and $\mathbf{v} = a + bi \in \mathbb{C}$. Then

$$(r + s)\mathbf{v} = (r + s)(a + bi) = r(a + bi) + s(a + bi),$$

because multiplication in a field enjoys the distributive property. Thus, we have shown that $(r+s)\mathbf{v} = r\mathbf{v}+s\mathbf{v}$, as desired. The remaining properties are just as straightforward to show.

▷ **Quick Exercise.** Show that properties (vi) through (viii) hold here. ◁

We will generalize the idea of this example in Exercise 9.

## Example 39.2

The set of complex numbers $\mathbb{C}$ forms a vector space over itself! Here, the vector space axioms are just properties that hold for addition and multiplication in a field.

▷ **Quick Exercise.** What does axiom (vi) mean in this context? ◁

In the next example we provide a natural generalization of the vector space $\mathbb{R}^3$ (over the field $\mathbb{R}$).

## Example 39.3

The set $F^n$ of $n$-tuples with coordinates from the arbitrary field $F$ is a vector space over $F$. Here, a typical vector is of the form $(a_1, a_2, \ldots, a_n)$ where $a_i \in F$ and scalars are elements of $F$. The addition is coordinate-wise, and multiplication by a scalar merely multiplies each coordinate by that scalar. In Exercise 3, you will verify the details of this.

To be more concrete, the set $\mathbb{Z}_3 \times \mathbb{Z}_3$ of ordered pairs with coordinates from the field $\mathbb{Z}_3$ is a vector space, over the field $\mathbb{Z}_3$.

▷ **Quick Exercise.** List all the elements of this vector space. ◁

**Example 39.4**

The set of polynomials $\mathbb{Q}[x]$ is a vector space over $\mathbb{Q}$. The vectors are polynomials and the scalars are rational numbers. (See Exercise 1.)

**Example 39.5**

Let $\mathbb{Q}_n[x]$ be the set of polynomials of degree no more than $n$ with coefficients from $\mathbb{Q}$. This is a vector space over $\mathbb{Q}$. Here, vectors are polynomials of degree no more than $n$ and scalars are rational numbers. This example illustrates the fact that in a vector space, there need be no multiplication of vectors, but only addition of vectors: $\mathbb{Q}_n[x]$ is closed under addition but not multiplication. (See Exercise 2.)

**Example 39.6**

$\mathbb{Q}(\sqrt{2}) = \{a + b\sqrt{2} : a, b \in \mathbb{Q}\}$, is a vector space over $\mathbb{Q}$. Here, vectors are elements from $\mathbb{Q}(\sqrt{2})$ and scalars are from $\mathbb{Q}$.

▷ **Quick Exercise.** This example bears certain similarities with Example 39.1; explain this. *Hint*: What kind of ring is $\mathbb{Q}(\sqrt{2})$? ◁

The point we've made above in Example 39.5 bears repeating: In many of these examples of vector spaces it is natural also to define multiplication of vectors, as in our first example $\mathbb{R}^3$. Indeed, in many cases the vector space is a ring or field in its own right. But when we look at these as vector spaces over some field, we look at their structure in a different way and multiplication of vectors is not one of the operations considered. This change of point of view is important to keep in mind.

We now prove some basic properties of vector spaces. These properties should seem familiar, both from your previous experience with ring and group theory, and also any previous experience you have had with vectors, either in calculus or linear algebra.

**Theorem 39.1** *Let $V$ be a vector space over $F$ with $\mathbf{v} \in V$ and $r \in F$. Then, the additive inverse $-\mathbf{v}$ of $\mathbf{v}$ is unique. Also, $0\mathbf{v} = \mathbf{0}$, $r\mathbf{0} = \mathbf{0}$, and*

$$(-r)\mathbf{v} = r(-\mathbf{v}) = -(r\mathbf{v}).$$

**Proof:** Because $(V, +)$ is a group, the uniqueness of the additive inverse follows from Theorem 20.1.d.

Note the difference between the scalar 0 and the zero vector $\mathbf{0}$. The first identity says that multiplying any vector by the zero scalar yields the zero vector. But,

$$0\mathbf{v} = (0 + 0)\mathbf{v} = 0\mathbf{v} + 0\mathbf{v},$$

and so adding $-(0\mathbf{v})$, the additive inverse of $0\mathbf{v}$, to both sides, we get that $\mathbf{0} = 0\mathbf{v}$, as desired.

Note that in proving that $0\mathbf{v} = \mathbf{0}$, we used the fact that 0 was the additive identity of $F$ and one of the distributive laws (v). To prove the next identity, we will use the fact that $\mathbf{0}$ is the additive identity of $V$ and the other distributive law (vii). So,

$$r\mathbf{0} = r(\mathbf{0} + \mathbf{0}) = r\mathbf{0} + r\mathbf{0},$$

and so $\mathbf{0} = r\mathbf{0}$, as before.

To show that $(-r)\mathbf{v} = -(r\mathbf{v})$, we will show that $(-r)\mathbf{v}$ is the additive inverse of $r\mathbf{v}$. That is, it must be equal to $-(r\mathbf{v})$. We do this by adding $(-r)\mathbf{v}$ to $r\mathbf{v}$ and showing that the sum is $\mathbf{0}$. Doing so, we have

$$(-r)\mathbf{v} + r\mathbf{v} = (-r + r)\mathbf{v} = 0\mathbf{v} = \mathbf{0},$$

as desired. You can show in a similar manner that $r(-\mathbf{v}) = -(r\mathbf{v})$.

▷ **Quick Exercise.** Show that $r(-\mathbf{v}) = -(r\mathbf{v})$. ◁ □

The proofs of the results of Theorem 39.1 might seem familar, and should indeed be compared to the solutions to Exercises 6.1 and 6.2.

---

# Chapter Summary

In this chapter we defined *vector space*, looked at a number of examples of vector spaces, and examined some elementary properties.

---

# Warm-up Exercises

a. Consider the vectors $\mathbf{v} = (2, -1)$ and $\mathbf{w} = (2, 3)$ in $\mathbb{R}^2$. Compute the following, and draw diagrams to interpret these computations geometrically:

$$\mathbf{v} + \mathbf{w}, \quad \mathbf{v} - \mathbf{w}, \quad 2\mathbf{w}, \quad \frac{1}{2}\mathbf{v}, \quad -\mathbf{v}$$

b. Explain the difference between $0$ and $\mathbf{0}$.

c. Let $V = \{f \in \mathbb{Z}_3[x] : \deg(f) \le 2\}$; explain why this is a vector space over $\mathbb{Z}_3$. List all vectors and all scalars in this case.

d. Let $V = \{a + b\pi : a, b \in \mathbb{Q}\}$. Is this a vector space over $\mathbb{Q}$? Over $\mathbb{R}$?

e. Let $V$ be a vector space over $\mathbb{R}$, and let $\mathbf{v} \in V$. Explain what $-\mathbf{v}$ and $(-1)\mathbf{v}$ mean, according to the definitions of our notation; why are they equal?

f. Is $\mathbb{C}$ a vector space over $\mathbb{Q}$? Is $\mathbb{Q}$ a vector space over $\mathbb{C}$?

g. Is $\mathbb{Z}[x]$ a vector space over $\mathbb{Z}$?

---

# Exercises

1. Check Example 39.4: That is, prove that $\mathbb{Q}[x]$ is a vector space over $\mathbb{Q}$.

2. Check Example 39.5: That is, prove that $\mathbb{Q}_n[x]$ is a vector space over $\mathbb{Q}$, although it is not a subring of $\mathbb{Q}[x]$.

3. Check Example 39.3: That is, let $F$ be a field, and let $F^n$ be the set of $n$-tuples with entries from $F$. Prove that $F^n$ is a vector space over $F$.

4. Let $V$ be a vector space over the field $F$. Suppose that $r, s \in F$ and $\mathbf{0} \ne \mathbf{v} \in V$. Prove that if $r\mathbf{v} = s\mathbf{v}$, then $r = s$.

5. Let $V$ be a vector space over the field $F$. Suppose that $0 \neq r \in F$. Define the function

$$\varphi_r : V \to V$$

by $\varphi_r(\mathbf{v}) = r\mathbf{v}$. Prove that $\varphi_r$ is a bijection that preserves addition. That is, $\varphi_r$ is an *additive group isomorphism* from $(V, +)$ onto itself.

6. Let $V$ be a vector space over the field $F$. Suppose that $r, s \in F$ and $\mathbf{v} \in V$. Prove that $r(s\mathbf{v}) = s(r\mathbf{v})$.

7. Show that $M_2(\mathbb{R})$, the two-by-two matrices with entries from $\mathbb{R}$, is a vector space over $\mathbb{R}$.

8. Show that $M_{m,n}(F)$, the $m$-by-$n$ matrices with entries from the field $F$, is a vector space over $F$.

9. If $F$ and $E$ are fields with $F \subseteq E$, show that $E$ is a vector space over $F$. This is an important example of a vector space in subsequent chapters.

10. Let $\mathbb{Q}(\sqrt[3]{2}) = \{a + b\sqrt[3]{2} + c\sqrt[3]{4} : a, b, c \in \mathbb{Q}\}$. (In Exercise 37.5 you showed this is a field.) Show that $\mathbb{Q}(\sqrt[3]{2})$ is a vector space over $\mathbb{Q}$.

11. Let $V$ be the set of real-valued functions defined on $\mathbb{R}$, with addition defined by

$$(f + g)x = f(x) + g(x).$$

For $c \in \mathbb{R}$ and $f \in V$, the scalar multiple of $f$ by $c$ is $cf(x) = c(f(x))$. Show $V$ is a vector space over $\mathbb{R}$.

12. Prove that $\mathbb{Z}$ is not a vector space over $\mathbb{Z}_p$, where $p$ is a positive prime integer.

13. Prove that $\mathbb{Z}$ is not a vector space over $\mathbb{Q}$.

# Chapter 40

## Vector Spaces II

In the previous chapter we defined vector spaces. In this chapter we examine subsets of vector spaces that in some way generate the entire vector space. This idea gives rise to a way of measuring the size of a vector space.

---

### 40.1  Spanning Sets

If $V$ is a vector space over $F$ and

$$\{\mathbf{v}_1, \mathbf{v}_2, \ldots, \mathbf{v}_n\} \subseteq V,$$

then a **linear combination** of $\mathbf{v}_1, \mathbf{v}_2, \ldots, \mathbf{v}_n$ is the vector

$$a_1\mathbf{v}_1 + a_2\mathbf{v}_2 + \cdots + a_n\mathbf{v}_n,$$

where the $a_i$ are scalars. The collection of vectors that can be written as linear combinations of a given set of vectors $\mathcal{V}$ is the set **spanned** by $\mathcal{V}$.

#### Example 40.1

Consider the vectors $\mathbf{v} = (1, 0, 0)$ and $\mathbf{w} = (0, 1, 0)$ in $\mathbb{R}^3$. The vector

$$\left(\frac{1}{3}, -1, 0\right) = \frac{1}{3}\mathbf{v} + (-1)\mathbf{w},$$

and so is a linear combination of $\mathbf{v}$ and $\mathbf{w}$. Evidently, the set of all vectors in $\mathbb{R}^3$ spanned by $\mathbf{v}$ and $\mathbf{w}$ is exactly $\{(x, y, 0) : x, y \in \mathbb{R}\}$.

▷ **Quick Exercise.** Why is the set of vectors spanned by $\mathbf{v}$ and $\mathbf{w}$ equal to $\{(x, y, 0) : x, y \in \mathbb{R}\}$? ◁

Note that if we include the vector $(0, 0, 1)$, we then obtain all vectors in $\mathbb{R}^3$ as linear combinations of these three vectors.

#### Example 40.2

It is not so evident which vectors in $\mathbb{R}^3$ are spanned by $(1, 0, -1)$ and $(3, 2, 1)$. We can certainly conclude that the set spanned by them consists of all vectors of the form

$$(x + 3y, 2y, -x + y), \quad \text{where} \quad x, y \in \mathbb{R},$$

but what vectors are these? For example, does the vector $(1, 1, -1)$ belong to this set? If it did, there would be a simultaneous solution to the three equations

$$
\begin{aligned}
x + 3y &= 1 \\
2y &= 1 \\
-x + y &= -1.
\end{aligned}
$$

It's pretty easy to see that there is no such solution.

▷ **Quick Exercise.** Verify that there is no solution to this system of equations. ◁

Thus, $(1, 1, -1)$ is not a linear combination of $(1, 0, -1)$ and $(3, 2, 1)$.

If every vector of $V$ can be written as a linear combination of vectors from a subset $\mathcal{V}$ of $V$, we say that $\mathcal{V}$ **spans** (or **generates**) $V$. Note that $\mathcal{V}$ may be infinite or finite, but any given linear combination of vectors from $\mathcal{V}$ involves only a finite number of vectors.

## Example 40.3

For instance, from our discussion above it is clear that

$$\{(1, 0, 0), (0, 1, 0), (0, 0, 1)\}$$

spans $\mathbb{R}^3$.

## Example 40.4

Similarly,

$$\{(1, 0, 0), (2, 3, 0), (0, 1, 0), (-1, 0, -1), (0, 0, 1)\}$$

spans $\mathbb{R}^3$ because

$$
\begin{aligned}
(x, y, z) ={}& x(1, 0, 0) + 0(2, 3, 0) + y(0, 1, 0) + \\
& 0(-1, 0, -1) + z(0, 0, 1).
\end{aligned}
$$

Note that this spanning set contains more vectors than necessary. Furthermore, there may be more than one way of expressing a given vector as a linear combination of vectors in the spanning set. For instance, we can express $(-2, -2, -5)$ as

$$-2(1, 0, 0) + 0(2, 3, 0) - 2(0, 1, 0) + 0(-1, 0, -1) - 5(0, 0, 1),$$

or as

$$2(1, 0, 0) - 1(2, 3, 0) + 1(0, 1, 0) + 2(-1, 0, -1) - 3(0, 0, 1).$$

## Example 40.5

Now let's consider the set

$$\{(1, 0, 0), (2, 3, 0), (-1, 0, -1)\}.$$

It is not so obvious that this also spans $\mathbb{R}^3$. To show this, we consider a arbitrary vector $(x, y, z)$ from $\mathbb{R}^3$ and show that we can express it as a linear combination of the three vectors in the set. That is, we wish to find scalars $a_1, a_2$, and $a_3$ such that

$$(x, y, z) = a_1(1, 0, 0) + a_2(2, 3, 0) + a_3(-1, 0, -1).$$

This vector equation is equivalent to the following three linear equations, which we must solve simultaneously:

$$
\begin{aligned}
x &= a_1 + 2a_2 - a_3 \\
y &= \phantom{a_1 +{}} 3a_2 \\
z &= \phantom{a_1 + 2a_2} - a_3
\end{aligned}
$$

By the usual techniques we find the solutions for $a_1, a_2,$ and $a_3$ are

$$a_1 = x - \frac{2}{3}y - z$$
$$a_2 = \frac{1}{3}y$$
$$a_3 = -z.$$

So, for instance,

$$(3, 1, -2) = \frac{13}{3}(1, 0, 0) + \frac{1}{3}(2, 3, 0) + 2(-1, 0, -1).$$

**Example 40.6**

Consider the vector space $\mathbb{Q}[x]$ over $\mathbb{Q}$. Note that $\{1, x, x^2, x^3, \ldots\}$ spans $\mathbb{Q}[x]$.

▷ **Quick Exercise.** Show that $\{1, x, x^2, x^3, \ldots\}$ spans $\mathbb{Q}[x]$. ◁

In this case, we have an *infinite* spanning set.

A vector space is said to be **finite dimensional** if there is a finite set of vectors that spans the vector space. So, $\mathbb{R}^3$ is finite dimensional over $\mathbb{R}$.

The vector space $\mathbb{C}$ over $\mathbb{R}$ is finite dimensional because $\{1, i\}$ spans $\mathbb{C}$.

▷ **Quick Exercise.** Verify that the vector space $\mathbb{C}$ over $\mathbb{R}$ is finite dimensional. ◁

Likewise, $\mathbb{Q}(\sqrt{2})$ over $\mathbb{Q}$ is finite dimensional because $\{1, \sqrt{2}\}$ spans $\mathbb{Q}(\sqrt{2})$.

▷ **Quick Exercise.** Verify that $\mathbb{Q}(\sqrt{2})$ over $\mathbb{Q}$ is finite dimensional. ◁

However, $\mathbb{Q}[x]$ is not finite dimensional over $\mathbb{Q}$. In Example 40.6 we gave an infinite spanning set for $\mathbb{Q}[x]$; this alone does not suffice in showing that $\mathbb{Q}[x]$ is not finite dimensional—it may be that there exists some finite set of vectors that does indeed span $\mathbb{Q}[x]$. To show that $\mathbb{Q}[x]$ is not finite dimensional, we must show that *every* finite subset of $\mathbb{Q}[x]$ fails to span $\mathbb{Q}[x]$. This is Exercise 7. For the most part, we will confine our study here to finite dimensional vector spaces.

## 40.2   A Basis for a Vector Space

In the remainder of this chapter we will examine spanning sets that are minimal in the sense that removing any one vector from the set will result in a set that does not span the entire vector space.

The following is an important definition for us. A set of vectors

$$\{\mathbf{v}_1, \mathbf{v}_2, \ldots, \mathbf{v}_n\}$$

is **linearly independent** if

$$a_1\mathbf{v}_1 + a_2\mathbf{v}_2 + \cdots + a_n\mathbf{v}_n = \mathbf{0}$$

implies that $a_1 = a_2 = \cdots = a_n = 0$. A set of vectors that is not linearly independent is **linearly dependent**.

### Example 40.7

We can easily see that $\{(1,0,0),(0,1,0),(0,0,1)\}$ is linearly independent in $\mathbb{R}^3$ because if

$$a_1(1,0,0) + a_2(0,1,0) + a_3(0,0,1) = (0,0,0), \text{ then}$$

$(a_1, a_2, a_3) = (0,0,0)$, or $a_1 = a_2 = a_3 = 0$. The set

$$\{(1,0,0), \ (2,3,0), \ (-1,0,-1)\}$$

is also linearly independent, which we can verify by showing that the resulting system of equations has a unique solution of $a_1 = a_2 = a_3 = 0$.

▷ **Quick Exercise.** Show that the above system of equations has a unique solution of $a_1 = a_2 = a_3 = 0$. ◁

### Example 40.8

By contrast,

$$\{(1,0,0), \ (2,3,0), (0,1,0), \ (-1,0,-1), \ (0,0,1)\}$$

is linearly dependent because

$$
\begin{aligned}
(0,0,0) \ = \ & -2(1,0,0) + 1(2,3,0) - 3(0,1,0) + \\
& 0(-1,0,-1) + 0(0,0,1).
\end{aligned}
$$

(There are other possible values here for the $a_i$.)

▷ **Quick Exercise.** Find a different set of values for the $a_i$. ◁

### Example 40.9

The set of vectors $\{1 + \sqrt{2}, \ 2 - \sqrt{2}\}$ is linearly independent in the vector space $\mathbb{Q}(\sqrt{2})$ over $\mathbb{Q}$. But if we include the third vector $6\sqrt{2}$ in this set, it becomes dependent, because

$$2(1 + \sqrt{2}) + (-1)(2 - \sqrt{2}) - \frac{1}{2}(6\sqrt{2}) = 0.$$

The following theorem conveniently characterizes linearly independent sets of vectors:

**Theorem 40.1** $\{v_1, v_2, \ldots, v_n\}$ *is a set of linearly dependent vectors in the vector space* $V$ *if and only if one of the vectors from this set can be written as a linear combination of the others.*

**Proof:** If $\{v_1, v_2, \ldots, v_n\}$ is linearly dependent then we can write

$$a_1 v_1 + a_2 v_2 + \cdots + a_n v_n = \mathbf{0},$$

where the $a_i$'s are not all 0. By renumbering the vectors, if necessary, let's assume that $a_1$ is not zero. But then,

$$v_1 = -\frac{a_2}{a_1}v_2 - \frac{a_3}{a_1}v_3 - \cdots - \frac{a_n}{a_1}v_n,$$

as desired. Note that division by scalars is permissible, because they come from a field.

Conversely, suppose one of the vectors (say, $v_1$) can be expressed as a linear combination of the others:

$$v_1 = a_2 v_2 + a_3 v_3 + \cdots + a_n v_n.$$

But then,

$$0 = -1\mathbf{v}_1 + a_2\mathbf{v}_2 + a_3\mathbf{v}_3 + \cdots + a_n\mathbf{v}_n.$$

And so the set $\{\mathbf{v}_1, \mathbf{v}_2, \ldots, \mathbf{v}_n\}$ is linearly dependent.                    □

Looking at this another way, this theorem says that a set of vectors is linearly independent if and only if no vector in the set can be expressed as a linear combination of the others. Thus, if a set of vectors $\mathcal{V}$ is linearly independent and spans the vector space $V$, no proper subset of $\mathcal{V}$ can span $V$. In particular, any vector removed from $\mathcal{V}$ cannot be written as a linear combination of the remaining vectors. In other words, $\mathcal{V}$ is a *minimal* spanning set for $V$. We have a name for such a spanning set: A **basis** for a vector space $V$ is a linearly independent set of vectors that spans $V$.

▷ **Quick Exercise.**   We have thus argued above that any *basis* is a *minimal spanning set*. The converse is also true: Any minimal spanning set is linearly independent, and so a basis. What is the argument for this?   ◁

## Example 40.10

$\{(1,0,0), (0,1,0), (0,0,1)\}$ is a basis for $\mathbb{R}^3$, as is

$$\{(1,0,0), (2,3,0), (-1,0,-1)\}.$$

## Example 40.11

$\{1, i\}$ is a basis for $\mathbb{C}$ over $\mathbb{R}$. Note in this case, where our basis has two elements, linear independence is particularly easy to check: By Theorem 40.1 we need only verify that neither of the vectors is a scalar (in this case, real) multiple of the other. This is certainly true for 1 and $i$.

## Example 40.12

$\{1+i, \sqrt{2}+4i\}$ is a basis for $\mathbb{C}$ over $\mathbb{R}$. To check this requires a solution of two equations in two unknowns that we leave as Exercise 12. This example should make evident the fact that there are infinitely many distinct bases for $\mathbb{C}$ over $\mathbb{R}$.

## Example 40.13

$\{1, \sqrt{2}\}$ is a basis for $\mathbb{Q}(\sqrt{2})$ over $\mathbb{Q}$.

In the Quick Exercise above, we emphasized the fact that being a linearly independent spanning set (that is, a basis) is actually equivalent to being a minimal spanning set. We can change perspective a bit, and also prove that being a basis is equivalent to being a *maximal independent set*; you will pursue this in Exercises 16 and 17.

The following is another important equivalent characterization of a basis, which you will prove in Exercise 8.

**Theorem 40.2** $\mathcal{V} = \{\mathbf{v}_1, \mathbf{v}_2, \ldots, \mathbf{v}_n\}$ *is a basis for a vector space $V$ if and only if every element of $V$ can be uniquely written as a linear combination of the $\mathbf{v}_i$.*

The last important goal of this chapter is to show that for a finite dimensional vector space, all bases have the same number of vectors. (This is also true for infinite dimensional vector spaces, but the proof requires more sophisticated set theory than we wish to use here.) The number of elements in a basis for a vector space will then become an *invariant* of the space; that is, the number of elements in a basis is *independent* of the particular basis determined. The number of basis vectors consequently provides us with a good measure of the 'size' of the space. It is exactly this measure of size that we will need when we return to the theory of field extensions.

But first, we make a simple observation about linear combinations: If $\mathbf{v}$ is a linear combination of $\mathbf{v}_1, \mathbf{v}_2, \ldots, \mathbf{v}_n$, and each $\mathbf{v}_i$ is a linear combination of $\mathbf{w}_1, \mathbf{w}_2, \ldots, \mathbf{w}_m$, then $\mathbf{v}$ is a linear combination of $\mathbf{w}_1, \mathbf{w}_2, \ldots, \mathbf{w}_m$. To show this, simply write $\mathbf{v}$ as a linear combination of the $\mathbf{v}_i$ and substitute for each $\mathbf{v}_i$ the appropriate linear combination of the $\mathbf{w}_j$. (This is Exercise 15.)

---

## 40.3   Finding a Basis

Now suppose $V$ is a finite dimensional vector space spanned by $\mathcal{V} = \{\mathbf{v}_1, \mathbf{v}_2, \ldots, \mathbf{v}_n\}$. Then we will argue below that we can find a subset of $\mathcal{V}$ that is a basis for $V$. That is, *any* finite spanning set for a vector space contains a linearly independent spanning set, which is then necessarily a basis.

We build this set in the following way. We may as well assume that $V$ does not have zero dimension and none of the $\mathbf{v}_i$ is the zero vector. First, we start with $\mathbf{v}_1$. We add each $\mathbf{v}_i$ in succession until we find a $\mathbf{v}_i$ that is a linear combination of the previous $\mathbf{v}_j$, $j < i$. We discard this $\mathbf{v}_i$. We continue in this manner, adding or discarding vectors until we've run through the vectors in $\mathcal{V}$. Call the set of vectors that remains $\mathcal{B}$. This set also spans $V$ because we have removed only those vectors that can be expressed as linear combinations of the vectors in $\mathcal{V}$ that come before it, and so, by Exercise 15, $\mathcal{B}$ also spans $V$. It remains to show that $\mathcal{B}$ is linearly independent.

By way of contradiction, suppose $\mathcal{B} = \{\mathbf{v}_{i_1}, \mathbf{v}_{i_2}, \ldots, \mathbf{v}_{i_k}\}$ is linearly dependent where $i_1, i_2, \ldots, i_k$ are indices of the vectors in $\mathcal{V}$ with $i_1 < i_2 < \cdots < i_k$. Then

$$0 = a_1 \mathbf{v}_{i_1} + a_2 \mathbf{v}_{i_2} + \cdots + a_k \mathbf{v}_{i_k},$$

where not all the $a_k$ are zero. Pick the largest $j$ with $a_j \neq 0$. Then, solving the above for $\mathbf{v}_{i_j}$, we can express $\mathbf{v}_{i_j}$ as a linear combination of vectors in $\mathcal{V}$ with smaller indices, which is a contradiction. Therefore, $\mathcal{B} \subseteq \mathcal{V}$ is linearly independent, and thus a basis for $\mathcal{V}$. We summarize this in the following theorem.

**Theorem 40.3** *Every finite spanning set of a vector space contains a subset that is a basis for the vector space.*

We demonstrate this technique by finding a subset of

$$\{(1,0,0), (2,3,0), (0,1,0), (-1,0,-1), (0,0,1)\}$$

that is a basis for $\mathbb{R}^3$. We have already shown in Examples 40.4 and 40.8 that this set is linearly dependent and spans $\mathbb{R}^3$.

We start with the set $\{(1,0,0)\}$. Now we add the vector $(2,3,0)$, if $(2,3,0)$ is not a linear combination of $\{(1,0,0)\}$. Because this is clearly not the case (a linear combination of a single vector is a scalar multiple of that vector), we add it to our set, which is now $\{(1,0,0),(2,3,0)\}$.

The next vector is $(0,1,0)$. To see if $(0,1,0)$ is a linear combination of $(1,0,0)$ and $(2,3,0)$, we attempt to solve

$$(0,1,0) = a(1,0,0) + b(2,3,0),$$

and find that there is a solution of $a = -2/3$ and $b = 1/3$. Thus, $(0,1,0)$ is discarded and our set remains $\{(1,0,0),(2,3,0)\}$.

We next see if the vector $(-1,0,-1)$ is a linear combination of $(1,0,0)$ and $(2,3,0)$ by attempting to solve

$$(-1,0,-1) = a(1,0,0) + b(2,3,0).$$

But we easily see that there is no solution to this equation. (There is no way to make the third coordinate of the right-hand side non-zero.) Hence, we add this vector, making our set

$$\{(1,0,0),(2,3,0),(-1,0,-1)\}.$$

Finally, we consider the last vector and try to solve

$$(0,0,1) = a(1,0,0) + b(2,3,0) + c(-1,0,-1).$$

We find that there is a solution of $a = c = -1$ and $b = 0$. Thus, $(0,0,1)$ is discarded, leaving the set $\{(1,0,0),(2,3,0),(-1,0,-1)\}$ as a basis for $\mathbb{R}^3$. Note that a reordering of the vectors in our original set might produce a different basis for $\mathbb{R}^3$. You will see this explicitly in Exercise 14.

---

## 40.4  Dimension of a Vector Space

We are now ready to prove that any two bases for a finite dimensional vector space have the same number of vectors.

**Theorem 40.4** *Suppose that $V$ is a finite dimensional vector space with basis $\mathcal{W} = \{\mathbf{w}_1, \mathbf{w}_2, \ldots, \mathbf{w}_m\}$ and $\mathcal{V} = \{\mathbf{v}_1, \mathbf{v}_2, \ldots, \mathbf{v}_n, \ldots\}$ is another linearly independent spanning set. Then $\mathcal{V}$ has no more than $m$ elements. Thus every basis for $V$ has exactly $m$ elements.*

**Proof:**  We start by considering the spanning set

$$\{\mathbf{v}_1, \mathbf{w}_1, \mathbf{w}_2, \ldots, \mathbf{w}_m\}.$$

This set is linearly dependent, because $\mathbf{v}_1$ is a linear combination of the $\mathbf{w}_i$. So, it contains a subset that is a basis. (Of course, $\mathcal{W}$ is such a subset, but we're going to force $\mathbf{v}_1$ into a basis.) We proceed to find this subset using the method described above. Here, we first include $\mathbf{v}_1$ and then proceed through the $\mathbf{w}_i$ and will discard at least one of the $\mathbf{w}_i$. We claim that only one is discarded. Suppose that $\mathbf{w}_i$ and $\mathbf{w}_j$ are both discarded, where $i < j$. Then,

$$\mathbf{w}_i = a_0\mathbf{v}_1 + a_1\mathbf{w}_1 + \cdots + a_{i-1}\mathbf{w}_{i-1}.$$

Now $a_0 \neq 0$, else we've expressed $\mathbf{w}_i$ as a linear combination of the vectors $\mathbf{w}_1, \mathbf{w}_2, \ldots, \mathbf{w}_{i-1}$, contradicting the linear independence of $\mathcal{W}$. So we can write

$$\mathbf{v}_1 = -\frac{a_1}{a_0}\mathbf{w}_1 - \frac{a_2}{a_0}\mathbf{w}_2 - \cdots - \frac{a_{i-1}}{a_0}\mathbf{w}_{i-1} + \frac{1}{a_0}\mathbf{w}_i. \tag{40.1}$$

Likewise, we can write $\mathbf{w}_j$ as

$$\mathbf{w}_j = b_0\mathbf{v}_1 + b_1\mathbf{w}_1 + \cdots + b_{j-1}\mathbf{w}_{j-1}. \tag{40.2}$$

But if we substitute expression (40.1) for $\mathbf{v}_1$ into equation (40.2), we will have expressed $\mathbf{w}_j$ as a linear combination of the other vectors in $\mathcal{W}$, again contradicting the linear independence of $\mathcal{W}$.

Thus, the linearly independent subset of

$$\{\mathbf{v}_1, \mathbf{w}_1, \mathbf{w}_2, \ldots, \mathbf{w}_m\}$$

has precisely one fewer vector in it, and that discarded vector was one of the $\mathbf{w}_i$. By renumbering, if necessary, let's suppose that the discarded vector was $\mathbf{w}_1$. So, the new basis we obtained is $\{\mathbf{v}_1, \mathbf{w}_2, \mathbf{w}_3, \ldots, \mathbf{w}_m\}$.

Now add $\mathbf{v}_2$ to this basis so that our set is now

$$\{\mathbf{v}_1, \mathbf{v}_2, \mathbf{w}_2, \ldots, \mathbf{w}_m\}.$$

Again, this is a spanning set that is linearly dependent and so contains a linearly independent subset which we find by the now familiar method. Note that neither $\mathbf{v}_1$ nor $\mathbf{v}_2$ will be discarded in the process, because $\mathbf{v}_1, \mathbf{v}_2$ is a linearly independent set. So, the discarded vector will be one of the $\mathbf{w}_i$. (There will be only one discarded by an argument similar to the one above.) Again, by renumbering if necessary, we assume that $\mathbf{w}_2$ was the one discarded, leaving the set $\{\mathbf{v}_1, \mathbf{v}_2, \mathbf{w}_3, \mathbf{w}_4, \ldots, \mathbf{w}_m\}$ as the basis.

We repeat this process, successively adding $\mathbf{v}_3, \mathbf{v}_4, \ldots,$ and $\mathbf{v}_n$ and each time discarding one of the $\mathbf{w}_i$ to get yet another basis for $V$. We will never run out of the $\mathbf{w}_i$'s before exhausting the $\mathbf{v}_j$'s, because then we would have a basis for $V$ consisting of a proper subset of $\mathcal{V}$. (This can't be, because by the Quick Exercise in Section 40.2 a basis is a *minimal* spanning set.) This means that $\mathcal{V}$ must actually be a finite set, with no more than $m$ elements.

By interchanging the roles of $\mathcal{V}$ and $\mathcal{W}$, and repeating the above argument, we obtain that these two sets have the same (finite) number of elements. □

The **dimension** of a given vector space is the number of vectors in any basis (this definition for our purposes applies only for finite-dimensional vector spaces, although it can be extended to the infinite case).

### Example 40.14

The dimension of the vector space $\mathbb{R}^3$ over $\mathbb{R}$ is 3.

### Example 40.15

The vector space $\mathbb{C}$ over $\mathbb{R}$ has dimension 2, and the vector space $\mathbb{Q}(\sqrt{2})$ over $\mathbb{Q}$ has dimension 2.

### Example 40.16

Consider the vector space

$$V = \{f \in \mathbb{Z}_3[x] : \deg(f) \le 2\}$$

over $\mathbb{Z}_3$, which we considered in Exercise 39.c. This vector space has dimension 3, because $\{1, x, x^2\}$ is a basis.

▷ **Quick Exercise.** Check this. ◁

We can easily obtain two important corollaries from Theorem 40.4:

**Corollary 40.5** *Suppose that $V$ is a finite dimensional vector space, and $\mathcal{U}$ is a linearly independent subset of $V$. Then $\mathcal{U}$ has finitely many elements, and there is a basis $\mathcal{W}$ for $V$ with $\mathcal{U} \subseteq \mathcal{W}$.*

**Proof:** Choose a basis $\mathcal{V} = \{\mathbf{v}_1, \mathbf{v}_2, \ldots, \mathbf{v}_n\}$ for $V$ with $n$ elements. Now if $\mathcal{U}$ is already a spanning set, then it is a basis, and by the previous theorem it has exactly $n$ elements.

If $\mathcal{U}$ is not a spanning set, we run through the vectors of $\mathcal{V}$ one at a time as follows, to build the basis $\mathcal{W}$ containing $\mathcal{U}$: Begin by letting $\mathcal{W} = \mathcal{U}$. For step $i$, check whether $\mathbf{v}_i$ is spanned by $\mathcal{W}$. If it is, we add nothing to $\mathcal{W}$. If, however, $\mathbf{v}_i$ is not in the span of $\mathcal{W}$, we add it to $\mathcal{W}$; the set $\mathcal{W}$ remains linearly independent. After $n$ steps, we will have augmented $\mathcal{U}$, building a larger set $\mathcal{W}$ that remains linearly independent but now must span $V$ (because $\mathcal{V}$ does). But the previous theorem asserts that this set is finite, with exactly $n$ elements. Thus the original set $\mathcal{U}$ must have been finite, and it is indeed a subset of a basis for $V$. $\square$

**Corollary 40.6** *Every set in an $n$-dimensional vector space with more than $n$ elements is linearly dependent.*

**Proof:** Any linearly independent set can by the previous corollary be augmented to obtain a basis for the vector space. But any basis for the vector space has no more than $n$ elements. $\square$

---

# Chapter Summary

In this chapter we defined *linear combination, spanning set, linear independence, linear dependence,* and *basis*.

We showed that any spanning set contains a subset that is a basis and that all bases for a given finite dimensional vector space have the same number of vectors. The number of vectors in a basis for a vector space is the *dimension* of the vector space.

---

# Warm-up Exercises

a. Give examples of the following, or explain why such an example cannot exist. Be sure to specify explicitly the vector space, the field of scalars, and the vectors required:

   (a) A spanning set that isn't a basis.

   (b) A basis that isn't a spanning set.

   (c) Two distinct bases for the same vector space.

   (d) Two distinct bases for the same vector space, one of which is a proper subset of the other.

   (e) A set of three linearly dependent vectors, any two of which are independent.

   (f) A set of four linearly dependent vectors, any three of which are independent.

b. Why could the zero vector never belong to a basis (for a vector space with more than one element)?

c. What is the dimension of $\mathbb{C}$ as a vector space over $\mathbb{R}$? What is the dimension of $\mathbb{C}$ as a vector space over $\mathbb{C}$?

## Exercises

1. Prove that $\sqrt{2}$ and $\sqrt{3}$ are linearly independent elements of the vector space $\mathbb{R}$, over $\mathbb{Q}$.

2. (Here, we extend Exercise 1.) Prove that $\{1, \sqrt{2}, \sqrt{3}\}$ is an independent subset of the vector space $\mathbb{R}$ over $\mathbb{Q}$. Is it a basis for $\mathbb{R}$?

3. Consider the set $V$, consisting of all polynomials of degree 0 or 1, with coefficients from $\mathbb{C}$. Prove that this is both a vector space over $\mathbb{C}$, and over $\mathbb{R}$. Find a basis in each case. Does its dimension stay the same, when we change the scalar field?

4. Find a basis for the vector space given in Exercise 39.7.

5. Find a basis for the vector space given in Exercise 39.8.

6. Give a basis for the vector space given in Exercise 39.10.

7. Show that $\mathbb{Q}[x]$ over $\mathbb{Q}$ is not finite dimensional by showing that no finite subset of $\mathbb{Q}[x]$ is a spanning set.

8. Prove Theorem 40.2.

9. (a) Define a **subspace** of a vector space.

   (b) Let $S$ be a subset of the vector space $V$. Show that the set of linear combinations of vectors from $S$ is a subspace of $V$. (Note that this subspace is all of $V$ if and only if $S$ spans $V$.)

10. Determine which of the following are subspaces of the vector space $M_2(\mathbb{R})$ discussed in Exercise 39.7.

    (a) Matrices with determinant 1.

    (b) Matrices with determinant 0.

    (c) The diagonal matrices.

    (d) Matrices with integer entries.

    (e) Matrices of the form
    $$\begin{pmatrix} a & a+b \\ b & 0 \end{pmatrix}.$$

    (f) Matrices with first row, first column entry of 0.

11. Determine which of the following are subspaces of the vector space of all real-valued functions discussed in Exercise 39.11.

    (a) All differentiable functions.

    (b) All functions $f$ with $f(2) = 0$.

    (c) All linear functions. (That is, functions of the form $f(x) = a + bx$.)

    (d) All polynomial functions.

12. Check Example 40.12: that is, prove that $\{1 + i, \sqrt{2} + 4i\}$ is a spanning set for $\mathbb{C}$ over $\mathbb{R}$.

13. Find a subset of the set

$$\{(1,1,0),(2,3,-1),(0,1,-1),(1,4,2),(0,0,3)\}$$

that is a basis for $\mathbb{R}^3$ using the technique described in this chapter.

14. Find a subset of the set

$$\{(-1,0,-1),(0,0,1),(1,0,0),(2,3,0),(0,1,0)\}$$

that is a basis for $\mathbb{R}^3$ using the technique in this chapter; this is the same set we used in Section 40.3 to demonstrate this technique. Take the vectors in the order given here, to see explicitly that we obtain a different subset than in the text.

15. Let $V$ be a vector space. Show that if $\mathbf{v}$ is a linear combination of the vectors $\mathbf{v}_1,\mathbf{v}_2,\ldots,\mathbf{v}_n \in V$ and each $\mathbf{v}_i$ is in turn a linear combination of $\mathbf{w}_1,\mathbf{w}_2,\ldots,\mathbf{w}_m$, then $\mathbf{v}$ is a linear combination of $\mathbf{w}_1,\mathbf{w}_2,\ldots,\mathbf{w}_m$.

16. Prove that a basis for a vector space $V$ is a *maximal independent set*: That is, it is a set $\mathcal{B} \subseteq V$ of independent vectors, so that $\mathcal{B} \cup \{w\}$ is a dependent set, whenever $w \in V\backslash\mathcal{B}$.

17. Prove that a maximal independent set of vectors of a vector space is necessarily a basis (that is, prove the converse of Exercise 16).

18. Use the technique in the proof of Corollary 40.5 to build a basis for $\mathbb{R}^4$ that contains the linearly independent set

$$\{(1,0,-1,0),\ (-2,1,2,0)\}.$$

# Chapter 41

## Field Extensions and Kronecker's Theorem

Consider the polynomial $x^2 + 1 \in \mathbb{R}[x]$. This polynomial clearly has no roots in $\mathbb{R}$, but we know that there exists a larger field—namely, $\mathbb{C}$—in which this polynomial does have roots. In this chapter we develop a general method for constructing bigger fields, with the aim of being able to further factor polynomials.

## 41.1   Field Extensions

A field $E$ is an **extension field** of a field $F$ if $E \supseteq F$. That is, $E$ is an extension field of $F$ exactly if $F$ is a subfield of $E$; in this situation, we call the field $F$ the **base field**. For example, $\mathbb{R}$ is an extension field of $\mathbb{Q}$, $\mathbb{C}$ is an extension field of both $\mathbb{R}$ and $\mathbb{Q}$, and $\mathbb{Q}(\sqrt{2})$ is an extension field of $\mathbb{Q}$.

If $E$ is an extension field of a field $F$ and $\alpha \in E$ is a root of a polynomial in $F[x]$, we say $\alpha$ is **algebraic over** $F$. Otherwise, $\alpha$ is **transcendental over** $F$.

### Example 41.1

Note first that if $E$ is an extension field of a field $F$ and $\alpha \in F$, then trivially $\alpha$ is algebraic over $F$ because we can easily find a polynomial $p \in F[x]$ such that $p(\alpha) = 0$.

▷ **Quick Exercise.**   Find such a polynomial. ◁

### Example 41.2

The complex field $\mathbb{C}$ is an extension field of $\mathbb{Q}$, and because $\sqrt{2}$ is a root of $x^2 - 2 \in \mathbb{Q}[x]$, $\sqrt{2}$ is algebraic over $\mathbb{Q}$. Also, $i$ is algebraic over $\mathbb{Q}$ because $i$ is a root of $x^2 + 1 \in \mathbb{Q}[x]$.

### Example 41.3

The real field $\mathbb{R}$ is an extension field of $\mathbb{Q}$. As we mentioned in Chapter 38 (see Theorem 38.5), the numbers $e$ and $\pi$ are transcendental over $\mathbb{Q}$; that is, neither $e$ nor $\pi$ are roots of any polynomial in $\mathbb{Q}[x]$. These famous and difficult results are due to Hermite and Lindemann, respectively.

## 41.2   Kronecker's Theorem

We now prove an extremely important theorem due to Leopold Kronecker, an eminent German mathematician of the latter part of the 19th century. Given a polynomial $f$ in

$F[x]$, is there a field extension of the field $F$ that contains a root of $f$? Kronecker's Theorem provides us with an affirmative answer to this question.

Note that this theorem will apply to any polynomial over any field. Furthermore, while the description of the extension field will seem somewhat abstract—we will find that it is a ring of cosets of $F[x]$—later theorems will allow us to explicitly describe some of these extension fields in a way that is more palatable.

**Theorem 41.1    Kronecker's Theorem**    *Let $F$ be a field and $f$ a polynomial in $F[x]$ of degree at least 1. Then there exists an extension field $E$ of $F$ and an $\alpha \in E$ such that $f(\alpha) = 0$.*

**Proof:**    If $f$ has a root in the field $F$, then $F$ itself is the required extension field. So we may as well assume that $f$ has no roots in $F$. From Chapter 10 we know that $f$ can be factored into irreducible polynomials in $F[x]$; note that each of these polynomials has degree at least two. Let $p$ be one of these irreducible  factors. We will show that there is an extension field $E$ of $F$ and an $\alpha \in E$ with $p(\alpha) = 0$.

From Theorem 15.1 we know that because $p$ is irreducible in $F[x]$, the principal ideal $\langle p \rangle$ is maximal in the principal ideal domain $F[x]$. Hence, from Theorem 15.2, $F[x]/\langle p \rangle$ is a field. This is our field $E$.

We now need to show two things. First, we must show that $E$ can be viewed as an extension field of $F$; that is, we will obtain an injection from $F$ into $E$. Second, we need to find an element of $E$ that is a root of $p$.

We show that there is an injection from $F$ into $F[x]/\langle p \rangle$ by considering the function $\psi : F \to F[x]/\langle p \rangle$ defined by $\psi(a) = \langle p \rangle + a$, where $a \in F$. That is, an element in $F$ is mapped to its coset in $F[x]/\langle p \rangle$. (Note this function makes a subtle shift in viewing $a$. First, we think of $a$ as an element of $F$ while the image of $a$ treats $a$ as a polynomial in $F[x]$.) It is evident that $\psi$ is a ring homomorphism, because of the way the ring operations are defined on the ring of cosets.

▷ **Quick Exercise.**    Check that $\psi$ is a ring homomorphism. ◁

We now claim that $\psi$ is one-to-one; to show this, we will verify that its kernel is $\{0\}$. For that purpose, suppose that

$$\psi(a) = \langle p \rangle + a = \langle p \rangle + 0.$$

Then $a \in \langle p \rangle$. But the ideal $\langle p \rangle$ is precisely the set of multiples of $p$ and because $p$ has degree at least 2, so do all the non-zero elements of $\langle p \rangle$. But $a$ (viewed as a polynomial in $F[x]$, of course) has degree 0, and so $a$ must be the zero polynomial. That is, the kernel of $\psi$ is trivial, and so $\psi$ is injective. We consequently may as well assume that $F$ is a subfield of the field $F[x]/\langle p \rangle$.

Finally, we must find an element of $F[x]/\langle p \rangle$ that is a root of our polynomial $p$. Consider the element $\alpha = \langle p \rangle + x$. Suppose $p = a_0 + a_1 x + \cdots + a_n x^n$. Then

$$p(\alpha) = a_0 + a_1(\langle p \rangle + x) + \cdots + a_n(\langle p \rangle + x)^n \in F[x]/\langle p \rangle.$$

But we do arithmetic in $F[x]/\langle p \rangle$ by choosing any coset representative we wish. If we pick $x$ as the coset representative of $\langle p \rangle + x$, then

$$p(\alpha) = \langle p \rangle + a_0 + a_1 x + \cdots + a_n x^n = \langle p \rangle + p = \langle p \rangle + 0.$$

But $\langle p \rangle + 0$ is the zero element of the field $F[x]/\langle p \rangle$, and so we have found our desired element of $F[x]/\langle p \rangle$.    □

Let's perform the construction of the theorem in two specific cases.

**Example 41.4**

Consider the polynomial $f = x^3 - 5 \in \mathbb{Q}[x]$; it has no roots in $\mathbb{Q}$.

▷ **Quick Exercise.** Verify that $f = x^3 - 5$ has no roots in $\mathbb{Q}$. ◁

Thus, $f$ is irreducible in $\mathbb{Q}[x]$, and so $\langle f \rangle$ is a maximal ideal. This makes $E = \mathbb{Q}[x]/\langle f \rangle$ a field. Furthermore, we may identify $\mathbb{Q}$ with the subfield

$$\{\langle x^3 - 5\rangle + q : q \in \mathbb{Q}\}$$

of $E$. The element $\langle f \rangle + x$ of $E$ is then the required root for the polynomial $f$, because

$$f(\langle f \rangle + x) = (\langle f \rangle + x)^3 - 5 = \langle f \rangle + x^3 - 5 = \langle f \rangle + f = \langle f \rangle + 0.$$

**Example 41.5**

Consider the polynomial $p = x^2 + x + 1$ in $\mathbb{Z}_2[x]$. It has no roots in $\mathbb{Z}_2$, because the only possibilities are 0 and 1, and neither works. Thus, $p$ is irreducible in $\mathbb{Z}_2[x]$, and so $\langle p \rangle$ is a maximal ideal. This makes $E = \mathbb{Z}_2[x]/\langle p \rangle$ a field. Furthermore, we may identify $\mathbb{Z}_2$ with the subfield

$$\{\langle p \rangle + 0, \langle p \rangle + 1\}$$

of $E$. The element $\langle p \rangle + x$ of $E$ is then the required root for the polynomial $p$.

In the next chapter, we prove a theorem that gives a much more explicit description of $F[x]/\langle p \rangle$. For example, $x^2 + 1$ is irreducible in $\mathbb{R}[x]$, and so $x^2 + 1$ has a root in the field $\mathbb{R}[x]/\langle x^2 + 1 \rangle$. But we also know that

$$\mathbb{R}(i) = \{a + bi : a, b \in \mathbb{R}\} = \mathbb{C},$$

is a field in which $i$ is a root of $x^2 + 1$. We will see that these two fields are in fact isomorphic.

---

## 41.3   The Characteristic of a Field

It turns out that every field has a unique smallest subfield, which we call the *prime* subfield. This fact provides us with many nice examples of field extensions.

To show that this is true, suppose that $K$ is an arbitrary field, with multiplicative identity 1. Consider what happens when we repeatedly add 1 to itself: We get the elements

$$1, \quad 1 + 1 = 2, \quad 1 + 2 = 3 \quad \cdots.$$

In the case of a field like $\mathbb{R}$, this process goes on forever, and we obtain in this way all the positive integers. But in the case of a field like $\mathbb{Z}_{13}$, we obtain 0, once we have added 1 to itself thirteen times. The **characteristic** of the field $K$ is the least positive integer $n$ so that $n \cdot 1 = 0$ (if such exists). If no such $n$ exists, we set the characteristic to 0. (If you did Exercise 8.14, you have already encountered this concept, in a more general context.)

**Example 41.6**

The characteristic of such fields as $\mathbb{Q}$, $\mathbb{R}$, $\mathbb{C}$, and $\mathbb{Q}(\sqrt{2})$ is zero, while the characteristic of the field $\mathbb{Z}_p$ is $p$.

**Example 41.7**

Consider now the field $E$ constructed in Example 41.5. What is the characteristic of this field? The multiplicative identity in $E$ is the coset $\langle p \rangle + 1$, and clearly

$$(\langle p \rangle + 1) + (\langle p \rangle + 1) = \langle p \rangle + 0.$$

This means that the characteristic of $E$ is 2.

Although we defined the characteristic in terms of the number of times 1 could be added to itself, notice that the properties of arithmetic in a field means that we can in fact define characteristic in terms of *any* non-zero element. For if $n \cdot 1 = 0$, then

$$n \cdot r = (r + r + \cdots + r) = r(1 + 1 + \cdots + 1) = r(n \cdot 1),$$

and so $n \cdot 1 = 0$ if and only if $n \cdot r = 0$.

▷ **Quick Exercise.** Why is the fact that a field has no zero divisors relevant to this remark? ◁

Notice that the only values for the characteristic we have obtained in our examples are 0 and positive prime integers. It is easy to prove that these are the only possible cases:

**Theorem 41.2** *The characteristic of any field is either 0 or a positive prime integer.*

**Proof:** Suppose that a field $F$ has characteristic $n > 0$, and that $n$ has a non-trivial factorization $n = rs$. Then

$$(r1)(s1) = (1 + \cdots + 1)(1 + \cdots + 1) = (1 + 1 + \cdots + 1) = (rs)1 = 0$$

(where we have added 1 to itself $r$ times, $s$ times, and then $rs$ times, in the previous computation). But because a field has no zero divisors, this means that either $r1$ or $s1$ is zero, and this contradicts the minimality of $n$.                                              □

Suppose now that $F$ is a field with characteristic zero. We define a function $\iota : \mathbb{Q} \to F$ as follows: Given $a/b \in \mathbb{Q}$ (where $a$ and $b$ are integers, with $b \neq 0$), let $\iota(a/b) = (a \cdot 1)/(b \cdot 1)$. Notice first that the quotient makes sense, because in a field of characteristic zero, $b \cdot 1 \neq 0$. It is now easy to prove that this function preserves both addition and multiplication, and so is a ring homomorphism.

▷ **Quick Exercise.** Check these two facts. ◁

But the kernel of $\iota$ is evidently $\{0\}$, because if $a \in \mathbb{Z}$ and $a \neq 0$, then $a \cdot 1 \neq 0$ (because the field has characteristic zero).

Thus, any field of characteristic zero contains (an isomorphic copy of) the field $\mathbb{Q}$. And so, such a field is necessarily a vector space over the rational numbers. This copy of $\mathbb{Q}$ is the unique smallest subfield of $F$. (You will prove this fact in Exercise 7a.) This copy of $\mathbb{Q}$ is called the **prime subfield** of $F$.

**Example 41.8**

In Example 41.4, we identified the prime subfield of the field $E$ as the set $\{\langle f \rangle + q : q \in \mathbb{Q}\}$.

What about fields with non-zero characteristic? If a field $F$ has characteristic $p$, where $p$ is a positive prime integer, then we define $\iota : \mathbb{Z}_p \to F$ by setting $\iota([1]) = 1$, and extending this function additively. That is, we set $\iota([2]) = 1 + 1$, and so forth. This function is also an injective ring homomorphism (see Exercise 6), and so any field with characteristic $p$ contains (an isomorphic copy of) the field $\mathbb{Z}_p$. This copy of $\mathbb{Z}_p$ is the unique smallest subfield of $F$. (Again: You will prove this fact in Exercise 7b.) This copy of $\mathbb{Z}_p$ is called the **prime subfield** of $F$.

**Example 41.9**

In Example 41.5, we identified the prime subfield of the field $E$ as the set $\{\langle p \rangle + 0, \langle p \rangle + 1\}$.

▷ **Quick Exercise.** What can you say about a field which has dimension 1 as a vector space over its prime field? ◁

We can view the prime subfield of a field as the result of applying all the field operations to the multiplicative identity 1. If the field has characteristic zero, this gives (an isomorphic copy of) $\mathbb{Q}$; if the field has characteristic $p$, this gives (an isomorphic copy of) $\mathbb{Z}_p$. Note that if we know the characteristic of a field $F$ and $E \supseteq F$ is any extension field, then $E$ will automatically have the same characteristic as $F$, because its prime subfield will be the same.

Alternatively, we could do the following: Given a field $K$, consider the set of all subfields $E \subseteq K$; this set is non-empty, because $K$ itself is such a subfield. Now consider the intersection of all such subfields. The result is certainly non-empty, because it necessarily contains the element 1. It is also a field (as you prove in Exercise 8). It is necessarily the smallest subfield of $K$, which is then the prime subfield.

---

## Chapter Summary

In this chapter we defined the phrase *E is an extension field of a field F*. In this case we also defined what it means for an element of $E$ to be *algebraic over F* and *transcendental over F*.

Next, we proved *Kronecker's Theorem* which says that for a field $F$, if $f \in F[x]$, then there exists an extension field of $F$ that contains a root of $f$.

Finally, we defined the *characteristic* of a field and showed that all fields have characteristic 0, or $p$, where $p$ is a positive prime integer.

---

## Warm-up Exercises

a. In each of the following cases, show that the given element is algebraic over the given base field, by finding an appropriate polynomial with the element as a root.

   (a) $\sqrt[3]{2}$ is algebraic over $\mathbb{Q}$.

   (b) $\sqrt[3]{2}$ is algebraic over $\mathbb{R}$.

   (c) $1 + \sqrt{2}$ is algebraic over $\mathbb{Q}$.

   (d) $\sqrt{1 + \sqrt{2}}$ is algebraic over $\mathbb{Q}$.

   (e) $1 + 2i$ is algebraic over $\mathbb{Q}$.

   (f) $\pi + i$ is algebraic over $\mathbb{R}$.

b. If $F$ is a field, is $F$ an extension field of $F$? If so, does $F$ have any transcendental elements over $F$?

c. What extension field do we get if we apply Kronecker's Theorem 41.1 to a degree 1 polynomial?

d. Recall the Fundamental Theorem of Algebra 10.1, which asserts that all non-constant polynomials in $\mathbb{C}[x]$ can be factored completely into linear factors. What does this say about finding elements outside $\mathbb{C}$ that are algebraic over $\mathbb{C}$?

e. Give examples of the following, or else explain why no such example exists:

    (a) An element in $\mathbb{C}$ transcendental over $\mathbb{R}$.

    (b) An element in $\mathbb{C}$ transcendental over $\mathbb{Q}$.

    (c) An element in $\mathbb{R}$, but not in $\mathbb{Q}(\sqrt{2})$, which is algebraic over $\mathbb{Q}(\sqrt{2})$.

f. For each of the following fields, identify its characteristic, and determine its prime subfield:

$$\mathbb{C}, \quad \mathbb{Q}, \quad \mathbb{Q}(\sqrt{2}), \quad \mathbb{Z}_{11}, \quad \mathbb{Q}[x]/\langle x^3 - 2 \rangle, \quad \mathbb{Z}_3[x]/\langle x^2 + x + 2 \rangle.$$

g. "Every polynomial of degree at least 1 has a root." Discuss the truth of this statement.

h. Why is $\cos 20°$ algebraic over $\mathbb{Q}$? *Hint:* Find the appropriate result in Chapter 38.

i. Give some examples of real numbers that are algebraic over $\mathbb{Q}$, but not constructible.

---

# Exercises

1. Let $F$ be a field and $f \in F[x]$ a polynomial of degree at least 1. Prove that there exists a field extension $E$ of $F$ in which $f$ can be factored into linear factors. Such a field extension is called a **splitting extension for** $f$.

2. In the proof of Kronecker's Theorem 41.1, we prove that the function $\psi : F \to F[x]/\langle p \rangle$ is an injective ring homomorphism. Show that $\psi$ is onto if and only if the irreducible polynomial $p$ is linear.

3. If $R$ is a commutative ring with unity, and $R$ is a subring of a field $F$, we can speak of the elements in $F$ that are **algebraic** over $R$, meaning simply those elements of the field which are roots of polynomials from $R[x]$.

    (a) Explain why all rational numbers are algebraic over $\mathbb{Z}$.

    (b) Prove that the set of elements of $\mathbb{R}$ that are algebraic over $\mathbb{Z}$ is the same as the set of elements of $\mathbb{R}$ which are algebraic over $\mathbb{Q}$.

4. Suppose that $\alpha \in \mathbb{C}$ is an algebraic number over $\mathbb{Q}$, and $r \in \mathbb{Q}$. Prove that $\alpha + r$ and $r\alpha$ are also algebraic over $\mathbb{Q}$. (Actually, this exercise is a special case of Exercise 43.10, where you will show that the set $\mathbb{A}$ of complex numbers algebraic over $\mathbb{Q}$ is a field, called the **field of algebraic numbers**. Why is our exercise a special case?)

5. Prove that $\sin 1°$ is algebraic over $\mathbb{Q}$.

6. Let $F$ be a field with characteristic $p$. Prove that the function $\iota : \mathbb{Z}_p \to F$ discussed in Section 41.3 is an injective ring homomorphism.

7. (a) For a field $F$ with characteristic zero, we defined in Section 41.3 an injective ring homomorphism $\iota : \mathbb{Q} \to F$. Prove that $\iota(\mathbb{Q})$ is the smallest subfield of $F$.

(b) For a field $F$ with characteristic $p$, we defined in Section 41.3 an injective ring homomorphism $\iota : \mathbb{Z}_p \to F$. (You completed the proof of this in Exercise 6.) Prove that $\iota(\mathbb{Z}_p)$ is the smallest subfield of $F$.

8. Let $K$ be a field, and consider the set of all subfields of $K$ (this set may well have infinitely many elements). Consider the intersection of all these subfields. Prove that the result is a field, by checking explicitly the field axioms.

9. Consider

$$f = x^4 - x^2 - 6 \in \mathbb{Q}[x].$$

Use Theorem 41.1 to construct an extension field $E$ of $\mathbb{Q}$ for which $f$ has a root in $E$. Specify the root explicitly as an element of the field $E$ you construct.

10. Consider

$$f = x^3 + 2x + 1 \in \mathbb{Z}_3[x].$$

Use Theorem 41.1 to construct an extension field $E$ of $\mathbb{Z}_3$ in which $f$ has a root. Now the root is a coset, while $E$ is a field. Thus the root must have a multiplicative inverse. Compute its multiplicative inverse, expressed as a coset.

# Chapter 42

## Algebraic Field Extensions

In this chapter we give a nicer description for certain field extensions. This will in particular allow us to exhibit some new finite fields. We will focus our attention on field extensions in which the additional elements are algebraic over the base field.

### 42.1   The Minimal Polynomial for an Element

As we've seen, $\mathbb{R}$ is an extension field of $\mathbb{Q}$ and $\sqrt{2} \in \mathbb{R}$ is algebraic over $\mathbb{Q}$ because $\sqrt{2}$ is a root of $x^2 - 2 \in \mathbb{Q}[x]$. Of course, $\sqrt{2}$ is a root of many other polynomials in $\mathbb{Q}[x]$; for instance, $3x^2 - 6$, $(x^2 - 2)(x + 1)$, and $(x^2 - 2)(x^5 + 2x - 3)$. But every polynomial listed is a multiple of the irreducible polynomial $x^2 - 2$. This is in fact true in general, as the next important theorem shows.

**Theorem 42.1** *If $E$ is an extension field of a field $F$ and $\alpha \in E$ is algebraic over $F$, then there exists an irreducible polynomial $p \in F[x]$ such that $p(\alpha) = 0$. Furthermore, if $f \in F[x]$ and $f(\alpha) = 0$, then $p$ divides $f$.*

**Proof:**   Because $\alpha$ is algebraic over $F$, there is at least one polynomial in $F[x]$ with $\alpha$ as a root. Let $p \in F[x]$ be a non-zero polynomial of minimal degree such that $p(\alpha) = 0$. We claim that $p$ is irreducible in $F[x]$. For suppose that $p = tq$, where $t$ and $q$ are polynomials in $F[x]$ with degree at least one. Then $0 = p(\alpha) = t(\alpha)q(\alpha)$, and so $\alpha$ would be a root of at least one of $t$ and $q$. Let's assume $t(\alpha) = 0$. But $p$ is of minimal degree among those polynomials with $\alpha$ as a root. Hence, $\deg(t) = \deg(p)$, and so $q$ is a constant polynomial, as required.

It remains to show that $p$ divides every polynomial $f \in F[x]$, where $f(\alpha) = 0$. By the Division Theorem we can write $f = pq + r$ where $q, r \in F[x]$ and $\deg(r) < \deg(p)$. But

$$0 = f(\alpha) = p(\alpha)q(\alpha) + r(\alpha) = 0q(\alpha) + r(\alpha) = r(\alpha).$$

Therefore, $\alpha$ is a root of $r$. But because $\deg(r) < \deg(p)$ and $p$ was of minimal degree among those polynomials with $\alpha$ as a root, $r$ must be the zero polynomial. Thus, $p$ divides $f$.   □

Note that the above proof shows that the irreducible polynomial in $F[x]$ with $\alpha$ as a root is unique up to a constant multiple. So, there exists a unique *monic* polynomial in $F[x]$ that is irreducible with $\alpha$ as a root. (Recall that a monic polynomial is one whose leading coefficient is 1.) We call this polynomial the **minimal polynomial of $\alpha$ over** $F$, and the degree of this polynomial the **degree of $\alpha$ over** $F$.

**Example 42.1**

   The minimal polynomial for $\sqrt{2}$ over $\mathbb{Q}$ is $x^2 - 2$. Thus, $\sqrt{2}$ is of degree 2 over $\mathbb{Q}$.

**Example 42.2**

The minimal polynomial for $i$ over $\mathbb{R}$ is $x^2 + 1$. Thus, $i$ is of degree 2 over $\mathbb{R}$. (The same can be said if $\mathbb{R}$ is replaced by $\mathbb{Q}$, because $x^2 + 1 \in \mathbb{Q}[x]$.)

**Example 42.3**

The minimal polynomial for $\sqrt[3]{2}$ over $\mathbb{Q}$ is $x^3 - 2$. Thus, $\sqrt[3]{2}$ is of degree 3 over $\mathbb{Q}$.

**Example 42.4**

We claim that $\sqrt{2 + \sqrt{2}}$ has minimal polynomial $x^4 - 4x^2 + 2$ over $\mathbb{Q}$ and so is of degree 4 over $\mathbb{Q}$.

▷ **Quick Exercise.** Check that $\sqrt{2 + \sqrt{2}}$ is indeed a root of $x^4 - 4x^2 + 2$. ◁

But $x^4 - 4x^2 + 2$ is clearly irreducible over $\mathbb{Z}[x]$, by Eisenstein's criterion 5.7, and so is irreducible over $\mathbb{Q}[x]$, by Gauss's Lemma 5.5.

---

## 42.2   Simple Extensions

Suppose $E$ is an extension field of the field $F$ and $\alpha \in E$. We wish to obtain the smallest subfield of $E$ that contains both $\alpha$ and all the elements of $F$. To do this, we consider the set of all such subfields of $E$. There is at least one such subfield—namely, $E$ itself; there may be infinitely many others. We now consider

$$F(\alpha) = \cap\{\text{fields } K : F \subseteq K \subseteq E, \alpha \in K\}.$$

This set clearly contains the elements of $F$ and also $\alpha$. But is it a subfield? To show that this set is closed under subtraction, choose $a, b \in F(\alpha)$. Then $a, b \in K$, for each of the subfields $K$ containing $F$ and $\alpha$. But then because $K$ is a subfield, $a - b \in K$, for every such $K$. Thus, $a - b$ must be an element of the intersection of all such $K$.

▷ **Quick Exercise.** The rest of the proof that $F(\alpha)$ is a subfield follows the same lines; complete this proof. Note that this argument is essentially the same as that for Exercise 41.8. ◁

Because $F(\alpha)$ is contained in all subfields of $E$ that contain both $F$ and $\alpha$, it is certainly the *smallest* such subfield. If $E = F(\alpha)$, we then say $E$ is a **simple extension** of $F$. If $\alpha$ is algebraic over $F$, we say that $F(\alpha)$ is an **algebraic simple extension** of $F$.

**Example 42.5**

One example of an algebraic simple extension we've seen before is $\mathbb{Q}(\sqrt{2}) = \{a + b\sqrt{2} : a, b \in \mathbb{Q}\}$. In Section 37.1 we argued directly that every element of this field belongs to *any* subfield of $\mathbb{C}$ that contains the rational numbers and $\sqrt{2}$; furthermore, the set of such elements forms a field. This was a more concrete way of seeing that this field is the smallest field extension of $\mathbb{Q}$ containing $\sqrt{2}$.

In the previous example we have an explicit description of the elements of a simple extension; of course, this explicit description depended on knowing the specific arithmetic properties of $\sqrt{2}$. Our goal now is to find such a description for the general algebraic simple extension $F(\alpha)$ of a field $F$. We get the answer in the next two theorems. Here, the minimal polynomial of $\alpha$ over $F$ will provide us with the arithmetic information we need about $\alpha$.

**Theorem 42.2** *Let $F$ be a subfield of a field $E$ and $\alpha$ an element of $E$ that is algebraic over $F$; let $p \in F[x]$ be the minimal polynomial for $\alpha$. Then the simple extension $F(\alpha)$ is isomorphic to the field $F[x]/\langle p \rangle$. Consequently, any two algebraic simple extensions of $F$ by a root of $p$ are isomorphic.*

**Proof:** Consider the evaluation homomorphism $\psi : F[x] \to E$ defined by $\psi(f) = f(\alpha)$. It is now evident that the image of $\psi$ is contained in $F(\alpha)$. From the proof of Kronecker's Theorem 41.1 it is clear that $p$ divides every polynomial in the kernel of $\psi$; it follows that the kernel of the homomorphism $\psi$ is exactly $\langle p \rangle$. The Fundamental Isomorphism Theorem for Rings 14.1 then tells us that $F[x]/\langle p \rangle$ is isomorphic to the image of $\psi$ in $F$. Because $p$ is irreducible, $F[x]/\langle p \rangle$ is a field (Theorems 15.1 and 15.2). The isomorphic image of this field in $E$ contains both $F$ and $\alpha$. Hence, the image of $\psi$ is exactly $F(\alpha)$, the smallest subfield of $E$ containing $F$ and $\alpha$.

Because this is true regardless of the field $E$, this means that any two algebraic simple extensions of this form are isomorphic to $F[x]/\langle p \rangle$, and hence to each other. $\qquad\square$

Apparently the algebraic simple extension $F(\alpha)$ depends heavily on the nature of the larger field $E$ in which we are doing our computations. However, this is not actually the case: The theorem (and its proof) asserts that $F(\alpha)$ is entirely determined by $F$, and by the minimal polynomial for $\alpha$.

The previous theorem now allows us to describe the elements of an algebraic simple extension quite explicitly.

**Theorem 42.3** *Consider the simple extension $F(\alpha)$ of $F$ where $\alpha$ is algebraic over $F$. Let $n \geq 1$ be the degree of $\alpha$ over $F$. Then every element $\beta$ of $F(\alpha)$ can be uniquely written as*

$$\beta = b_0 + b_1\alpha + b_2\alpha^2 + \cdots + b_{n-1}\alpha^{n-1},$$

*where $b_i \in F$.*

**Proof:** Consider the evaluation homomorphism $\psi_\alpha$ from $F[x]$ to $F(\alpha)$: If $f \in F[x]$, then $\psi_\alpha(f) = f(\alpha)$. (The previous theorem shows us that this is onto.) So, if $f = a_0 + a_1x + \cdots + a_mx^m$, then $\psi_\alpha(f) = a_0 + a_1\alpha + a_2\alpha^2 + \cdots + a_m\alpha^m$, an element of $F(\alpha)$. Let $p = c_0 + c_1x + c_2x^2 + \cdots + x^n$ be the minimal polynomial of $\alpha$ over $F$. (Recall that the minimal polynomial is monic.) Because $p(\alpha) = 0$, we have $\alpha^n = -c_0 - c_1\alpha - c_2\alpha^2 - \cdots - c_{n-1}\alpha^{n-1}$. Note that we can use this equation to write every $\alpha^m$ for $m \geq n$ in terms of powers of $\alpha$ less than $n$. For example,

$$\begin{aligned} \alpha^{n+1} &= \alpha\alpha^n = -c_0\alpha - c_1\alpha^2 - \cdots - c_{n-1}\alpha^n \\ &= -c_0\alpha - c_1\alpha^2 - \cdots - c_{n-1}(-c_0 - c_1\alpha - \cdots - c_{n-1}\alpha^{n-1}). \end{aligned}$$

Higher powers of $\alpha$ are whittled down in this manner. So, every element $\beta$ of $F(\alpha)$ can be written

$$\beta = b_0 + b_1\alpha + b_2\alpha^2 + \cdots + b_{n-1}\alpha^{n-1}.$$

We need only show that this expression is unique. So, suppose

$$\beta = b_0 + b_1\alpha + \cdots + b_{n-1}\alpha^{n-1} = d_0 + d_1\alpha + \cdots + d_{n-1}\alpha^{n-1}$$

for $b_i, d_i \in F$. Consider the polynomial

$$f = (b_0 - d_0) + (b_1 - d_1)x + \cdots + (b_{n-1} - d_{n-1})x^{n-1}.$$

Then $f \in F[x]$ and $f(\alpha) = 0$. Furthermore, the degree of $f$ is less than the degree of $p$. But we know that $p$ is of minimal degree among those polynomials with $\alpha$ as a root. Thus, $f$ must be the zero polynomial. In other words, $b_i = d_i$ for $i = 0, 1, \ldots, n - 1$, as desired.    □

## Example 42.6

Consider the real number $\alpha = \sqrt{2} + \sqrt[3]{2}$. It turns out that this element has degree 6 over $\mathbb{Q}$, with minimal polynomial $x^6 - 6x^4 - 4x^3 + 12x^2 - 24x - 4$. (See Exercise 3 and also Exercise 38.2.) This means that

$$\mathbb{Q}(\alpha) = \{b_0 + b_1\alpha + b_2\alpha^2 + b_3\alpha^3 + b_4\alpha^4 + b_5\alpha^5 \in \mathbb{R} : b_i \in \mathbb{Q}\}.$$

That is, every element of $\mathbb{Q}(\alpha)$ can be expressed as a fifth-degree (or smaller) polynomial from $\mathbb{Q}[x]$, evaluated at $\alpha$.

But consider the real number $\frac{1}{\alpha}$. This is obviously an element of the field $\mathbb{Q}(\alpha)$, and so we must be able to express it in terms of a fifth degree polynomial evaluated at $\alpha$. But we know that $\alpha^6 - 6\alpha^4 - 4\alpha^3 + 12\alpha^2 - 24\alpha = 4$. If we divide by $\alpha$, we then have that

$$\frac{4}{\alpha} = -24 + 12\alpha - 4\alpha^2 - 6\alpha^3 + \alpha^5.$$

Division by the integer 4 gives us the required expression for $\frac{1}{\alpha}$. In Exercise 5 you will do some more computations of this sort in this field.

## Example 42.7

Let's return to Example 41.5: The polynomial $p = x^2 + x + 1$ is irreducible in $\mathbb{Z}_2[x]$. Kronecker's Theorem 41.1 says that there is an extension field $E$ of $\mathbb{Z}_2$ that contains a root $\alpha$ of $p$. But $p$ is degree 2, and so Theorem 42.3 says that $\mathbb{Z}_2(\alpha) = \{a + b\alpha : a, b \in \mathbb{Z}_2\}$. That is,

$$\mathbb{Z}_2(\alpha) = \{0 + 0\alpha, 0 + 1\alpha, 1 + 0\alpha, 1 + 1\alpha\}.$$

Thus, $\mathbb{Z}_2(\alpha)$ is a field with four elements—something we have not seen before (unless you did Exercise 8.15). It is easy to give the addition and multiplication tables for this field. The critical fact when computing these tables is that $\alpha^2 + \alpha + 1 = 0$. We leave the actual computations as Exercise 1.

## Example 42.8

As another example, we return to $\mathbb{R}[x]/\langle x^2 + 1 \rangle$, a field extension of $\mathbb{R}$ that contains a root for $x^2 + 1$. Specifically, the root is the coset $\alpha = \langle x^2 + 1 \rangle + x$. Then

$$\mathbb{R}(\alpha) = \mathbb{R}[x]/\langle x^2 + 1 \rangle$$

and

$$\mathbb{R}(\alpha) = \{a + b\alpha : a, b \in \mathbb{R}\},$$

where $\alpha^2 + 1 = 0$. Of course,

$$\mathbb{C} = \{a + bi : a, b \in \mathbb{R}\},$$

where $i^2 + 1 = 0$. It is easy to show that these two fields are isomorphic via the function $\varphi : \mathbb{R}(\alpha) \to \mathbb{C}$ given by $\varphi(a + b\alpha) = a + bi$. (This is because $\alpha$ and $i$ play the same roles in their respective fields. Specifically, $\alpha^2 = -1$ and $i^2 = -1$.)

▷ **Quick Exercise.**  Show that the map $\varphi$ given above is an isomorphism. ◁

**Example 42.9**

Consider the irreducible polynomial $x^3 - 2 \in \mathbb{Q}[x]$. This polynomial has three distinct roots in $\mathbb{C}$, namely $\sqrt[3]{2}$, $\sqrt[3]{2}\zeta$, $\sqrt[3]{2}\zeta^2$, where

$$\zeta = e^{\frac{2\pi i}{3}} = -\frac{1}{2} + \frac{\sqrt{3}i}{2}$$

is the primitive cube root of unity (see Exercise 10.25).

$\triangleright$ **Quick Exercise.** Verify that $\zeta$ and $\zeta^2$ are cube roots of 1, and so the three given numbers are in fact roots of $x^3 - 2$. $\triangleleft$

Now consider the three simple extensions

$$\mathbb{Q}(\sqrt[3]{2}), \quad \mathbb{Q}(\sqrt[3]{2}\zeta), \quad \mathbb{Q}(\sqrt[3]{2}\zeta^2)$$

of $\mathbb{Q}$. According to Theorem 42.2, all three of these fields are isomorphic, even though the first is a subfield of the real numbers $\mathbb{R}$, while the latter two fields obviously include complex numbers.

It is clear that $\sqrt[3]{2}\zeta, \sqrt[3]{2}\zeta^2 \notin \mathbb{Q}(\sqrt[3]{2})$, but we claim that it is also the case that $\sqrt[3]{2} \notin \mathbb{Q}(\sqrt[3]{2}\zeta)$. If it were, then by Theorem 42.3,

$$\sqrt[3]{2} = a + b(\sqrt[3]{2}\zeta) + c(\sqrt[3]{2}\zeta)^2,$$

for some rational numbers $a, b, c$. By setting real and imaginary parts of this equation equal, we obtain

$$\sqrt[3]{2} = a - \frac{1}{2}\sqrt[3]{2}b - \frac{1}{2}\sqrt[3]{4}c$$

and

$$0 = \frac{\sqrt{3}}{2}\sqrt[3]{2}b - \frac{\sqrt{3}}{2}\sqrt[3]{4}c.$$

The second equation implies that $b = c = 0$, because otherwise we could infer that $\sqrt[3]{2}$ is rational. But then the first equation implies that $\sqrt[3]{2}$ is rational. By similar arguments, you can check (see Exercise 9) that none of the three simple extensions

$$\mathbb{Q}(\sqrt[3]{2}), \quad \mathbb{Q}(\sqrt[3]{2}\zeta), \quad \mathbb{Q}(\sqrt[3]{2}\zeta^2)$$

of $\mathbb{Q}$ contain either of the other two roots for $x^3 - 2$.

Let's actually find an explicit isomorphism $\varphi$ between $\mathbb{Q}(\sqrt[3]{2})$ and $\mathbb{Q}(\sqrt[3]{2}\zeta)$. Since 1 must be sent to 1, it is easy to see that $\varphi(q) = q$ for all rational numbers $q$. Now if $\alpha$ is a root of $x^3 - 2$, then $\alpha^3 - 2 = 0$, and so $\varphi(\alpha)^3 - \varphi(2) = \varphi(0)$, or in other words, $\varphi(\alpha)$ must also be a root of $x^3 - 2$. But because we have argued that $\sqrt[3]{2}\zeta$ is the *only* root of $x^3 - 2$ belonging to $\mathbb{Q}(\sqrt[3]{2}\zeta)$, it then follows that $\varphi(\sqrt[3]{2}) = \sqrt[3]{2}\zeta$. Thus, the isomorphism is easily described in terms of the unique representation of Theorem 42.3 as follows: Suppose that $\beta = a + b\sqrt[3]{2} + c(\sqrt[3]{2})^2$ is an arbitrary element of $\mathbb{Q}(\sqrt[3]{2})$; then $\varphi(\beta) = a + b\sqrt[3]{2}\zeta + c(\sqrt[3]{2}\zeta)^2$.

In Chapters 44 and 46 we will return to this example and examine this isomorphism (and others), from a more sophisticated point of view.

---

## 42.3   Simple Transcendental Extensions

Notice that Theorem 42.3 describing simple extensions applies only to *algebraic* extensions. To see how nice this description is, let's examine a *transcendental* simple extension:

**Example 42.10**

Consider the simple extension $\mathbb{Q}(\pi)$ of $\mathbb{Q}$ by the transcendental element $\pi$. The general intersection argument we gave above for obtaining the simple extension still applies in this case, and so we do have the smallest subfield $\mathbb{Q}(\pi)$ of $\mathbb{R}$, which contains the rational numbers and the transcendental $\pi$ as well. Thus, such real numbers as

$$\frac{1}{\pi} \quad \text{and} \quad \frac{1/2 + 3\pi^2}{2 + 3\pi + (5/4)\pi^2}$$

must belong to this field because they are obtained just by field operations. But in this case we *cannot* express such elements as a rational polynomial in $\pi$. Indeed,

$$\mathbb{Q}(\pi) = \left\{ \frac{f(\pi)}{g(\pi)} \in \mathbb{R} : f, g \in \mathbb{Q}[x],\ g \neq 0 \right\}.$$

In Exercise 6 you will prove that this set of real numbers is actually a field and consequently is the simple extension required.

For our purposes, algebraic simple extensions are much more important than transcendental ones, which is fortunate, considering how much nicer our description is in the algebraic case.

## 42.4   Dimension of Algebraic Simple Extensions

Notice that the description of the elements of algebraic simple extensions in Theorem 42.3 has a distinctive vector space flavor. Indeed, the theorem says that when the algebraic simple extension $F(\alpha)$ is viewed as a vector space over $F$, the set

$$\{1, \alpha, \alpha^2, \ldots, \alpha^{n-1}\}$$

is a basis for this vector space. Let us restate this important fact: *If the degree of the algebraic element $\alpha$ over a field F is n, then n is the dimension of $\mathrm{F}(\alpha)$ as a vector space over F.*

**Example 42.11**

We now rephrase Example 42.6: $\mathbb{Q}(\sqrt[3]{2} + \sqrt{2})$ is a vector space of dimension 6 over $\mathbb{Q}$. A basis consists of

$$1,\ \sqrt[3]{2} + \sqrt{2},\ (\sqrt[3]{2} + \sqrt{2})^2,\ \cdots,\ (\sqrt[3]{2} + \sqrt{2})^5.$$

**Example 42.12**

The field with four elements described in Example 42.7 is a vector space of dimension 2 over $\mathbb{Z}_2$. A basis consists of 1 and $\alpha$.

The following is an interesting corollary to the fact just stated.

**Corollary 42.4** *If $\alpha$ is algebraic over $F$ and $F(\alpha)$ has dimension $n$ over $F$ and $\beta \in F(\alpha)$, then $\beta$ is also algebraic over $F$. Furthermore, the degree of $\beta$ over $F$ is at most $n$.*

**Proof**: Because $F(\alpha)$ has dimension $n$ over $F$, any collection of more than $n$ elements of $F(\alpha)$ is linearly dependent (Corollary 40.6). Specifically, consider the $n+1$ elements

$$1, \beta, \beta^2, \ldots, \beta^n.$$

Because this set is linearly dependent, there exist $b_0, b_1, \ldots, b_n \in F$, such that

$$b_0 + b_1 \beta + b_2 \beta^2 + \cdots + b_n \beta^n = 0.$$

That is, $\beta$ is a root of the polynomial $b_0 + b_1 x + b_2 x^2 + \cdots + b_n x^n \in F[x]$. So, $\beta$ is algebraic over $F$ of degree no more than $n$. $\qquad \square$

To rephrase: Every element of an algebraic simple extension is algebraic over the base field.

What about transcendental simple extensions? The reason we have no nice element-wise description for such a field is exactly this: The vector space dimension of such a field over its base field is infinite! You will prove this fact in Exercise 8.

## Chapter Summary

For an element that is algebraic over a field we defined the *minimal polynomial* for that element over the field. The degree of this polynomial is the *degree* of the element over the field. We showed that such a polynomial always exists.

We also examined *simple extensions* and showed that in the case of a simple extension of a field by an algebraic element, the extension is in fact a vector space over the base field with dimension equal to the degree of the element. We showed this by explicitly displaying a basis for this vector space. Using this theorem, we were then able to construct a field with four elements.

## Warm-up Exercises

a. Return to Exercise a in the previous chapter. In each case, determine a basis for the simple extension obtained by adjoining the element to the field in question.

b. Express the following field elements as a linear combination of the basis elements you determined in Exercise a.

   (a) $\dfrac{1}{2 + \sqrt{2}} \in \mathbb{Q}(1 + \sqrt{2})$.

   (b) $\dfrac{\sqrt[3]{2}}{1 + 2\sqrt[3]{2}} \in \mathbb{Q}(\sqrt[3]{2})$.

c. Explain why the following statements are true, or else give a counterexample. In each case $F$ is a subfield of the field $E$, and $\alpha \in E \backslash F$:

   (a) $F(\alpha)$ is finite dimensional over $F$.

   (b) Let $\alpha$ be algebraic over $F$. Then $F(\alpha)$ is finite dimensional over $F$.

   (c) Every element of $F(\alpha)$ is algebraic over $F$.

   (d) Every element of $F(\alpha)$ is algebraic over $E$.

   (e) Suppose that $\alpha$ is algebraic over $F$. Then every element of $F(\alpha)$ is algebraic over $F$.

d. Explain why the minimal polynomial is unique.

## Exercises

1. (a) Compute the addition and multiplication tables for the field $\mathbb{Z}_2(\alpha)$ described in Example 42.7. Use your table to determine explicitly the multiplicative inverse of each non-zero element of this field. (This exercise is essentially a repeat of Exercise 8.15.)

   (b) In Exercise 12.18, you considered the Frobenius isomorphism, defined on any finite field. Compute this homomorphism explicitly in this case.

2. Consider the polynomial

$$f = 1 + x^2 + x^3 \in \mathbb{Z}_2(\alpha)[x],$$

   where $\mathbb{Z}_2(\alpha)$ is the field considered in the previous exercise.

   (a) Use the Root Theorem to show that this polynomial is irreducible.

   (b) By Kronecker's Theorem 41.1 we can construct an extension field of $\mathbb{Z}_2(\alpha)$ so that $f$ has a root $\beta$. How many elements does this field have?

   (c) If you're not doing anything this weekend, construct a multiplication table for this field.

3. In this problem you will (almost) prove that

$$f = x^6 - 6x^4 - 4x^3 + 12x^2 - 24x - 4$$

   is the minimal polynomial for $\alpha = \sqrt{2} + \sqrt[3]{2}$, as stated in Example 42.6. (We will return to this example in Example 43.3.)

   (a) Check that $f(\alpha) = 0$.

   (b) It can be shown that $f$ is irreducible in $\mathbb{Q}[x]$, but you will not do so here. However, show that $f$ can be factored into irreducibles in $\mathbb{R}[x]$ as follows:

$$\left( x - \sqrt{2} - \sqrt[3]{2} \right) \left( x^2 + (2\sqrt{2} + \sqrt[3]{2})x + (2 + \sqrt{2}\sqrt[3]{2} + \sqrt[3]{4}) \right)$$
$$\left( x + \sqrt{2} - \sqrt[3]{2} \right) \left( x^2 - (2\sqrt{2} - \sqrt[3]{2})x + (2 - \sqrt{2}\sqrt[3]{2} + \sqrt[3]{4}) \right).$$

   (c) How would you factor $f$ into irreducibles in $\mathbb{C}[x]$? (You need not feel obliged to actually carry this out!)

4. Determine the minimal polynomial of $\sqrt{2} + \sqrt{3}$ over the following three fields: $\mathbb{Q}$, $\mathbb{Q}(\sqrt{2})$, $\mathbb{R}$.

5. Let $\alpha = \sqrt{2} + \sqrt[3]{2}$, and $f$ be its minimal polynomial, as in Example 42.6, and in Exercise 3. Show explicitly that the following elements of $\mathbb{Q}(\alpha)$ are linear combinations of

$$1, \alpha, \alpha^2, \cdots, \alpha^5.$$

   (a) $\dfrac{1}{\alpha + 1}$.

   (b) $\dfrac{\alpha}{1 + \alpha^2}$.

6. Let

$$K = \left\{ \frac{f(\pi)}{g(\pi)} \in \mathbb{R} : f, g \in \mathbb{Q}[x], \; g \neq 0 \right\},$$

as given in Example 42.10.

  (a) Why do the quotients defined above always make sense? (That is, why are the denominators always non-zero?)

  (b) Show that $K$ is a subfield of $\mathbb{R}$.

  (c) Argue that $K = \mathbb{Q}(\pi)$.

  (d) Are the elements in $K$ uniquely represented, as we've presented them in the definition of $K$?

7. Let $F$ be a subfield of the field $E$, and suppose that $\alpha \in E$ is transcendental over $F$. Using Exercise 6 as a model, formulate an explicit description of the members of the simple extension $F(\alpha)$, and prove that your formulation works.

8. Let $F$ be a subfield of the field $E$, and suppose that $\alpha \in E$ is transcendental over $F$.

  (a) Show that

$$\{1, \alpha, \alpha^2, \alpha^3, \cdots\}$$

   is a linearly independent set of the vector space $F(\alpha)$ (over $F$).

  (b) Why does part (a) mean that the dimension of $F(\alpha)$ over $F$ is infinite?

  (c) Show that the set in part (a) is *not* a basis for $F(\alpha)$ over $F$.

9. In this exercise we perform some additional verifications related to Example 42.9. Show that $\sqrt[3]{2}, \sqrt[3]{2}\zeta \notin \mathbb{Q}(\sqrt[3]{2}\zeta^2)$ and $\sqrt[3]{2}\zeta^2 \notin \mathbb{Q}(\sqrt[3]{2}\zeta)$. Thus, in each case the field isomorphisms between the three simple extensions of Example 42.9 are very simply described. How?

10. Consider $\zeta$, the primitive cube root of unity used in Example 42.9. What is the minimial polynomial for $\zeta$ over $\mathbb{Q}$? What is the dimension of $\mathbb{Q}(\zeta)$ over $\mathbb{Q}$? Justify your answers.

11. Find the minimal polynomial for $\sqrt{2} + i$ over $\mathbb{Q}$. Be sure you justify the assertion that the degree is minimal. What are the other roots of this minimal polynomial?

# Chapter 43

## Finite Extensions and Constructibility Revisited

In the last chapter we obtained a good description for a simple extension of a field by an algebraic element. Specifically, if $\alpha$ is algebraic over $F$, we showed that $F(\alpha)$ is a vector space over $F$ with dimension equal to the degree of $\alpha$ over $F$. Furthermore, every element of $F(\alpha)$ is algebraic over $F$. In this chapter we are interested in field extensions that are not necessarily simple, but in which every element is algebraic. We'll then use our further results to provide more elegant proofs of the impossibility of two of the classical construction problems of the ancient Greeks.

## 43.1 Finite Extensions

We say that a field extension $E$ of a field $F$ is an **algebraic extension** if every element of $E$ is algebraic over $F$. That is, every element of $E$ is a root of some polynomial in $F[x]$. We proved in the last chapter that every element of a simple extension by an algebraic element is in fact an algebraic element itself; using this new terminology, this means that every such extension is algebraic.

We say $E$ is a **finite extension** of the field $F$ if it has finite dimension as a vector space over $F$. We shall use $[E : F]$ to stand for the dimension of $E$ over $F$, also called the **degree of $E$ over $F$**.

**Example 43.1**

> We know that $\mathbb{Q}(\sqrt[3]{3})$ is a finite extension of $\mathbb{Q}$ and $[\mathbb{Q}(\sqrt[3]{3}) : \mathbb{Q}] = 3$. Likewise, $\mathbb{C}$ is a finite extension of $\mathbb{R}$, and $[\mathbb{C} : \mathbb{R}] = 2$.

We showed in the last chapter that any simple extension by an algebraic element is a finite extension (Section 42.4). The following theorem asserts that every finite extension is algebraic; its proof uses an argument similar to that we used in Corollary 42.4 when we proved that elements of algebraic simple extensions are algebraic.

**Theorem 43.1** *A finite extension of a field is an algebraic extension.*

**Proof:**   Suppose that $E$ is a finite extension of $F$ and $[E : F] = n$. Let $\alpha \in E$. Then the $n + 1$ elements

$$1, \alpha, \alpha^2, \ldots, \alpha^n$$

cannot be linearly independent, by Corollary 40.6. Therefore, there exist

$$a_0, a_1, \ldots, a_n \in F$$

such that

$$a_0 + a_1\alpha + a_2\alpha^2 + \cdots + a_n\alpha^n = 0.$$

In other words, $\alpha$ is a root of the polynomial $a_0 + a_1 x + a_2 x^2 + \cdots + a_n x^n$ in $F[x]$. □

What about the converse of this theorem? In Exercise 5 you will explore an example of a field extension that is algebraic but not finite, thus showing that the converse is false.

The next theorem is the most important result of this chapter; it is this theorem that will make our new approach to the constructibility problems quite easy. It is a counting theorem, relating the dimensions of field extensions to one another. Counting theorems are always important. Despite the theorem's importance, its proof is easy. In fact, before reading the proof, you might attempt it yourself.

**Theorem 43.2** *If $K$ is a finite extension of a field $E$ and $E$ is a finite extension of a field $F$, then $K$ is a finite extension of the field $F$. Furthermore,*

$$[K : F] = [K : E][E : F].$$

**Proof:**   Let $\{\alpha_i : i = 1, \ldots, n\}$ be a basis for $E$ over $F$ and $\{\beta_j : j = 1, \ldots, m\}$ be a basis for $K$ over $E$. We will show that

$$\{\alpha_i\beta_j : i = 1, \ldots, n, \ j = 1, \ldots, m\}$$

is a basis for $K$ over $F$. Note that there are $mn$ distinct elements in this set.

First, we show that the $mn$ elements $\alpha_i\beta_j$ span $K$. Accordingly, let $k \in K$. Because the $\beta_j$ span $K$ as a vector space over $E$, there exist elements $a_1, \ldots, a_m$ of $E$ such that

$$k = \sum_{j=1}^{m} a_j\beta_j.$$

Likewise, because the $\alpha_i$ span $E$ as a vector space over $F$, for each $a_j$ there exist elements $b_{1j}, \ldots, b_{nj}$ of $F$ such that

$$a_j = \sum_{i=1}^{n} b_{ij}\alpha_i.$$

Substituting these into the sum for $k$, we get

$$k = \sum_{j=1}^{m} \left( \sum_{i=1}^{n} b_{ij}\alpha_i \right) \beta_j = \sum_{i,j} b_{ij}(\alpha_i\beta_j).$$

Thus, the $mn$ elements $\alpha_i\beta_j$ span $K$. (This argument is really a repeat of Exercise 40.15.)

We now show that the $\alpha_i\beta_j$ are linearly independent. So, suppose that

$$\sum_{i,j} c_{ij}(\alpha_i\beta_j) = 0, \quad c_{ij} \in F.$$

We will show that $c_{ij} = 0$ for all $i$ and $j$. But,

$$0 = \sum_{i,j} c_{ij}(\alpha_i\beta_j) = \sum_{j=1}^{m} \left( \sum_{i=1}^{n} c_{ij}\alpha_i \right) \beta_j.$$

The $\beta_j$ form a basis for $K$ over $E$; therefore,

$$\sum_{i=1}^{n} c_{ij}\alpha_i = 0$$

for each $j$. But the $\alpha_i$ form a basis for $E$ over $F$, so $c_{ij} = 0$ for each $i$ and $j$. Thus, the $mn$ elements $\alpha_i \beta_j$ are linearly independent, and so form a basis for $K$ as a vector space over $F$. $\square$

A rough paraphrase of this theorem is this: A finite extension of a finite extension is a finite extension. Notice that we have explicitly constructed a basis for $K$ over $F$ from the bases given for $K$ over $E$ and $E$ over $F$. We will explore some specific examples of this procedure below. The theorem is pictured in the following diagram:

Before looking at examples, we will sharpen our result for the case of simple algebraic extensions. Suppose that $E$ is an extension field of $F$ and $\alpha \in E$ is algebraic over $F$. Let $\beta \in F(\alpha)$. Thus, $F(\beta) \subseteq F(\alpha)$. We have seen from Corollary 42.4 that $\beta$ is algebraic over $F$, and so it follows from the last theorem that $[F(\beta) : F]$ must divide $[F(\alpha) : F]$. That is, the degree of $\beta$ over $F$ must divide the degree of $\alpha$ over $F$. We restate this in the following corollary.

**Corollary 43.3** *Suppose $E$ is an extension field of the field $F$ and $\alpha \in E$ is algebraic over $F$. If $\beta \in F(\alpha)$, then the degree of $\beta$ over $F$ divides the degree of $\alpha$ over $F$.*

We now illustrate the theorem with a couple of examples.

**Example 43.2**

Consider $\mathbb{Q}(\sqrt{2}, \sqrt{3})$, the smallest subfield of $\mathbb{R}$ containing $\mathbb{Q}$, $\sqrt{2}$, and $\sqrt{3}$, as a vector space over $\mathbb{Q}$. Now, $\sqrt{3} \notin \mathbb{Q}(\sqrt{2})$ (see Exercise 1 or Example 37.2), and so $x^2 - 3$ is the minimal polynomial for $\sqrt{3}$ over $\mathbb{Q}(\sqrt{2})$. Therefore, $[\mathbb{Q}(\sqrt{2}, \sqrt{3}) : \mathbb{Q}(\sqrt{2})] = 2$ and $\{1, \sqrt{3}\}$ is a basis for $\mathbb{Q}(\sqrt{2}, \sqrt{3})$ over $\mathbb{Q}(\sqrt{2})$. Because $[\mathbb{Q}(\sqrt{2}) : \mathbb{Q}] = 2$ and $\{1, \sqrt{2}\}$ is a basis for $\mathbb{Q}(\sqrt{2})$ over $\mathbb{Q}$, we have that $[\mathbb{Q}(\sqrt{2}, \sqrt{3}) : \mathbb{Q}] = 4$ and $\{1, \sqrt{2}, \sqrt{3}, \sqrt{6}\}$ is a basis for $\mathbb{Q}(\sqrt{2}, \sqrt{3})$ over $\mathbb{Q}$. Notice that here we are thinking of $\mathbb{Q}(\sqrt{2}, \sqrt{3})$ as a simple extension of $\mathbb{Q}(\sqrt{2})$, which in turn is a simple extension of $\mathbb{Q}$.

Alternatively, we could think of $\mathbb{Q}(\sqrt{2}, \sqrt{3})$ as an extension of $\mathbb{Q}(\sqrt{3})$. Then, $[\mathbb{Q}(\sqrt{2}, \sqrt{3}) : \mathbb{Q}(\sqrt{3})] = 2$ and $[\mathbb{Q}(\sqrt{3}) : \mathbb{Q}] = 2$. These are two ways of stepping up from $\mathbb{Q}$ to $\mathbb{Q}(\sqrt{2}, \sqrt{3})$: First, we adjoin $\sqrt{2}$ to $\mathbb{Q}$ and then adjoin $\sqrt{3}$, or first adjoin $\sqrt{3}$ to $\mathbb{Q}$ and then adjoin $\sqrt{2}$.

The following diagram illustrates this situation.

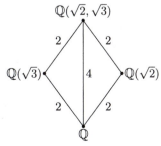

You should refer to Example 37.2 for an account of this example that does not use the theory of vector spaces.

**Example 43.3**

Now consider $\mathbb{Q}(\sqrt{2}, \sqrt[3]{2})$. Because $[\mathbb{Q}(\sqrt[3]{2}) : \mathbb{Q}] = 3$ and $[\mathbb{Q}(\sqrt{2}) : \mathbb{Q}] = 2$, it follows from Corollary 43.3 that $\sqrt[3]{2} \notin \mathbb{Q}(\sqrt{2})$. Thus, $[\mathbb{Q}(\sqrt{2}, \sqrt[3]{2}) : \mathbb{Q}(\sqrt{2})] = 3$. So,

$$[\mathbb{Q}(\sqrt{2}, \sqrt[3]{2}) : \mathbb{Q}] = [\mathbb{Q}(\sqrt{2}, \sqrt[3]{2}) : \mathbb{Q}(\sqrt{2})][\mathbb{Q}(\sqrt{2}) : \mathbb{Q}] = 3 \cdot 2 = 6.$$

Again, we could build $\mathbb{Q}(\sqrt{2}, \sqrt[3]{2})$ another way. Because $\sqrt{2} \notin \mathbb{Q}(\sqrt[3]{2})$ (from Corollary 43.3), we have that $[\mathbb{Q}(\sqrt{2}, \sqrt[3]{2}) : \mathbb{Q}(\sqrt[3]{2})] = 2$, and so

$$[\mathbb{Q}(\sqrt{2}, \sqrt[3]{2}) : \mathbb{Q}] = [\mathbb{Q}(\sqrt{2}, \sqrt[3]{2}) : \mathbb{Q}(\sqrt[3]{2})][\mathbb{Q}(\sqrt[3]{2}) : \mathbb{Q}] = 2 \cdot 3 = 6.$$

The following diagram illustrates this.

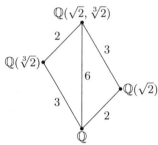

But now consider the simple algebraic extension $\mathbb{Q}(\alpha)$, where $\alpha = \sqrt{2} + \sqrt[3]{2}$. It is evident that $\alpha \in \mathbb{Q}(\sqrt{2}, \sqrt[3]{2})$, and so $\mathbb{Q}(\alpha) \subseteq \mathbb{Q}(\sqrt{2}, \sqrt[3]{2})$; we claim that this last field extension is trivial. In Exercise 42.3 we suggested that $\mathbb{Q}(\alpha)$ is of degree 6 over $\mathbb{Q}$, by providing a monic polynomial in $\mathbb{Q}[x]$ with $\alpha$ as a root. So by the theorem, we have that $[\mathbb{Q}(\sqrt{2}, \sqrt[3]{2}) : \mathbb{Q}(\alpha)] = 1$, as we require.

But let's prove that $\mathbb{Q}(\sqrt{2}, \sqrt[3]{2}) = \mathbb{Q}(\alpha)$, without making use of Exercise 42.3. We know that $\sqrt[3]{2} = \alpha - \sqrt{2}$. Cubing both sides of this equation gives

$$2 = \alpha^3 - 3\sqrt{2}\alpha^2 + 6\alpha - 2\sqrt{2}.$$

Solving this equation for $\sqrt{2}$ gives us

$$\sqrt{2} = \frac{\alpha^3 + 6\alpha - 2}{3\alpha^2 + 2}.$$

▷ **Quick Exercise.**  Check this.  ◁

But this means that $\sqrt{2} \in \mathbb{Q}(\alpha)$; but then $\sqrt[3]{2} = \alpha - \sqrt{2} \in \mathbb{Q}(\alpha)$ too, and so $\mathbb{Q}(\sqrt{2}, \sqrt[3]{2}) \subseteq \mathbb{Q}(\alpha)$. Because we've already noted the reverse inclusion, we have proved again that $\mathbb{Q}(\alpha) = \mathbb{Q}(\sqrt{2}, \sqrt[3]{2})$. This means of course that the minimal polynomial for $\alpha$ must be of degree 6.

Note furthermore that the finite extension $\mathbb{Q}(\sqrt{2}, \sqrt[3]{2})$ over $\mathbb{Q}$ is actually simple. It turns out that *all* finite extensions over the rational field are simple; we will prove this important fact in Theorem 44.5.

The calculations carried out above can be taken one step further to get the minimal polynomial for $\alpha$ over $\mathbb{Q}$. If you square both sides of the equation we have for $\sqrt{2}$ above, then a bit of algebra yields the polynomial for which $\alpha$ is a root. (See Exercise 11.)

It is clear that an induction argument extends Theorem 43.2 to any number of steps. For completeness, we include this as a corollary. (The proof of this is left as Exercise 2.)

**Corollary 43.4** *If $F_i$, $i = 1, \ldots, n$ are fields with $F_{i+1}$ a finite extension of $F_i$, for $i = 1, \ldots, n-1$, then $F_n$ is a finite extension of $F_1$ and*

$$[F_n : F_1] = [F_n : F_{n-1}][F_{n-1} : F_{n-2}] \cdots [F_2 : F_1].$$

## 43.2 Constructibility Problems

Let's start by recalling the three Greek construction problems: doubling the cube, trisecting an arbitrary angle, and squaring the circle. The proofs that these constructions are impossible depend in an essential way on the Constructible Number Theorem 37.3, which we re-state here:

**Constructible Number Theorem**     *The number $\alpha$ is constructible if and only if there exists a finite sequence of fields*

$$\mathbb{Q} = F_0 \subset F_1 \subset \cdots \subset F_N,$$

*with $\alpha \in F_N$ and $F_{i+1} = F_i(\sqrt{k_i})$ for some $k_i \in F_i$, with $k_i > 0$ for $i = 0, \ldots, N-1$.*

Let's re-consider this theorem, in light of our theory of field extensions. We now know that the degree of each of the extensions in the Constructible Number Theorem is 2. Thus, $[F_N : \mathbb{Q}]$ must be a power of 2. Now clearly the extension $\mathbb{Q}(\alpha) \subseteq F_N$, and so by Theorem 43.2, we must have that $[\mathbb{Q}(\alpha) : \mathbb{Q}]$ is also a power of 2. We thus obtain the following corollary of the Constructible Number Theorem:

**Corollary 43.5** *If a number $\alpha$ is constructible, then $[\mathbb{Q}(\alpha) : \mathbb{Q}] = 2^n$, for some positive integer $n$.*

Note that the converse to this corollary is false; $[\mathbb{Q}(i) : \mathbb{Q}] = 2$ is a counterexample, because constructible numbers are necessarily real. But the converse is also false for real numbers. In Exercise 48.7 we will present an irreducible fourth degree polynomial in $\mathbb{Q}[x]$ with a real root $\alpha$ that is not constructible, even though $[\mathbb{Q}(\alpha) : \mathbb{Q}] = 4$.

Let's now return to the impossibility proofs of the construction problems. Recall that to double the cube, we are to construct an edge of a cube whose volume is twice that of a given cube. (Actually, we are given the *edge* of the original cube.) If we consider the edge of the original cube to be length 1, then we are required to construct a line segment of length $\sqrt[3]{2}$. But $[\mathbb{Q}(\sqrt[3]{2}) : \mathbb{Q}] = 3$, and so Corollary 43.5 tells us that $\sqrt[3]{2}$ is not constructible. Thus, it is impossible to double the cube with only a compass and straightedge.

In our impossibility proof for the trisection problem, we showed that it is in fact impossible to trisect a 60° angle. If we could do so, we could construct an angle of 20° which in turn implies the construction of a root of the polynomial $x^3 - 3x - 1$, which is irreducible in $\mathbb{Q}[x]$. But any root of this polynomial has degree 3 over $\mathbb{Q}$; once again, Corollary 43.5 shows that this is not constructible.

We are not able to use the theory of this chapter to more easily prove the impossibility of squaring the circle, because this problem involves considering the *transcendental* simple extension $\mathbb{Q}(\sqrt{\pi})$. We must still rely on Lindemann's difficult Theorem 38.5 asserting that $\pi$ is transcendental over $\mathbb{Q}$.

## Chapter Summary

In this chapter we showed that every *finite extension* is an *algebraic extension*. If $K$ is a finite extension of $E$, which is a finite extension of $F$, then $K$ is a finite extension of $F$ with $[K : F] = [K : E][E : F]$.

We then used this theorem to give alternate proofs for the impossibility of doubling the cube and trisecting an arbitrary angle.

# Warm-up Exercises

a. If $[K : F] = 1$, what can you say about the fields $K$ and $F$?

b. If $[K : F]$ is prime and $E$ is a field where $F \subseteq E \subseteq K$, what can you say about the field $E$?

c. If $F \subseteq E \subseteq K$ is a sequence of finite field extensions and $[K : F] = [E : F]$, what can you say about the fields $K$ and $E$?

d. Explain why the following statements are true, or else give a counterexample.

   (a) Every finite extension is algebraic.

   (b) Every simple extension is algebraic.

   (c) Every simple extension is finite.

   (d) Every algebraic extension is simple.

   (e) Every algebraic extension is finite.

e. How many elements belong to a field that is a degree 2 extension of a degree 3 extension of $\mathbb{Z}_7$?

f. Give examples of the following, or else explain why the example does not exist:

   (a) A degree 2 extension of $\mathbb{Q}(\sqrt{2})$.

   (b) A degree 3 extension of $\mathbb{R}$.

   (c) A degree 3 extension of $\mathbb{Z}_2$.

---

# Exercises

1. Show that $\sqrt{3} \notin \mathbb{Q}(\sqrt{2})$, using Theorem 43.2. (See Example 37.2 for a more elementary solution of this exercise.)

2. Prove Corollary 43.4. That is, suppose that each $F_i$ is a field, and $F_1 \subseteq F_2 \subseteq \cdots \subseteq F_n$ is a sequence of finite extensions. Show that

$$[F_n : F_1] = [F_n : F_{n-1}][F_{n-1} : F_{n-2}] \cdots [F_2 : F_1].$$

3. Prove that $\mathbb{Q}(\sqrt{2} + \sqrt{3}) = \mathbb{Q}(\sqrt{2}, \sqrt{3})$.

4. (a) Determine $[\mathbb{Q}(\sqrt[3]{2}, i) : \mathbb{Q}]$; include a proof of your result.

   (b) Find $\alpha \in \mathbb{C}$ so that $\mathbb{Q}(\alpha) = \mathbb{Q}(\sqrt[3]{2}, i)$.

5. In this exercise you will show that not every algebraic extension is a finite extension by considering the field $F$, constructed as follows. Let

$$F_1 = \mathbb{Q}(\sqrt{2}), \ F_2 = F_1(\sqrt[3]{2}), \ F_3 = F_2(\sqrt[4]{2}), \ F_4 = F_3(\sqrt[5]{2}), \ \cdots,$$

and continue inductively. Then let

$$F = \bigcup_{n=1}^{\infty} F_n.$$

Argue that $F$ is field. Then prove that $F$ is not a finite extension, thus showing that the converse of Theorem 43.1 is false.

6. Give an example to show that if $[F(a) : F] = n$ and $[F(b) : F] = m$, it does not necessarily follow that $[F(a, b) : F] = mn$.

7. Prove that an algebraic extension of an algebraic extension is an algebraic extension.

8. Consider the primitive cube root of unity

$$\zeta = -\frac{1}{2} + i\frac{\sqrt{3}}{2} = e^{\frac{2\pi i}{3}} = \cos\left(\frac{2\pi}{3}\right) + i\sin\left(\frac{2\pi}{3}\right),$$

which we used in Example 42.9. In Exercise 42.10 you concluded that $[\mathbb{Q}(\zeta) : \mathbb{Q}] = 2$. Now draw a complete field extension diagram for the field extension $\mathbb{Q}(\sqrt[3]{2}, \zeta)$ of $\mathbb{Q}$, computing and justifying all of the degrees in your diagram. (We will return to this example in Example 44.4 and Example 46.7.)

9. Consider the field $F = \mathbb{Z}_2(\alpha, \beta)$ constructed in Exercise 42.2.

    (a) Use Theorem 43.2 to determine the possible degrees $[F : K]$, where $K$ is a subfield of $F$. Find a subfield with each of these degrees, and draw a field extension diagram similar to that in Examples 43.2 and 43.3.

    (b) Prove that $\mathbb{Z}_2(\alpha + \beta) = \mathbb{Z}_2(\alpha, \beta)$.

10. Prove that $\mathbb{A}$, the set of all complex numbers algebraic over $\mathbb{Q}$, is a field.

11. Finish the calculations started in Example 43.3 to find the minimal polynomial for $\sqrt{2} + \sqrt[3]{2}$ over $\mathbb{Q}$; compare your result with Exercise 42.3.

12. Consider

$$\zeta = e^{\frac{2\pi i}{5}},$$

the primitive fifth root of unity. Use Exercise 5.17 to compute $[\mathbb{Q}(\zeta) : \mathbb{Q}]$.

13. Consider the field $\mathbb{A}$ of algebraic numbers over $\mathbb{Q}$, discussed in Exercise 10.

    (a) Show that $\mathbb{A}$ is not a finite extension of $\mathbb{Q}$.

    (b) Suppose $K$ is a field, and $\mathbb{Q} \subset \mathbb{A} \subset K \subset \mathbb{C}$. Show that $K$ is not a finite extension of $\mathbb{A}$.

# Section IX in a Nutshell

This section presents some more sophisticated mathematical ideas that allow us to prove more elegantly the impossibility of the constructibility problems of the previous section. This more powerful approach to field theory is interesting in its own right and also plays a vital role in Section X.

First, we define a *vector space* $V$ over a field $F$. This is a set of objects (called *vectors*) equipped with a binary operation called addition, making the set an additive group. In addition a vector space has a *scalar multiplication* whereby vectors are multipied by elements (*scalars*) from the field $F$. The complete set of axioms follows, where $\mathbf{v}, \mathbf{w}, \mathbf{u} \in V$, and $r, s \in F$:

1. $\mathbf{v} + \mathbf{w} = \mathbf{w} + \mathbf{v}$,

2. $\mathbf{v} + (\mathbf{w} + \mathbf{u}) = (\mathbf{v} + \mathbf{w}) + \mathbf{u}$,

3. there exists a **zero vector 0**, with the property that $\mathbf{v} + \mathbf{0} = \mathbf{v}$, and

4. every vector $\mathbf{v}$ has an **additive inverse** $-\mathbf{v}$, with the property that $\mathbf{v} + (-\mathbf{v}) = \mathbf{0}$.

5. $(r + s)\mathbf{v} = r\mathbf{v} + s\mathbf{v}$,

6. $(rs)\mathbf{v} = r(s\mathbf{v})$,

7. $r(\mathbf{v} + \mathbf{w}) = r\mathbf{v} + r\mathbf{w}$, and

8. $1\mathbf{v} = \mathbf{v}$.

Classic examples of vector spaces are the set $F^n$ of $n$-tuples from the field $F$, and the polynomial ring $F[x]$. If $F$ and $E$ are fields and $F \subseteq E$, then the field $E$ is actually a vector space over $F$. (In this case we say $E$ is an *extension field* of the field $F$.) This eventually allows us to apply vector space theory to field theory.

A *basis* for a vector space $V$ is a linearly independent set of vectors that spans $V$. The number of vectors in a basis is the *dimension* of the vector space. (See Theorem 40.4.) Thus $F^n$ has dimension $n$ and $F[x]$ has infinite dimension over $F$. If $B$ is a basis for vector space $V$ then every element of $V$ can be uniquely written as a linear combination of elements in $B$ (Theorem 40.2). Every spanning set contains a subset that is a basis (Theorem 40.3).

*Kronecker's Theorem 41.1* says that if $F$ is a field and $f \in F[x]$ of degree at least 1, we can always construct an extension field $E$ of $F$ such that $f(\alpha) = 0$ for some $\alpha \in E$. The proof of this theorem reveals that $F[x]/\langle p \rangle$ is the desired field, where $p$ is an irreducible factor of $f$ in $F[x]$.

If $E$ is an extension field of $F$ and $\alpha \in E$ is a root of a polynomial in $F[x]$, then $\alpha$ is algebraic over $F$. The smallest subfield $F(\alpha)$ of $E$ containing $F$ and $\alpha$ is essentially unique, and isomorphic to the field $F[x]/\langle p \rangle$, where $p$ is an irreducible polynomial $p \in F[x]$ with $p(\alpha) = 0$ (Theorem 42.2). Then $p$ divides all $f \in F[x]$ where $f(\alpha) = 0$ (Theorem 42.1), and if we further require that $p$ be monic, it is unique and called the *minimal polynomial of $\alpha$ over $F$*.

Fields of the form $F(\alpha)$, are called *simple extensions* of $F$. If $\alpha$ happens to be algebraic over $F$, then $F(\alpha)$ is called an *algebraic simple extension* of $F$. Furthermore, if $\deg(p) = n$, then $F(\alpha)$ is a vector space over $F$ of dimension $n$ with basis $\{1, \alpha, \alpha^2, \ldots, \alpha^{n-1}\}$. It follows that every element in $F(\alpha)$ is algebraic over $F$ (Corollary 42.4).

If $E$ is an extension field of $F$ and has finite dimension as a vector space over $F$, we call $E$ a *finite extension* of $F$ and denote the degree of the extension by $[E : F]$. Every finite extension is algebraic (Theorem 43.1). Degrees multiply: If $F_1 \subseteq F_2 \subseteq F_3$ form a sequence of finite extensions, then $[F_3 : F_1] = [F_3 : F_2][F_2 : F_1]$. This extends to any finite sequence of finite extensions. If $E$ is an extension field of $F$ and $\alpha \in E$ is algebraic over $F$ and $\beta \in F(\alpha)$, then the degree of $\beta$ over $F$ divides the degree of $\alpha$ over $F$ (Corollary 43.3).

It follows from the Constructible Number Theorem that if $\alpha$ is constructible, then $[\mathbb{Q}(\alpha) : \mathbb{Q}] = 2^n$. This allows us to show that it is impossible to double the cube (because $[\mathbb{Q}(\sqrt[3]{2}) : \mathbb{Q}] = 3$), it is impossible to trisect a $60°$ angle (because to do so means we could construct a root of $x^3 - 3x - 1$, which is irreducible in $\mathbb{Q}[x]$ of degree 3), and it is impossible to square the circle (because $\mathbb{Q}(\sqrt{\pi})$ is a transcendental simple extension of $\mathbb{Q}$).

# Part X

# Galois Theory

# Chapter 44

## *The Splitting Field*

We are all familiar with the way in which the quadratic formula gives us an explicit formula for the roots of any degree two polynomial in $\mathbb{Q}[x]$ (and, this formula works in $\mathbb{C}[x]$ too – see Exercises 10.1 and 10.2). In Exercises 10.12–10.19, you can explore the cubic formula that performs the same task as the quadratic formula; it is a good bit more complicated. In the case of the quadratic formula, we need to perform the field operations from $\mathbb{Q}$, and in addition extract a square root. In the case of the cubic formula, we need to perform the field operations from $\mathbb{Q}$, extract cube roots, and also have to extract one or more square roots. Can this process be extended to higher and higher degree polynomials from $\mathbb{Q}[x]$? The answer for quartic (that is, fourth degree) equations is yes; it is a nightmare of a formula, which involves not only the extraction of a fourth root, but also cube and square roots as well. In essence, the cubic problem reduces finally to a quadratic, and the quartic in turn finally reduces to a cubic! You can explore the quartic formula in Exercise 10.20.

It seems natural to suppose that this project could be carried on indefinitely (at the cost of more and more complicated calculations), to obtain formulas solving fifth, sixth and higher degree polynomial equations over $\mathbb{C}$. This supposition is false: It was one of the triumphs of nineteenth century abstract algebra to prove that such formulas for degree 5 and higher are impossible. This result can and should be compared to the results we have encountered in Chapters 38 and 43, showing that the classical construction problems are impossible.

In order to prove the impossibility of solving all polynomial equations by radicals, we will need some more field theory, which we will begin exploring in this chapter. To complete our project, we will eventually make use of the group theory we encountered in Chapter 30. The exciting interplay between groups and fields makes this subject an outstanding example of the power of abstract algebra.

## 44.1 The Splitting Field

In Chapter 41 we encountered Kronecker's Theorem 41.1: Given a field $F$ and a non-constant polynomial $f \in F[x]$, we can always build a (potentially) bigger field than $F$ in which $f$ has at least one root. By an easy inductive argument (Exercise 41.1) on the degree of $f$, we can then build a (potentially) larger field than $F$, in which $f$ can be completely factored into linear polynomials; such a field is called a **splitting extension** over $F$ for $f$. In this chapter we will show that this can be done minimally in essentially only one way.

We make a definition to capture this idea. Suppose that $F$ is a field and $f \in F[x]$, with $\deg(f) > 0$. Then the field $K$ is a **splitting field for $f$ over** $F$, if $F \subseteq K$, $f$ factors into linear polynomials in $K[x]$, and if $L$ is any other field with $F \subseteq L \subset K$, then $f$ cannot be factored into linear polynomials in $L[x]$. Any field containing $K$ is a splitting extension of $f$ over $F$. The splitting field of $f$ over $F$ is a minimal splitting extension. When the polynomial $f$ does factor into linear polynomials in a field $K$, we say $f$ **splits in** $K$.

### Example 44.1

If the polynomial $f \in F[x]$ can already be factored into linear polynomials in $F[x]$, then $F$ itself is a splitting field for $f$.

### Example 44.2

The field $\mathbb{C}$ is a splitting field for the polynomial $x^2 + 1 \in \mathbb{R}[x]$ over $\mathbb{R}$.

### Example 44.3

Let $f \in \mathbb{Q}[x]$ be any quadratic polynomial that is irreducible in $\mathbb{Q}[x]$, and $\alpha \in \mathbb{C}$ be a root of $f$ given by the quadratic formula. Then the quadratic extension $\mathbb{Q}(\alpha)$ is a splitting field for $f$ over $\mathbb{Q}$. This is true because the second root for $f$ in $\mathbb{C}$ is the complex conjugate $\bar{\alpha}$, and $\bar{\alpha} \in \mathbb{Q}(\alpha)$.

▷ **Quick Exercise.** Why is $\bar{\alpha} \in \mathbb{Q}(\alpha)$?

### Example 44.4

Consider the polynomial $f = x^3 - 2 \in \mathbb{Q}[x]$ that we looked at in Example 42.9. The field $\mathbb{Q}\left(\sqrt[3]{2}\right)$ certainly contains a root for $f$, namely $\sqrt[3]{2}$. But it is not a splitting field, because there are two other cube roots of 2 not belonging to this field, namely $\sqrt[3]{2}\zeta$ and $\sqrt[3]{2}\zeta^2$, where $\zeta = e^{\frac{2\pi i}{3}} = -\frac{1}{2} + \frac{\sqrt{3}i}{2}$ is the primitive cube root of unity. Thus, a splitting field for $f$ is

$$\mathbb{Q}\left(\sqrt[3]{2}\right)(\zeta) = \mathbb{Q}\left(\sqrt[3]{2}\right)(\sqrt{3}i).$$

### Example 44.5

Consider the polynomial $f = x^3 + x + 1 \in \mathbb{Z}_2[x]$. It is evident that $f$ is an irreducible polynomial. By Kronecker's Theorem 41.1 we can build the extension field

$$
\begin{aligned}
\mathbb{Z}_2[x]/\langle f \rangle &= \mathbb{Z}_2(\alpha) \\
&= \{0,\ 1,\ \alpha,\ 1+\alpha,\ \alpha^2,\ 1+\alpha^2,\ \alpha+\alpha^2,\ 1+\alpha+\alpha^2\}
\end{aligned}
$$

where $\alpha = \langle f \rangle + x$. This is a splitting field for $f$, because

$$f = (x + \alpha)\left(x + \alpha^2\right)\left(x + \alpha + \alpha^2\right).$$

You will check this factorization in Exercise 44.1 and also show that

$$\mathbb{Z}_2(\alpha) = \mathbb{Z}_2(\alpha^2) = \mathbb{Z}_2(\alpha + \alpha^2).$$

We will now prove that a non-constant polynomial over a field always has a splitting field; furthermore, such a field is unique (up to isomorphism, of course). We will prove existence (and more) in Theorem 44.1 and uniqueness in Theorem 44.2.

**Theorem 44.1** *Let $F$ be a field and $f \in F[x]$ a non-constant polynomial. Then there exists a splitting field $K$ for $f$ over $F$. Furthermore, if the linear factorization of $f$ in $K[x]$ is given by*

$$f = \beta(x - \alpha_1)(x - \alpha_2)\cdots(x - \alpha_n),$$

*Then $K$ is equal to the finite extension $F(\alpha_1, \alpha_2, \cdots, \alpha_n)$.*

**Proof:** Let $F$ be a field and $f \in F[x]$ a non-constant polynomial. We know by Exercise 41.1 that we can construct a splitting extension $M$ of $F$ for $f$. Now consider the finite field extension $K = F(\alpha_1, \alpha_2, \cdots, \alpha_n)$, where the elements $\alpha_i$ are the roots of $f$ in $M$. It is then evident that $K$ is a splitting extension of $F$ for $f$. But if $L$ is any other splitting extension of $F$ contained in $M$, then $L$ must contain the roots of $f$, and so must contain $K$. This means that $K$ is minimal among such extensions, and so is a splitting field. $\square$

We now prove that splitting fields are unique, up to isomorphism. The following theorem carefully sets up what we mean by this uniqueness:

**Theorem 44.2** *Suppose that $F$ is a field and $f \in F[x]$ is a non-constant polynomial. Suppose that $K$ and $\overline{K}$ are two splitting fields of $F$ for $f$. Then there exists an isomorphism $\varphi : K \to \overline{K}$ that leaves $F$ fixed.*

**Proof:** Assume that $F$, $f$, $K$, and $\overline{K}$ are as in the hypotheses of the theorem. We know from Theorem 44.1 that $K = F(\alpha_1, \alpha_2, \alpha_3, \cdots, \alpha_n)$, where the $\alpha_i$ are the (not necessarily distinct) roots of $f$ in $K$. Furthermore, $\overline{K} = F(\beta_1, \beta_2, \cdots, \beta_n)$, where the $\beta_i$ are the (not necessarily distinct) roots of $f$ in $\overline{K}$. We will proceed by induction on the degree $n$. To facilitate this, define the fields $K_k$ inductively, by letting $K_0 = F$, and $K_{k+1} = K_k(\alpha_{k+1})$. We will define the fields $\overline{K}_k$ in the same manner; we will let $\overline{K}_0 = F$ and define the subfields of $\overline{K}$ at each inductive step, by reordering the roots $\beta_1, \cdots, \beta_n$, which we will describe below. We will extend the field isomorphism $\varphi$ from each of the extension fields $K_k$ onto the extension fields $\overline{K}_k$, one step at a time.

We start with $\varphi : K_0 \to \overline{K}_0$ being the identity isomorphism (since $K_0 = \overline{K}_0 = F$). We now assume by induction that $\varphi$ has in fact been defined and is an isomorphism from $K_k$ onto $\overline{K}_k$. Furthermore, this isomorphism leaves $F$ fixed and has been constructed so that $\varphi(\alpha_i) = \beta_i$, for $i = 1, 2, \cdots, k$.

Now consider the root $\alpha_{k+1}$ of the polynomial $f \in F[x] \subseteq K_k[x]$. If $\alpha_{k+1} \in K_k$, then $f \in F[x] \subseteq \overline{K}_k[x]$ must have a root in $\overline{K}_k$ other than $\beta_1, \beta_2, \cdots, \beta_k$, and, by renumbering, we may assume that this is the root $\beta_{k+1}$. We consequently have $\varphi$ already defined on the field $K_{k+1} = K_k$ onto the field $\overline{K}_{k+1} = \overline{K}_k$.

We thus may assume that the root $\alpha_{k+1} \notin K_k$. We then know by Theorem 42.2 that $K_k(\alpha_{k+1})$ is isomorphic to the field $K_k[x]/\langle p \rangle$, where $p$ is the irreducible factor of $f$ in $K_k[x]$ for which $\alpha_{k+1}$ is a root. Now because the fields $K_k$ and $\overline{K}_k$ are isomorphic, we know that the polynomial rings $K_k[x]$ and $\overline{K}_k[x]$ are isomorphic and there is an irreducible factor $\overline{p}$ of $f$ in $\overline{K}_k[x]$ corresponding to $p$. (See Exercise 14.17.) Since $\overline{p}$ divides $f$ in $\overline{K}[x]$, the roots of $\overline{p}$ are also roots of $f$ in $\overline{K}$. So choose any root of $\overline{p}$ in $\overline{K}$, which we can (by renumbering) call $\beta_{k+1}$ and let $\overline{K}_{k+1} = \overline{K}_k(\beta_{k+1})$. Then $\overline{K}_{k+1}$ is isomorphic to $\overline{K}_k[x]/\langle \overline{p} \rangle$. But by Exercise 14.20, it is clear that $K_k[x]/\langle p \rangle$ and $\overline{K}_k[x]/\langle \overline{p} \rangle$ are isomorphic, and the isomorphism between them extends the assumed isomorphism for $K_k$ to $\overline{K}_k$, and it takes $\langle p \rangle + x$ to $\langle \overline{p} \rangle + x$. This means that we have an isomorphism from $K_{k+1} = K_k(\alpha_{k+1})$ to $\overline{K}_{k+1} = \overline{K}_k(\beta_{k+1})$, which takes $\alpha_{k+1}$ to $\beta_{k+1}$.

Thus, by induction, we have that there is an isomorphism from $K$ to $\overline{K}$ that extends the isomorphism $\varphi$ from $F$ to $\overline{F}$, as required. $\square$

## Example 44.6

Suppose that $f \in \mathbb{Q}[x] \subseteq \mathbb{C}[x]$. Recall the Fundamental Theorem of Algebra 10.1, which implies that every polynomial with complex coefficients can be factored completely into linear factors. That is, we have in this case that all of the roots $\alpha_1, \cdots, \alpha_n$ of $f$ belong to $\mathbb{C}$. Consequently, $\mathbb{C}$ contains a splitting field for $f$ over $\mathbb{Q}$ (that is, $\mathbb{C}$ is a splitting extension for $f$ over $\mathbb{Q}$), and it consists precisely of $\mathbb{Q}(\alpha_1, \cdots, \alpha_n)$. By Theorem 44.2, we know that this is essentially the only splitting field.

The next theorem will give us some insight into just how nice splitting fields are. It says that if $K$ is a splitting field over $F$ for $f$ and $K$ contains *any* root for an irreducible polynomial $g \in F[x]$, then $K$ will contain *all* roots for $g$. That is, $g$ will also split in $K[x]$. The proof of this theorem relies heavily on the uniqueness of splitting fields.

**Theorem 44.3** *Let $K$ be the splitting field over the field $F$ for the polynomial $f$. Suppose that $g \in F[x]$ is an irreducible polynomial over $F$ and $\gamma \in K \backslash F$ is a root for $g$. Then $g$ factors completely into linear factors in $K[x]$.*

**Proof**: Suppose that $K$ is the splitting field over the field $F$ for the polynomial $f$. We know that $K = F(\alpha_1, \alpha_2, \cdots, \alpha_n)$, where the $\alpha_i$ are the roots of $f$ in $K$. Consider an irreducible polynomial $g \in F[x]$, and suppose that $\gamma \in K \backslash F$ with $g(\gamma) = 0$. Let's assume by way of contradiction that $g$ does not completely factor into linear factors inside $K[x]$. But then by Kronecker's Theorem 41.1 we can construct a strictly larger extension field $K(\beta)$ of $K$, for which $\beta$ is a root of $g$. Now because $\gamma$ and $\beta$ are both roots of the irreducible polynomial $g$, we know by Theorem 42.2 that there is an isomorphism between the fields $F(\gamma)$ and $F(\beta)$ that leaves $F$ fixed. Now $K(\beta) = F(\alpha_1, \cdots, \alpha_n)(\beta) = F(\alpha_1, \cdots, \alpha_n, \beta) = F(\beta)(\alpha_1, \cdots, \alpha_n)$, and this is the splitting field of $f$ over $F(\beta)$. But because $F(\beta)$ and $F(\gamma)$ are isomorphic, there is an isomorphism between $K(\beta)$ and $K = K(\gamma) = F(\gamma, \alpha_1, \cdots, \alpha_n)$, since the latter field is the splitting field for $f$ over $F(\gamma)$.

But now we count degrees. By our isomorphism between $K$ and $K(\beta)$ as splitting fields over the isomorphic fields $F(\gamma)$ and $F(\beta)$ we have that $[K : F(\gamma)] = [K(\beta) : F(\gamma)]$. But then by Theorem 43.2

$$[K : F] = [K : F(\gamma)][F(\gamma) : F] = [K(\beta) : F(\gamma)][F(\gamma) : F] = [K(\beta) : F],$$

and this means that $[K(\beta) : K] = 1$, and so $\beta \in K(\beta) = K$, which is as we wish. □

Suppose that $K$ is an extension field of the field $F$, and that whenever an irreducible polynomial $f \in F[x]$ has one root in $K$, then it splits in $K$. In this case we say that $K$ is a **normal extension** of $F$. With this terminology, we can rephrase Theorem 44.3 by saying that if $K$ is a splitting field over the field $F$ for the polynomial $f$, then $K$ is in fact a normal extension. Later in this chapter we will encounter Theorem 44.6, which asserts that for a field with characteristic zero, being a splitting field is *equivalent* to being a finite normal extension.

In Chapter 47 we will discover that the concept of normal extension is actually closely related to the concept of normal subgroup!

---

## 44.2   Fields with Characteristic Zero

Theorems 44.1 and 44.2 together say that given any field $F$ and a non-constant polynomial $f$, we can always build a unique minimal extension field, in which $f$ completely factors into linear factors. So, for any polynomial over a field, we obtain the unique (up to isomorphism) splitting field for the polynomial merely by adjoining the roots of the polynomial, by repeated application of Kronecker's Theorem 41.1. In general, some of these roots may already belong to the base field, and some of the roots may be repeated. However, if the polynomial is irreducible, the situation is particularly nice (for fields of characteristic zero).

**Theorem 44.4** *Let $F$ be a field of characteristic zero and $f$ an irreducible polynomial in $F[x]$. Then the roots of $f$ in its splitting field are all distinct.*

**Proof**: Suppose that $f$ has a repeated root $\alpha$ in its splitting field $K$. This means that $f = (x - \alpha)^k g$, where $k$ is an integer greater than one, and $g$ is a polynomial over $K$. We will now make use of the formal derivative of this polynomial. We first encountered this idea in Exercise 4.7 in the polynomial ring $\mathbb{Q}[x]$, and in Exercise 3 you will check that the appropriate results hold, for any field (and in fact for any commutative ring). In particular, the product rule like you encountered in calculus works here, and so the formal derivative $f'$ can be computed thus:

$$f' = k(x - \alpha)^{k-1} g + (x - \alpha)^k g'.$$

Now because $K$ has characteristic zero, this is necessarily a non-zero polynomial. We then have that $x - \alpha$ is a factor of both $f$ and $f'$. But if we use term-by-term differentiation instead, it is clear that $f' \in F[x]$. Since $f$ is irreducible, we know that in $F[x]$ we have $\gcd(f, f') = 1$, and so by the GCD identity for $F[x]$ we can find polynomials $a, b \in F[x]$ for which $1 = af + bf'$. But if we evaluate this polynomial at $\alpha$ we obtain the following:

$$1 = 1(\alpha) = a(\alpha)f(\alpha) + b(\alpha)f'(\alpha) = 0 + 0 = 0,$$

a contradiction. Thus $f$ must not have any repeated roots in the splitting field $K$. $\square$

Theorem 44.4 remains true for all finite fields, and you will prove this fact in Exercise 45.8. The theorem is false, however, for infinite fields with characteristic $p$ (although we will not pursue this topic in this book). We will find Theorem 44.4 of considerable use when we return to the problem of determining when polynomial equations can be solved with radicals.

We close this section by proving another important result about fields with characteristic zero. It is surprisingly the case that for such fields any finite extension is simple. Now we know by Theorem 43.1 that any finite extension is algebraic. Thus to assert that a finite extension is simple, we are really saying that if we adjoin finitely many algebraic numbers to a field with characteristic zero, we only need to adjoin a single element. In particular, over such a field the splitting field is always a simple extension.

### Example 44.7

An interesting example of this is Example 43.3, where we rather laboriously proved that $\mathbb{Q}(\sqrt{2}, \sqrt[3]{2})$ is a simple extension $\mathbb{Q}(\alpha)$ of $\mathbb{Q}$, where $\alpha = \sqrt{2} + \sqrt[3]{2}$. As we shall see in the proof of the theorem, it is no accident that $\alpha$ is a linear combination of the elements $\sqrt{2}, \sqrt[3]{2}$.

**Theorem 44.5** *Let $F$ be a field with characteristic zero, and $K$ a finite extension of $F$. Then $K$ is an algebraic simple extension of $F$.*

**Proof**: Suppose that $\alpha$ and $\beta$ are algebraic elements over a field $F$, which has characteristic zero. We will show that $F(\alpha, \beta) = F(\mu)$, for some algebraic element $\mu$. This reduction from two generators to one is clearly the induction step needed to show that any finite extension is simple (see Exercise 4).

Now $\alpha$ and $\beta$ are roots of irreducible polynomials $f, g \in F[x]$ of degrees $n$ and $m$, respectively. We may as well do all of our computations in a field $K$ in which both $f$ and $g$ split into linear factors. Because $F$ has characteristic zero, Theorem 44.4 tells us that the roots $\alpha = \alpha_1, \cdots, \alpha_n$ of $f$ in $K$ are all distinct from one another, and that the roots $\beta = \beta_1, \cdots, \beta_m$ of $g$ in $K$ are also all distinct from one another.

Consider the finite set of elements in the field $K$:

$$\left\{ \frac{\alpha_j - \alpha}{\beta - \beta_i} \right\},$$

where $i > 1$ and $j > 1$. Since $F$ is of characteristic zero, it is an infinite field, and so we can choose a non-zero element $a \in F$ not equal to any of these quotients.

We will now define $\mu = \alpha + a\beta$. It is quite evident that $F(\mu) \subseteq F(\alpha, \beta)$. We must show the reverse inclusion is true, by arguing that $\alpha, \beta \in F(\mu)$. For that purpose, consider the polynomial $h \in F(\mu)[x]$, defined by $h(x) = f(\mu - ax)$. Notice that

$$h(\beta) = f(\mu - a\beta) = f(\alpha) = 0.$$

Thus $h$ and $g$ share the root $\beta$.

However, $h$ and $g$ can share no other root. For if they did, it would be some $\beta_i$, with $i > 1$. But then

$$0 = h(\beta_i) = f(\mu - a\beta_i) = f(\alpha + a\beta - a\beta_i).$$

This would mean that $\alpha + a(\beta - \beta_i) = \alpha_j$, for some $j$, and we chose $a$ precisely so that this cannot be true.

Because $g$ factors into linear factors in $K$, it is now evident that the gcd of $h$ and $g$ in $K[x]$ is $x - \beta$. But $h, g \in F(\mu)[x]$, and so when we perform Euclid's algorithm to compute $\gcd(h, g)$, we will remain in the ring $F(\mu)[x]$. This means that $x - \beta \in F(\mu)[x]$, or in other words, $\beta \in F(\mu)$. But because $\alpha = \mu - a\beta$, $\alpha \in F(\mu)$, too. Thus, $F(\alpha, \beta) = F(\mu)$.   □

Examples of this theorem appear in Example 43.3, and Exercises 43.3 and 43.4.

We can now use this theorem to show that for fields of characteristic zero, a finite extension is normal exactly if it is the splitting field for some irreducible polynomial.

**Theorem 44.6** *Suppose that $F$ is a field of characteristic zero, with finite extension $K$. Then $K$ is the splitting field for some irreducible $f \in F[x]$ if and only if $K$ is a normal extension of $F$.*

**Proof:**   Suppose that $K$ is a finite normal extension of $F$, a field with characteristic zero. Then by Theorem 44.5 $K = F(\alpha)$, where $\alpha$ is algebraic over $F$. Let $f \in F[x]$ be the minimal polynomial for $\alpha$. Since $K$ is normal, $f$ splits in $K$. But clearly no smaller subfield of $K$ contains all the roots of $f$, because $K = F(\alpha)$. Thus $K$ is the splitting field for $f$ over $F$.

The converse is just Theorem 44.3.   □

Theorems 46.2 and 47.2 will give other conditions equivalent to being a finite normal extension.

---

# Chapter Summary

In this chapter we proved that every polynomial over a base field admits a unique minimal field extension of that field, in which the polynomial factors into linear factors; this is called its *splitting field*. A splitting field for a polynomial actually has the stronger property that if an irreducible polynomial over the base field has even one root in the splitting field, it has all of its roots. We call such field extensions *normal*. We also showed that for a field of characteristic zero, finite extensions are necessarily simple extensions.

# Warm-up Exercises

a. Answer the following true or false; if your answer is false, give a counterexample. Assume that $F$, $K$, and $L$ are fields.

   (a) A splitting field for $f$ over $F$ is an algebraic extension of $F$.

   (b) All finite extensions of $\mathbb{Q}$ are simple.

   (c) A normal extension of $\mathbb{Q}$ is the splitting field for some $f \in \mathbb{Q}[x]$.

   (d) A splitting field for $f$ over $F$ is a normal extension.

   (e) If $f \in \mathbb{Q}[x]$ is irreducible, then all of the roots of $f$ in $\mathbb{C}$ are distinct.

   (f) Suppose that $f \in F[x]$ is irreducible and has degree bigger than 1. To construct the splitting field for $f$ over $F$, it is always possible to use the field $F[x]/\langle f \rangle$ constructed by Kronecker's Theorem 41.1.

   (g) Suppose that $f \in F[x]$ is irreducible and has degree bigger than 1. To construct the splitting field for $f$ over $F$, it is never possible to use the field $F[x]/\langle f \rangle$ constructed by Kronecker's Theorem 41.1.

   (h) All splitting fields are finite extensions.

   (i) If $F \subseteq L \subseteq K$, and $K$ is the splitting field for $f$ over $F$, then $K$ is the splitting field for $f$ over $L$.

   (j) Suppose $K$ is a proper finite extension of $F$. Then $K$ is the splitting field for some polynomial $f$ over $F$.

b. Consider the polynomial $x^3 - 2$. Describe its splitting field over the following fields:

$$\mathbb{Q}, \quad \mathbb{R}, \quad \mathbb{Q}(i), \quad \mathbb{C}$$

c. Let $F$ be a field, and suppose that $f \in F[x]$ has degree 1. What is the splitting field of $f$ over $F$?

# Exercises

1. In this problem we check the claims made in Example 44.5.

   (a) Show that the polynomial $f$ factors as

$$f = (x + \alpha)\left(x + \alpha^2\right)\left(x + \alpha + \alpha^2\right).$$

   (b) Prove that $\mathbb{Z}_2(\alpha) = \mathbb{Z}_2(\alpha^2)$.

   (c) Prove that $\mathbb{Z}_2(\alpha) = \mathbb{Z}_2(\alpha + \alpha^2)$.

2. In this problem you will follow the proof of Theorem 44.5 in the particular case where $\alpha = \sqrt{2}$, $\beta = \sqrt{7}$, and $F = \mathbb{Q}$.

   (a) Make a complete list of the elements of $\mathbb{Q}(\sqrt{2}, \sqrt{7})$ that are not allowed to be the element $a$. Choose your own value of $a$ (there are many choices) and find an appropriate value for $\mu$.

   (b) Determine the polynomial $h \in \mathbb{Q}(\mu)[x]$ and check that $\beta$ is a root.

(c) Compute the gcd of $h$ and $g$ in $\mathbb{Q}(\mu)[x]$ using Euclid's algorithm, and thus verify the claim about this gcd in the proof.

3. Suppose that $R$ is a commutative ring, and $f \in R[x]$. Then we can write $f$ as $f = a_n x^n + a_{n-1} x^{n-1} + a_1 x + a_0$, where the coefficients $a_i$ are elements of $R$. Define the formal derivative of $f$ as

$$f' = na_n x^{n-1} + (n-1)a_{n-1} x^{n-2} + \cdots + a_1,$$

which is just the formula we expect from calculus. (Note that we explored this notion for $R = \mathbb{Q}$ in Exercise 4.7.)

(a) Suppose that $a \in R$ and $f \in R[x]$. Prove that $(af)' = af'$. (The *constant multiple rule*.)

(b) Suppose that $f, g \in R[x]$. Prove that $(f + g)' = f' + g'$. (The *sum rule*.)

(c) Suppose that $f, g \in R[x]$. Then $(fg)' = f'g + fg'$. (The *product rule*.)

(d) Suppose that $a \in R$, and $n$ is a positive integer. Prove that if $f = (x - a)^n$, then $f' = n(x - a)^{n-1}$. (The *power rule*.)

4. Complete the inductive proof for Theorem 44.5: Suppose $K = F(\alpha_1, \cdots, \alpha_n)$, for algebraic elements $\alpha_i$. Show that there is an algebraic element $\mu$ so that $K = F(\mu)$.

5. Consider the polynomial $f = x^4 - 4x^2 + 2 \in \mathbb{Q}[x]$.

(a) Argue that $f$ is irreducible over $\mathbb{Q}$.

(b) Factor $f$ completely into linear factors in $\mathbb{C}[x]$.

(c) Describe the splitting field of $f$ over $\mathbb{Q}$ in $\mathbb{C}$.

(d) This splitting field can actually be described in the form $\mathbb{Q}(\alpha)$, where $\alpha$ is one of the roots of $f$. Do so if you haven't done so already, justifying your assertion rigorously.

6. Consider the polynomial $f = x^4 - 2 \in \mathbb{Q}[x]$.

(a) Argue that $f$ is irreducible over $\mathbb{Q}$.

(b) Factor $f$ completely into linear factors in $\mathbb{C}[x]$.

(c) Describe the splitting field $K$ of $f$ over $\mathbb{Q}$ in $\mathbb{C}$. (The easiest such description is of the form $\mathbb{Q}(\alpha, \beta)$, where $\alpha, \beta$ are appropriately chosen elements of $\mathbb{C}$.)

(d) Prove that $K$ *cannot* be described in the form $\mathbb{Q}(\alpha)$, where $\alpha$ is one of the roots of $f$.

(e) Why does Theorem 44.5 imply that this splitting field is a simple extension of $\mathbb{Q}$? Give such a description, justifying your assertion rigorously.

7. Suppose that $F$ is a field, and $K$ is a field extension of $F$ that is *algebraically closed* (see Exercise 10.26 for a definition). Suppose that $f \in F[x]$. Argue that $K$ contains a copy of the splitting field of $f$ over $F$. Describe this splitting field more explicitly.

8. This exercise follows up on Exercise 6. Let $\beta = \sqrt[4]{2}(1 + i)$.

(a) Show that $\beta$ is a root of $g = x^4 + 8$.

(b) Factor $g$ completely into linear factors in $\mathbb{C}[x]$. Then use your factorization to argue that $g$ is the minimal polynomial for $\beta$ over $\mathbb{Q}$.

(c) Let $F$ be the splitting field of $g$ over $\mathbb{Q}$. Use Exercise 6b above to argue that $F \subseteq K$.

(d) What is $[\mathbb{Q}(\beta) : \mathbb{Q}]$? Use this fact to argue that $[\mathbb{Q}(\beta) : \mathbb{Q}(\sqrt{2}i)] = 2$.

(e) Use (b) to argue that $\sqrt[4]{2}, i \in F$, and so $F = K$.

(f) What is $[\mathbb{Q}(\sqrt[4]{2}) : \mathbb{Q}]$? Then argue that $[F : \mathbb{Q}] = 8$.

# Chapter 45

## Finite Fields

In the previous chapter we proved that any irreducible polynomial $f$ over a field $F$ has a unique splitting field: A minimal field extension of $F$ in which $f$ can be factored into linear factors. This provides a field inside of which we can explore whether or not the roots of $f$ are obtainable by elementary algebraic operations. We will pursue this goal in the remaining chapters in this book.

But a wonderful bonus flows from the existence and uniqueness of splitting fields, and we will take a small detour from our goal to explore this bonus in the present chapter. We are now able to completely describe all finite fields. We have for a long time been familiar with the finite fields $\mathbb{Z}_p$, where $p$ is a prime integer. We have also encountered various finite fields as finite extensions of such fields; for example, consider Example 41.5, Exercise 42.2 and Example 44.5. In this chapter we will be able to place these examples in a beautiful general context.

## 45.1  Existence and Uniqueness

Theorem 41.2 says that every field has characteristic zero or $p$, where $p$ is a prime integer. Fields with characteristic zero have a subfield isomorphic to $\mathbb{Q}$ and so are infinite. Thus, any finite field has characteristic $p$, for some prime $p$. This means that every finite field contains (an isomorphic copy of) one of the fields $\mathbb{Z}_p$ as a subfield (recall that this is called the *prime subfield*). We use these considerations to prove our first result about arbitrary finite fields.

**Theorem 45.1** *A finite field of characteristic $p$ has $p^n$ elements, for some positive integer $n$.*

**Proof**:    With our knowledge of group theory, this becomes quite an easy theorem. Let $F$ be a finite field, and consider its additive group structure. Saying that its characteristic is $p$ means exactly that every non-zero element has (additive) order $p$. But then Cauchy's Theorem 24.5 says that $p$ must be the only prime integer dividing the order of $F$. That is, every finite field has $p^n$ elements, for some positive integer $n$. (We've seen this proof before, and we could instead have cited Theorem 28.1 here.)    □

But do such fields exist, for every prime $p$, and every positive integer $n$? The next theorem says that they do. Here, field theory comes to the rescue.

**Theorem 45.2** *For every prime integer $p$ and every positive integer $n$, there exists a field with $p^n$ elements.*

**Proof**:    Consider the polynomial $f = x^{p^n} - x \in \mathbb{Z}_p[x]$. By Theorem 44.1 we know that there exists a splitting field $F$ of $\mathbb{Z}_p$ for $f$. Because $F$ is a field extension of $\mathbb{Z}_p$, it must be a field of characteristic $p$. We will show that $F$ is exactly the required field of $p^n$ elements.

Pick two roots $r, s \in F$ of the polynomial $f$. We claim that $r - s$ is also a root of $f$. To check this, we must evaluate $f(r - s) = (r - s)^{p^n} - (r - s)$. But in a field of characteristic $p$, we know that $(r - s)^{p^n} = r^{p^n} - s^{p^n}$ (see Exercise 1c). Thus,

$$f(r - s) = r^{p^n} - s^{p^n} - r + s = \left(r^{p^n} - r\right) - \left(s^{p^n} - s\right) = 0 - 0 = 0,$$

and so $r - s$ is a root of $f$, as claimed. That is, the set of roots of $f$ is an (additive) group.

Now pick two roots $r, s \in F$ of the polynomial $f$, with $s \neq 0$. We claim that $rs^{-1}$ is also a root of $f$. But

$$f(rs^{-1}) = r^{p^n} s^{-p^n} - rs^{-1} = rs^{-1} - rs^{-1} = 0.$$

Thus, the set of roots of $f$ actually forms a *subfield* of $F$. But because a splitting field is minimal among all fields containing the roots, $F$ consists exactly of the roots of $f$.

To show that $F$ contains exactly $p^n$ elements, it remains to check that $f$ has no repeated roots. Pick any root $r$ of $f$. Then

$$
\begin{aligned}
f(x - r) &= (x - r)^{p^n} - (x - r) \\
&= x^{p^n} - x - (r^{p^n} - r) = f(x) - 0 = f(x).
\end{aligned}
$$

Thus,

$$f = (x - r)((x - r)^{p^n - 1} - 1).$$

But clearly $x - r$ is not a factor of $(x - r)^{p^n - 1} - 1$ (because $x - r$ does not divide 1). Thus, $r$ is not a repeated root. Consequently, $f$ has $p^n$ roots, and so the set of roots of $f$ is equal to the splitting field $F$ and is the required field with $p^n$ elements. $\square$

We shall now use Theorem 44.2 (the uniqueness of splitting fields) to show that any two finite fields with $p^n$ elements are isomorphic:

**Theorem 45.3** *All fields with $p^n$ elements are isomorphic.*

**Proof:**   Any finite field $F$ with $p^n$ elements has characteristic $p$, and so is a finite extension of $\mathbb{Z}_p$. Consider again the polynomial $f = x^{p^n} - x \in \mathbb{Z}_p[x]$. We will now show that the elements of $F$ are exactly the distinct roots of $f$, and so is exactly the splitting field of $f$ over $\mathbb{Z}_p$.

We know that the multiplicative group $F^*$ has $p^n - 1$ elements. Now pick $a \in F^*$. By Lagrange's Theorem we know that the order of $a$ divides $p^n - 1$; thus, $a^{p^n - 1} = 1$. But then $a^{p^n} = a$, and so $a$ is a root of $x^{p^n} - x$. Thus, every element of $F$ is a root of $x^{p^n} - x$. The field $F$ has $p^n$ elements and $x^{p^n} - x$ can have no more than $p^n$ roots. Hence, $F$ consists of exactly those roots.

But then the uniqueness of splitting fields (Theorem 44.2) means that all fields with $p^n$ elements are isomorphic. $\square$

Because the field with $p^n$ elements is unique (up to isomorphism), we can unambiguously denote it by $GF(p^n)$. This stands for **Galois field of order** $p^n$, after the Frenchman Evariste Galois who was the first mathematician to consider finite fields. Of course, $GF(p)$ is merely a different notation for the familiar field $\mathbb{Z}_p$.

The uniqueness of finite fields was proved in 1893 by the American mathematician E. H. Moore, who was by that time at the University of Chicago. Moore played a crucial part in bringing American mathematics to the level of European mathematics. He was the founder of the mathematics department at the University of Chicago and was one of the founders of the American Mathematical Society. But most important was his work as an admired mathematician and as a teacher and mentor for the next generation of American mathematicians.

## 45.2 Examples

**Example 45.1**

Consider the polynomial $x^2 + 1 \in \mathbb{Z}_3[x]$. It is easy to see that this polynomial has no roots in $\mathbb{Z}_3$, by checking the three possibilities 0, 1, and 2. Thus, by the Root Theorem, $x^2 + 1$ is irreducible, and so the ideal $\langle x^2 + 1 \rangle$ is maximal (Theorems 15.1 and 15.2). Thus, $\mathbb{Z}_3[x]/\langle x^2 + 1 \rangle$ is a field. This field has degree 2 over $\mathbb{Z}_3$ and so has 9 elements. It is consequently the Galois field $GF(9)$. We should really write these elements as (additive) cosets. For simplicity's sake, we will replace $\langle x^2 + 1 \rangle + x$ by $\alpha$. With this convention, the elements of this field are

$$\{0, \ 1, \ 2, \ \alpha, \ \alpha + 1, \ \alpha + 2, \ 2\alpha, \ 2\alpha + 1, \ 2\alpha + 2\},$$

where the multiplication is determined by the law $\alpha^2 + 1 = 0$; that is, $\alpha^2 = 2$.

Let's look at the multiplicative structure of this field. That is, what sort of group is $GF(9)^*$? It is certainly an abelian group with 8 elements, and up to isomorphism there are three such groups. It turns out that this group is cyclic, and $\alpha + 1$ is a generator. Let's verify this, by brute calculation:

$$(\alpha + 1)^1 = \alpha + 1, \quad (\alpha + 1)^2 = \alpha^2 + 2\alpha + 1 = 2\alpha,$$

$$(\alpha + 1)^3 = 2\alpha(\alpha + 1) = 2\alpha^2 + 2\alpha = 2\alpha + 1,$$

$$(\alpha + 1)^4 = (\alpha + 1)(2\alpha + 1) = 2\alpha^2 + 3\alpha + 1 = 2,$$

$$(\alpha + 1)^5 = 2(\alpha + 1) = 2\alpha + 2,$$

$$(\alpha + 1)^6 = (\alpha + 1)(2\alpha + 2) = 2\alpha^2 + 4\alpha + 2 = \alpha,$$

$$(\alpha + 1)^7 = (\alpha + 1)\alpha = \alpha^2 + \alpha = \alpha + 2,$$

$$(\alpha + 1)^8 = (\alpha + 1)(\alpha + 2) = \alpha^2 + 3\alpha + 2 = 1.$$

It turns out that the group of units of *any* finite field is cyclic! This is called the Primitive Root Theorem:

**Theorem 45.4    Primitive Root Theorem**    *The group of units of any finite field* $GF(p^n)$ *is cyclic.*

**Proof:**    A freestanding proof of this theorem is difficult, but we can obtain it by using the Fundamental Theorem for Finite Abelian Groups 29.12 (which we did not prove). Obviously $GF(p^n)^*$ is a finite abelian group, and therefore by the Fundamental Theorem it is isomorphic to a direct product of cyclic groups of the form

$$\mathbb{Z}_{q_1}^{k_1} \times \mathbb{Z}_{q_2}^{k_2} \times \cdots \times \mathbb{Z}_{q_r}^{k_r}$$

with each $q_i$ a prime, and where the primes need not be distinct. Now if all of the primes $q_i$ are distinct, this group is cyclic of order $p^n - 1 = q_1^{k_1} \cdots q_r^{k_r}$, by the Chinese Remainder Theorem 16.1. So we wish to show that there are no repetitions in the primes $q_i$. We proceed by way of contradiction. We may assume without loss of generality that $q_1 = q_2$, and $k_1 \leq k_2$. But then every element in the direct product of cyclic groups has order no more than $m = q_2^{k_2} \cdots q_r^{k_r} < p^n - 1$.

▷ **Quick Exercise.**  Why is this true? ◁

But this means that every non-zero element in the field $GF(p^n)$ is a root of the polynomial $x^m - 1$; that is, this polynomial has $p^n - 1$ roots. But by Corollary 4.4 (true for any field), such a polynomial can have at most $m$ roots. Since $m < p^n - 1$, this is a contradiction, and so $GF(p^n)^*$ is a cyclic group as claimed. □

**Example 45.2**

We now know that groups $\mathbb{Z}_{11}^*$ and $\mathbb{Z}_{17}^*$ are cyclic. You will check this in Exercise 11.

We now consider when finite fields are subfields of one another. The following example will help motivate this:

**Example 45.3**

In order to construct the field $GF(8)$, we need only find an irreducible polynomial $f$ in $\mathbb{Z}_2[x]$ of degree three; then form $\mathbb{Z}_2[x]/\langle f \rangle$. We actually performed this calculation in Example 44.5 (see also Exercise 44.1), using the polynomial $f = x^3 + x + 1$. In Exercise 9 you will use the polynomial $x^3 + x^2 + 1$ instead, and show explicitly that the resulting field is isomorphic to the field constructed in Example 44.5.

Now consider the polynomial $g = x^2 + x + 1 \in GF(8)[x]$. We claim that this is an irreducible polynomial in $GF(8)[x]$. To show this, let's try by brute force to see whether any of the 8 elements of $GF(8)$ are roots of the polynomial. For example,

$$g(\alpha^2 + \alpha) = (\alpha^2 + \alpha)^2 + (\alpha^2 + \alpha) + 1 =$$

$$\alpha^4 + \alpha^2 + \alpha^2 + \alpha + 1 = \alpha^4 + \alpha + 1 =$$

$$\alpha(\alpha + 1) + \alpha + 1 = \alpha^2 + 1 \neq 0.$$

Thus, $\alpha^2 + \alpha$ is not a root.

▷ **Quick Exercise.** Try the other seven elements of the field, and thus see that this polynomial is irreducible. ◁

But then we can construct the field

$$GF(8)/\langle x^2 + x + 1 \rangle.$$

This is a degree 2 extension of $GF(8)$, which has 8 elements, and so is a field with 64 elements altogether. Thus we have constructed the essentially unique field with 64 elements $GF(2^6) = GF(64)$. Note of course that we could have also constructed this field, had we been able to find an irreducible polynomial in $\mathbb{Z}_2[x]$ of degree 6. You will find such a polynomial in Exercise 10.

The previous example shows that the unique field with $2^3$ elements can be considered a subfield of the unique field with $2^6$ elements. The full story is given by the following theorem, which you prove in Exercise 5.

**Theorem 45.5** *The field $GF(p^m)$ is a subfield of $GF(p^n)$ if and only if $m$ divides $n$.*

---

# Chapter Summary

In this chapter we prove that for every positive prime integer $p$ and positive integer $n$, there exists a unique finite field with exactly $p^n$ elements. This is a nice consequence of the uniqueness of splitting fields.

# Warm-up Exercises

a. Describe how you would construct the field $GF(25)$; specify a polynomial $p \in \mathbb{Z}_5[x]$ that you could use.

b. Consider the multiplicative group $GF(32)^*$. Why is this group obviously cyclic? (You need not refer to the Primitive Root Theorem.)

c. Give three non-isomorphic commutative rings with nine elements. How many are fields? Are they isomorphic as abelian groups, or not?

d. Consider $GF(343)$. Give a field of characteristic 7 in which this field is a proper subfield. Give a field of characteristic 7 with fewer than 343 elements, which cannot be found as a subfield of this field.

# Exercises

1. Suppose that $F$ is a field with characteristic $p$, and $a, b \in F$.

    (a) Prove that $(a + b)^p = a^p + b^p$. *Hint:* Use Exercise 2.19 and the binomial theorem Exercise 6.17.

    (b) For any positive integer $n$, prove that $(a + b)^{p^n} = a^{p^n} + b^{p^n}$.

    (c) For any positive integer $n$, prove that $(a - b)^{p^n} = a^{p^n} - b^{p^n}$.

2. Explicitly write out the elements of $GF(25)$, as constructed along the lines of Warm-up Exercise a, above. Find a generator for the cyclic multiplicative group $GF(25)^*$.

3. Define the function

$$\varphi : GF(p^n) \to GF(p^n)$$

by $\varphi(a) = a^p$. Prove that this is a ring isomorphism, called the **Frobenius** isomorphism (this is actually a repeat of Exercise 12.18).

4. Describe how to construct $GF(81)$ by using a specific polynomial $p \in GF(9)[x]$.

5. Prove Theorem 45.5. That is, show that $GF(p^m)$ is a subfield of $GF(p^n)$ if and only if $m$ divides $n$.

6. Prove that every element in a finite field with characteristic $p$ has a $p$th root.

7. Suppose that $f \in GF(p^n)[x]$ is a non-constant polynomial and suppose that its formal derivative is zero. Prove that $f$ is not irreducible.

8. Suppose that $F$ is a finite field and $f$ is an irreducible polynomial in $F[x]$. Prove that the roots of $f$ in its splitting field are all distinct. (That is, prove Theorem 44.4, replacing the hypothesis that the field is characteristic zero, with the hypothesis that it is finite.)

9. In this exercise we follow up on Example 45.3, where we constructed $GF(8)$ as $\mathbb{Z}_2[x]/\langle f \rangle$, with $f = x^3 + x + 1$. Then $GF(8) = \mathbb{Z}_2(\alpha)$, where $\alpha = \langle f \rangle + x$.

    (a) We know from the proof of Theorem 45.2 that $GF(8)$ consists exactly of the 8 distinct roots of the polynomial $x^8 - x \in \mathbb{Z}_2[x]$. Explain why $f$ is a factor of this polynomial.

    (b) Factor $x^8 - x$ completely in $\mathbb{Z}_2[x]$ as a product of irreducible polynomials. (You should obtain a cubic irreducible polynomial $g$ other than $f$ as a factor.)

    (c) Factor $g$ into linear factors in $\mathbb{Z}_2(\alpha)[x]$.

    (d) Why are $\mathbb{Z}_2[x]/\langle f \rangle$ and $\mathbb{Z}_2[x]/\langle g \rangle$ isomorphic as fields? (Your answer should just involve a citation of the appropriate theorem(s).)

    (e) Let $\beta = \langle g \rangle + x$. Construct an explicit isomorphism between the fields $\mathbb{Z}_2(\alpha)$ and $\mathbb{Z}_2(\beta)$. You should specify a function from one of these fields to the other, including precisely where each of the eight elements goes.

10. In Example 45.3 we constructed $GF(64)$ by building $GF(8)$ first, and then constructing a degree two extension of that field. The alternative is to find a degree six irreducible polynomial $h \in \mathbb{Z}_2[x]$, and then construct the field $\mathbb{Z}_2[x]/\langle h \rangle$.

    (a) Give a complete list (with justification) of all degree 1, 2, or 3 irreducible polynomials in $\mathbb{Z}_2[x]$.

    (b) Determine all fourth degree irreducible polynomials in $\mathbb{Z}_2[x]$.

    (c) Determine all sixth degree irreducible polynomials in $\mathbb{Z}_2[x]$.

11. Show by direct computation that $\mathbb{Z}_{11}$ and $\mathbb{Z}_{17}$ have primitive roots: That is, verify explicitly that 2 (and 3, respectively) generate their groups of units.

# Chapter 46

## Galois Groups

We now return to our goal of understanding whether the roots of an irreducible polynomial over a field can be obtained by elementary algebraic computations. In Chapter 44 we constructed the unique splitting field for such a polynomial, inside of which such computations must occur. In the present chapter we will look closely at what sort of field extension the splitting field must be. We will use group theory to do this.

In Chapter 17 we saw how geometry could be illuminated by considering functions leaving geometric properties fixed; we thus obtained groups of symmetries. Here we will illuminate field extensions (and splitting fields in particular) by considering functions leaving field properties fixed; we will thus obtain groups of automorphisms called Galois groups.

## 46.1 The Galois Group

To better understand a field $F$, it is useful to consider all functions that preserve essential algebraic structure. Such functions are one-to-one and onto ring homomorphisms from the field to itself; such homomorphisms are called **automorphisms**. We have denoted the set of all such automorphisms $\mathrm{Aut}(F)$; this set is a group under functional composition (see Example 19.21).

Suppose that $E$ and $F$ are fields, and $F \subseteq E$. We may now consider the following subset of $\mathrm{Aut}(E)$:

$$\mathrm{Gal}(E|F) = \{\varphi \in \mathrm{Aut}(E) : \varphi(f) = f, \text{for all } f \in F\}.$$

That is, we are considering only those automorphisms of the field $E$ that leave all elements of the subfield $F$ *fixed*. It is easy to check that $\mathrm{Gal}(E|F)$ is a subgroup of $\mathrm{Aut}(E)$, using the Subgroup Theorem 20.2.

▷ **Quick Exercise.** Show that $\mathrm{Gal}(E|F)$ is a subgroup of $\mathrm{Aut}(E)$. ◁

We call $\mathrm{Gal}(E|F)$ the **Galois Group of the field $E$ over $F$**. Let's consider some examples.

#### Example 46.1

$\mathrm{Gal}(\mathbb{R}|\mathbb{Q})$ is the trivial group, because $\mathrm{Aut}(\mathbb{R})$ itself is the group with only one element (see Exercise 19.13).

#### Example 46.2

Let $K$ be any field, with prime subfield $F$. The prime subfield is the field generated from 1 by field operations, and so any automorphism must leave it fixed. This means that $\mathrm{Gal}(K|F) = \mathrm{Aut}(K)$.

## Example 46.3

Gal($\mathbb{C}|\mathbb{R}$) is the two element group $\{\iota, \varphi\}$, where $\iota$ is the identity automorphism, and $\varphi$ is the complex conjugation map. This is true because there are only two automorphisms of $\mathbb{C}$ that fix $\mathbb{R}$ (Exercise 19.14).

## Example 46.4

Let's compute the Galois group Gal($\mathbb{Q}(\sqrt{2})|\mathbb{Q}$). Because $\mathbb{Q}$ is the prime subfield of $\mathbb{Q}(\sqrt{2})$, it is left fixed by any automorphism (Example 46.2). Since any element of $\mathbb{Q}(\sqrt{2})$ is of the form $a + b\sqrt{2}$, where $a, b \in \mathbb{Q}$, it is clear than any automorphism $\varphi$ of this field is determined by $\varphi(\sqrt{2})$. But

$$2 = \varphi(2) = \varphi\left(\sqrt{2}^2\right) = (\varphi(\sqrt{2}))^2,$$

and so $\varphi(\sqrt{2})$ must be a square root of 2. There are only two choices: $\varphi(\sqrt{2}) = \sqrt{2}$ and $\varphi(\sqrt{2}) = -\sqrt{2}$. The first of these choices leads to the identity automorphism. The second leads to $\varphi(a + b\sqrt{2}) = a - b\sqrt{2}$. You can check that this is an automorphism of $\mathbb{Q}(\sqrt{2})$ leaving $\mathbb{Q}$ fixed, and so Gal($\mathbb{Q}(\sqrt{2})|\mathbb{Q}$) is a two element group.

▷ **Quick Exercise.** Check explicitly that the set of these two automorphisms forms a group. ◁

We will generalize this example in Theorem 46.1 below.

## Example 46.5

Consider the field $\mathbb{Q}(\sqrt{2}, \sqrt{3})$. Now every element of this field can be expressed as $a + b\sqrt{2} + c\sqrt{3} + d\sqrt{6}$ (see Example 43.2). Thus every automorphism $\varphi$ of $\mathbb{Q}(\sqrt{2}, \sqrt{3})$ is determined by what it does to $\sqrt{2}$, $\sqrt{3}$, $\sqrt{6}$. Since $\sqrt{6} = (\sqrt{2})(\sqrt{3})$, we actually need only know $\varphi(\sqrt{2})$ and $\varphi(\sqrt{3})$. By an argument similar to that in Example 46.4, we have two choices for each of these. These lead to exactly four possibilities:

$$
\begin{aligned}
\iota(a + b\sqrt{2} + c\sqrt{3} + d\sqrt{6}) &= a + b\sqrt{2} + c\sqrt{3} + d\sqrt{6} \\
\varphi_1(a + b\sqrt{2} + c\sqrt{3} + d\sqrt{6}) &= a - b\sqrt{2} + c\sqrt{3} - d\sqrt{6}, \\
\varphi_2(a + b\sqrt{2} + c\sqrt{3} + d\sqrt{6}) &= a - b\sqrt{2} - c\sqrt{3} + d\sqrt{6}, \\
\varphi_3(a + b\sqrt{2} + c\sqrt{3} + d\sqrt{6}) &= a + b\sqrt{2} - c\sqrt{3} - d\sqrt{6}.
\end{aligned}
$$

You can check directly that these are all automorphisms. We thus have a group with four elements. It is easy to check that each of these elements has order 2, and so this group is (up to isomorphism) the Klein Four Group (see the discussion in Section 24.3).

▷ **Quick Exercise.** Check that each $\varphi_i$ is in fact an automorphism, and that each such has order 2. ◁

We will now obtain a theorem generalizing the arguments made in Examples 46.4 and 46.5:

**Theorem 46.1** *Let* $F \subseteq K$ *be fields, and* $f \in F[x]$ *an irreducible polynomial, and* $\alpha \in K \backslash F$ *a root of* $f$. *Suppose that* $\varphi \in$ Gal($F(\alpha)|F$). *Then* $\varphi$ *is entirely determined by* $\varphi(\alpha)$. *Furthermore,* $\varphi(\alpha)$ *must be a root of* $f$ *in* $K$, *and so*

$$|\mathrm{Gal}(F(\alpha)|F)| \leq \deg(f) = [F(\alpha) : F].$$

**Proof**: Let $\deg(f) = n$. We know from Theorem 42.3 that every element of $F(\alpha)$ can be written (uniquely) in the form

$$\beta = b_0 + b_1\alpha + \cdots + b_{n-1}\alpha^{n-1},$$

where $b_i \in F$. But then

$$\varphi(\beta) = \varphi(b_0) + \varphi(b_1)\varphi(\alpha) + \cdots + \varphi(b_{n-1})\varphi\left(\alpha^{n-1}\right) =$$

$$b_0 + b_1\varphi(\alpha) + \cdots + b_{n-1}\varphi(\alpha)^{n-1};$$

this follows because $\varphi$ is a ring homomorphism, and $\varphi(b_i) = b_i$ since $b_i \in F$. It then follows that $\varphi$ is entirely determined by what it does to $\alpha$.

But $\alpha$ is a root of $f = a_0 + a_1x + \cdots + a_nx^n$, and so by a similar argument

$$f(\varphi(\alpha)) = a_0 + a_1\varphi(\alpha) + \cdots + a_n\varphi(\alpha)^n = \varphi(f(\alpha)) = 0.$$

This means that $\varphi(\alpha)$ is also a root of $f$. Since $\deg(f) = n$, there are at most $n$ choices for $\varphi(\alpha)$. It then follows that $|\mathrm{Gal}(F(\alpha)|F)| \leq \deg(f)$. The irreducibility of $f$ implies that $\deg(f) = [F(\alpha) : F]$. $\square$

This theorem provides us with a practical approach for computing Galois groups of simple algebraic extensions. Because each element of the Galois group takes each root of $f$ to a root of $f$, and since an automorphism is of course an injection, we can (up to isomorphism) view the Galois group as a subgroup of the group of permutations of the roots of $f$. In the examples that follow, we shall usually take this point of view.

▷ **Quick Exercise.** Review the previous two examples in light of this theorem and the observations that follow it. ◁

**Example 46.6**

Let's reconsider Example 44.4 in light of this theorem. Consider the irreducible polynomial $x^3 - 2 \in \mathbb{Q}[x]$ and the field extension $\mathbb{Q}\left(\sqrt[3]{2}\right) \subset \mathbb{R}$ of the rational numbers $\mathbb{Q}$. Theorem 46.1 says that $\mathrm{Gal}(\mathbb{Q}(\sqrt[3]{2})|\mathbb{Q})$ has at most three elements. But in this case the Galois group is trivial, because only one of the three roots of $x^3 - 2$ is a real number, and so is the only root in $\mathbb{Q}(\sqrt[3]{2}) \subset \mathbb{R}$.

---

## 46.2   Galois Groups of Splitting Fields

In light of the previous example and theorem, it seems natural to ask when the Galois group of an algebraic simple field extension is as large as possible (namely, is equal to the degree of the field extension). It turns out that we already have the appropriate concept at hand: This takes place exactly when the extension is normal (at least for fields of characteristic zero). The result is really not all that surprising as Theorem 46.1 tells us that any automorphism in $\mathrm{Gal}(F(\alpha)|F)$ maps a root of $f$ to another root of $f$ and there are exactly $\deg(f)$ roots of $f$ available to us in a normal extension. This is the content of the next theorem:

**Theorem 46.2** *Let $K$ be a finite extension of the field $F$, which is of characteristic zero. Then $|\mathrm{Gal}(K|F)| = [K : F]$ if and only if $K$ is a normal extension of $F$.*

**Proof:**   Suppose first that $K$ is a normal extension of the field $F$ with characteristic zero. We know from Theorem 44.5 that there exists an algebraic element $\beta$ so that $K = F(\beta)$. Because $K$ is a normal extension, $K$ is the splitting field for the minimal polynomial $g$ for $\beta$ over $F$; let's suppose that the degree of $g$ is $m$. We know from Theorem 42.3 that $[K : F] = m$. Suppose that

$$\beta = \beta_1, \ \beta_2, \ \cdots, \ \beta_m$$

are the roots of $g$ in $K$; these roots are distinct, because $F$ has characteristic zero (Theorem 44.4).

Consider now an element $\varphi \in \mathrm{Gal}(K|F)$. By Theorem 46.1 it is entirely determined by $\varphi(\beta)$, and we have only $m$ distinct choices $\beta_i$ to consider. Now by the proof of Exercise 14.20, the function

$$a_0 + a_1\beta + \cdots + a_{m-1}\beta^{m-1} \mapsto a_0 + a_1\beta_i + \cdots + a_{m-1}\beta_i^{m-1}$$

is an isomorphism between $F(\beta)$ and $F(\beta_i)$ that leaves $F$ fixed. Furthermore, $F(\beta) = F(\beta_i) = K$, since $K$ is the unique splitting field for $g$ and a normal extension of $F$. This means that each choice $\beta_i$ leads to a distinct element of $\mathrm{Gal}(K|F)$. Thus $|\mathrm{Gal}(K|F)| \geq m$. But $m = \deg(g)$ and Theorem 46.1 tells us that $|\mathrm{Gal}(K|F)| \leq m$, and so $|\mathrm{Gal}(K|F)| = m = [K : F]$.

Conversely, suppose that $[K : F] = |\mathrm{Gal}(K|F)|$. Since our fields are of characteristic zero, $K = F(\alpha)$, a simple extension of $F$, where $\alpha$ is a root of an irreducible polynomial $f \in F[x]$ and $\deg(f) = [K : F]$. But any $\varphi \in \mathrm{Gal}(K|F)$ is determined by its value $\varphi(\alpha)$, and $\varphi(\alpha)$ is a root of $f$. Since there are by assumption $[K : F]$ distinct elements of $\mathrm{Gal}(K|F)$, this means that all of the roots of $f$ already belong to $K$. That is, $K$ is the splitting field for $f$ over $F$ and so is a normal extension, by Theorem 44.6.   □

Under the conditions of Theorem 46.2, the group $\mathrm{Gal}(K|F)$ can be thought of as a subgroup of $S_m$, where $K = F(\beta)$, $g \in F[x]$ is the minimal polynomial for $\beta$ over $F$, and $m = \deg(g)$. Furthermore, by Theorem 46.1, each of these permutations is completely determined by which of the $m$ roots of $g$ the root $\beta$ is mapped to.

But notice that Theorem 46.2 applies to any finite normal extension, whether we express it explicitly as a simple extension or not. The following example illustrates this:

### Example 46.7

As we saw in Example 44.4, the splitting field for $x^3 - 2$ over $\mathbb{Q}$ is $\mathbb{Q}(\sqrt[3]{2}, \zeta)$, and so is a normal extension of $\mathbb{Q}$. We note that $[\mathbb{Q}(\sqrt[3]{2}) : \mathbb{Q}] = 3$ and $[\mathbb{Q}(\zeta) : \mathbb{Q}] = 2$. This means that $[\mathbb{Q}(\sqrt[3]{2}, \zeta) : \mathbb{Q}] = 6$, and so by Theorem 46.2 the Galois group $G = \mathrm{Gal}(\mathbb{Q}(\sqrt[3]{2}, \zeta)|\mathbb{Q})$ has six elements. Since each element of the group $G$ permutes the roots of $x^3 - 2$, we can (and should) think of $G$ as a group of permutations of these roots. But since $3! = 6$, $G$ consists of all such permutations. We shall thus label the three roots $\sqrt[3]{2}, \sqrt[3]{2}\zeta, \sqrt[3]{2}\zeta^2$ by the three integers $1, 2, 3$ when thinking of the elements of the Galois group as permutations. It is actually easiest to determine the elements $\varphi$ of $G$ by choosing both $\varphi(\sqrt[3]{2})$ and $\varphi(\zeta)$, since $\varphi$ must also permute the two roots $\zeta, \zeta^2$ of the polynomial $x^2 + x + 1 = \frac{x^3 - 1}{x - 1}$.

| $\varphi(\sqrt[3]{2})$ | $\varphi(\zeta)$ | perm |
|:---:|:---:|:---:|
| $\sqrt[3]{2}$ | $\zeta$ | $\iota$ |
| $\sqrt[3]{2}$ | $\zeta^2$ | $(23)$ |
| $\sqrt[3]{2}\zeta$ | $\zeta$ | $(123)$ |
| $\sqrt[3]{2}\zeta$ | $\zeta^2$ | $(12)$ |
| $\sqrt[3]{2}\zeta^2$ | $\zeta$ | $(132)$ |
| $\sqrt[3]{2}\zeta^2$ | $\zeta^2$ | $(13)$ |

▷ **Quick Exercise.** Check that these six permutations make sense. ◁

We could instead consider $\mathbb{Q}(\sqrt[3]{2}, \zeta)$ as a simple extension of $\mathbb{Q}$, and in particular as $\mathbb{Q}(\sqrt[3]{2} + \zeta)$; you can check the proof of Theorem 44.5 to verify that the element $\sqrt[3]{2} + \zeta$ works to make $\mathbb{Q}(\sqrt[3]{2}, \zeta)$ a simple extension. Since we know our field extension is degree 6, we can infer that $\sqrt[3]{2} + \zeta$ must be a root of an irreducible sixth degree polynomial $f \in \mathbb{Q}[x]$. We can then think of the elements of the Galois group as permutations of the roots of $f$. You will check these observations explicitly in Exercise 10; in Exercise 11 you will look at $\mathbb{Q}(\sqrt[3]{2}, \zeta)$ as a simple extension using a different field element and a different minimal polynomial.

## Example 46.8

The argument in Example 46.7 can be made more generally. Suppose that $f \in \mathbb{Q}[x]$ is any irreducible cubic polynomial with one real root and two complex roots. Let $K$ be the splitting field for this polynomial over the rational field. Then $K$ contains a degree 3 field extension of $\mathbb{Q}$, and so 3 divides $[K : \mathbb{Q}]$. But complex conjugation is a non-trivial element of $\mathrm{Gal}(K|\mathbb{Q})$. Since this has order two as a group element, 2 divides $|\mathrm{Gal}(K|\mathbb{Q})| = [K : \mathbb{Q}]$. But then the Galois group consists of all 6 permutations of the three roots of $f$ in $K$ and so is isomorphic to $S_3$.

## Example 46.9

In Exercise 44.6 you factored the polynomial $x^4 - 2 \in \mathbb{Q}$ over the complex numbers as follows:

$$
\begin{aligned}
x^4 - 2 &= \left(x^2 - \sqrt{2}\right)\left(x^2 + \sqrt{2}\right) \\
&= \left(x - \sqrt[4]{2}\right)\left(x + \sqrt[4]{2}\right)\left(x - \sqrt[4]{2}i\right)\left(x + \sqrt[4]{2}i\right).
\end{aligned}
$$

You then argued that the splitting field of this polynomial is $\mathbb{Q}(\sqrt[4]{2}, i)$. Let's compute the Galois group $G = \mathrm{Gal}(\mathbb{Q}(\sqrt[4]{2}, i)|\mathbb{Q})$ of this splitting field. Now

$$[\mathbb{Q}(\sqrt[4]{2}, i) : \mathbb{Q}] = [\mathbb{Q}(\sqrt[4]{2}, i) : \mathbb{Q}(\sqrt[4]{2})][\mathbb{Q}(\sqrt[4]{2}) : \mathbb{Q}] = 2 \cdot 4 = 8,$$

and so the Galois group has eight elements. We can view the elements of $G$ as permutations of the four roots

$$\sqrt[4]{2}, \quad -\sqrt[4]{2}, \quad \sqrt[4]{2}i, \quad -\sqrt[4]{2}i;$$

we will label these roots by the integers 1 through 4. Notice that in this case we will not obtain the entire permutation group, which has 24 elements. As in Example 46.7, in practice the elements $\varphi$ of this Galois group are determined by $\varphi(\sqrt[4]{2})$, which is equal to one of the four roots, and $\varphi(i)$, which is equal to one of the two roots of $x^2 + 1$, namely $\pm i$. We thus obtain the following elements of the Galois group $G$.

| $\varphi(\sqrt[4]{2})$ | $\varphi(i)$ | perm |
|:---:|:---:|:---:|
| $\sqrt[4]{2}$ | $i$ | $\iota$ |
| $\sqrt[4]{2}$ | $-i$ | $(34)$ |
| $-\sqrt[4]{2}$ | $i$ | $(12)(34)$ |
| $-\sqrt[4]{2}$ | $-i$ | $(12)$ |
| $\sqrt[4]{2}i$ | $i$ | $(1324)$ |
| $\sqrt[4]{2}i$ | $-i$ | $(13)(24)$ |
| $-\sqrt[4]{2}i$ | $i$ | $(1423)$ |
| $-\sqrt[4]{2}i$ | $-i$ | $(14)(23)$ |

We recognize this group of order 8 as $D_4$, the group of symmetries of a square.

**Example 46.10**

We consider next the Galois group for the splitting field of $x^7 - 1$ over $\mathbb{Q}$. Now $x^7 - 1 = (x-1)\Phi_7(x) = (x-1)(x^6 + \cdots + x + 1)$, where $\Phi_7(x)$ is the cyclotomic polynomial, which is irreducible by Exercise 5.17. Furthermore, its six distinct roots are exactly $\zeta = e^{\frac{2\pi i}{7}}, \zeta^2, \cdots, \zeta^6$. So in this case the splitting field is exactly $\mathbb{Q}(\zeta)$, and $[\mathbb{Q}(\zeta) : \mathbb{Q}] = 6$. There are thus 6 elements in $\text{Gal}(\mathbb{Q}(\zeta)|\mathbb{Q})$, and they are determined by which seventh root of unity $\zeta^k$ that $\zeta$ is sent to. We thus obtain the following elements of the Galois group:

| $\varphi(\zeta)$ | perm |
|:---:|:---:|
| $\zeta$ | $\iota$ |
| $\zeta^2$ | $(124)(365)$ |
| $\zeta^3$ | $(132645)$ |
| $\zeta^4$ | $(142)(356)$ |
| $\zeta^5$ | $(154623)$ |
| $\zeta^6$ | $(16)(25)(34)$ |

We recognize this group as a cyclic group of order 6.

The previous example admits a natural generalization that we will later find quite useful.

**Theorem 46.3** *Let $p$ be a prime integer. Suppose that $F$ is a subfield of the complex numbers $\mathbb{C}$. Let $\zeta = e^{\frac{2\pi i}{p}}$ be the primitive $p$th root of unity; suppose that $\zeta \notin F$. Then $F(\zeta)$ is the splitting field for the polynomial $\Phi_p(x) = x^{p-1} + x^{p-2} + \cdots + x + 1$ over $F$. Furthermore, $\text{Gal}(F(\zeta)|F)$ is abelian (and in fact cyclic of order $p - 1$).*

**Proof:**　We know that the splitting field for $\Phi_p$ can be considered a subfield of $\mathbb{C}$, and its roots are exactly the powers $\zeta, \zeta^2, \zeta^3, \cdots \zeta^{p-1}$ of the primitive $p$th root of unity $\zeta$. Consequently, given an element $\varphi \in \text{Gal}(F(\zeta)|F)$, it is certainly determined by its value applied to $\zeta$, and that value must be a root of $\Phi_p$, namely, one of the powers $\zeta^k$. So if $\varphi(\zeta) = \zeta^k$, we will denote $\varphi$ by $\varphi_k$.

Consider now the group $\mathbb{Z}_p^*$ of units of the field $\mathbb{Z}_p$. This is clearly a finite abelian group, and it is cyclic by the Primitive Root Theorem 45.4. Define the function

$$\Psi : \mathbb{Z}_p^* \to \text{Gal}(F(\zeta)|F)$$

by setting $\Psi([k]) = \varphi_k$. We claim first that this function is well-defined. But this follows, because

$$\varphi_{k+pr}(\zeta) = \zeta^{k+pr} = \zeta^k(\zeta^p)^r = \zeta^k = \varphi_k(\zeta).$$

We now claim that this is a group homomorphism; that is, we claim that $\Psi(jk) = \varphi_{jk}$ is the same as $\Psi(j) \circ \Psi(k)$. But

$$\Psi(jk)(\zeta) = \varphi_{jk}(\zeta) = \zeta^{jk} = (\varphi_k(\zeta))^j = \varphi_j \circ \varphi_k(\zeta) = \Psi(j) \circ \Psi(k)(\zeta).$$

Now clearly $\Psi$ is bijective.

▷ **Quick Exercise.**　Why is this so?　◁

Thus $\text{Gal}(F(\zeta)|F)$ is isomorphic to the cyclic group $\mathbb{Z}_p^*$, and hence isomorphic to $\mathbb{Z}_{p-1}$.　□

Note that when we use this theorem in Chapter 48, we will actually only need the weaker conclusion that $\text{Gal}(F(\zeta)|F)$ is abelian; this follows from the proof above, without actually using the Primitive Root Theorem 45.4.

Let's summarize what we have learned from our previous examples. Suppose that $K$ is a splitting field over the field $F$ for an irreducible polynomial $f \in F[x]$. Then $[K : F] = |\text{Gal}(K|F)|$ must at least be $n = \deg(f)$, because $K$ contains an isomorphic copy of the field $F[x]/\langle f \rangle$ constructed by Kronecker's Theorem, and this field has degree $n$ over $F$. And since $\text{Gal}(K|F)$ can and should be viewed as a group of permutations of the roots of $f$, it can never have more than $|S_n| = n!$ elements. In Example 46.10, the Galois group is as small as possible; it has $n$ elements. In Example 46.7, the Galois group is as large as possible; it has $n!$ elements. And in Example 46.9, the Galois group has a number of elements intermediate between $n$ and $n!$.

## Example 46.11

The splitting field of the polynomial $x^7 - 2 \in \mathbb{Q}[x]$ is quite evidently $\mathbb{Q}(\sqrt[7]{2}, \zeta)$, where $\zeta$ is the seventh root of unity with smallest positive argument, which we considered in Example 46.10.

▷ **Quick Exercise.** What are the roots of this polynomial, in terms of $\zeta$ and $\sqrt[7]{2}$? ◁

Let's now compute $G = \text{Gal}(\mathbb{Q}(\sqrt[7]{2}, \zeta)|\mathbb{Q}(\zeta))$. Any element $\varphi \in G$ clearly leaves $\zeta$ fixed and is determined by what it does to $\sqrt[7]{2}$. There are thus seven possible choices, and since our extension is normal, all lead to automorphisms. Thus $G$ is a group of order seven, and so is necessarily isomorphic to a cyclic group of order seven.

You should note that if $a \in \mathbb{Q}$, $p$ prime, and $\zeta = e^{\frac{2\pi i}{p}}$, the primitive $p$th root of unity, then $\mathbb{Q}(\sqrt[p]{a}, \zeta)$ is the splitting field of $x^p - a \in \mathbb{Q}[x]$. Thus the field $\mathbb{Q}(\sqrt[p]{a}, \zeta)$ contains all the $p$th roots of $a$, namely $\sqrt[p]{a}, \sqrt[p]{a}\zeta, \ldots, \sqrt[p]{a}\zeta^{p-1}$.

## Example 46.12

Care must be taken when trying to apply the technique of the last example. Consider the polynomial $x^6 - 4 \in \mathbb{Q}[x]$. The six roots are $\sqrt[3]{2}, \sqrt[3]{2}\zeta, \ldots, \sqrt[3]{2}\zeta^5$, where $\zeta = e^{\frac{2\pi i}{6}} = e^{\frac{\pi i}{3}}$, a primitive sixth root of unity. (Note $\sqrt[6]{4} = \sqrt[3]{2}$.) Now we compute $G = \text{Gal}(\mathbb{Q}(\sqrt[3]{2}, \zeta)|\mathbb{Q}(\zeta))$. Again an element $\varphi \in G$ is determined by what it does to $\sqrt[3]{2}$, since $\zeta$ is fixed. There are six possible roots. However some of these mappings are problematic. For if $\varphi(\sqrt[3]{2}) = \sqrt[3]{2}\zeta$, then $\varphi(\sqrt[3]{2^3}) = \varphi(2) = 2$ but $(\varphi(\sqrt[3]{2}))^3 = (\sqrt[3]{2}\zeta)^3 = 2\zeta^3 = -2$. So $\varphi$ is not a homomorphism and so not in $G$. We similarly must eliminate the cases if $\varphi(\sqrt[3]{2}) = \sqrt[3]{2}\zeta^5$ or $\varphi(\sqrt[3]{2}) = \sqrt[3]{2}\zeta^3 = -\sqrt[3]{2}$.

▷ **Quick Exercise.** Check that $\varphi(\sqrt[3]{2}) = \sqrt[3]{2}\zeta^5$ and $\varphi(\sqrt[3]{2}) = \sqrt[3]{2}\zeta^3$ are not homomorphisms and so not in $G$. ◁

This leaves us with three possible roots to map $\sqrt[3]{2}$, namely $\sqrt[3]{2}, \sqrt[3]{2}\zeta^2$, and $\sqrt[3]{2}\zeta^4$. The resultant Galois group is therefore isomorphic to a cyclic group of order 3.

The explanation for these difficulties is that $x^6 - 4 = (x^3 + 2)(x^3 - 2)$ is *not* irreducible in $\mathbb{Q}[x]$. Indeed, we really have been working with the splitting field of $x^3 - 2$ over $\mathbb{Q}(\zeta)$. Note that $[\mathbb{Q}(\sqrt[3]{2}, \zeta) : \mathbb{Q}(\zeta)] = 3$, with the Galois group having the same number of elements. Note that we could alternatively have been considering this field extension as the splitting field for $x^3 + 2$ over $\mathbb{Q}(\zeta)$.

▷ **Quick Exercise.** Why? ◁

You should compare this with Example 46.6.

Example 46.11 is an important one for us, and we will make use of it in Chapter 48 when we prove that it is impossible to solve the quintic by radicals. And so we generalize the conditions of Example 46.11 (and put Example 46.12 in an appropriate context):

**Theorem 46.4** *Suppose that $p$ is prime and $F$ a subfield of $\mathbb{C}$. Suppose that $r \in F$, but $r$ has no pth root in $F$ while $\alpha \in \mathbb{C}$ and $\alpha^p = r$. Suppose that $\zeta = e^{\frac{2\pi i}{p}} \in F$. Then $F(\alpha)$ is the splitting field of $x^p - r \in F[x]$ over $F$, and $\mathrm{Gal}(F(\alpha)|F)$ is a cyclic group of order $p$.*

**Proof:**    We may again suppose that the splitting field of $x^p - r$ is a subfield of $\mathbb{C}$. But the roots of $x^p - r$ in $\mathbb{C}$ are evidently the elements $\alpha,\ \alpha\zeta,\ \alpha\zeta^2,\ \cdots\ \alpha\zeta^{p-1}$. Because we are assuming that $\zeta \in F$, all these elements belong to the field $F(\alpha)$, and so the latter field is evidently the splitting field, as required.

Given an element $\varphi \in \mathrm{Gal}(F(\alpha)|F)$, it is clearly determined by what it does to $\alpha$, and obviously $\varphi(\alpha) = \alpha\zeta^k$, for some integer $k$, with $0 \le k \le p-1$. Because $F(\alpha)$ is normal over $F$, each such choice leads to a distinct automorphism. So the Galois group has exactly $p$ elements. The only group with $p$ elements is a cyclic group of that order.    □

### Example 46.13

Consider now the polynomial $x^5 - 6x + 3 \in \mathbb{Q}[x]$. This is clearly irreducible, by Eisenstein's Criterion 5.7. We will now calculate the Galois group of its splitting field over the rational field. We will do this rather less directly than in our previous examples.

Let's consider the polynomial function $f(x) = x^5 - 6x + 3$ and use some calculus. Now $f'(x) = 5x^4 - 6$, which has exactly two real roots $\pm\sqrt[4]{\frac{6}{5}}$. The negative root corresponds to a local maximum for $f$, and the positive root corresponds to a local minimum. Furthermore, $f$ takes on a positive value at the local maximum and a negative value at the local minimum. Since $\lim_{x \to \pm\infty} f(x) = \pm\infty$, $f$ has exactly three real roots. It must consequently have two complex roots, and these form a complex conjugate pair. The graph of $f$ in the picture below illustrates its properties.

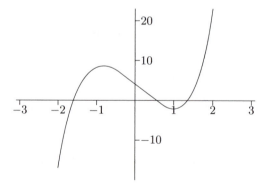

▷ **Quick Exercise.**    Verify the details regarding the graph of this function.    ◁

Now let $K$ be the splitting field for $f$ over $\mathbb{Q}[x]$. We shall as usual view the Galois group $\mathrm{Gal}(K|\mathbb{Q})$ as a group of permutations of the 5 distinct roots $\alpha_1, \alpha_2, \alpha_3, \alpha_4, \alpha_5$ of $f$ in $K$. We may as well label our roots so that the first two are the complex roots. Since $f$ is irreducible and of degree 5 and its Kronecker field is a subfield of the splitting field, 5 divides $[K : \mathbb{Q}] = |\mathrm{Gal}(K|\mathbb{Q})|$. Now Cauchy's Theorem 24.5 implies that the group $\mathrm{Gal}(K|\mathbb{Q})$ has an element of order 5, and as a permutation of a set of five elements, the element must be a 5-cycle. As usual, we will think of the elements of our Galois group as permutations on the root subscripts. So we may suppose that this 5-cycle is $\alpha = (1abcd)$, where $a, b, c, d$ are the integers $2, 3, 4, 5$ in some order. But some power of $\alpha$ is then of the form $(12abc)$, and we may as well relabel the three real roots so that the permutation $(12345)$ belongs to the Galois group.

But complex conjugation is clearly a field automorphism of $K$ leaving $\mathbb{Q}$ fixed and so belongs to this Galois group. Complex conjugation leaves the three real roots fixed and interchanges the two conjugate roots, and so as a permutation of the roots it is simply (12).

But now Exercise 27.9 says that (12) and (12345) together generate the entire group $S_5$. Consequently, we have shown that the Galois group of the splitting field $K$ of the polynomial $x^5 - 6x + 3 \in \mathbb{Q}[x]$ is the full permutation group $S_5$.

This example will be of considerable importance in Chapter 48, where we will show that the Galois group being $S_5$ will imply that it is impossible to solve for the roots of the polynomial $x^5 - 6x + 3$ using ordinary field arithmetic, and the extraction of roots!

---

## Chapter Summary

For any finite field extension $K \supseteq F$ we defined the Galois group $\text{Gal}(K|F)$ as the group of all automorphisms of the field $K$ that leave $F$ fixed. When $K$ is the splitting field for some polynomial $f \in F[x]$, then the Galois group can be viewed as a group of permutations of the roots of $f$ in $K$, whose size is equal to the degree of the field extension $K$ over $F$. This size is always at least $n = \deg(f)$ and can in principle be as large as $n!$.

---

## Warm-up Exercises

a. Answer the following true or false; if your answer is false, give a counterexample. Assume that $F \subseteq K$ are fields.

   (a) $\text{Gal}(F|F)$ is the trivial group.

   (b) Let $\alpha \in K \backslash F$ and $\alpha$ is algebraic over $F$. Then $\text{Gal}(F(\alpha)|F)$ is a finite group.

   (c) $\text{Gal}(\mathbb{Q}(\sqrt{5})|\mathbb{Q})$ is a two element group.

   (d) A field automorphism preserves addition but need not preserve multiplication.

   (e) The group $\text{Gal}(\mathbb{Q}(\sqrt[5]{2})|\mathbb{Q})$ has at least five elements.

b. Suppose that $\zeta$ is the primitive 11th root of unity. What can you say about Galois group $\text{Gal}(\mathbb{Q}(\sqrt[11]{2}, \zeta)|\mathbb{Q}(\sqrt[11]{2}))$?

c. Suppose that $\zeta$ is the primitive 11th root of unity. What can you say about Galois group $\text{Gal}(\mathbb{Q}(\sqrt[11]{2}, \zeta)|\mathbb{Q}(\zeta))$?

d. Give examples of the following, or else argue that such an example does not exist:

   (a) A Galois group with exactly 4 elements.

   (b) A field $F$ and an element $\alpha$ that is algebraic over $F$, where $\text{Gal}(F(\alpha)|F)$ has infinitely many elements.

   (c) A non-cyclic Galois group.

## Exercises

1. Compute the Galois groups

$$\mathrm{Gal}(\mathbb{Q}(\sqrt{2}, i)|\mathbb{Q}) \text{ and } \mathrm{Gal}(\mathbb{Q}(\sqrt{2}, i)|\mathbb{Q}(\sqrt{2})).$$

2. Consider the splitting field $\mathbb{Z}_2(\alpha)$ of the polynomial $x^3 + x + 1 \in \mathbb{Z}_2[x]$ that we looked at in Example 44.5. Explicitly describe the elements of the Galois group $\mathrm{Gal}(\mathbb{Z}_2(\alpha)|\mathbb{Z}_2)$ as permutations of the three roots $\alpha, \alpha^2, \alpha + \alpha^2$.

3. Compute the Galois group $\mathrm{Gal}(K|\mathbb{Q})$, where $K$ is the splitting field for the polynomial $x^4 - 4x^2 + 2$. You computed this splitting field in Exercise 44.5. Now compute $\mathrm{Gal}(K|\mathbb{Q}(\sqrt{2}))$.

4. Suppose that $\zeta$ is the primitive cube root of unity. Let K be the splitting field of $x^9 - 1 \in \mathbb{Q}[x]$. Compute $\mathrm{Gal}(K|\mathbb{Q}(\zeta))$.

5. Consider the polynomial $f = x^3 - 3x - 1 \in \mathbb{Q}[x]$. This is the polynomial we considered in Section 38.2 (and Section 43.2) when we proved that it is impossible to trisect a $60°$ angle with ruler and compass. In this exercise we will compute the Galois group of the splitting field for $f$ over the rational field.

   (a) Show that $f$ is irreducible over $\mathbb{Q}$.

   (b) In Section 38.2 we used a trigonometric identity to show that $\alpha = 2\cos 20°$ is a root of this polynomial. Use similar arguments to show that $\beta = -2\cos 40°$ and $\gamma = -2\sin 10°$ are the other two roots of $f$.

   (c) Use elementary trigonometry to show that $\mathbb{Q}(\beta)$ is the splitting field for $f$ over $\mathbb{Q}$. (That is, show that $\alpha, \gamma \in \mathbb{Q}(\beta)$.)

   (d) We can think of $\mathrm{Gal}(\mathbb{Q}(\beta)|\mathbb{Q})$ as a group of permutations of the roots $\alpha, \beta, \gamma$. Argue that in this case this group is a cyclic group of order three.

   (e) In this problem we solved for the roots of $f$ by taking advantage of trigonometry. Instead, use the Cardano-Tartaglia formula to obtain a root of $f$. Show that the root you obtain is one of the roots given above.

6. Apply the same reasoning as in Example 46.13 to conclude that the Galois group of the splitting field for $x^5 - 14x^2 + 7$ over the rational numbers is the permutation group $S_5$.

7. Repeat the previous exercise for the polynomial $x^5 - 4x^4 + 2x + 2 \in \mathbb{Q}[x]$.

8. Generalize the reasoning from Example 46.13 to prove the following: Consider the irreducible $f \in \mathbb{Q}[x]$ with $\deg(f) = p$, where $p$ is a prime integer. Suppose that $f$ has exactly two non-real roots. Prove that the Galois group of the splitting field of $f$ is the permutation group $S_p$.

9. Prove the following modest variation on Theorem 46.2: Let $K$ be a simple algebraic extension of the finite field $F$. Then $|\mathrm{Gal}(K|F)| = [K : F]$ if and only if $K$ is a normal extension of $F$.

10. In this exercise you will perform some addition calculations related to Example 46.7, where we compute $\mathrm{Gal}(\mathbb{Q}(\sqrt[3]{2}, \zeta)|\mathbb{Q})$, where $\zeta = -\frac{1}{2} + \frac{\sqrt{3}}{2}i$. Let $\beta = \sqrt[3]{2} + \zeta$.

(a) Show that $\mathbb{Q}(\sqrt[3]{2}, \zeta) = \mathbb{Q}(\beta)$.

(b) Derive the minimal polynomial $f$ for $\beta$ over $\mathbb{Q}$ using the technique from Example 43.3 and Exercise 43.11. (This should be of degree 6.)

(c) We know that the elements $\varphi \in \mathrm{Gal}(\mathbb{Q}(\sqrt[3]{2}, \zeta)|\mathbb{Q})$ are entirely determined by $\varphi(\beta)$. Use the description of the Galois group elements in Example 46.7 to obtain a complete list of the roots of $f$.

(d) Pick an element of the list of roots other than $\beta$ and verify explicitly that it is a root of $f$ (this exercise is painful indeed if performed by hand).

11. In this exercise we continue computations related to Example 46.7. Recall from Example 44.4 that $\mathbb{Q}(\sqrt[3]{2}, \zeta) = \mathbb{Q}(\sqrt[3]{2}, \sqrt{3}i)$.

(a) Show that if $\alpha = \sqrt[3]{2} + \sqrt{3}i$, then $\mathbb{Q}(\sqrt[3]{2}, \zeta) = \mathbb{Q}(\alpha)$. That is, the choice of the element adjoined to make our field extension a simple extension is by no means unique.

(b) Now find the minimal polynomial $g$ for $\alpha$ over $\mathbb{Q}$. This will be different from the minimal polynomial you found in the previous exercise, but will still be of degree 6, of course.

(c) Apply the same technique as in the previous exercise to obtain a complete list of all roots of $g$.

12. Consider the splitting field $\mathbb{Q}(\sqrt[6]{2}, \zeta)$ of $x^6 - 2 \in \mathbb{Q}[x]$. Compute the Galois group $\mathrm{Gal}(\mathbb{Q}(\sqrt[6]{2}, \zeta)|\mathbb{Q}(\zeta))$, where $\zeta = e^{\frac{2\pi i}{6}}$.

13. In Example 46.10 we discovered that if $\zeta$ is the primitive seventh root of unity, then $\mathrm{Gal}(\mathbb{Q}(\zeta)|\mathbb{Q})$ is isomorphic to $\langle \varphi_3 \rangle$, the cyclic group of order 6 generated by the automorphism determined by $\varphi(\zeta) = \zeta^3$. And note that Theorem 46.3 asserts this is true. In this exercise you will verify this theorem in a couple of cases.

(a) Let $\zeta$ be the primitive eleventh root of unity. Show that $\mathrm{Gal}(\mathbb{Q}(\zeta)|\mathbb{Q})$ is isomorphic to $\langle \varphi_2 \rangle$, the cyclic group of order 10 generated by the automorphism determined by $\varphi_2(\zeta) = \zeta^2$.

(b) Let $\zeta$ be the primitive seventeenth root of unity. Show that $\mathrm{Gal}(\mathbb{Q}(\zeta)|\mathbb{Q})$ is isomorphic to $\langle \varphi_3 \rangle$, the cyclic group of order 16 generated by the automorphism determined by $\varphi_3(\zeta) = \zeta^3$.

(c) What do these results have to do with Exercise 45.11?

# Chapter 47

## The Fundamental Theorem of Galois Theory

We saw in the last chapter that the Galois group of a finite extension of a field provides a lot of information about the structure of the extension field. In fact, if the extension is normal, then the degree of the extension is equal to the number of automorphisms belonging to the Galois group. In this chapter we encounter the Fundamental Theorem of Galois Theory, which shows that this connection between field extensions and groups carries even more information than that. In Chapter 48 we will be able to exploit this connection between field theory and group theory to address our goal of better understanding the solution of polynomial equations by field arithmetic and the extraction of roots.

## 47.1 Subgroups and Subfields

Suppose that we have fields $F \subseteq E \subseteq K$. Then it is easy to see that $\mathrm{Gal}(K|E)$ is a subgroup of $\mathrm{Gal}(K|F)$, because automorphisms of $K$ that fix $E$ clearly also fix $F \subseteq E$.

▷ **Quick Exercise.** Check that $\mathrm{Gal}(K|E)$ is not only a subset of $\mathrm{Gal}(K|F)$ but also a subgroup. ◁

### Example 47.1

Referring to Example 46.7, we have that

$$\mathrm{Gal}(\mathbb{Q}(\sqrt[3]{2}, \zeta)|\mathbb{Q}) = \{\iota, (23), (123), (12), (132), (13)\},$$

viewed as a group of permutations of the roots of $x^3 - 2$. But $\mathrm{Gal}(\mathbb{Q}(\sqrt[3]{2}, \zeta)|\mathbb{Q}(\zeta))$ is quite evidently the subgroup $\{\iota, (123), (132)\}$.

▷ **Quick Exercise.** What subgroup is $\mathrm{Gal}(\mathbb{Q}(\sqrt[3]{2}, \zeta)|\mathbb{Q}(\sqrt[3]{2}))$? ◁

### Example 47.2

Referring back to Example 46.5, we have that

$$\mathrm{Gal}(\mathbb{Q}(\sqrt{2}, \sqrt{3})|\mathbb{Q}) = \{\iota, \varphi_1, \varphi_2, \varphi_3\}.$$

Note that $\mathrm{Gal}(\mathbb{Q}(\sqrt{2}, \sqrt{3})|\mathbb{Q}(\sqrt{3}))$ is quite evidently the subgroup $\{\iota, \varphi_1\}$.

▷ **Quick Exercise.** What subgroup is $\mathrm{Gal}(\mathbb{Q}(\sqrt{2}, \sqrt{3})|\mathbb{Q}(\sqrt{2}))$? How about $\mathrm{Gal}(\mathbb{Q}(\sqrt{2}, \sqrt{3})|\mathbb{Q}(\sqrt{6}))$? ◁

So for a field $E$ between the fields $F \subseteq K$ we always obtain a subgroup $\mathrm{Gal}(K|E)$ of the Galois group $\mathrm{Gal}(K|F)$. But we can also proceed in the reverse direction and obtain fields from subgroups. Suppose that $H$ is a subgroup of the Galois group $\mathrm{Gal}(K|F)$. We define

$$\mathrm{Fix}(H) = \{k \in K : \eta(k) = k, \text{ for all } \eta \in H\}.$$

It is obvious that $F \subseteq \mathrm{Fix}(H) \subseteq K$, because all elements of $H \subseteq \mathrm{Gal}(K|F)$ fix elements from $F$, by definition. But more is true:

**Theorem 47.1** *Suppose that $F \subseteq K$ are fields, and $H$ is a subgroup of $\mathrm{Gal}(K|F)$. Then* $\mathrm{Fix}(H)$ *is a subfield of $K$ containing $F$.*

**Proof:** We leave the straightforward details to Exercise 5. □

We thus call $\mathrm{Fix}(H)$ the **fixed field** of the subgroup $H$.

**Example 47.3**

Returning again to Example 46.7, we have that $\mathrm{Gal}(\mathbb{Q}(\sqrt[3]{2}, \zeta)|\mathbb{Q})$ is (up to isomorphism) the permutation group $S_3$. This group has the following subgroups:

$$\{\iota\}, \ G_1 = \{\iota, (23)\}, \ G_2 = \{\iota, (13)\}, \ G_3 = \{\iota, (12)\}, \ A_3, \ S_3$$

It is easy to compute the fixed fields:

$$\begin{aligned}
\mathrm{Fix}(\{\iota\}) &= \mathbb{Q}(\sqrt[3]{2}, \zeta) \\
\mathrm{Fix}(G_1) &= \mathbb{Q}(\sqrt[3]{2}) \\
\mathrm{Fix}(G_2) &= \mathbb{Q}(\sqrt[3]{2}\zeta) \\
\mathrm{Fix}(G_3) &= \mathbb{Q}(\sqrt[3]{2}\zeta^2) \\
\mathrm{Fix}(A_3) &= \mathbb{Q}(\zeta) \\
\mathrm{Fix}(S_3) &= \mathbb{Q}
\end{aligned}$$

▷ **Quick Exercise.** Check these fixed fields. ◁

Notice that in Example 47.3 we actually have a one-to-one correspondence between all subgroups of $S_3$ and all fields intermediate between $\mathbb{Q}$ and $\mathbb{Q}(\sqrt[3]{2}, \zeta)$. It turns out that this is the case precisely because we have a normal extension. We shall discover that this (and more) is the content of the Fundamental Theorem of Galois Theory 47.3 that we prove below.

---

## 47.2 Symmetric Polynomials

Before we can tackle the problem of understanding the correspondence we have begun establishing between intermediate fields and subgroups of the Galois group, we have a technical matter we need to discuss, regarding the coefficients of a polynomial that splits in a given field.

Suppose that $f \in F[x]$ is a monic polynomial with roots $\alpha_1, \alpha_2, \cdots, \alpha_n$ that exist in the splitting field for $f$ over $F$. We then have that

$$f = (x - \alpha_1)(x - \alpha_2) \cdots (x - \alpha_n).$$

If we multiply this out by using the distributive law, we will obtain

$$f = \sum_{k=0}^{n} (-1)^{n-k} a_{n-k} x^k = a_o x^n - a_1 x^{n-1} + a_2 x^{n-1} - \cdots + (-1)^n a_n,$$

where the coefficient $a_k$ consists of the sum of all products of exactly $k$ of the $\alpha_j$ with distinct subscripts (where we interpret this for $k = 0$ as $a_0 = 1$). For example, if $n = 4$, then

$$a_0 = 1$$

$$a_1 = \alpha_1 + \alpha_2 + \alpha_3 + \alpha_4$$

$$a_2 = \alpha_1\alpha_2 + \alpha_1\alpha_3 + \alpha_1\alpha_4 + \alpha_2\alpha_3 + \alpha_2\alpha_4 + \alpha_3\alpha_4$$

$$a_3 = \alpha_1\alpha_2\alpha_3 + \alpha_1\alpha_2\alpha_4 + \alpha_1\alpha_3\alpha_4 + +\alpha_2\alpha_3\alpha_4$$

$$a_4 = \alpha_1\alpha_2\alpha_3\alpha_4$$

Notice that for convenience of description we have reversed the usual subscript convention: $a_k$ is (up to plus or minus) the coefficient on the $x^{n-k}$ term.

We can justify these formulas with a combinatorial argument. After all multiplications have been distributed out, there will be $2^n$ terms obtained by making all possible choices, taking one of the two terms in each binomial factor $x - \alpha_i$. To obtain a contributor to the $x^k$ term, we need to choose exactly $k$ $x$-terms, and $n-k$ $\alpha_j$ terms, each with a distinct subscript $j$. When we add up each of these terms, we get exactly the $a_{n-k}$ terms described above. Note also that each $\alpha_j$ term selected contributes a factor of $-1$ as well. It is also possible to construct a careful proof by induction; such a proof is tedious but straightforward, and we will leave it as Exercise 7.

We call these coefficients the **symmetric polynomials** in the constants $\alpha_1, \cdots, \alpha_n$. What is important for our purposes to observe is that these polynomials are symmetric in the $\alpha_i$'s. This means that if we have any field automorphism that permutes the roots of $f$, it will leave these coefficients fixed. This observation will be important in our proof of the Fundamental Theorem of Galois Theory 47.3 below.

## 47.3   The Fixed Field and Normal Extensions

Given fields $F \subseteq K$, we have a way to obtain a subgroup $\mathrm{Gal}(K|E)$ of $\mathrm{Gal}(K|F)$, for each intermediate field $E$. And for each subgroup $H$ of $\mathrm{Gal}(K|F)$, we have a way to obtain an intermediate field $\mathrm{Fix}(H)$. When $K$ is a finite *normal* extension of $F$, it turns out that these two processes are inverses of one another: Group theory will perfectly mirror field theory, and vice versa.

If $E$ is a field with field extension $K$, then it is clear from the definition of the fixed field and the Galois group that $\mathrm{Fix}(\mathrm{Gal}(K|E)) \supseteq E$.

▷ **Quick Exercise.**   Check this.   ◁

The following theorem asserts that the reverse inclusion holds only in case that $K$ is normal over $E$:

**Theorem 47.2** *Suppose that $E \subseteq K$ are fields of characteristic zero, and $K$ is a finite extension of $E$. Then $\mathrm{Fix}(\mathrm{Gal}(K|E)) = E$ if and only if $K$ is a normal extension of $E$.*

**Proof:** Suppose first that $K$ is a normal extension of $E$; we may as well assume that $K$ is strictly larger than $E$. Since $K$ is finite over $E$ and of characteristic zero, this means by Theorem 44.5 that $K = E(\alpha)$, for some algebraic $\alpha \in K \backslash E$. We shall assume that the minimal polynomial for $\alpha$ over $E$ is $f$, and $\deg(f) = n$. Then $1, \alpha, \alpha^2, \cdots, \alpha^{n-1}$ is a basis for $K$ over $E$. Furthermore, because $K$ is normal over $E$, $K$ contains all $n$ roots $\alpha = \alpha_1, \cdots, \alpha_n$ of $f$; these roots are distinct, by Theorem 44.4.

Let $\beta \in K$, and suppose that $\beta$ is fixed by all elements of $\mathrm{Gal}(K|E)$. We will complete the proof if we can conclude that $\beta \in E$. (As we observed above, the reverse inclusion is always true.)

Now we can express $\beta$ in terms of our basis as $\beta = \sum_{k=0}^{n-1} a_k \alpha^k$. But $\beta$ is left fixed by all $n$ elements of $\mathrm{Gal}(K|E)$, and for each $i$ there is such a Galois group element $\varphi_i$ that takes $\alpha$ to $\alpha_i$ (see the proof of Theorem 46.2). We thus have

$$\beta = \varphi_i(\beta) = \varphi_i \left( \sum_{k=0}^{n-1} a_k \alpha^k \right) = \sum_{k=0}^{n-1} a_k \alpha_i^k.$$

But now consider the polynomial $g \in K[x]$ defined by

$$g = a_{n-1} x^{n-1} + \cdots + a_1 x + (a_0 - \beta).$$

By the previous equation, we see that $g$ has $n$ distinct roots $\alpha_i$. But this is too many roots for a polynomial with degree no more than $n - 1$. Thus, $g$ must be the identically zero polynomial. This means that $\beta = a_0 \in E$.

For the converse, suppose that $\mathrm{Fix}(\mathrm{Gal}(K|E)) = E$. We again may assume that $K$ is strictly larger than $E$, and furthermore that it is a simple extension $K = E(\alpha)$ with minimal polynomial $f$, with $\deg(f) = n$. We will show that $K$ contains all $n$ of the roots of $f$ and thus is the splitting field for $f$ over $E$, and so is normal, by Theorem 44.3.

Since $\alpha \in K \backslash E$ and $\mathrm{Fix}(\mathrm{Gal}(K|E)) = E$, there must be some $\varphi \in \mathrm{Gal}(K|E)$ so that $\varphi(\alpha) = \alpha_2 \neq \alpha$. Now $\alpha_2$ is clearly a root of $f$, and $\alpha_2 \in K$. If $\deg(f) = n = 2$, then $f$ splits in $K$ as required.

If $n > 2$ we must continue by induction. So we will suppose that $\alpha = \alpha_1, \alpha_2, \cdots, \alpha_k$ are all roots of $f$ belonging to $K$. Let

$$g = (x - \alpha_1)(x - \alpha_2) \cdots (x - \alpha_k).$$

This is clearly a factor of $f$ in $K[x]$. If $k < n$ then $g$ is a non-trivial factor of $f$. Since $f$ is irreducible in $E[x]$, this means that $g \notin E[x]$. But the coefficients of $g$ are precisely the symmetric polynomials in $\alpha_1, \alpha_2, \cdots, \alpha_k$. So if each element of $\mathrm{Gal}(K|E)$ merely permutes these $k$ roots, the coefficients must belong to the fixed field of the Galois group, which is by assumption $E$. Thus it must be the case that at least one of the $\alpha_i$ is moved by an element of $\mathrm{Gal}(K|E)$ to a root of $f$ not on our current list. This provides us with another root of $f$ that belongs to $K$. By induction, we then obtain all the roots of $f$ in $K$. Thus $K$ is a normal extension of $E$. $\qquad\square$

### Example 47.4

We return to Example 46.6. Since $\mathrm{Gal}(\mathbb{Q}(\sqrt[3]{2})|\mathbb{Q})$ is trivial, the fixed field is $\mathbb{Q}(\sqrt[3]{2})$. Because this is strictly larger than the rational field, it is not a normal extension. (Of course, it is easy to see that this extension is not normal, directly from the definition.)

## 47.4    The Fundamental Theorem

We are now ready to state and prove the Fundamental Theorem of Galois Theory.

**Theorem 47.3    The Fundamental Theorem of Galois Theory, Part One**    *Suppose that $K$ is a finite normal extension of the field $F$, which is of characteristic zero. There is a one-to-one order reversing correspondence between fields $E$ with $F \subseteq E \subseteq K$ and the subgroups $H$ of $\mathrm{Gal}(K|F)$. We can describe this correspondence by two maps that are inverses of one another; namely, we have*

$$E \longmapsto \mathrm{Gal}(K|E)$$

*and*

$$H \longmapsto \mathrm{Fix}(H).$$

**Proof:**    It is clear from the definition of the Galois group and the fixed field that if we have fields with $E_1 \subseteq E_2$ then $\mathrm{Gal}(K|E_1) \supseteq \mathrm{Gal}(K|E_2)$. Also, if $H_1 \subseteq H_2$ are subgroups of the Galois group, then $\mathrm{Fix}(H_1) \supseteq \mathrm{Fix}(H_2)$. These two maps are thus order-reversing. We need only show that the maps are inverses of one another. To do this, we will compose them in both directions.

Since $K$ is normal over $F$, it is clearly normal over $E$ (see Exercise 44.a(i)). But then $\mathrm{Fix}(\mathrm{Gal}(K|E)) = E$, as required.

We'd now like to show that for any subgroup $H$ of $\mathrm{Gal}(K|F)$, we have that $H = \mathrm{Gal}(K|\mathrm{Fix}(H))$. Because our fields are of characteristic zero, we have that $K$ is a simple extension $\mathrm{Fix}(H)(\alpha)$ of $\mathrm{Fix}(H)$, where $\alpha$ is algebraic over $\mathrm{Fix}(H)$.

Suppose that the subgroup $H$ has $h$ elements, which we may specify as $\iota = \eta_1, \eta_2, \cdots, \eta_h$. We can now consider the polynomial

$$f = (x - \eta_1(\alpha))(x - \eta_2(\alpha)) \cdots (x - \eta_h(\alpha)) \in K[x].$$

The coefficients for the polynomial $f$ are then the symmetric polynomials in the constants

$$\eta_1(\alpha), \ \eta_2(\alpha), \ \cdots, \ \eta_h(\alpha).$$

But if we apply any element of $H$ to the elements of this set, we will just permute them. Consequently, the coefficients for the polynomial $f$ belong to $\mathrm{Fix}(H)$. Thus $\alpha$ is a root of a polynomial with degree $h$ in $\mathrm{Fix}(H)[x]$. This means that $[K : \mathrm{Fix}(H)] = [\mathrm{Fix}(H)(\alpha) : \mathrm{Fix}(H)] \leq h$.

But $K$ is normal over $\mathrm{Fix}(H)$, and so

$$[K : \mathrm{Fix}(H)] = |\mathrm{Gal}(K|\mathrm{Fix}(H))| \geq |H| = h.$$

Putting these inequalities together, we have that

$$h \leq |\mathrm{Gal}(K|\mathrm{Fix}(H))| \leq h,$$

and so $|\mathrm{Gal}(K|\mathrm{Fix}(H))| = h$ and thus $\mathrm{Gal}(K|\mathrm{Fix}(H)) = H$, as required.    $\square$

We will at the end of this chapter give numerous examples of the Fundamental Theorem in action, but we will first provide some important additional information about the correspondence between subfields and subgroups.

We first make a few observations about counting. On the group side, we can count by making use of Lagrange's Theorem 24.2, while on the field side our primary counting tool is Theorem 43.2 (a finite extension of a finite extension is finite). So if $K$ is a (characteristic zero) normal field extension of the field $F$, and $E$ is an intermediate field, we have that $|\text{Gal}(K|F)| = [K : F]$ and $|\text{Gal}(K|E)| = [K : E]$, and

$$[K : F] = [K : E][E : F].$$

But on the group side we have that

$$|\text{Gal}(K|F)| = |\text{Gal}(K|E)|[\text{Gal}(K|F) : \text{Gal}(K|E)].$$

This means that the degree $[E : F]$ of the field extension $E$ over $F$ is precisely equal to the index $[\text{Gal}(K|F) : \text{Gal}(K|E)]$ of the subgroup $\text{Gal}(K|E)$ in the Galois group $\text{Gal}(K|F)$. In the examples in the next section this will be useful for us, because in practice it is often easier to count group elements than it is to calculate the degree of field extensions.

There is another more profound refinement we can add to the one-to-one correspondence we have between intermediate fields and subgroups of the Galois group. It turns out that normal subgroups correspond exactly to normal extensions. This is the reason why such extensions are called normal! This is important enough that we will call this result the second part of the Fundamental Theorem:

**Theorem 47.4    Fundamental Theorem of Galois Theory, Part Two**    *Suppose that $K$ is a finite normal extension of the field $F$, with characteristic zero. An intermediate field $E$ is a normal extension of $F$ if and only if the Galois group $\text{Gal}(K|E)$ is a normal subgroup of $\text{Gal}(K|F)$. Furthermore, the Galois group $\text{Gal}(E|F)$ is isomorphic to*

$$\text{Gal}(K|F)/\text{Gal}(K|E).$$

**Proof:**    Suppose that $E$ is a field intermediate between $K$ and $F$. Then $\text{Gal}(K|E)$ is normal in $\text{Gal}(K|F)$ if and only if $\varphi^{-1}\psi\varphi \in \text{Gal}(K|E)$, for all $\varphi \in \text{Gal}(K|F)$ and $\psi \in \text{Gal}(K|E)$. But because $K$ is a normal extension of $E$, this is equivalent to asserting that $\varphi^{-1}\psi\varphi(e) = e$, for all $e \in E$. But this is true exactly if $\psi\varphi(e) = \varphi(e)$, for all $e \in E$. But this means precisely that $\varphi(e) \in \text{Fix}(\text{Gal}(K|E)) = E$. And this says that all roots of the minimal polynomial for $e$ over $F$ actually belong to $E$. This means exactly that $E$ is a normal extension of $F$.

To show the group isomorphism, we shall define a group homomorphism $\Gamma$ from $\text{Gal}(K|F)$ onto $\text{Gal}(E|F)$ with the appropriate kernel. Given $\varphi \in \text{Gal}(K|F)$, we shall define $\Gamma(\varphi) = \varphi|_E$, the restriction of $\varphi$ to the subfield $E$. Because $E$ is a normal extension of $F$, we have that $\varphi(e) \in E$, for all $e \in E$. This means that this map is well defined.

▷ **Quick Exercise.**    Why is the map well defined? ◁

It clearly preserves the group operation (functional composition).

Now suppose that $\Gamma(\varphi) = \iota$, the identity automorphism. This means exactly that $\varphi$ leaves all elements of $E$ fixed. In other words, $\varphi \in \text{Gal}(K|E)$. So the Fundamental Isomorphism Theorem for Groups 26.4 asserts the isomorphism we require.    □

## 47.5    Examples

We shall conclude this chapter by looking at a number of examples illustrating the full strength of the Fundamental Theorem of Galois Theory.

## Example 47.5

Let's examine the normal field extension $\mathbb{Q}(\sqrt{2}, \sqrt{3})$ of $\mathbb{Q}$ that we considered in Examples 46.5 and 47.2. The Galois group $G = \{\iota, \varphi_1, \varphi_2, \varphi_3\}$ is (isomorphic to) the Klein Four Group, which clearly has three two element subgroups $\langle\varphi_1\rangle$, $\langle\varphi_2\rangle$ and $\langle\varphi_3\rangle$. These subgroups correspond exactly to the three intermediate fields

$$\mathbb{Q}(\sqrt{3}), \; \mathbb{Q}(\sqrt{6}), \; \mathbb{Q}(\sqrt{2}),$$

respectively. Since these are the only proper, nontrivial subgroups, the Fundamental Theorem guarantees that these are the only proper intermediate subfields. The index of each of these subgroups in $G$ is two, which corresponds precisely to the fact that each of these extensions is of degree two over $\mathbb{Q}$. The group $G$ is abelian, and so each of these subgroups is normal. This corresponds to the fact that the three quadratic extensions are of course normal. Pictured below is the order-reversing correspondence between subgroups and intermediate fields:

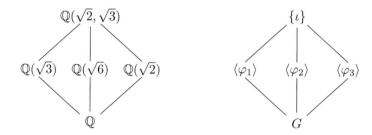

## Example 47.6

In Example 47.3 we had already computed the fixed fields corresponding to all of the subgroups of $\mathrm{Gal}(\mathbb{Q}(\sqrt[3]{2}, \zeta)|\mathbb{Q})$ (see also Example 46.7, where we actually computed the Galois group). Only one of the subgroups is normal, namely $A_3$. Its fixed field is $\mathbb{Q}(\zeta)$ which is a quadratic normal extension; this is the only intermediate field that is normal over $\mathbb{Q}$. Here is the picture of the subgroups and corresponding intermediate fields:

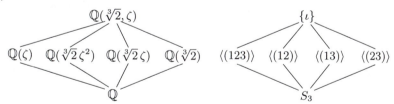

## Example 47.7

Let's return to Example 46.10, where we considered the splitting extension of $x^7 - 1$ over $\mathbb{Q}$. We know that the splitting extension is $\mathbb{Q}(\zeta)$, where $\zeta = e^{\frac{2\pi i}{7}}$. The Galois group $G = \mathrm{Gal}(\mathbb{Q}(\zeta)|\mathbb{Q})$ is a cyclic group of order six represented as a permutation group as

$$\{\iota, (132645), (124)(365), (16)(25)(34), (142)(356), (154623)\}.$$

A cyclic group of order six has exactly two proper, non-trivial subgroups, which in this case are $G_1 = \{\iota, (16)(25)(34)\}$ and $G_2 = \{\iota, (124)(365), (142)(356)\}$. We thus have exactly two proper intermediate fields, $\mathrm{Fix}(G_1)$ and $\mathrm{Fix}(G_2)$. The first must be a cubic extension of $\mathbb{Q}$, while the second is a quadratic extension. It is evident that such field elements as $\zeta + \zeta^6$, $\zeta^2 + \zeta^5$, $\zeta^3 + \zeta^4$ are left fixed by $G_1$, while elements $\zeta + \zeta^2 + \zeta^4$ and $\zeta^3 + \zeta^5 + \zeta^6$ are left fixed by $G_2$. These are consequently candidates for field elements that might produce the appropriate intermediate fields. However, we cannot immediately rule out the possibility that these elements might belong to $\text{Fix}(G) = \mathbb{Q}$.

To rigorously determine the fixed fields for these subgroups requires us to inquire more carefully into the arithmetic in $\mathbb{Q}(\zeta)$. We first should remember that $\zeta$ (and its powers) are the roots of the irreducible cyclotomic polynomial $1 + x + x^2 + \cdots + x^6$. For more insight into this field, it is helpful to look at the picture of the seventh roots of unity as complex numbers.

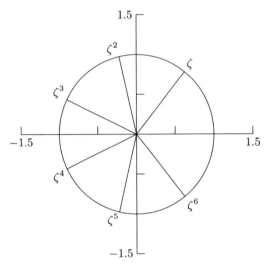

It is now quite evident that $\bar{\zeta} = \zeta^6$, $\bar{\zeta^2} = \zeta^5$, and $\bar{\zeta^3} = \zeta^4$.

▷ **Quick Exercise.**  Check the above example by examining these complex numbers in trigonometic form.  ◁

Consequently, $\zeta + \zeta^6$, $\zeta^2 + \zeta^5$, $\zeta^3 + \zeta^4$ are real numbers, and we can in fact express them trigonometrically (we will freely use the double angle and sum formulas for the cosine function in the calculations below):

$$\zeta + \zeta^6 = 2\cos\frac{2\pi}{7}$$

$$\zeta^2 + \zeta^5 = 2\cos\frac{4\pi}{7} = 4\cos^2\frac{2\pi}{7} - 2$$

$$\zeta^3 + \zeta^4 = 2\cos\frac{6\pi}{7} = 8\cos^3\frac{2\pi}{7} - 6\cos\frac{2\pi}{7}$$

▷ **Quick Exercise.**  Verify these trigonometric calculations.  ◁

We can now show that $\zeta + \zeta^6$ actually satisfies an irreducible cubic polynomial in $\mathbb{Q}[x]$. We begin with the cyclotomic polynomial (with terms reordered):

$$0 = 1 + \left(\zeta + \zeta^6\right) + \left(\zeta^2 + \zeta^5\right) + \left(\zeta^3 + \zeta^4\right) =$$

$$1 + 2\cos\frac{2\pi}{7} + 4\cos^2\frac{2\pi}{7} - 2 + 8\cos^3\frac{2\pi}{7} - 6\cos\frac{2\pi}{7}.$$

This means that $\zeta + \zeta^6$ is a root of $x^3 + x^2 - 2x - 1 \in \mathbb{Q}[x]$, which is clearly irreducible by the Rational Root Theorem 5.6.

Consequently, $\mathbb{Q}\left(\zeta + \zeta^6\right)$ is a cubic extension of $\mathbb{Q}$ and is thus necessarily the fixed field of $G_1$. In Exercise 6 you will check that the other two roots of $x^3 + x^2 - 2x - 1$ are precisely $\zeta^2 + \zeta^5$ and $\zeta^3 + \zeta^4$.

Now $\zeta + \zeta^2 + \zeta^4$ is clearly not a real number, and so

$$\mathbb{Q}\left(\zeta + \zeta^2 + \zeta^4\right)$$

must be the other intermediate field. To actually show that it is a quadratic extension, we square it by brute force:

$$\left(\zeta + \zeta^2 + \zeta^4\right)^2 = \zeta^8 + 2\zeta^6 + 2\zeta^5 + \zeta^4 + 2\zeta^3 + \zeta^2$$

$$= \left(\zeta + \zeta^2 + \zeta^4\right) + 2\left(\zeta^3 + \zeta^5 + \zeta^6\right).$$

But $\zeta^3 + \zeta^5 + \zeta^6 = -1 - \left(\zeta + \zeta^2 + \zeta^4\right)$, and so $\zeta + \zeta^2 + \zeta^4$ is a root of the polynomial $x^2 + x + 2$, and its conjugate $\zeta^3 + \zeta^5 + \zeta^6$ is the other root. By the quadratic formula we obtain these roots as $-\frac{1}{2} \pm \frac{\sqrt{7}i}{2}$.

We thus have the following diagram of the fields and subgroups:

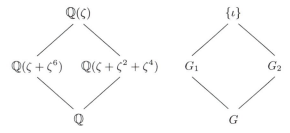

## Example 47.8

In Example 46.13 we showed that the Galois group of the splitting field $K$ of the irreducible quintic $x^5 - 6x + 3 \in \mathbb{Q}[x]$ consists of the full permutation group $S_5$. As before, we will denote the roots by $\alpha_1, \alpha_2, \alpha_3, \alpha_4, \alpha_5$, where $\alpha_1$ and $\alpha_2$ are the two complex conjugate roots.

Now $S_5$ has a large number of subgroups, and we will be unable to analyze all of them! However, it will be useful to look at a couple of examples that are easy to handle.

Consider first the five stabilizer subgroups $\mathrm{Stab}(k)$ (for $k = 1, 2, 3, 4, 5$). The subgroup $\mathrm{Stab}(k)$ is precisely the set of elements of $S_5$ that leave the root $\alpha_k$ fixed (see Exercise 23.14 for information about these subgroups). Each of these groups is isomorphic to $S_4$, and so has 24 elements. It is evident that the fixed field of $\mathrm{Stab}(k)$ is precisely $\mathbb{Q}(\alpha_k)$. These five fields are necessarily distinct, because the subgroups are. Furthermore, they are all isomorphic as field extensions of $\mathbb{Q}$, by Kronecker's Theorem 41.1. Kronecker's Theorem says these are degree 5 field extensions; equivalently we can see this because

$$[S_5 : \mathrm{Stab}(k)] = 120/24 = 5.$$

These are of course not normal subgroups, and the corresponding fields $\mathbb{Q}(\alpha_k)$ are not normal extensions.

We can generalize to other stabilizer subgroups, as in Exercise 23.15. For example, if $J = \{3, 4\}$, then the stabilizer subgroup $\mathrm{Stab}(J)$ is isomorphic to $S_3$ and its fixed field is $\mathbb{Q}(\alpha_3, \alpha_4)$, which is necessarily a field extension of degree $[S_5 : \mathrm{Stab}(J)] = 120/6 = 20$.

An obvious element of the Galois group to consider is complex conjugation. It leaves the three real roots $\alpha_3, \alpha_4, \alpha_5$ fixed, and interchanges the conjugate pair $\alpha_1$ and $\bar{\alpha}_1 = \alpha_2$. The 2 element subgroup $\{\iota, (12)\}$ has as its fixed field $\mathbb{R} \cap K$, which is necessarily a field extension of $\mathbb{Q}$ of degree 60.

We have left out an obvious subgroup of $S_5$ to consider. What about its unique non-trivial normal subgroup $A_5$? Its fixed field is necessarily a quadratic extension of $\mathbb{Q}$ (which will be a normal extension, of course). It turns out that this intermediate field can be found – our polynomial has an invariant integer $d$ called its *discriminant*, and the appropriate field extension of $\mathbb{Q}$ inside $K$ is $\mathbb{Q}(\sqrt{d})$. We will not pursue this topic here.

## Chapter Summary

For a finite normal extension of a field of characteristic zero, we prove the Fundamental Theorem of Galois Theory. This theorem provides a one-to-one order-reversing correspondence between the subgroups of the Galois group and the intermediate fields. Normal subgroups correspond to normal extensions in this correspondence.

## Warm-up Exercises

a. Suppose that $f = (x - \alpha_1) \cdots (x - \alpha_5)$. Calculate the coefficient on the $x^3$ term that results when we multiply $f$ out.

b. Suppose that we have fields $F \subseteq E \subseteq K$. Which of the following groups is a subgroup of which?

$$\mathrm{Gal}(K|E) \quad \mathrm{Gal}(K|F) \quad \mathrm{Gal}(E|F)$$

c. What further relationship holds among the groups in Exercise b, if $K$ is a finite normal extension of $F$, and $E$ is a normal extension of $F$?

d. Suppose that we have the following containments among groups: $H_1 \subseteq H_2 \subseteq \mathrm{Gal}(K|F)$. What containment relationship always holds for the fixed fields $\mathrm{Fix}(H_1)$ and $\mathrm{Fix}(H_2)$?

e. Suppose that $K$ is a finite normal extension of the field $F$, and $|\mathrm{Gal}(K|F)| = p$, where $p$ is a positive prime integer. What can you say about intermediate fields $E$, where $F \subset E \subset K$?

f. Suppose that $K$ and $F$ are fields of characteristic zero, $K$ is a normal extension, and $[K : F] = 4$. Why do we know there is an intermediate field $E$ for which $[K : E] = 2$?

## Exercises

1. Consider the splitting field $K$ of $x^5 - 3$ over $\mathbb{Q}(\zeta)$, where $\zeta$ is the primitive fifth root of unity. Compute the Galois group $\mathrm{Gal}(K|\mathbb{Q}(\zeta))$. Then illustrate the correspondence between subfields and subgroups given by the Fundamental Theorem of Galois Theory, by drawing a diagram illustrating all of the subgroups of the Galois group of this splitting field and the corresponding order-reversed picture of all the fields between $\mathbb{Q}(\zeta)$ and the splitting field. What are the orders of each of the groups you have in your diagram?

2. Consider the cyclotomic polynomial $\Phi_5 = x^4 + x^3 + x^2 + x + 1 \in \mathbb{Q}[x]$. Then we know that the splitting field for $\Phi_5$ is just $\mathbb{Q}(\zeta)$, where $\zeta$ is the primitive fifth root of unity, and the Galois group of this splitting field is a cyclic group of order 4 (see Theorem 46.3). Draw a diagram illustrating all of the subgroups of the Galois group of this splitting field and the corresponding order-reversed picture of all the fields between $\mathbb{Q}$ and the splitting field. What are the orders of each of the groups you have in your diagram?

3. Consider again the splitting field of $x^4 - 2 \in \mathbb{Q}[x]$, as in Exercise 44.6 and Example 46.9. As in the previous exercise, draw a diagram illustrating all of the subgroups of the Galois group of this splitting field and the corresponding order-reversed picture of all the fields between $\mathbb{Q}$ and the splitting field. You should explicitly verify that each intermediate field is the fixed field of the appropriate subgroup. What are the orders of each of the groups in your diagram?

4. Which of the subgroups of the Galois group in the previous exercise are normal in the Galois group of the splitting field? What property do the corresponding fields have?

5. Prove Theorem 47.1.

6. In this exercise you check a computation from Example 47.7. In that example we showed that $\zeta + \zeta^6$ is a root of the polynomial $x^3 + x^2 - 2x - 1$, where $\zeta$ is the primitive seventh root of unity. Show that $\zeta^2 + \zeta^5$ and $\zeta^3 + \zeta^4$ are the other two roots.

7. In this exercise you provide an inductive proof for the formula for the symmetric polynomials, as discussed in Section 47.2. Suppose that $f = (x - \alpha_1)(x - \alpha_2) \cdots (x - \alpha_n) \in F[x]$, where $F$ is a field. Let $(-1)^{n-k} a_{n-k}$ be the coefficient on $x^k$ when we multiply $f$ out using the distributive law. Use induction on $n$ to prove that

$$a_{n-k} = \Sigma\{\alpha_{j_1} \alpha_{j_2} \cdots \alpha_{j_k}\},$$

where the sum is over all possible choices of $k$ distinct $\alpha_i$.

8. Suppose that $f \in F[x]$ is an irreducible polynomial of degree $n$ over the field $F$, where $F$ is a field of characteristic zero. Let $K$ be the splitting field of $f$ over $F$. Then we can think of $G = \text{Gal}(K|F)$ as a subgroup of the permutation group $S_n$. Prove that $G$ is a transitive subgroup of this permutation group (see Exercise 27.13 for a definition).

9. Consider the field $K = \mathbb{Q}(\sqrt{2}, \sqrt{3}, \sqrt{5})$. This is a normal extension of $\mathbb{Q}$. Find the values of each of the following.

   (a) $[K : \mathbb{Q}]$
   (b) $[K : \mathbb{Q}(\sqrt{2})]$
   (c) $[K : \mathbb{Q}(\sqrt{5})]$
   (d) $[K : \mathbb{Q}(\sqrt{2}, \sqrt{5})]$
   (e) $[\mathbb{Q}(\sqrt{2}, \sqrt{5}) : \mathbb{Q}]$
   (f) $[K : \mathbb{Q}(\sqrt{2} + \sqrt{3})]$
   (g) $[\mathbb{Q}(\sqrt{2} + \sqrt{3}) : \mathbb{Q}]$

10. Draw the diagrams of the subgroups of the Galois group $\text{Gal}(\mathbb{Q}(\sqrt{2}, \sqrt{3}, \sqrt{5})|\mathbb{Q})$ and the corresponding one of all the fields between $\mathbb{Q}(\sqrt{2}, \sqrt{3}, \sqrt{5})$ and $\mathbb{Q}$.

11. Consider the splitting field $K$ for $x^5 - 6x + 3$ over $\mathbb{Q}$, as considered in Example 47.8, and identify $\mathrm{Gal}(K|\mathbb{Q})$ with $S_5$, as done in the example. Consider the subgroup $H = \langle(125)\rangle$ of the Galois group. Determine $[K : \mathrm{Fix}(H)]$. Then answer the following questions about the field extension $K$ over $\mathrm{Fix}(H)$ (with justification):

   (a) Why is this extension simple?

   (b) What is the degree of the minimal polynomial for this simple extension?

   (c) Why is this extension normal?

   (d) Why is $\mathrm{Fix}(H)$ not a normal extension of $\mathbb{Q}$?

   (e) Would the answers to any of the previous questions be different, if we chose a different 3-cycle from $S_5$ than $(125)$?

   (f) Give a stabilizer subgroup of $H$ of the form $\mathrm{Stab}(J)$, where $J$ is a subset of $\{1, 2, 3, 4, 5\}$ (see Exercise 23.15). Choose $J$ to be as large as possible.

   (g) Use the previous answer to obtain two fields intermediate between $K$ and $\mathbb{Q}$, which are subfields of one another.

   (h) For the two fields in the previous answer, what is the degree of the larger field over the smaller?

   (i) Why is the field extension from the previous answer normal?

   (j) According the Fundamental Theorem, Part Two, what corresponding Galois group is then necessarily normal in what other group? To what Galois group does the quotient group have to be isomorphic?

12. Let $\zeta$ be the primitive eleventh root of unity. Then by Theorem 46.3 $\mathrm{Gal}(\mathbb{Q}(\zeta)|\mathbb{Q})$ is a cyclic group of order 10. Specify the generator of this group using permutation notation, and analyze all of the subgroups, and the corresponding fields, following the exposition of Example 47.7. Include explicit descriptions for the fixed fields.

# Chapter 48

## Solving Polynomials by Radicals

We are now ready to focus our attention on the problem of whether it is possible to solve all polynomial equations over a subfield of the complex numbers, by ordinary field arithmetic, together with the extraction of roots. We are able to do this in the quadratic case (using the quadratic formula Exercise 10.1), in the cubic case (using the Cardano-Tartaglia approach, Exercise 10.12), and in the quartic case (using the Ferrari approach, Exercise 10.20).

In this chapter we will recast this problem in terms of field extensions, just as we did for the notion of constructible numbers, in Chapter 37. In that context we simplified matters considerably, by focusing our attention on the sequence of ever larger fields necessary to obtain the constructible number. In that case the larger fields were built as quadratic extensions. In our present case, we will need to build larger and larger fields, but allow extensions by higher power roots instead.

## 48.1   Field Extensions by Radicals

Let's consider a single step in the process of building larger fields by extraction of roots. Given a field $F$ and an integer $n \geq 2$, a **radical extension** of $F$ is a simple algebraic field extension of the form $F(\beta)$, where $\beta^n \in F$. Thus every element of the radical extension can be expressed in terms of field arithmetic on $\beta$, together with elements of $F$. But we can view $\beta$ as the result of an extraction of an $n$th root of $\beta^n \in F$. That is, every element of $F(\beta)$ can be obtained by a formula involving only ordinary field arithmetic, together with a root extraction.

We then generalize this inductively, supposing that we have a finite sequence of radical extensions

$$F = F_0 \subset F_1 = F_0(\beta_1) \subset F_2 = F_1(\beta_2) \subset \cdots \subset F_N = F_{N-1}(\beta_N),$$

where at each stage $\beta_i^{n_i} \in F_{i-1}$. We call the field $F_N$ an **extension of $F$ by radicals**; for every element in $F_N$ there is a (perhaps very complicated!) formula, involving only field operations and extractions of roots, applied iteratively to the elements of the field $F$. We'll call the finite sequence of fields a **root tower over $F$**.

Now, if $f \in F[x]$ is a polynomial, and its splitting field $K$ is contained in an extension of $F$ by radicals, we say that $f$ is **solvable by radicals**.

This description of solvability by radicals can and should be compared to our description of constructible numbers in the Constructible Number Theorem 37.3.

Let's look at some examples.

**Example 48.1**

Any quadratic polynomial $f = ax^2 + bx + c \in \mathbb{Q}[x]$ is solvable by radicals over $\mathbb{Q}$, because the splitting field for $f$ is contained in (and is in this case equal to) the field extension $\mathbb{Q}\left(\sqrt{b^2 - 4ac}\right)$. Here the root tower consists of just two fields: $\mathbb{Q} \subseteq \mathbb{Q}\left(\sqrt{b^2 - 4ac}\right)$. Obviously the quadratic formula expression is our explicit solution by radicals.

## Example 48.2

Consider the polynomial $f = x^6 - 2x^3 - 1 \in \mathbb{Q}[x]$. Since this is a quadratic equation in $x^3$, it is evident by the quadratic formula that it factors as

$$f = (x^3 - (1 + \sqrt{2}))(x^3 - (1 - \sqrt{2})).$$

But the roots in $\mathbb{C}$ of these two factors are in turn just cube roots. If as usual we let $\zeta = -\frac{1}{2} + \frac{\sqrt{3}}{2}i$ be the primitive cube root of unity, we then have $f$ completely factored as

$$f = \left(x - \sqrt[3]{1 + \sqrt{2}}\right)\left(x - \sqrt[3]{1 + \sqrt{2}}\zeta\right)\left(x - \sqrt[3]{1 + \sqrt{2}}\zeta^2\right)$$
$$\left(x - \sqrt[3]{1 - \sqrt{2}}\right)\left(x - \sqrt[3]{1 - \sqrt{2}}\zeta\right)\left(x - \sqrt[3]{1 - \sqrt{2}}\zeta^2\right).$$

So each of these six roots can be expressed in terms of a formula involving only field operations, together with both cube and square roots. But we can also look at this in terms of the corresponding root tower:

$$\mathbb{Q} \subset \mathbb{Q}(\sqrt{2}) \subset \mathbb{Q}(\sqrt{2})\left(\sqrt[3]{1 + \sqrt{2}}\right) \subset$$
$$\mathbb{Q}(\sqrt{2})\left(\sqrt[3]{1 + \sqrt{2}}\right)(\zeta) = F_3.$$

It may not be immediately obvious that all six roots belong to the field $F_3$, but we can conclude this if we note that

$$\left(\sqrt[3]{1 + \sqrt{2}}\right)\left(\sqrt[3]{1 - \sqrt{2}}\right) = \sqrt[3]{1 - 2} = -1,$$

and so

$$\sqrt[3]{1 - \sqrt{2}} = \frac{-1}{\sqrt[3]{1 + \sqrt{2}}} \in \mathbb{Q}(\sqrt{2})\left(\sqrt[3]{1 + \sqrt{2}}\right).$$

In this case $F_3$ is again equal to the splitting field for $f$ over $\mathbb{Q}$, and $[F_3 : \mathbb{Q}] = 12$.

▷ **Quick Exercise.** Why are the assertions in the last statement true? Note that by Theorem 46.2 the Galois group $\text{Gal}(F_3|\mathbb{Q})$ is a group of 12 elements — you will determine which group in Exercise 1. ◁

## Example 48.3

Let's now consider the irreducible cubic polynomial $f = x^3 + 3x + 1 \in \mathbb{Q}[x]$. We can use the Cardano-Tartaglia formula to obtain a real root for this polynomial. In Exercise 10.16 you do this to obtain

$$\alpha_1 = \sqrt[3]{\frac{-1 + \sqrt{5}}{2}} - \sqrt[3]{\frac{1 + \sqrt{5}}{2}}.$$

Furthermore in that exercise you verify that if $\zeta$ is the primitive cube root of unity, then the other two roots for $f$ in $\mathbb{C}$ are a conjugate pair of complex numbers:

$$\alpha_2 = \sqrt[3]{\frac{-1 + \sqrt{5}}{2}}\zeta - \sqrt[3]{\frac{1 + \sqrt{5}}{2}}\zeta^2$$

and

$$\alpha_3 = \sqrt[3]{\frac{-1+\sqrt{5}}{2}}\zeta^2 - \sqrt[3]{\frac{1+\sqrt{5}}{2}}\zeta.$$

It is now quite evident that the splitting field $K$ for $f$ can be reached by the following root tower:

$$\mathbb{Q} \subset \mathbb{Q}(\zeta) \subset \mathbb{Q}(\zeta, \sqrt{5}) \subset \mathbb{Q}\left(\zeta, \sqrt{5}, \sqrt[3]{\frac{-1+\sqrt{5}}{2}}\right) = F_3.$$

That is, we have that $K$ is a subfield of $F_3$.

▷ **Quick Exercise.** Why does $\sqrt[3]{\frac{1+\sqrt{5}}{2}} \in F_3$? ◁

In Exercise 2 you will check that each of these field extensions is in fact proper, and so $[F_3 : \mathbb{Q}] = 12$. But from Example 46.8 we know that $[K : \mathbb{Q}] = |\mathrm{Gal}(K|\mathbb{Q})| = 6$. Consequently, in this example our root tower reaches a field extension strictly larger than the splitting field. We will have to account for this possibility in the general theory we develop later in this chapter.

## 48.2 Refining the Root Tower

Suppose again that we have a field $F$, a polynomial $f \in F[x]$, and a root tower over $F$ that enables us to solve $f$ by radicals. Our eventual goal is to show that the existence of such a root tower places restrictions on the sort of group we can get as the Galois group of the splitting field for $f$. This restriction on the group will enable us to conclude that not all polynomials are solvable by radicals, which is the goal of this chapter.

To accomplish this, we will modify our root tower in a couple of ways. First of all, we'd like each step along the tower to be small enough that it can be well understood. So now consider a single step in our root tower. We suppose that we have the field extension $E \subseteq E(\beta)$, where $\beta^n \in E$, and $E$ is one of the fields reached at some stage of our root tower for $F$. We now claim that we can assume that $n$ is a prime integer. For if $n = p_1 p_2 \cdots p_m$ is the prime factorization of $n$ into (not necessarily distinct) prime factors, we can replace the single radical extension from $E$ to $E(\beta)$ by a sequence of extensions as follows:

$$E \subseteq E(\beta^{p_2 p_3 \cdots p_m}) \subseteq E(\beta^{p_3 \cdots p_m}) \subseteq \cdots \subseteq E(\beta).$$

We obviously arrive at the same spot in our root tower, at the expense of taking finitely many extra steps in the construction. When we pass to the group side later, it will be more convenient for us to assume we need only extract roots of prime order.

Our second modification of the root tower requires us to impose a serious restriction on the arguments that follow. We will assume not only that our field $F$ is of characteristic zero, but that it is actually a subfield of the complex numbers $\mathbb{C}$. The examples of fields of characteristic zero that we have explored in this and earlier chapters have all satisfied this criterion, although there certainly do exist fields of characteristic zero that do not (see Exercise 8). Our desire in this chapter is to recover Abel's result that there exist fifth degree polynomials over $\mathbb{Q}$ that are not solvable by radicals, and so this restrictive assumption does not effect that goal.

The advantage of this restriction is that in $\mathbb{C}$ we have a very concrete understanding of the $p$th roots of unity, where $p$ is a prime integer. Theorem 46.3 tells us that the splitting field of the polynomial $x^p - 1 \in \mathbb{Q}[x]$ is just $\mathbb{Q}(\zeta)$, where

$$\zeta = e^{\frac{2\pi i}{p}}$$

is the primitive $p$th root of unity.

With this restriction in place, we are now ready to return to our project of refining our root tower. Let's suppose that the root tower we now have looks like this:

$$F = F_0 \subset F_1 = F(\beta_1) \subset F_2 = F(\beta_2) \subset \cdots \subset F_m = F_{m-1}(\beta_m),$$

where for each $i$ there is a prime integer $p_i$ with $\beta_i^{p_i} \in F_{i-1}$. Now consider the set of all these primes $p_i$ appearing as the degrees of the radical extensions in our root tower (of course, some of these primes may occur more than once in this list). We will now extend our root tower further, by first adjoining each of the $p_i$th primitive roots of unity $\zeta_i$, one at a time. The first terms of our root tower will then look like this:

$$F \subseteq E_1 = F(\zeta_1) \subseteq E_2 = E_1(\zeta_2) \subseteq \cdots \subseteq E_m = F(\zeta_1, \zeta_2, \cdots, \zeta_m).$$

We note that by Theorem 46.3 it is clear that each of these fields $E_i$ is normal over $F$, because evidently $E_i$ is the splitting field over $F$ for the polynomial

$$(x^{p_1} - 1)(x^{p_2} - 1) \cdots (x^{p_i} - 1).$$

Note that in case we have repetitions in our list of primes, some of these extensions may be trivial; this is harmless for what follows.

So from our original root tower, we have obtained a new tower, where we only extract roots of prime order, and where we have first included all of the primitive roots of unity, for each of the primes we need to use. Of course, by including all these extra elements we may well have arrived at a much larger field extension of the original field $F$, but that is harmless, because the new larger field will still allow for the solution of the polynomial $f$ by radicals. We record our progress so far in refining our root tower:

**Lemma 48.1** *Suppose that $F$ is a subfield of the complex numbers, and $f \in F[x]$ is solvable by radicals over $F$. Then there exists a root tower of the following form, where the splitting field of $f$ is contained in the largest field $F_m$ of the tower:*

$$F \subseteq E_1 = F(\zeta_1) \subseteq E_2 = E_1(\zeta_2) \subseteq \cdots \subseteq E_m = F(\zeta_1, \zeta_2, \cdots, \zeta_m) = F_0$$

$$\subseteq F_1 = F_0(\beta_1) \subseteq F_2 = F_1(\beta_2) \subseteq \cdots \subseteq F_m = F_{m-1}(\beta_m),$$

*where $p_i$ are prime integers, $\zeta_i$ is the primitive $p_i$th root of unity, and $\beta_i^{p_i} \in F_{i-1}$.*

Notice that the index $m$ is a count of the number of (not necessarily distinct) primes used to perform the original root extractions, one prime at a time. Consequently, many of the extensions in the tower above may be trivial; this does not affect our goal. We have thrown in many additional roots, precisely so that each individual extension in the Lemma is fairly easy to understand: In each case we have adjoined a single prime order root.

Unfortunately, this is still not the root tower we will use, in general. Our next major step is to be able to translate our field extension problem into a problem of group theory, by using the Fundamental Theorem of Galois Theory (Theorems 47.3 and 47.4). These theorems do not yet apply, because we presently do not have any control over whether our extensions are normal over $F$.

To be specific, consider the extension $F_i = F_{i-1}(\beta_i)$; since $\beta_i^{p_i} \in F_{i-1}$, $\beta_i$ is a root of the polynomial $x^{p_i} - \beta_k^{p_i}$. But we know that the primitive $p_i$-th root of unity $\zeta_i$ belongs to $F_{i-1}$, and so the roots of our polynomial are just $\beta_i$, $\beta_i\zeta_i$, $\beta_i\zeta_i^2$, ..., $\beta_i\zeta_i^{p_i-1}$, which all belong to $F_i$. However, this only makes $F_i$ normal over $F_{i-1}$, and not over $F$. What follows is an example showing that we cannot in general expect normality over the base field $F$.

## Example 48.4

Let $\beta = \sqrt[4]{2}(1+i)$. As you can check directly, this is a root of $x^4 + 8$; see Exercise 44.8. That is, $\beta$ is a fourth root of $-8$. Now suppose that $\mathbb{Q} \subset \mathbb{Q}(\beta)$ is a step in our original root tower. Since 4 is not prime, we must insert an additional field in our tower. But $\beta^2 = 2\sqrt{2}i$, and this latter element is evidently a square root of $-8$. Our refined root tower is now

$$\mathbb{Q} \subset \mathbb{Q}(2\sqrt{2}i) \subset \mathbb{Q}(2\sqrt{2}i, \beta) = \mathbb{Q}(\beta).$$

The next step we took above was to first insert all of the appropriate $p$th roots of unity. But in this case the only prime involved is 2, and so we should first adjoin the primitive second root of unity. But this is just $-1$, which already belongs to $\mathbb{Q}$. We now claim that $\mathbb{Q}(\beta)$ is *not* a normal extension over $\mathbb{Q}$. In Exercises 44.6 and 44.8 you show that the splitting field of $x^4 + 8$ over $\mathbb{Q}$ is $\mathbb{Q}(\sqrt[4]{2}, i)$, which has degree 8 over the rationals, but $[\mathbb{Q}(\beta) : \mathbb{Q}] = 4$. (This example was pointed out by Thomas Hungerford in a recent paper in the *American Mathematical Monthly*.)

Our goal now is to replace each of the fields $F_i$ above by a larger field $K_i$, with the property that $K_i$ is in fact a normal extension of $F$. We will do this by including enough elements at each stage that $K_i$ will be the splitting field of some polynomial $g_i$ over $F$. We will proceed inductively. Our starting point for the induction is the normal field extension $F \subseteq K_0 = F(\zeta_1, \zeta_2, \cdots, \zeta_m)$. Here $K_0$ is exactly the splitting field of the polynomial

$$g_0 = (x^{p_1} - 1)(x^{p_2} - 1) \cdots (x^{p_m} - 1).$$

So suppose that we have already constructed our new sequence of fields

$$F \subseteq K_0 \subseteq K_1 \subseteq K_2 \cdots \subseteq K_i,$$

so that for every $1 \leq j \leq i$, we have that $\beta_j \in K_j$, and each $K_j$ is an extension of $K_{j-1}$ by radicals, and is a normal extension of $F$; in fact, we will assume that $K_j$ is the splitting field of a polynomial $g_j \in F[x]$. Note that because $F_i = F(\beta_1, \beta_2, \cdots, \beta_i)$, it is evident that $F_j \subseteq K_j$, for all $1 \leq j \leq i$.

We shall now provide the inductive argument to construct $K_{i+1}$. We know that $\beta_{i+1}^{p_{i+1}} \in F_i \subseteq K_i$; for notational convenience, let's call this element $e$, and set $p = p_{i+1}$. We can enumerate the elements of the Galois group $\mathrm{Gal}(K_i|F)$ as $\{\iota = \varphi_1, \varphi_2, \cdots, \varphi_k\}$. Consider the polynomial

$$h = (x^p - \varphi_1(e))(x^p - \varphi_2(e)) \cdots (x^p - \varphi_k(e)) \in K_i[x].$$

We will now build the splitting field of $h$ over $K_i$ by radical extensions, one factor at a time. Notice that the coefficients of the polynomial $h$ are symmetric polynomials in the roots

$$\varphi_1(e), \cdots, \varphi_k(e);$$

see Section 47.2. Furthermore, we know that the elements of $\mathrm{Gal}(K_i|F)$ permute the roots of $h$. But symmetric polynomials in the roots are thus left fixed by the elements of $\mathrm{Gal}(K_i|F)$, and so these coefficients belong to the fixed field of the Galois group. By the induction hypothesis, we know that $K_i$ is normal over $F$, and so the fixed field of the Galois group is equal to $F$. Consequently, the coefficients belong to $F$, and so $h \in F[x]$.

Now we know that the primitive $p$th root of unity $\zeta$ is an element of the field $K_i$. Consequently, in order to assure that the factor $x^p - \varphi_i(e)$ splits, we need only adjoin a single root $\rho_i$ to the previous field (Theorem 46.4); here $\rho_i$ is a root of the polynomial $x^p - \varphi_i(e)$. We thus have

$$K_i \subset K_i(\rho_1) \subset K_i(\rho_1, \rho_2) \subset \cdots \subset K_i(\rho_1, \cdots, \rho_k).$$

Note that at each stage in this sequence we have a splitting field for a polynomial

$$(x^p - \varphi_1(e))(x^p - \varphi_2(e)) \cdots (x^p - \varphi_j(e)) \in K_i[x],$$

and so each of these fields is normal over $K_i$. (They may not be normal over $F$, because these intermediate polynomials need not belong to $F[x]$.) We shall now let $K_{i+1} = K_i(\rho_1, \cdots, \rho_k)$; evidently $K_{i+1}$ is the splitting field over $F$ of the polynomial $g_{i+1} = g_i h \in F[x]$, where by the induction hypothesis $g_i$ is the polynomial in $F[x]$ for which $K_i$ is the splitting field over $F$.

Let's summarize the situation we now have.

**Lemma 48.2** *Suppose that $F \subset \mathbb{C}$ is a field, and $f \in F[x]$ is a polynomial that can be solved by radicals over $F$. Then there exists a sequence of field extensions*

$$F \subseteq E_1 = F(\zeta_1) \subseteq E_2 = E_1(\zeta_2) \subseteq \cdots \subseteq E_m = K_0 = F(\zeta_1, \zeta_2, \cdots, \zeta_m)$$
$$\subseteq K_1 \subseteq K_2 \subseteq \cdots \subseteq K_m,$$

*so that the following conditions hold:*

1. *For each $i$, $\zeta_i$ is the $p_i$th primitive root for unity, for the prime integer $p_i$.*

2. *Each $K_i$ is normal over $F$.*

3. *Each $K_{i+1}$ is obtained from $K_i$ by finitely many radical extensions of degree $p_i$; these extensions are all normal over $K_i$.*

4. *$K_m$ contains the splitting field of $f$ over $F$.*

We have now put all the technical details in place and are now ready to apply Galois theory to this lemma, which we shall do in the next section.

---

## 48.3   Solvable Galois Groups

In the last section we showed that whenever a polynomial $f \in F[x] \subseteq \mathbb{C}[x]$ is solvable by radicals, we can construct a root tower of a particular type, as specified in Lemma 48.2. This gives us the information we need to translate our root tower into a sequence of groups. This translation from field theory into group theory gives us the following:

**Theorem 48.3** *Suppose that $f \in F[x] \subseteq \mathbb{C}[x]$ is solvable by radicals. Then the Galois group of the splitting field for $f$ is solvable.*

**Proof**: We will now apply the Fundamental Theorem of Galois Theory to the field extensions guaranteed by Lemma 48.2. Let

$$G = G_0 = \mathrm{Gal}(K_m|F),\ G_1 = \mathrm{Gal}(K_m|E_1),\ \cdots,\ G_m = \mathrm{Gal}(K_m|E_m)$$

$$G_{m+1} = \mathrm{Gal}(K_m|K_1),\ \cdots,\ G_{2m} = \mathrm{Gal}(K_m|K_m).$$

It is evident that

$$G_{2m} = \{\iota\} \subseteq G_{2m-1} \subseteq \cdots \subseteq G_m \subseteq \cdots \subseteq G_1 \subseteq G$$

and each of these subgroups is normal in $G$, because each of the fields $E_i$ and $K_i$ are normal extensions of $F$.

We would like to conclude that $G$ is a solvable group; we will use Theorem 30.4. For $0 < i \leq m$, we have that

$$G_{i-1}/G_i = \mathrm{Gal}(K_m|E_{i-1})/\mathrm{Gal}(K_m|E_i) \approx \mathrm{Gal}(E_i|E_{i-1}).$$

The isomorphism follows from the Fundamental Theorem of Galois Theory. But by Theorem 46.3 we know that this is an abelian group, and hence solvable, of course.

For $0 < i \leq m$, we have that

$$G_{m+i-1}/G_{m+i} = \mathrm{Gal}(K_m|K_{i-1})/\mathrm{Gal}(K_m|K_i) \approx \mathrm{Gal}(K_i|K_{i-1}).$$

We would like to argue that this last group is a solvable group.

But note that we obtain $K_i$ from $K_{i-1}$ by finitely many radical extensions of degree $p = p_i$, a prime integer. That is, we have that

$$K_{i-1} \subseteq L_1 \subseteq L_2 \cdots \subseteq L_h \subseteq K_i,$$

where each of the fields $L_j$ is obtained from its predecessor by adjoining a $p$th root of an element from the previous field. Furthermore, we know by construction that the primitive $p$th root of unity $\zeta \in K_{i-1}$. Consequently, each of these extensions is normal over $K_{i-1}$, because it is the splitting field over $K_{i-1}$ for a polynomial of the form $x^p - e$ (see Lemma 48.2 above).

We thus have the subnormal series

$$\{1\} \subseteq \mathrm{Gal}(K_i|L_h) \subseteq \mathrm{Gal}(K_i|L_{h-1}) \subseteq \cdots \subseteq \mathrm{Gal}(K_i|L_1) \subseteq \mathrm{Gal}(K_i|K_{i-1}).$$

When we compute the quotient group of two of these groups we have

$$\mathrm{Gal}(K_i|L_{j-1})/\mathrm{Gal}(K_i|L_j) \approx \mathrm{Gal}(L_j|L_{j-1}).$$

But this last group is cyclic by Theorem 46.4. This means that our subnormal series actually makes $\mathrm{Gal}(K_i|K_{i-1})$ a solvable group.

Putting this all together, we have now concluded by Theorem 30.4 that $G = \mathrm{Gal}(K_m|F)$ is also a solvable group.

This is almost the conclusion that we want. However, $K_m$ is in practice probably much larger than the splitting field for $f$ over $F$. But if we let $K$ be that splitting field, then $K$ is a normal extension of $F$, and so another application of the Fundamental Theorem of Galois Theory gives us that

$$\mathrm{Gal}(K|F) \approx \mathrm{Gal}(K_m|F)/\mathrm{Gal}(K_m|K).$$

But this means that $\mathrm{Gal}(K|F)$ is a homomorphic image of the solvable group $\mathrm{Gal}(K_m|F)$ and so is itself solvable, by Theorem 30.1. $\qquad\square$

It is actually the case that the converse of Theorem 48.3 is true — that is, if the Galois group for the splitting field of a polynomial $f \in F[x] \subseteq \mathbb{Q}[x]$ is solvable, then $f$ is actually solvable by radicals over $F$. To prove this, we would have to start with the subnormal series with abelian quotients for the Galois group, and build a sequence of radical field extensions that eventually contain the splitting field. We will not actually carry out this project here.

The next example is what we've been looking for! It is a polynomial of degree five that is not solvable by radicals.

### Example 48.5

Consider again the polynomial $f = x^5 - 6x + 3 \in \mathbb{Q}[x]$. In Example 46.13 we showed that the Galois group of the splitting field for $f$ over $\mathbb{Q}$ is $S_5$. In Example 30.4 we asserted that $S_5$ is not a solvable group. Theorem 48.3 now means that $f$ is *not solvable by radicals!*

### Example 48.6

In Example 48.2 we considered the polynomial $f = x^6 - 2x^3 - 1 \in \mathbb{Q}[x]$ and showed explicitly that it is solvable by radicals. In Exercise 1 you will compute the Galois group for the splitting field of this polynomial and show that it is a solvable group.

We have thus solved (in the negative) a problem that bedeviled European mathematicians for several centuries. It is in principle impossible to extend the progressively more complicated algebraic formulas we have for second, third, and fourth degree equations to the fifth degree or higher. The mathematics involved is more complicated, but the situation is the same as for the classical constructibility problems: Field theory (in this case with an important assist from the theory of groups) has solved an important mathematical problem that at first blush does not seem to require the abstract approach.

The abstract algebraic approach was successful in settling the construction problems of classical antiquity and in solving the solution by radicals problems of the Italian renaissance; this is only the beginning of the story. Powerful algebraic techniques have played an important role in bringing profound insights into many difficult and important problems, over the course of the nineteenth, twentieth, and twenty-first centuries. You are invited to learn more about the successes and beauty of modern mathematics.

---

## Chapter Summary

In this chapter we prove that if a polynomial over a subfield of the complex numbers can be solved by radicals, then the Galois group of its splitting field over the base field is necessarily a solvable group. We can then easily exhibit a fifth degree polynomial over the rational numbers that *cannot* be solved by radicals.

---

## Warm-up Exercises

a. Give a root tower whose last field contains the splitting field for $x^4 - 2$ over the rational numbers. (This is Example 46.9.)

b. Give a root tower whose last field contains the splitting field for $x^3 - 2$ over the rational numbers. (This is Example 46.7.)

c. Give an example of a root tower whose last field is *not* normal over the base field.

d. Suppose that $\zeta$ is the primitive $p$th root of unity, $F$ is a subfield of the complex numbers, $\zeta \in F$, but $F$ contains no $p$th root of $a \in F$. Let $\alpha$ be a $p$th root of $a$ in $\mathbb{C}$. Give as much information as you can about the field extension $F(\alpha)$ over $F$.

e. Is the sequence of fields given by the Constructible Number Theorem 37.3 a root tower?

f. Give an example of an irreducible polynomial $f \in \mathbb{Q}[x]$ that can be solved by radicals, none of whose roots are constructible numbers. Give such an example where all the roots are real numbers.

g. Suppose that $K$ is the splitting field for $f \in \mathbb{Q}[x]$, and $\mathrm{Gal}(K|\mathbb{Q}) \approx S_7$. What can you say about this situation?

---

# Exercises

1. Consider the polynomial $x^6 - 2x^3 - 1 \in \mathbb{Q}[x]$ from Example 48.2 and 48.6. Compute the Galois group for the splitting field of this polynomial over $\mathbb{Q}$, and show explicitly that this is a solvable group. Find the subgroups corresponding to the fixed fields

$$\mathbb{Q}(\sqrt{2}), \ \mathbb{Q}(\zeta), \ \mathbb{Q}(\zeta, \sqrt{2}), \ \mathbb{Q}(\alpha), \ \mathbb{Q}(\alpha\zeta), \ \mathbb{Q}(\alpha\zeta^2).$$

2. Consider the root tower obtained in Example 48.3 to obtain a field containing the splitting field $K$ for $x^3 + 3x + 1 \in \mathbb{Q}[x]$. Show that each field extension in the root tower is proper, and so the field $F_3$ is a proper extension of $K$.

3. In this problem and its successor we will describe a method that helps determine the Galois group of the splitting field for a quartic polynomial. Let $F$ be a subfield of the complex numbers, and suppose that $f \in F[x]$ is an irreducible polynomial of degree four. We may as well assume that $f$ is monic, and so

$$f = (x - \alpha_1)(x - \alpha_2)(x - \alpha_3)(x - \alpha_4),$$

where the $\alpha_i$ are distinct elements of $\mathbb{C}$. Let $K = F(\alpha_1, \alpha_2, \alpha_3, \alpha_4)$ be the splitting field for $f$. We will consider $G = \mathrm{Gal}(K|F)$ as a subgroup of the group $S_4$ of permutations of these roots, in this order. By Section 47.2 we know that $f = x^4 - a_1x^3 + a_2x^2 - a_3x + a_4$, where the $a_i$ are the symmetric polynomials in the roots $\alpha_i$.

(a) Let
$$\beta_1 = \alpha_1\alpha_2 + \alpha_3\alpha_4, \quad \beta_2 = \alpha_1\alpha_3 + \alpha_2\alpha_4,$$
$$\beta_3 = \alpha_1\alpha_4 + \alpha_2\alpha_3.$$

Form the polynomial

$$g = (x - \beta_1)(x - \beta_2)(x - \beta_3) = x^3 - b_1x^2 + b_2x - b_3.$$

By making use of what we know about symmetric polynomials, prove that

$$b_1 = a_2, \quad b_2 = a_1a_3 - 4a_4, \quad b_3 = a_1^2a_4 + a_3^2 - 4a_2a_4.$$

Thus, $g \in F[x]$. *Note:* Pay heed to the signs involved. The polynomial $g$ is called the **resolvent polynomial** for $f$.

(b) Let $E = F(\beta_1, \beta_2, \beta_3)$ be the splitting field for $g$ over $F$; this is a subfield of $K$. Let

$$H = G \cap \{\iota, (12)(34), (13)(24), (14)(23)\}.$$

   Prove that $E = \text{Fix}(H)$.

(c) Why is $H$ normal in $G$?

(d) Prove that $\text{Gal}(E|F)$ is isomorphic to $G/H$.

4. In this exercise we use the same notation and terminology as in the previous exercise. By Exercise 47.8 the Galois group $G$ is necessarily a transitive subgroup of $S_4$. In Exercise 27.13 you compiled a list of all the transitive subgroups of $S_4$. We will in what follows think of $G/H = \text{Gal}(E|F)$ as a subgroup of the group of permutations $S_3$ of the roots of $g$. We now look at all possible cases for $G$:

(a) Suppose that $G = S_4$. Then show that $G/H$ is isomorphic to $S_3$.

(b) Suppose that $G = A_4$. Then show that $G/H$ is isomorphic to $\mathbb{Z}_3$.

(c) Suppose that $G = \{\iota, (12)(34), (13)(24), (14)(23)\}$. Then show that $G/H$ is the trivial group.

(d) Suppose that $G$ is a cyclic group of order 4. Then show that $G/H$ is a cyclic group of order 2.

(e) Suppose that

$$G = \{\iota, (12)(34), (13)(24), (14)(23), (12), (34), (1423), (1324)\},$$

   the subgroup of $S_4$ of order 8 (isomorphic to $D_4$). Then show that $G/H$ is a cyclic group of order 2.

*Note:* Since $H$ is the Galois group of the splitting field of a cubic polynomial, it is presumably easier to compute than is $G$. But the results above say that in most cases knowing $H$ actually determines $G$. We will illustrate this principle in the exercises that follow. The ambiguous case that remains (when $H$ is a cyclic group of order 2) can actually also be resolved by a refinement of the argument given here. For information about that case you can consult Nathan Jacobson's *Basic Algebra I* (W. H. Freeman and Company, 1974), from which these exercises have been adapted.

5. From Example 46.9 we know that the Galois group of the splitting field for $x^4 - 2$ is isomorphic to $D_4$. Thus if we form the resolvent polynomial $g$, we know from Exercise 4 that its Galois group is a cyclic group of order two. Compute $g$ and the Galois group explicitly, to check this computation.

6. Consider the polynomial $f = x^4 + 2x + 2 \in \mathbb{Q}[x]$.

(a) Why is $f$ irreducible over $\mathbb{Q}[x]$?

(b) Use calculus to argue that $f$ has no real roots.

(c) Show that the resolvent polynomial $g$ obtained as in Exercise 3 is $g = x^3 - 8x - 4$.

(d) Why is $g$ irreducible over $\mathbb{Q}[x]$?

(e) Use calculus to argue that $g$ has three real roots.

(f) Argue that the Galois group of $g$ over the rational field must contain an element of order 3. What does this mean about the Galois group of the splitting field for $f$?

7. Consider the polynomial $f = x^4 + 2x - 2 \in \mathbb{Q}[x]$. Notice that this is just a vertical shift of the polynomial in the previous exercise.

   (a) Why is $f$ irreducible over $\mathbb{Q}[x]$?

   (b) Use calculus to argue that $f$ has exactly two real roots.

   (c) Show that the resolvent polynomial $g$ obtained as in Exercise 3 is $g = x^3 + 8x - 4$.

   (d) Why is $g$ irreducible over $\mathbb{Q}[x]$?

   (e) Use Example 46.8 to argue that the Galois group for the splitting field for $g$ over the rational field is $S_3$.

   (f) What is the Galois group of the splitting field for $f$ over $\mathbb{Q}$?

   (g) Let $r$ be one of the real roots of $f$. What is $[\mathbb{Q}(r) : \mathbb{Q}]$?

   (h) We now argue that $r$ cannot be constructible; So we suppose by way of contradiction that it is. Why would there be a proper intermediate field $\mathbb{Q} \subset F \subset \mathbb{Q}(r)$? What kind of extension does that make $\mathbb{Q}(r)$

   (i) Use what you know about the roots of $f$ to conclude that $\mathbb{Q}(r)$ cannot be normal over $\mathbb{Q}$. This allows us to conclude that $r$ is not constructible. This example shows that the converse of Corollay 43.5 is false.

8. In this exercise we will sketch a construction of a field with characteristic zero, which is not (isomorphic to) a subfield of $\mathbb{C}$. Our method of proof for Theorem 48.3 relies on assuming our field is a subfield of $\mathbb{C}$, although there are actually more general proofs that remove this hypothesis. Consider $S_{\mathbb{C}}$, the ring of all complex-valued sequences, with addition and multiplication defined component-wise (this is a ring entirely analogous to the ring of real-valued sequences considered in Exercise 6.19). We can also consider the ideal $\Sigma_{\mathbb{C}}$ of this ring, consisting of all sequences with at most finitely many non-zero entries (this is an ideal entirely analogous to that in the real case, considered in Exercise 9.29). We can now apply the Maximal Ideal Theorem (see Exercise 15.31) to find a maximal ideal $M$ of $S_{\mathbb{C}}$, for which $\Sigma_{\mathbb{C}} \subset M$. Let's denote by $K$ the field $S_{\mathbb{C}}/M$.

   (a) Consider the element $M + (1, 1, 1, \ldots) \in K$. Why is this element non-zero in $K$?

   (b) Define a function $\varphi : \mathbb{C} \to K$ defined by $\varphi(\alpha) = M + (\alpha, \alpha, \alpha, \ldots)$. Prove that this is an injective ring homomorphism, which is not onto.

   (c) Now suppose by way of contradiction that there is an injective ring homomorphism $\psi$ of $K$ into $\mathbb{C}$. Use the composition $\psi\varphi$ and suitable slight generalizations of Exercises 19.13 and 19.14 to derive a contradiction.

# Section X in a Nutshell

This section applies field theory and group theory to prove that there is no general method to solve fifth degree equations using only field operations and extractions of roots.

We start by defining a *splitting field* for a polynomial $f$ over a field $F$, which is a minimal field extension of $F$ where $f$ factors into linear polynomials. There is always a splitting field and a splitting field is a finite extension of $F$ (Theorem 44.1). Furthermore, any two splitting fields for a given $f \in F[x]$ are isomorphic (Theorem 44.2). Splitting fields have the surprising property that if $K$ is a splitting field over $F$ for $f$, and $g \in F[x]$ is irreducible over $F$ with a root in $K$, then $g$ also factors into linear polynomials in $K$ (Theorem 44.3).

An extension field $K$ of $F$ is a *normal extension* of $F$ if whenever an irreducible $f \in F[x]$ has one root in $K$, then it splits in $K$. If $F$ has characteristic zero, then $K$ is a finite normal extension of $F$ if and only if $K$ is a splitting field for some irreducible polynomial in $F[x]$ (Theorem 44.6). Furthermore, the roots of an irreducible polynomial over $F$ are all distinct in its splitting field (Theorem 44.4), as long as the field is of characteristic zero. Another important property of fields of characteristic zero is this: any finite extension is actually simple (Theorem 44.5).

The section digresses a bit with a chapter on finite fields where we use the uniqueness of the splitting field to prove that every finite field of characteristic $p$ (prime) has $p^n$ elements (Theorem 45.1), and, indeed, a unique such field exists of order $p^n$, for every $p$ and $n$. We denote this *Galois field* by $GF(p^n)$. Also, $GF(p^m)$ is a subfield of $GF(p^n)$ if and only if $m$ divides $n$.

If $E$ is an extension field of $F$, then the set of automorphisms of $E$ that fix elements in $F$ forms a group, denoted by $\mathrm{Gal}(E|F)$ and called the *Galois group of the field $E$ over $F$*. We then explore the relationship between the order of these Galois groups and the degrees of the field extensions and discover that these counts are equal in the case of normal extensions:

- (Theorem 46.1) Let $F \subseteq K$ be fields, and $f \in F[x]$ an irreducible polynomial, and $\alpha \in K \backslash F$ a root of $f$. Suppose that $\varphi \in \mathrm{Gal}(F(\alpha)|F)$. Then $\varphi$ is entirely determined by $\varphi(\alpha)$. Furthermore, $\varphi(\alpha)$ must be a root of $f$ in $K$, and so

$$|\mathrm{Gal}(F(\alpha)|F)| \leq \deg(f) = [F(\alpha) : F].$$

- (Theorem 46.2) Let $K$ be a finite extension of the field $F$, which is of characteristic zero. Then $|\mathrm{Gal}(K|F)| = [K : F]$ if and only if $K$ is a normal extension of $F$.

If $H$ is a subgroup of $\mathrm{Gal}(K|F)$, then the elements of $K$ fixed by $H$ (called the *fixed field* of $H$ and denoted $\mathrm{Fix}(H)$) is a subfield of K containing $F$ (Theorem 47.1). Once again, in the case of normal extensions, the situation is particularly nice: $\mathrm{Fix}(\mathrm{Gal}(K|F)) = F$ if and only if $K$ is a normal extension of $F$ (Theorem 47.2).

The theory culminates in the Fundamental Theorem of Galois Theory, which we presented in two parts (Theorems 47.3 and 47.4) but consolidate here. It shows how group theory and field theory mirror one another:

Suppose that $K$ is a finite normal extension of the field $F$, which is of characteristic zero. There is a one-to-one order reversing correspondence between fields $E$ with $F \subseteq E \subseteq K$ and the subgroups $H$ of $\mathrm{Gal}(K|F)$. We can describe this correspondence by two maps that are inverses of one another; namely, we have

$$E \longmapsto \mathrm{Gal}(K|E)$$

and

$$H \longmapsto \mathrm{Fix}(H).$$

Furthermore, an intermediate field $E$ is a normal extension of $F$ if and only if the Galois group $\mathrm{Gal}(K|E)$ is a normal subgroup of $\mathrm{Gal}(K|F)$. And in that case, the Galois group $\mathrm{Gal}(E|F)$ is isomorphic to

$$\mathrm{Gal}(K|F)/\mathrm{Gal}(K|E).$$

We now have the theory in place to prove that the general fifth degree polynomial equation cannot be solved by radicals. A *radical extension* of a field $F$ is a simple algebraic extension $F(\beta)$ where $\beta^n \in F$, for some integer $n \geq 2$. A sequence of radical extensions

$$F = F_0 \subset F_1 = F_0(\beta_1) \subset F_2 = F_1(\beta_2) \subset \cdots \subset F_N = F_{N-1}(\beta_N),$$

where at each stage $\beta_i^{n_i} \in F_{i-1}$ is called a *root tower* over $F$ and the last field $F_N$ is an *extension of $F$ by radicals*. If $f \in F[x]$ has a splitting field contained in an extension of $F$ by radicals, we say that $f$ is *solvable by radicals*. Notice how this is similar to our description of constructible numbers in Section VII. After some delicate adjustments to this root tower, we obtain a careful refinement fully described in Lemma 48.2. We then translate this lemma into group theory using the Fundamental Theorem, thus obtaining the theorem we need (Theorem 48.3):

Suppose that $f \in F[x] \subseteq \mathbb{C}[x]$ is solvable by radicals. Then the Galois group of the splitting field for $f$ is solvable.

Example 48.4 considers the polynomial $x^5 - 6x + 3 \in \mathbb{Q}[x]$, which has Galois group $S_5$ (as we showed in Example 46.13). But $S_5$ is not solvable (Example 30.4) and so this polynomial is *not* solvable by radicals.

# Hints and Solutions

We provide here various short answers or hints for the warm-up and odd numbered exercises. Normally we will give hints for all but the simplest proofs or on occasion a brief outline. Please note that if an exercise asks for an example we provide one, although there are usually many other possibilities.

## Chapter 1 — The Natural Numbers

**a.** $\mathbb{Z}$ is closed under subtraction (has additive inverses). $\mathbb{Q}$ is closed under division (has multiplicative inverses).

**b.** $\mathbb{Z}$ and $\mathbb{Q}$ do not have a least element. The subset $\{x : 0 < x \le 1\}$ does not have a least element.

**c.** Number the dominoes with the natural numbers in the order they stand. If you tip over domino $n$, the domino $n + 1$ will tip over also. So, if I knock over domino one, by induction all dominoes will knock over.

**d.** Any finite subset of an ordered set has a least element.

**1.** *Hint:* This follows by straightforward induction.

**3.** *Hint:* Use mathematical induction and the triangle inequality.

**5.** *Hint:* Note that here your base case for induction is when $n = 4$. You can easily check that this statement is false for $n = 1$, $2$, and $3$.

**7.** *Hint:* This follows by straightforward induction. You can also first show (by induction) this is true for $a = 1$, and then just multiply by $a$.

**9.** 4 (see Theorem 2.4).

**11.** *Hint:* Note the base case is when $n = 1$ and when $n = 2$. For the induction step, assume $n > 2$.

**13.** *Hint:* Again, base case is when $n = 1, 2$. For induction step, assume $n > 2$ and use the defining formula $a_{n+2} = a_{n+1} + a_n$.

**15.** The "induction" step is false when $n = 2$.

**17.** *Hint:* It's probably easier to prove that the Strong Principle is equivalent to the Well-ordering Principle. This works because of Theorem 1.1 and Exercise 16.

## Chapter 2 — The Integers

**a.**
$$-120 = 13(-10) + 10;$$
$$120 = (-13)(-9) + 3;$$
$$-120 = (-13)(10) + 10.$$

**b.** $0, 1, 2.$ $\{\cdots, -5, -2, 1, 4, 7, \cdots\}$.

**c.** 1 and $|p|$.

**d.** The multiples of $m$.

**e.**
$$92 = 2^2 \cdot 23; 100 = 2^2 5^2; 101; 102 = 2 \cdot 3 \cdot 17;$$
$$502 = 2 \cdot 251; 1002 = 2 \cdot 3 \cdot 167.$$

**f.** 1/30.

**g.** One method is algorithmic, relying on Euclid's Algorithm. The other is existential, proving the existence by the well-ordering principle. The later is (relatively) quick to describe. The former allows one to compute the GCD identity.

**1.** $(13)(21) + (-8)(34) = 1, (157)(772) + (-50)(2424) = 4, (-53)(2007) + (524)(203) = 1$, and $(4)(3604) + (-3)(4770) = 106$.

**3.** *Hint:* There are two things to prove here: Each linear combination of $a$ and $b$ is a multiple of $\gcd(a, b)$ and each multiple of $\gcd(a, b)$ can be written as a linear combination of $a$ and $b$. The latter part is easy to show. For the former, note that a common divisor of $a$ and $b$ also divides $ax + by$.

**5.** *Hint:* The base case is trivial, since $n = 1$. For induction, assume $p|a_1 a_2 \cdots a_n = (a_1 a_2 \cdots a_{n-1})a_n$, for $n > 1$.

**7.** *Hint:* Consider $(n + 1)! + 2, (n + 1)! + 3, \cdots, (n + 1)! + (n + 1)$.

**9.** *Hint:* Use the GCD identity: Let $g = bx + cy$ be least such linear combination of $b$ and $c$. Now suppose $h = \gcd(ab, ac) = abx' + acy' = a(bx' + cy')$. If $bx' + cy'$ is not smallest such linear combination then ....

**11.** $2^2 \cdot 3 \cdot 5 \cdot 19, 2 \cdot 5 \cdot 3^2 \cdot 7$.

**13.** *Hint:* By way of contradiction, assume $\text{lcm}(a, b) = x$ does not divide $m$. By the Division Theorem, $x = dm + r$, where $0 < r < x$. Argue $x$ is a common multiple of $a$ and $b$.

**15.** *b. Hint:* Consider the square with three vertices $E, P$, and $C$, and use part (a). Why does this mean that the algorithm never halts?

**17.** *a.* Note the first recursive call is $\gcd(772, 108)$.

**19.** *Hint:* We know from Exercise 1.14 that $\binom{p}{k}$ is an integer.

# Chapter 3 — Modular Arithmetic

**a.** $[0]_3 = \{\cdots, -6, -3, 0, 3, \cdots\}, [1]_3 = \{\cdots, -5, -2, 1, 4, \cdots\}, [2]_3 = \{\cdots, -4, -1, 2, 5, \cdots\}$. $[1]_3^{-1} = [1]_3, [2]_3^{-1} = [2]_3$.

**b.** No. Yes. Yes.

**c.** $[15]$.

**d.** Arithmetic modulo 12.

**e.** 7 o'clock. Wednesday.

**f.** $[1]$. No solution. $[8]$. $[3]$.

**1.** $[0]_8 = \{\ldots, -16, -8, 0, 8, \ldots\}, [1]_8 = \{\ldots, -16, -7, 1, 9, \ldots\}, [2]_8 = \{\ldots, -14, -6, 2, 10, \ldots\}$, $[3]_8 = \{\ldots, -13, -5, 3, 11, \ldots\}, [4]_8 = \{\ldots, -12, -4, 4, 12, \ldots\}, [5]_8 = \{\ldots, -11, -3, 5, 11, \ldots\}$, $[6]_8 = \{\ldots, -10, -2, 6, 14, \ldots\}, [7]_8 = \{\ldots, -9, -1, 7, 15, \ldots\}$. $[1]_8^{-1} = [1]_8, [3]_8^{-1} = [3]_8, [5]_8^{-1} = [5]_8, [7]_8^{-1} = [7]_8$.

**3.** *Hint:* You need to have $13x = 1 + 28y$; you'll use the GCD identity.

**5.** $[1]_m = [au + mv]_m = [au]_m + [mv]_m = [a]_m[u]_m + [0]_m$.

**7.** $[1], [5], [7], [11], [13], [17], [19], [23]$. Each is its own inverse.

**9.** *Hint:* Suppose $[a] = [b]$ and $[c] = [d]$. You need to show that $[a][c] = [b][d]$.

**11.** *Hint:* To show that two sets are equal, show that each is a subset of the other.

**13.** $[0], [1], [4], [2]. [0], [1], [4]. [0], [1], [4], [7]$.

# Chapter 4 — Polynomials with Rational Coefficients

**a.** The sum is $3 - 2x + 2x^2 - (1/2)x^3 - (2/3)x^4$. The difference is $-1 - 2x - 2x^2 + (5/2)x^3 - (2/3)x^4$. The product is $2 - 4x + 2x^2 - (7/2)x^3 + (5/3)x^4 + 2x^5 - (17/6)x^6 + x^7$.

**b.** The quotient is $(3/2)x^3 - (5/4)x^2 + (5/8)x + (27/16)$; the remainder is $5/16$.

**c.** $1 + x, 1 - x$.

**d.** No. Yes. Yes. No.

**1.** $28(x^4 - x) = (5-x)(3x^6 + 4x^5 - 3x^3 - 4x^2) + (3x - 14)(x^6 + x^5 - 2x^4 - x^3 - x^2 + 2x)$.

**3.** $x^3 - 2$, $(x-1)^3$, $(x-1)^2 x$, $(x-1)(x-2)(x-3)$.

**5.** *Hint:* Show $-1$ is a root.

**7.** *Hint:* Write $f = (x - a)^2 g$ where $g$ has degree $n - 2$. Now find $f'$ using the product rule.

**9.** *Hint:* Your argument should closely parallel that for the integers.

**11.** *Hint:* Argue that the degree of $g$ must be zero.

**13.** For example, $x^4 + 2x^3 + 3x^2 + 2x$.

# Chapter 5 — Factorization of Polynomials

**a.** In any factorization, one factor must be of degree 0.

**b.** No roots.

**c.** $(x-2)(x-1)(x+1)(x+2)$.

**d.** For example, $(2x-4)(x-1)((1/2)x + (1/2))(x+2)$. Your factorizations will differ by associate factors.

**e.** No.

**f.** $kf$, for all $0 \neq k \in \mathbb{Q}$.

**g.** $(2x-1)(x^2 + 4x + 1)$.

**1.** *Hint:* Your proof should closely parallel the theorem for the integers.

**3.** *Hint:* Your proof should closely parallel the theorem for the integers.

**5.** *Hint:* First show that this polynomial can have no linear factors, and hence no cubic factors. Then consider what a quadratic factor could look like.

**7.** $2x^3 - 17x^2 - 10x + 9 = (x - 9)(x + 1)(2x - 1)$.

**9.** *Hint:* If $p/q$ is a root of $x^3 - 2$ ($p, q$ integers) then $p$ must divide what? And $q$ must divide what? Show all possibilities do not work.

**11.** *Hint:* Use Exercise 10 and the Division Theorem.

**15.** *a.* *Hint:* Use $p = 2$ and $p = 5$, respectively. *b.* *Hint:* After the substitution, Eisenstein will apply. *c.* *Hint:* Ditto. *d.* *Hint:* Show $f(x)$ factors into polynomials in $x$ if and only if $f(y + m)$ factors into polynomials in $y$.

# Chapter 6 — Rings

**a.** It is defined to be a function on all ordered pairs.

**b.** Yes. No. No.

**c.** Matrix multiplication; subtraction.

**d.** $\mathbb{Z}_6, M_2(\mathbb{Z})$.

**e.** Yes. No. No. Yes. Yes. Yes.

**f.** $2, 1/16; 0, 0; 8, 0; 2, 1; 4 + 12x^2, 81x^8 + 108x^6 + 54x^4 + 12x^2 + 1; \begin{pmatrix} 4 & 8 \\ -4 & 12 \end{pmatrix}, \begin{pmatrix} -31 & 48 \\ -24 & 17 \end{pmatrix}$

**1.** *Hint:* $0 + a0 = a0 = a(0 + 0) = a0 + a0$, now cancel. Similar for $0a$.

**3.** *Hint:* Show that you get 0 when $(-a)b$ or $a(-b)$ is added to $ab$.

**5.** *Hint:* Recall that $a - b = a + (-b)$ and use Exercise 3 for parts (b) and (c).

**7.** *Hint:* You'll need to compute $(AB)C$ and $A(BC)$ and see that these are equal. Same for $A(B + C)$ and $AB + AC$.

**9.** *Hint:* You'll need to tick off each rule of a ring – in particular find the 0 element and the additive inverses. Don't forget that you'll need to argue the desired functions are in $C[0, 1]$.

**11.** *Hint:* You'll need to verify the rules of a ring. The grubby parts will be to verify associativity and the distributive laws.

**13.** *Hint:* Same hint as for Exercise 11.

**15.** *Hint:* $(a, b) + (c, d) = (a + c, b + d)$ and $(a, b)(c, d) = (ac, bd)$. Then verify all the rules, which easily follow since $R$ and $S$ are rings. It should now be straightforward to generalize to longer direct products.

**17.** *Hint:* The same proof you used in Exercise 1.14 will work.

**19.** *Hint:* Since operations are coordinate-wise, the rules for rings should follow easily since $\mathbb{R}$ is a ring.

**21.** *Hint:* Compute the $x^k$ coefficient in $f(gh)$ and $(fg)h$. In each case these will be double summations. Argue that these double summations actually include the same terms and are therefore the same.

**23.** *Hint:* See Exercises 21 and 22.

**25.** *Hint:* No. Show $2\mathbb{Z} \cup 3\mathbb{Z}$ is not closed under addition.

**27.** *Hint:* This is difficult. Apply the hypothesis to $2a$ to conclude that $6a = 0$. Apply the hypothesis to $a + a^2$ to get that $3(a + a^2) = 0$. And apply the hypothesis to $a + b$ and $a - b$ to get that $2(ab - ba) = 0$. More algebra gives $ab - ba = 0$. Get that $2(ab - ba) = 0$.

# Chapter 7 — Subrings and Unity

**a.** $\pi\mathbb{Z}$ in $\mathbb{R}$; $\mathbb{Q}^*$ in $\mathbb{Q}$.

**b.** Yes. No. No. No. No. No.

**c.** $\{0\}, \mathbb{Z}_5; \{0\}, \{0, 2, 4\}, \{0, 3\}, \mathbb{Z}_6; \{0\}, \mathbb{Z}_7; \{0\}, \{0, 2, 4, 6, 8, 10\}, \{0, 3, 6, 9\}, \{0, 4, 8\}, \{0, 6\}, \mathbb{Z}_{12}$.

**d.** $D_2(\mathbb{Z})$ in $M_2(\mathbb{Z})$. Does not exist. $2\mathbb{Z}$ in $\mathbb{Z}$. $\mathbb{Z}_5$.

**e.** $X$.

**f.** $(1, 1)$.

**g.** $\varphi(3 \cdot 1) = \varphi(3)\varphi(1) = 12 \neq 6 = \varphi(3)$.

**h.** The map is $\varphi(n) = (n, 0)$. This is clearly a bijection. $\varphi(n + m) = (n + m, 0) = (n, 0) + (m, 0)\varphi(n) + \varphi(m, 0)$. Similarly $\varphi$ preserves multiplication.

**i.** $g$ and $h$ are injective. $g$ and $k$ are surjective.

**j.** $Y$ has at least as many elements as $X$.

**k.** $Y$ has no more elements than $X$.

**1.** *Hint:* Showing it is closed under subtraction is straightforward. To show closed under multiplication you must put your result into the form $a + b\sqrt{2}$.

**3.** *Hint:* Showing it is closed under subtraction is straightforward. To show closed under multiplication you must show your result can be simplified into the form $a + b\alpha + c\alpha^2$, using that $\alpha^3 = 5$.

**5.** *Hint:* Show set inclusion both directions.

**7.** *Hint:* To show $S$ is a subring, use the Subring Theorem. For part (b), the relationship is this: let $f = x$.

**9.** *Hint:* The intersection is closed under subtraction because each of the subrings are. Same for multiplication.

**11.** *a.* $1/2 \notin \mathbb{Z}_{\langle 2 \rangle}$. *b.* Use the Subring Theorem. *c.* Not closed under subtraction. *d.* Use the Subring Theorem.

**13.** $Z(M_2(\mathbb{Z})) = \left\{ \begin{pmatrix} a & 0 \\ 0 & a \end{pmatrix} : a \in \mathbb{Z} \right\}$.

**15.** *a.* Use the Subring Theorem. *b.* $\{0\}$. *c.* $\{0, 2, 4, 6\}$.

**17.** *Hint:* Let 1 and $e$ both be identities. Since 1 is an identity, $1e$ is equal to what? Since $e$ is an identity, $1e$ is equal to what?

**19.** *Hint:* Use Subring Theorem. No unity.

**21.** *Hint:* Use Subring Theorem. No – it would be $(1, 1, 1, \ldots) \notin I_1$.

**23.** *Hint:* Use Subring theorem.

**25.** *a.* In $\mathbb{Z}_6$, $4^2 = 4$. *b.* Compute $f^2$. *c.* *Hint:* Use the Subring Theorem. Need to show $ae = a$ for all $a \in Re$. Remember that $R$ is commutative here. *d.* Consider $ae = b(1 - e)$ and part (c) of this exercise. *e.* *Hint:* What must $a$ and $b$ be equal to?

**27.** *Hint:* Let $a = \psi(s)$ and $b = \psi(t)$. Now $\varphi$ is an isomorphism and $\varphi(a) = s$ and $\varphi(b) = t$. Apply to $\psi(s + t)$.

## Chapter 8 — Integral Domains and Fields

**a.** $\mathbb{Z} \times \mathbb{Z}$: units: $\{(1,1), (1,-1), (-1,1), (-1,-1)\}$; zero divisors: $\{(a,0) : a \neq 0\} \cup \{(0,b) : b \neq 0\}$. $\mathbb{Z}_{20}$: units: $\{1, 3, 7, 9, 11, 13, 17, 19\}$; zero divisors:$\{2, 4, 5, 6, 8, 10, 12, 14, 15, 16, 18\}$. $\mathbb{Z}_4 \times \mathbb{Z}_2$: units: $\{(1,1), (3,1)\}$; zero divisors: $\{(0,1), (1,0), (2,0), (3,0), (2,1)\}$. $\mathbb{Z}_{11}$: units: $(\mathbb{Z}_{11})^*$; no zero divisors. $\mathbb{Z}[x]$: units $\{1, -1\}$; no zero divisors.
**b.** Yes; if $a^{-1} = b$ then $(-a)(-b) = ab = 1$ .
**c.** Yes, No, Yes, No, Yes, No, Yes.
**d.** Sets of associates are $\{1, 2, 4, 7, 8, 11, 13, 14\}$, $\{3, 6, 9, 12\}$, $\{5, 10\}$.
**e.** All ordered pairs with both coordinates non-zero. All ordered pairs with both coordinates non-zero. All ordered pairs with second coordinate 0.
**f.** $\begin{pmatrix} 1 & 0 \\ 0 & 0 \end{pmatrix}$ and $\begin{pmatrix} 0 & 0 \\ 0 & 1 \end{pmatrix}$.
**g.** $\begin{pmatrix} 0 & 1 \\ 0 & 0 \end{pmatrix}$.
**h.** $(1,0)(0,1) = (0,0)$.
**i.** argument $-\tan^{-1}(4/3) \sim -.927$; modulus 5; inverse $(3/25) + (4/25)i$
**j.** 0 if $r > 0$; $\pi$ if $r < 0$.
**k.** The distance from the origin (equal to the modulus) is 1.
**l.** $|\alpha\beta| = |\alpha||\beta| = 1$.
**m.** $1 - \sqrt{3}i$: argument $-\pi/3$, modulus 2. $2 + 2i$: argument $\pi/4$, modulus $2\sqrt{2}$. Computations should now be straightforward. Product is $(2 - 2\sqrt{3}) - (2 - 2\sqrt{3})i$. $(1 - \sqrt{3}i)^{-1} = 1/4 + \sqrt{3}/4i)$. $(2 + 2i)^{-1} = 1/4 - 1/4i$.
**n.** $-1$; $\frac{1}{\sqrt{2}} + \frac{1}{\sqrt{2}}i$; $.5403 + .8415i$; $-1 + \sqrt{3}i$.
**o.** $\mathbb{Z}_5$; does not exist; does not exist; $\mathbb{Q}$; $\mathbb{Z}$.
**p.** No. All finite domains are fields. More directly, the only $m$ for which $\mathbb{Z}_m$ has no zero divisors are prime, which makes $\mathbb{Z}_m$ a field.
**q.** $9 = 4 \cdot 2 + 1 \rightarrow 1 = (-4)(2) + 9 \rightarrow [1] = [-4][2] \rightarrow [2]^{-1} = [-4] = [5]$.
**r.** $[2]^{(5-2)} = [8] = [3]$. So $[2]^{-1} = [3]$.
**1.** If $ba = 0$, then $bax = 0x = 0$.
**3.** Try $\begin{pmatrix} 2 & 1 \\ -1 & 1 \end{pmatrix}$ and $\begin{pmatrix} 1 & 1 \\ -1 & 2 \end{pmatrix}$.
**5.** If $1 = |\alpha\beta|^2 = |\alpha|^2|\beta|^2$, then $|\alpha| = 1$ and so $\alpha = \pm 1, \pm i$.
**7. a.** *Hint:* If $rs = 0$ in $R$, then what is a zero divisor of $\varphi(r)$ in $S$? **b.** *Hint:* If $rs = 1$ in $R$, then what is the inverse of $\varphi(r)$ in $S$? **c.** *Hint:* Use part (a). **d.** *Hint:* Use part (b).
**9.** $[6]^{-1} = [6]^{17} = [6][36]^8 = [6][17]^8 = [6][289]^4 = [6][4]^4 = [6][256] = [6][9] = [54] = [16]$.
**11.** $S = \{[3], [6], [9], [12], [15], [18] = [1], [21] = [4], [24] = [7], [27] = [10], [30] = [13], [33] = [16], [36] = [2], [39] = [5], [42] = [8], [45] = [11], [48] = [14]\} = T$.
**13.** The list $ad_1, ad_2, \ldots$ is now infinite and could be (is, in this case) a proper subset of $\mathbb{Z}$, and so no element in this list need be equal to 1.

**15. a.**

| $+$ | 0 | 1 | $\alpha$ | $1 + \alpha$ |
|---|---|---|---|---|
| 0 | 0 | 1 | $\alpha$ | $1 + \alpha$ |
| 1 | 1 | 0 | $1 + \alpha$ | $\alpha$ |
| $\alpha$ | $\alpha$ | $1 + \alpha$ | 0 | 1 |
| $1 + \alpha$ | $1 + \alpha$ | $\alpha$ | 1 | 0 |

| $*$ | 0 | 1 | $\alpha$ | $1 + \alpha$ |
|---|---|---|---|---|
| 0 | 0 | 0 | 0 | 0 |
| 1 | 0 | 1 | $\alpha$ | $1 + \alpha$ |
| $\alpha$ | 0 | $\alpha$ | $1 + \alpha$ | 1 |
| $1 + \alpha$ | 0 | $1 + \alpha$ | 1 | $\alpha$ |

**b.** Note each element has an inverse. **c.** Characteristic of $F$ is 2, of $\mathbb{Z}_4$ is 4.
**17.** *Hint:* If $\gcd(a, m) > 1$ then consider the GCD identity to find $b$ where $ab = 0$. In $\mathbb{Z}$, non-units are not zero divisors.

**19.** *Hint:* If $A$ not a unit, the $ab - cd = 0$. Consider three cases: $a \neq 0$, $c \neq 0$, $a = c = 0$.

**21.** *Hint:* Simply add the series for $\cos x$ and $i \sin x$.

## Chapter 9 — Ideals

**a.** No, Yes, No, No, No, Yes, Yes, No.

**b.** No, Yes, Yes, No, Yes.

**c.** $I = R$.

**d.** $\{a + b\sqrt{7} \in \mathbb{Z}[\sqrt{7}] : 7 \text{ divides } a\}$.

**e.** $\langle 2 \rangle$ in $\mathbb{Z}_8$. None. $\langle 1 + i \rangle$. $\langle x \rangle$.

**f.** $\langle 35, 15 \rangle = \langle 5 \rangle$. $\langle 12, 20, 15 \rangle = \mathbb{Z}$.

**g.** $\langle 4, x \rangle = \{a_0 + a_1 x + \cdots a_n x^n : 4 \text{ divides } a_0\}$.
$\langle 4, x^2 \rangle = \{a_0 + a_1 x + \cdots a_n x^n : 4 \text{ divides } a_0, a_1\}$.

**h.** $\langle x - 1 \rangle$.

**i.** Both are $\mathbb{Z} \times \mathbb{Z}$.

**j.** The only ideals of a field are $\langle 0 \rangle, \langle 1 \rangle$.

**k.** $\mathbb{Z}[x]$. $x^2, -x^2$. $g = x^2 + x, f = x$.

**1.** No, Yes, No, No, Yes, Yes, Yes, Yes, No, No.

**3.** *b.* $x^2 + 1$.

**5.** $(x - 3)(x^2 - 3)$.

**7.** $\langle \alpha \rangle = \{a + b\alpha + c\alpha^2 : 5 \text{ divides } a\}$. $\langle 2 \rangle = \{a + b\alpha + c\alpha^2 : 2 \text{ divides } a, b, c\}$.

**9.** *Hint:* An element of the ideal is $(3 + \sqrt{5})(a + b\sqrt{5}) = (3a + 5b) + (a + 3b)\sqrt{5}$. Now show this set is equal to the one given.

**11.** *Hint:* To say that it is the smallest such ideal merely means that $\langle a, b \rangle$ should be a subset of *every* ideal containing $a$ and $b$.

**13.** *Hint:* Check that if $x$ and $y$ are integers and $x + y$ is even, then $x - y$ is even too. Also, note that $2 = (1 + \sqrt{3})(-1 + \sqrt{3})$.

**15.** *Hint:* Show set inclusion both ways.

**17.** *Hint:* Absorption is hereditary.

**19.** *Hint:* First three parts are straightforward. *d.* It helps to think about part (c) and Exercise 11.

**21.** *b.* $I = \langle a \rangle$.

**23.** *Hint:* Since the ring is commutative, this should be straightforward.

**25.** *Hint:* To show it's a subring is straightforward. For absorption, you need to show if $s \in A(I)$, then so is $sa$, for $a \in R$.

**27.** *Hint:* From Exercise 24, we know this is an ideal. Use $1 - e$ as a generator. For one inclusion, if $es = 0$, then $s = (e + 1 - e)s = es + (1 - e)s = \cdots$. The other inclusion is straightforward.

**29.** *a.* This should be straightforward. *b.* $\Sigma = \cup I_n$. *c. Hint:* No matter which $\vec{s}_i \in \Sigma$ you pick, find a sequence in $\Sigma$ but not in $\langle \vec{s}_1, \vec{s}_2, \cdots, \vec{s}_n \rangle$.

**31.** *a.* $a_{0,0} = 0$. *b. Hint:* If $x$ and $y$ are to be multiples of the same polynomial in $\mathbb{Q}[x, y]$, then what must that polynomial be?

## Chapter 10 — Polynomials over a Field

**a.** $x^2 + 3x + 3; 0$. $3x^2 + x + 4; 2$. $-ix^2 + (-2 + 2i)x + (2 + 2i); 7 + 2i$. $(1/\pi)x^3 - (2/\pi + 1/\pi^2)x^2 + (2/\pi^2 + 1/\pi^3)x - (2/\pi^3 + 1/\pi^4); 1/3 + 2/\pi^3 + 1/\pi^4$.

**b.** $(x + 1)(x + 2)(x + 3)(x + 4)$.

**c.** $1 + i$ is a unit.

**d.** $2x^2 + 3x + 3, 4x^2 + x + 1, x^2 + 4x + 4 = (x + 2)^2, 3x^2 + 2x + 2$.

**e.** No roots.

**f.** In $\mathbb{Q}[x]$: $x^3 - 2$. In $\mathbb{R}[x]$: $(x - \sqrt[3]{2})(x^2 + \sqrt[3]{2}x + \sqrt[3]{4})$. In $\mathbb{C}[x]$: $(x - \sqrt[3]{2})(x - (-\sqrt[3]{2}/2 + \sqrt{8\sqrt[3]{2} - \sqrt[3]{4}}/2i))(x - (-\sqrt[3]{2}/2 - \sqrt{8\sqrt[3]{2} - \sqrt[3]{4}}/2i))$.

**g.** The modulus of this number is $\sqrt{12}$ and its argument is $5\pi/6$. Thus its square roots are $\pm\sqrt[4]{12}\,(\cos(5\pi/12) + i\sin(5\pi/12))$.

**h.** Never. $x$ can't have a multiplicative inverse.

**1.** The next step is $x^2 + \frac{b}{a}x + \left(\frac{b}{2a}\right)^2 = \left(\frac{b}{2a}\right)^2 - \frac{c}{a} \Rightarrow \cdots$

**3.** $x^5 + 4x + 1, 1$.

**5.** $1; 1; 1$.

**7.** $1 = (1/54)(x^2 - x - 9)(x^2 + x - 2) + (1/54)(-x + 4)(x^3 + 4x^2 + 4x + 9)$.

**9.** $2x^5 - 9x^4 + 16x^3 - 14x^2 + 14x - 5$.

**11.** $x, x + 1, x^2 + x + 1, x^3 + x^2 + 1, x^3 + x + 1, x^4 + x^3 + x^2 + x + 1, x^4 + x^3 + 1, x^4 + x + 1$. Note that $(x^2 + x + 1)^2 = x^4 + x^2 + 1$.

**13.** *Hint:* If you use the negative square root of $D$ instead, factor out $-1$ from both cube roots.

**15.** *Hint:* Use the identities in Exercise 12(b).

**17.** *a.* $g(y') > 0$ so $g(y)$ is increasing. *b.* $-\sqrt[3]{q}, -\sqrt[3]{q}\zeta, -\sqrt[3]{q}\zeta^2$, where $\zeta = e^{\frac{2\pi}{3}}$ (see Exercise 15). *c.* *Hint:* Recall $D = q^2 + \frac{4p^3}{27}$, which is $> 0$. Use this when evaluating $g$ at the two roots.

**19.** Use the fact that if $(c + di)^3 = a + bi$, then $(-c + di)^3 = -a + bi$.

**21.** The cubic you get is $2b^3 - b^2 + 6b - 7 = 0$; this has an integer root you can find by inspection.

**23.** *a.* *Hint:* Compute powers of complex numbers using DeMoivre's Theorem. *b.* Consider the arguments. *c.* Consider the Fundamental Theorem of Algebra.

**25.** *Hint:* Use DeMoivre's Theorem.

**27.** Try $x^2 + \alpha$.

**29.** *a.* 0, 1, 4, 2. Each non-zero element has two square roots. *b.* *Hint:* $2^{-1} = 4$, and so we need $(4b)^2$. *c.* $b = 1, c = 5$. *d.* $b = 1, c = 4$. *e.* $b = 6, c = 2$.

# Chapter 11 — Ring Homomorphisms

**a.** Example 11.1, 11.4, 11.5, 11.7, 11.7, none, 11.3 ($\pi(1,0) = 1$), 11.12 ($\varphi(1)$), 11.1 ($\varphi(2)$), 11.3 ($\pi(1,0)$).

**b.** No.

**c.** Does not preserve addition.

**d.** No, Yes.

**1.** Not unless $m = 1$.

**3.** No.

**5.** Yes.

**7.** Yes.

**9.** Yes.

**11.** No.

**13.** Yes.

**15.** No.

**17.** No.

**19.** Yes.

**21.** No.

**23.** No.

**25.** *a.* *Hint:* Use the Subring Theorem. *b.* Consider $\varphi : \mathbb{C} \to \mathbb{C}$ given by $\varphi(a + bi) = a - bi$, as in Example 11.4.

**27.** *Hint:* You need to show each element of $\varphi(Z(R))$ commutes with every element of $S$.

**29.** *Hint:* Use the Subring Theorem.

**31.** *a. Hint:* You will need to use some facts about convergent sequences when using the Subring Theorem. *b. Hint:* You will need some theorems from calculus.

## Chapter 12 — The Kernel

**a.** $\{0, 2, 4\}$.
**b.** $\{0, 2, 4\}, \{1, 3, 5\}$.
**c.** 1,2. All preimages for a homomorphism have the same number of elements.
**d.** Identity map. $\varphi : \mathbb{Z} \to \mathbb{Z}_4, \varphi(n) = [n]$. *Any* one-to-one ring homomorphism. Impossible.
**e.** $\mathbb{Z} \subseteq \mathbb{Q}$.
**f.** None.
**1.** $\{0\}$. It is injective.
**3.** $\{b \in P(X) : x \notin b\} = \langle X \backslash \{x\} \rangle$.
**5.** *Hint:* Proving one set inclusion is straightforward. By Division Theorem, $a + b\alpha + c\alpha^2 = (c\alpha + (b - 3c))(3 + \alpha) + a - 3b + 9c$. But $a - 3b + 9c = a + b + c - 4b + 8c$. But $a + b\alpha + c\alpha^2 \in \ker(\varphi)$ if and only if $4|(a + b + c)$.
**7.** $\{f \in C[0, 1] : f(0) = f(1) = 0\}$.
**9.** *Hint:* For each $r \in \mathbb{Q}$ find an $s \in \mathbb{Q}$ so that $g(2) = r$ for every $g \in \langle x - 2 \rangle + s$.
**11.** $\{(0, s) : s \in S\}$. When they have the same first component. $R$.
**13.** $\{a_0 + a_1 x + \cdots + a_n x^n : a_0 = a_1 = 0\} = \langle x^2 \rangle$.
**15.** *Hint:* Use $\varphi(rs) = \varphi(r)\varphi(s)$.
**17.** *a. Hint:* You'll use characteristic 2 in show $\varphi$ preserves addition. *b.* The identity map. *c.* Refer back to the multiplication tables you made for Exercise 8.15.

## Chapter 13 — Rings of Cosets

**a.** $4, \langle 4 \rangle + 0$, Yes, Yes.
**c.** $\langle x - 2 \rangle + (-4)$.
**d.** $\langle x - 2 \rangle + (1/4)$.
**e.** Cosets of $I$.
**f.** $I$.
**g.** Yes.
**h.** No. Only if $I = \{0\}$.
**i.** Yes.
**1.** $\langle x + 1 \rangle + 1/10$.
**3.** *Hint:* Consult Chapter 9 for a complete description of the ideals of $\mathbb{Z}$.
**5.** *a. Hint:* Consider part (c) of the Coset Theorem. *b.* You must find a function $g$ to be the inverse of $I + f$. So, $g(1/4) = ?$ .
**7.** *Hint:* $(J + (a, b))(J + (c, d)) = J$ implies $(ac, bd) \in J$ implies what about $ac$ and $bd$?
**9.** Note that $\mathbb{Z} + (3/2) = \mathbb{Z} + (1/2)$. But $(\mathbb{Z} + (1/2))^2 = \mathbb{Z} + (1/4)$, while $(\mathbb{Z} + (1/2))(\mathbb{Z} + (3/2)) = \mathbb{Z} + (3/4) \neq \mathbb{Z} + (1/4)$.
**11.** *a. Hint:* The key thing to show is absorption. *b. Hint:* If $N(R) + r^n = N(R)$, then $r^n \in N(R)$. What do you conclude about $r$? *c.* $\mathbb{Z}_2$.

## Chapter 14 — The Isomorphism Theorem for Rings

**a.** $\{0\}, R$.
**b.** Yes; what ideals does every ring have?
**c.** $\langle x^2 - 2 \rangle$. $\mathbb{Z}[x]/\langle x^2 - 2 \rangle$ and $\mathbb{Z}[\sqrt{2}]$.
**d.** 2: $\{0\}$ and $F$. So only possibilities for $\varphi(F)$ are $F$ and $\{0\}$.
**e.** $\{0\}, \langle n \rangle$, for $n \in \mathbb{N}$. $\mathbb{Z}, \mathbb{Z}_n$, and $\{0\}$.

**f.** $f = a_0 + x(a_1 + \cdots a_n x^{n-1})$, so $\langle 2, x \rangle + f = \langle 2, x \rangle + a_0$. But $a_0 = 2k + a$, where $a = 0$ or $1$. So $\langle 2, x \rangle + a_0 = \langle 2, x \rangle + a$. $\mathbb{Z}[x]/\langle 2, x \rangle$ is isomorphic to $\mathbb{Z}_2$.

**g.** Only if $\varphi$ is onto. $R/\ker(\varphi)$ is isomorphic to $\varphi(R)$, a subring of $S$.

**h.** There is a bijection between ideals and kernels of homomorphisms.

**1.** $\varphi(a + bi) = [a + b]_2$.

**3.** $\varphi(a, b) = [4a + 9b]_{12}$.

**5.** $\varphi(a_0 + a_1 x + \cdots a_n x^n) = (a_0, a_1)$.

**7.** *Hint:* Build an isomorphism between these rings (with domain $P(X)$) by making use of the homomorphisms considered in Exercise 11.10.

**9.** This is straightforward.

**11.** $\mathbb{R}$ and $C/I$, where $I$ is the ideal consisting of all sequences that converge to $0$.

**13.** $\varphi(1) = 1$; $\varphi(-1) = -1$; $-1 = \varphi(i^2) = (\varphi(i))^2$.

**15.** *Hint:* The same reasoning from Exercise 13 works here.

**17. a.** *Hint:* In order to extend $\varphi$, $\bar{\varphi}(a_0 + a_1 x + a_2 x^2 + \cdots + a_n x^n)$ should be what polynomial in $S[x]$? **b.** *Hint:* Consider $\bar{\varphi}(f(r))$. **c.** *Hint:* Suppose $f, g \in R[x]$, then $\bar{\varphi}(fg)$ factors into what in $S[x]$. The converse is similar.

**19.** *Hint:* Define a homomorphism from $R$ to $S/\varphi(I)$ with kernel $I$.

**21. b.** Use the Isomorphism Theorem. Define a ring homomorphism from $R/I$ onto $R/A$, and show that it has the appropriate kernel.

**23.** Define a homomorphism from $I$ to $(I + J)/J$ with the appropriate kernel.

# Chapter 15 — Maximal and Prime Ideals

**a.** No. Yes. Yes. Yes. No. No. No.

**b.** $\langle x^2 + 1 \rangle$ in $\mathbb{R}[x]$. $\langle x \rangle$ in $\mathbb{Z}[x]$. $\langle 3, x + 1 \rangle$ in $\mathbb{Z}[x]$. $\langle 4, x \rangle$ in $\mathbb{Z}[x]$.

**c.** A maximal ideal must be proper.

**d.** $\{0\}$ in $\mathbb{Z}$. Impossible: Every maximal ideal is prime. $\langle (0, 1) \rangle$ in $\mathbb{Z} \times \mathbb{Z}$.

**e.** $\langle x^2 - 2 \rangle \subset \langle x - \sqrt{2} \rangle$, but $x^2 - 2$ is irreducible in $\mathbb{Q}[x]$.

**f.** $\mathbb{Z}[x]/\langle x - 2 \rangle$ is isomorphic to the domain $\mathbb{Z}$.

**g.** Every field is a domain.

**h.** $\{(0, n) : n \neq 0\} \cup \{(m, 0) : m \neq 0\}$. $\{(0, n) : n \neq 0\} \cup \{(2, n) : n \neq 0\} \cup \{(m, 0) : m \neq 0\}$.

**i.** $R$ a domain; $R$ a field.

**j.** $R/R = \{0\}$ has no identity and so could not be a domain.

**3.** $2 = 4 \cdot 4$, $3 = 3 \cdot 3$, $4 = 2 \cdot 2$. Other elements are zero or units.

**5.** $\{(p, \pm 1)\} \cup \{(\pm 1, p)\}$ where $p$ is prime in $\mathbb{Z}$.

**7.** *Hint:* First show $a \in \langle c \rangle$ and $b \in \langle c \rangle$. For the other inclusion, use the GCD identity.

**9.** $\{0\}$ and $\langle p \rangle$, $p$ a prime integer. Prime $\equiv$ maximal.

**11.** $\langle 2 \rangle, \langle 5 \rangle$. Prime $\equiv$ maximal.

**13.** Prime $\equiv$ maximal: $\mathbb{Z}_2 \times \{0\}$, $\{0\} \times \mathbb{Z}_3$. Prime $\equiv$ maximal: $\mathbb{Z}_2 \times \{0, 2\}$, $\{0\} \times \mathbb{Z}_4$.

**15.** *Hint:* See Example 9.12.

**17.** *Hint:* See Example 9.12 to show not principal. Show $\langle 9, x \rangle \subset \langle 3, x \rangle$.

**19.** *Hint:* Consider $\langle x, y \rangle$.

**21.** The homomorphism is not onto.

**23. a:** *Hint:* Must show both a homomorphism and onto. *b.* $\ker(\varphi) = \langle 3 \rangle$.

**25.** *Hint:* Use the fact that if $f \notin \langle x^2 + 1 \rangle$, then $\langle x^2 + 1 \rangle + f$ has a multiplicative inverse in $\mathbb{R}[x]/\langle x^2 + 1 \rangle$.

**27.** *Hint:* What is the kernel of this homomorphism?

**29.** *Hint:* Define a homomorphism $R[x] \to (R/I)[x]$ by letting $\varphi(f)$ be the polynomial with coefficient $a$ replaced by $I + a$. $R/I$ is a domain, so $(R/I)[x]$ is also. The kernel of this map is $I[x]$. If $I$ is maximal ideal, consider $R = \mathbb{Q}$ and its maximal ideal $I = \{0\}$.

**31.** *a. Hint:* Use Theorem 15.2. *b. Hint:* Since $M$ is proper it can contain no units. Suppose $r \notin M$. Then the principal ideal $\langle r \rangle$ is proper if $r$ is not a unit and so is contained in a maximal ideal. But $M$ is the unique maximal ideal, so $r \in M$, a contradiction. Thus $r$ is a unit and $M$ must be all the non-units.

**33.** *Hint:* Take your cue from the proof of Theorem 2.7.

**35.** Given $r \in R$, consider the product $r(1 - r)$.

# Chapter 16 — The Chinese Remainder Theorem

**a.** $\mathbb{Z}_8 \times \mathbb{Z}_3, \mathbb{Z}_4 \times \mathbb{Z}_3 \times \mathbb{Z}_5, \mathbb{Z}_{11}, \mathbb{Z}_9$.

**b.** $[2]^4$ is in every prime ideal, so if $[2]^4$ is in a prime ideal then either $[2]$ is in the ideal or $[2]^3$ is. If $[2]^3$ is in a prime ideal, then either $[2]$ or $[2]^2$ is. If $[2]^2$ is then either $[2]$ or $[2]$ is. Regardless, $[2]$ is in every prime ideal.

**c.** $n$ divides $m$.

**d.** $x \equiv 5 \pmod 6$.

**e.** $x \equiv 2 \pmod 3, x \equiv 3 \pmod 5, x \equiv 2 \pmod 7$

**1.** $\mu([a]) = ([a]_3, [a]_4)$.

**3.** Prime $\equiv$ maximal; $\langle 4 \rangle, \langle 3 \rangle$.

**5.** $x \equiv 201 \pmod{252}$.

**9.** $\mathbb{Z}_3 \times \mathbb{Z}_3$ has characteristic 3, but $\mathbb{Z}_9$ doesn't.

**11.** *Hint:* Use the Exercise 10.

**13.** *a. Hint:* Find a zero divisor. *b.* $(0,1) \notin R$. *d. Hint:* Use the ideals $P_1$ and $P_2$. *e. Hint:* Note that $(3,0)(0,3) = (0,0) \in P$.

**15.** *Hint:* Domain corresponds to prime ideals. Fields correspond to maximal ideals.

# Chapter 17 — Symmetries of Geometric Figures

**a.** Every symmetry of an equilateral triangle has one that moves the triangle back to its original orientation.

**b.** A rotation followed by a rotation or a flip followed by a flip results in a rotation, while a flip followed by a rotation or vice versa results in a flip.

**c.** $\begin{pmatrix} \sqrt{2}/2 & -\sqrt{2}/2 \\ \sqrt{2}/2 & \sqrt{2}/2 \end{pmatrix} \begin{pmatrix} \sqrt{3}/2 & -1/2 \\ 1/2 & \sqrt{3}/2 \end{pmatrix}$

**d.** The result in all three cases should be $\varphi$.

**e.** Two.

**f.** One.

**g.** $\rho_2^2, \varphi_1, \rho_2^2$.

**h.** For example, any of the four symmetries in the set $F$ followed by any of the four symmetries in the set $R$ results in a symmetry in the set $R$. Similarly for the eight other block combinations.

**1.** You should find 8 symmetries of the square: the identity, 3 rotations, a flip about the vertical bisecting line, a flip about the horizontal bisecting line, and the 2 flips about the diagonals.

**3.** You should have 4 elements in your group: the identity, 1 rotation, and 2 flips.

**5.** *a.* $(x_2 - x_1)^2 + (y_2 - y_1)^2$. *b. Hint:* Use the matrix found in Section 17.3.

**7.** There are infinitely many rotations, one for each angle from 0 to $2\pi$, and an infinite number of flips, one about each line through the center.

**9.** *a. Hint:* Solve $\begin{pmatrix} a & b \\ c & d \end{pmatrix} \begin{pmatrix} x \\ y \end{pmatrix} = \begin{pmatrix} 0 \\ 0 \end{pmatrix}$.

**11.** There are four, all rotations of the base.

**13.** Flatlanders cannot understand flips.

**15.** There are 3 rotations about the triangular ends and a rotation of the base. There can be no flips of the triangular ends as that would "turn the tent inside out."

**17.** *a.* The relative orientation of all sides are fixed by symmetries in space. *b.* Should be 24 symmetries: Choice of 6 colors for the front and then any of the 4 adjacent colors for the right face.

## Chapter 18 — Permutations

**a.** $\alpha(5) = 1$, $\alpha(4) = 4$, $\alpha^{-1} = \begin{pmatrix} 1 & 2 & 3 & 4 & 5 & 6 & 7 \\ 5 & 3 & 1 & 4 & 7 & 6 & 2 \end{pmatrix}$.

**b.** $(1\ 3\ 2\ 7\ 5)$. $(1\ 5\ 7\ 2\ 3)$. $\{4, 6\}$. $\{1, 2, 3, 5, 7\}$. $\{1, 2, 3, 5, 7\}$. $\{6\}$.

**c.** $\beta = (14)(2670$. $\beta^2 = (276)$, $\beta\alpha = (136754)$, $\alpha\beta = (143265)$, $\beta\alpha^{-1} = (15234)$, $\alpha\beta\alpha = (127)(3654)$.

**d.** $(\alpha\beta)^{-1} = \begin{pmatrix} 1 & 2 & 3 & 4 & 5 & 6 & 7 \\ 5 & 3 & 4 & 1 & 6 & 2 & 7 \end{pmatrix}$.

**e.** 24, 120, 720.

**f.** Identity. One with only one cycle.

**g.** If both permutations are the same.

**h.** Both are unique factorization theorems. Cycles can be further factored into products of shorter cycles, although will not be disjoint.

**1.** $(123)^2$ is a cycle, but $(1234)^2$ is not.

**3.** *Hint:* Count number of 2-, 3-, 4-, and 5-cycles. Then count number of products of two disjoint 2-cycles and product of disjoint 2-cycle times 3-cycle.

**5.** *Hint:* The two 3-cycles may be disjoint or their supports overlap at 1, 2, or 3 points.

**7.** 2, 8, 20.

**9.** If $x \in \text{Fix}(\alpha) = \text{Fix}(\beta)$, then $\alpha\beta(x) = \alpha(x) = x$. For an example, consider $\beta = \alpha^{-1}$.

## Chapter 19 — Abstract Groups

**a.** $\mathbb{Z}, \mathbb{Z}_n, U(M_2(\mathbb{R}))$, symmetries of the square.

**b.** $[2], [2], \begin{pmatrix} -5 & -4 \\ 4 & 3 \end{pmatrix}, \varphi\rho, (165)(243), (-2, 4), (1/2, -1/4), 3/5 - 4/5i$.

**c.** $[4]$ in $\mathbb{Z}_5^*$. Let $a = \varphi, b = \rho$ in $D_3$. $[1]$ in $\mathbb{Z}_4$.

**d.** $R$ with operation addition and $U(R)$ with operation multiplication.

**e.** $A = 2I$. No.

**f.** No, No, Yes, No, Yes, No, No, No, No, Yes, No.

**g.** No identity. Also, operation is not associative.

**1.** $n^3$.

**3.** $I^{-1} = I$, $\begin{pmatrix} -1 & 0 \\ 0 & -1 \end{pmatrix}^{-1} = \begin{pmatrix} -1 & 0 \\ 0 & -1 \end{pmatrix}$, $\begin{pmatrix} 0 & 1 \\ -1 & 0 \end{pmatrix}^{-1} = \begin{pmatrix} 0 & -1 \\ 1 & 0 \end{pmatrix}$,

$\begin{pmatrix} 0 & i \\ i & 0 \end{pmatrix}^{-1} = \begin{pmatrix} 0 & -i \\ -i & 0 \end{pmatrix}$, $\begin{pmatrix} i & 0 \\ 0 & -i \end{pmatrix}^{-1} = \begin{pmatrix} -i & 0 \\ 0 & i \end{pmatrix}$.

**5.** *a.* *Hint:* You must first check that this is a binary operation on this set. *b.* *Hint:* You must show that the composition of two such function is also order-preserving.

**7.** This group will have $n$ elements.

**9.** *Hint:* Use associativity and cancellation.

**11.** *Hint:* $\mathbb{Z}$ is generated by 1. If $f \in \text{Aut}(\mathbb{Z})$, then $f(1) = ?$

**13.** *b.* *Hint:* $\varphi(a - b) = \varphi(x^2) = \varphi(x)^2$. *c.* If $\varphi \in \text{Aut}(\mathbb{R})$ then consider just restricting the domain to $\mathbb{Q}$. *d.* Suppose $\varphi(r) > r$ and $q$ is a rational between these two numbers, what is the relation of $q$ and $\varphi(q)$?

**15.** *Hint:* An automorphism of $F$ must fix 0 and 1, so $\alpha$ has only two possible images. One gives the identity. Verify the other is the Frobenius automorphism.

# Chapter 20 — Subgroups

**a.** Yes, No, No, Yes, No, No.

**b.** Yes.

**c.** No.

**d.** One identity, one inverse per element.

**e.** $\rho_1^3, \alpha_4$, 8, 14, (124567).

**f.** Yes; ignore the multiplication for the rings.

**g.** Yes.

**3.** *Hint:* Assume both $b$ and $c$ are inverses of $a$ and show $b = c$.

**5.** *Hint:* To show $H \cup K$ need not be a subgroup, try subgroups of $\mathbb{Z}$.

**7.** *Hint:* Multiply on correct side by the proper inverses.

**9.** *Hint:* Recall that $(a^{-1})^{-1} = a$.

**11.** $U(\mathbb{Z}_8), \{1\}, \{1, 3\}, \{1, 5\}, \{1, 7\}.$    $\mathbb{Z}_7^*, \{1\}, \{1, 2, 4\}, \{1, 6\}.$
$U(\mathbb{Z}_{15}), \{1\}, \{1, 4\}, \{1, 11\}, \{1, 14\}, \{1, 2, 4, 8\}, \{1, 7, 4, 13\}.$

**13.** Only the trivial subgroup is finite. There are infinitely many – consider elements of the form $e^{2\pi/n}$.

**15.** *a.* $C(\rho) = \{\iota, \rho, \rho^2\}$. *b.* $C(4) = \mathbb{Z}_7$. *c, d.* Straightforward using the Subgroup Theorem.

**17.** *Hint:* The intersection of two subgroups is a subgroup, so need only show this is a subset of $Z(H)$.

**19.** *Hint:* Need to find the inverse of a typical element in order to use the Subgroup Theorem.

**21.** *Hint:* This should be similar to the corresponding theorem for rings.

**23.** *Hint:* This should be similar to the corresponding theorem for rings.

**25.** *Hint:* Each element has a unique inverse.

**27.** *Hint:* The group operation is composition. The identity and inverses should be clear. All other properties should follow the fact that these are isomorphisms.

**29.** *a. Hint:* Refer back to the definition of the group operation. *b. Hint:* The notation $F$ and $R$ are meant to be suggestive of flip and rotation.

# Chapter 21 — Cyclic Groups

**a.** 3, 4, 2, $\infty, \infty$, 4, $\infty$, 8.

**b.** Yes, No, No, Yes, No, No, No, Yes, No, No, Yes.

**c.** Yes, $\mathbb{Z} \times \mathbb{Z}_2$.

**d.** Yes.

**e.** Consider $\langle g \rangle$.

**f.** 2 (1, -1), 4 (1, 3, 7, 9), 6 (1, 2, 3, 4, 5, 6), 2 (2, 5).

**h.** 14, 7, 2, or 1.

**i.** It divides 8.

**j.** Only $\mathbb{Z}_8$ has an element of order 8. $\mathbb{Z}_4 \times \mathbb{Z}_2$ is abelian. $D_4$ is not abelian.

**k.** If $b \neq 0$, then $n(a + bi)$ $(n \neq 0)$ can never be an element of $\mathbb{Z}$. If $b = 0$, then $na$ has no imaginary part.

**l.** Yes: Consider $\varphi : 2\mathbb{Z} \to 4\mathbb{Z}$ given by $\varphi(2a) = 4a$. If $G$ is finite it is impossible since all proper subgroups have fewer elements than $G$ and so the map cannot be one-to-one.

**m.** No; the group of units will never contain 0 and so must have fewer elements.

**1.** $\left\{ \begin{pmatrix} 1 & n \\ 0 & 1 \end{pmatrix} : n \in \mathbb{Z} \right\}, \left\{ \begin{pmatrix} 0 & 1 \\ 1 & 0 \end{pmatrix}, \begin{pmatrix} 1 & 0 \\ 0 & 1 \end{pmatrix} \right\}.$

**3.** Any of several small finite groups we have considered will do the job. Provide more than one example if you can.

**5.** *Hint:* Let $n_1, n_2, \ldots, n_k$ be the orders of the disjoint cycle factors $\varphi_1, \varphi_2, \ldots, \varphi_k$ of $\varphi$, and $n$ the lcm. Then $\varphi^n = \varphi_1^n \varphi_2^n \cdots \varphi_k^n = \iota$.

**7.** The intersection includes only the identity.

**9.** *Hint:* For $\mathbb{Z}_n$, 1 is a generator. If $n > 2$, what other element works?

**11.** *Hint:* Take care; $\mathbb{Z}_4$ is an additive group.

**13.** *a. Hint:* If $g$ is a generator, then find $g^k$ with order $n$. *b. Hint:* You can find example of both using direct products of some small groups.

**15.** Consider the rotation in the subgroup of smallest angle.

**17.** *Hint:* Think of the order of the cycle.

**19.** *Hint:* Think of ways their supports can overlap.

**21.** *Hint:* What are the inverses of $a^n$ and $ba^n$? Now use the Subgroup Theorem.

## Chapter 22 — Group Homomorphisms

**a.** No.

**b.** Yes.

**c.** No.

**e.** No.

**g.** No.

**h.** $\{2x + a : a \in \mathbb{R}\}$.

**i.** $\varphi([0]_2) = [0]_4$, $\varphi([1]_2) = [2]_4$.

**j.** $\varphi(g \circ h \circ k) = \varphi(g) * \varphi(h \circ k) = \varphi(g) * \varphi(h) * \varphi(k)$.

**k.** Yes.

**1.** This is a straightforward verification.

**3.** $\varphi(a, b) = a$, $\varphi(a, b) = b$, $\varphi(a, b) = a + b$. There are 4 homomorphisms, altogether.

**5.** *Hint:* Try a 'small' non-abelian group.

**7.** Let $h, k \in \varphi(G)$ with $\varphi(a) = h$ and $\varphi(b) = k$. Then $hk = \varphi(a)\varphi(b) = \varphi(ab) = \varphi(ba) = \varphi(b)\varphi(a) = kh$. Let $\varphi : \mathbb{Z} \to \mathbb{Z} \times D_3$ be defined by $\varphi(n) = (n, 0)$.

**9.** *Hint:* You will need an appropriate theorem from Chapter 11.

**11.** *Hint:* Consider a function that has a derivative, but no second derivative.

**13.** *Hint:* A homomorphism from a cyclic group with generator $g$ is entirely determined by the image of $g$. So a surjection must send the generator to another generator.

**15.** *a.* To show $\varphi$ preserves the group operation is straightforward. To show onto, if $k \in G$, what element gets mapped to $k$? *c. Hint:* Show $h^n = 1$ if and only if $(ghg^{-1})^n = 1$.

## Chapter 23 — Structure and Representation

**a.** 99.

**b.** Let $a, b \in G_1 G_2$, then $a = g_1 h_1$, $b = g_2 h_2$ for some $g_i \in G_1$ and $h_i \in G_2$. Then $ab^{-1} = g_1 h_1 h_2^{-1} g_2^{-1} = g_1 g_2^{-1} h_1 h_2^{-1} \in G_1 G_2$. So $G_1 G_2$ a subgroup by the Subgroup Theorem.

**c.** $(12)(13) = (132)$, but $(132)^{-1} = (123) = (13)(12)$ which is not in $G_1 G_2$.

**d.** It is a isomorphic to a subgroup of some $S_n$.

**e.** Every finite group is isomorphic to a subgroup of $S_n$, for some $n$, which is non-abelian. (For $n = 1, 2$, $S_n$ is a subgroup of $S_3$.)

**1.** *Hint:* Consider the cyclic subgroups $\langle 11 \rangle$ and $\langle 31 \rangle$ of $\mathbb{Z}_{341}$. (Be careful on which subgroup has which order.)

**3.** *Hint:* You need to check the three assertions stated in the proof.

**5.** *Hint:* Try $G_1 = \langle 1 \rangle$ and $G_2 = \langle \sqrt{2} \rangle$.

**7.** *Hint:* Not all criteria are satisfied; $D_3$ is not isomorphic to a direct product of these groups.

**9.** *Hint:* Need to first find elements in $\mathbb{Z}_{30}$ of order 2, 3, and 5.

**11.** 1 corresponds to $\begin{pmatrix} 1 & 2 & 3 & 4 \\ 1 & 2 & 3 & 4 \end{pmatrix}$, -1 corresponds to $\begin{pmatrix} 1 & 2 & 3 & 4 \\ 2 & 1 & 4 & 3 \end{pmatrix}$, i corresponds to $\begin{pmatrix} 1 & 2 & 3 & 4 \\ 3 & 4 & 2 & 1 \end{pmatrix}$, -i corresponds to $\begin{pmatrix} 1 & 2 & 3 & 4 \\ 4 & 3 & 1 & 2 \end{pmatrix}$.

**13.** For purposes of identifying elements of $S_6$, we label the permutations in $S_3$ as follows: $\iota \to 1$, $(12) \to 2$, $(13) \to 2$, $(23) \to 4$, $(123) \to 5$, and $(132) \to 6$. Then under the isomorphism of left multiplication, the elements of $S_3$ map to permutations of these 6 elements as follows:

$\iota \to \begin{pmatrix} 1 & 2 & 3 & 4 & 5 & 6 \\ 1 & 2 & 3 & 4 & 5 & 6 \end{pmatrix}$, $(12) \to \begin{pmatrix} 1 & 2 & 3 & 4 & 5 & 6 \\ 2 & 1 & 6 & 5 & 4 & 3 \end{pmatrix}$,

$(13) \to \begin{pmatrix} 1 & 2 & 3 & 4 & 5 & 6 \\ 3 & 5 & 1 & 6 & 2 & 4 \end{pmatrix}$, $(23) \to \begin{pmatrix} 1 & 2 & 3 & 4 & 5 & 6 \\ 4 & 6 & 5 & 1 & 3 & 2 \end{pmatrix}$,

$(123) \to \begin{pmatrix} 1 & 2 & 3 & 4 & 5 & 6 \\ 5 & 3 & 4 & 2 & 6 & 1 \end{pmatrix}$, $(132) \to \begin{pmatrix} 1 & 2 & 3 & 4 & 5 & 6 \\ 6 & 4 & 2 & 3 & 1 & 5 \end{pmatrix}$.

**15.** Stab$(K)$ has $(n - |K|)!$ elements. It is straightforward to show it is a subgroup.

**17.** *a. Hint:* Be sure to show $\mu$ is a permutation of $N$. *b. Hint:* Be sure to show that $\phi(\alpha) \in$ Stab$(J)$.

## Chapter 24 — Cosets and Lagrange's Theorem

**a.** $\langle 6 \rangle = \{6, 12, 18, 4, 10, 16, 2, 8, 14, 0\}$, $\langle 6 \rangle + 1 = \{7, 13, 19, 5, 11, 17, 3, 9, 15, 1\}$. $|\langle 6 \rangle| = 10$, $[\mathbb{Z}_{20} : \langle 6 \rangle] = 2$.

**b.** $U(\mathbb{Z}_{20}) = \{1, 3, 7, 9, 11, 13, 17, 19\}$, $\langle 7 \rangle = \{7, 9, 3, 1\}$, $\langle 7 \rangle \cdot 11 = \{17, 19, 13, 11\}$, $[U(\mathbb{Z}_{20}) : \langle 7 \rangle] = 2$, $|\langle 7 \rangle| = 4$.

**c.** $\{1, -1\} \cdot i = \{i, -i\}$, $\{1, -1\} \cdot j = \{j, -j\}$, $\{1, -1\} \cdot k = \{k, -k\}$, $\{1, -1\}$.

**d.** Those two cosets are in fact identical.

**e.** Yes, if $Ha = H$.

**f.** $Ha = Hb$.

**g.** Infinite.

**h.** Infinite.

**1.** $\langle (124) \rangle = \{(124), (142), \iota\}$, $\langle (124) \rangle (123) = \{(14)(23), (234), (123)\}$, $\langle (124) \rangle (132) = \{(134), (13)(24), (132)\}$, $\langle (124) \rangle (143) = \{(243), (12)(34), (143)\}$.

**3.** *Hint:* Everything should follow through easily, but take care when proving part (d).

**5.** *Hint:* Note that the map is a map between sets, no claim of group operations being preserved are being made here.

**7.** $K = \{\iota, (123), (132)\}$, $(12)K = \{(12), (23), (13)\}$. These are the same as the right cosets of $K$.

**9.** *Hint:* If $g \in G$ then what are the possible orders of $g$?

**11.** *a.* 1, 2, 4, or 8. *b. Hint:* Consider your answer to part (a).

**13.** *Hint:* If $|G|$ is prime, then apply Corollary 24.3. Now assume $|G|$ is not prime but every non-identity element generates $G$. In particular $G = \langle g \rangle$. Show this leads to a contradiction.

**15.** *a. Hint:* Corollary 24.3 says that the order of $a$ divides $|G|$. *b. Hint:* Apply part (a) to the group $\mathbb{Z}_p^*$ of units of $\mathbb{Z}_p$. *c. Hint:* Use the same idea as in part (b), by applying part (a) to the group $U(\mathbb{Z}_n)$ of units of $\mathbb{Z}_n$.

**17.** *Hint:* $\mathbb{Z} + r = \mathbb{Z} + s$ if and only if $r - s \in \mathbb{Z}$.

**19.** *a. Hint:* Every non-identity element of $H$ has order 2, 3, or 6. If an element of order 6, then find elements of order 2 and 3. Show $H$ can't have all elements of order 3 – these cyclic subgroups must pairwise intersection at the identity; now count how many elements in $H$ if there are 2 such elements, then 3. Likewise $H$ can't have all elements of order 2 – since

$A_4$ only has 3 elements of order 2. *b. Hint:* Look at the elements of $A_4$. *c. Hint:* Perform multiplications to generate elements in $H$.

# Chapter 25 — Groups of Cosets

**a.** $H\iota = \iota H$.

**b.** Because $ab = ba$ for all $a, b \in G$, then $aH = Ha$ for all $a \in G$.

**c.** $\mathbb{Z}_7$.

**d.** $\mathbb{Z}$ does not have the absorption property and so is not an ideal of $\mathbb{R}$. But the *group* $\mathbb{Z}$ (under addition) is a normal subgroup in the abelian group $\mathbb{R}$.

**e.** $\{1\}$ is certainly a subgroup. Also, $a\{1\} = \{a\} = \{1\}a$ for all $a \in G$.

**f.** No, $\mathcal{K}$ is not closed under the group operation.

**1.** *Hint:* From the proof of Theorem 25.2, we know that $gHg^{-1} \subseteq H$.

**3.** No, No.

**5.** *Hint:* Let $k = 2$ and consider a conjugate of $(13)$.

**7.** *Hint:* Define a function $\psi : \mathbb{Z}_n/\langle d \rangle \to \mathbb{Z}_d$, and prove that it works.

**9.** This is straightforward.

**11.** $\mathbb{Z}_2 \times \mathbb{Z}_2$.

**13.** *a. Hint:* Use the Subgroup Theorem. *b. Hint:* You must show every conjugate can be put in the form $hk$, $h \in H$, $k \in K$.

**15.** *Hint:* $A_4$ is a normal subgroup of $S_4$ by the Index 2 Theorem.

**17.** *a. Hint:* Use the Subgroup Theorem to establish that $gh^{-1}$ has finite order if $g$ and $h$ do. *b. Hint:* Suppose $t(G)h$ has finite order. Show $h \in t(G)$.

**19.** *Hint:* Certainly $H$ contains all finite products of commutators. Now show that this set is indeed a subgroup. Now if $H \subseteq K$, then $(gK)(hK) = ghK = gh(h^{-1}g^{-1}hg)K = hgK = (hK)(gK)$. If $G/K$ is abelian then $g^{-1}h^{-1}K = (g^{-1}K)(h^{-1}K) = (h^{-1}K)(g^{-1}K) = h^{-1}g^{-1}K$ which implies that $ghg^{-1}h^{-1} \in K$ and so $H \subseteq K$.

# Chapter 26 — The Isomorphism Theorem for Groups

**a.** Consider the natural map $G \to G/N$.

**b.** The commutative diagram given by the Fundamental Isomorphism Theorem allows us to determine the action of the homomorphism by the natural map.

**c.** No.

**d.** Yes.

**1.** The kernel is $\mathbb{Z}$. The groups $\mathbb{R}/\mathbb{Z}$ and $\mathbb{S}$ are isomorphic.

**3.** This is straightforward.

**5.** *a. Hint:* Need to show that if $\ker(\varphi)g = \ker(\varphi)h$, then $\varphi(g) = \varphi(h)$. *b. Hint:* That $\mu$ is a homomorphism follows from $\varphi$ being a group homomorphism. Now need to show onto.

**7.** *Hint:* To show $\varphi$ is a homomorphism is straightforward. Think of the possible normal subgroups of $\mathbb{Z}$ for kernels.

**9.** *Hint:* Define a homomorphism from $H$ to $HK/K$ with the appropriate kernel.

**11.** *a.* This is straightforward. *b.* $\langle R^2 \rangle$ *c.* The subgroup is the kernel of a homomorphism. *d.* $D_3/\langle R^2 \rangle$ and $\mathbb{Z}_2 \times \mathbb{Z}_2$.

**13.** *Hint:* First show that $H/K$ is actually a normal subgroup of $G/K$. Find a homomorphism from $G$ onto $(G/K)/(H/K)$ with the correct kernel.

**15.** *a. Hint:* Try two elements where $x_i$'s and $y_i$'s are all zero except for one coordinate. *b. Hint:* This is straightforward if you are careful with your notation. *c. Hint:* You need to find inverses and an identity element. *d. Hint:* Addition of two elements in $B$ amounts to just what? *e. Hint:* $B + g = B + h$ implies $g - h \in B$ implies what about $g$ and $h$? This should suggest a homomorphism from $W$ to a well-known group. *f.* You need a homomorphism $\varphi : \mathbb{Z} \to \text{Aut}(S)$. It will be a 'shift' operation.

## Chapter 27 — The Alternating Groups

**a.** For example, conjugating by $(123)$ yields $(127)(345)(69)$.

**b.** 60, 360.

**c.** The identity permutation is not odd.

**d.** The subgroup would then be normal by the Index 2 Theorem.

**e.** The subgroup $\{\iota, (12)\}$ in $S_3$.

**f.** Even.

**g.** Move to your sister's room, paint the north wall of her room, and move back to the room you started in.

**1.** Yes.

**3.** *a.* You should have the identity; 20 3-cycles; 24 5-cycles; 15 products of two disjoint 2-cycles. *b.* You should have the trivial subgroup; 10 distinct order 3 subgroups; 15 distinct order 2 subgroups; 6 distinct order 5 subgroups. *c. Hint:* Use Theorem 27.2.

**5.** $(1456)(29) = [(712)(3956)(48)](7895)(13)[(217)(6593)(84)]$.

**7.** *Hint:* Note that $(abc) = (1a)(1c)(1b)(1a)$.

**9.** Look at repeated conjugation of the transposition by the $n$-cycle; use Exercise 8.

**11.** *Hint:* Let $K = H \cap A_n$. What is $[H : K]$?

**13.** *a.* $S_4$, $H_1 =$ subgroup of disjoint 2-cycles (plus $\iota$), $A_4$ (since $H_1 \subseteq A_4$), $H_2 = \langle (1234) \rangle = \{\iota, (1234), (13)(24), (1432)\}$, $H_3 = \langle (1243) \rangle = \{\iota, (1243), (14)(23), (1342)\}$, $H_4 = \langle (1324) \rangle = \{\iota, (1324), (12)(34), (1423)\}$, $H_5 = H_1 \cup \{(13), (24), (12)(34), (14)(23)\}$, $H_6 = H_2 \cup \{(14), (23), (12)(34), (13)(24)\}$, $H_7 = H_4 \cup \{(12), (34), (13)(24), (14)(23)\}$.
*b. Hint:* It is easy to write down an element of $A_n$ taking $j$ to $k$ when $n \geq 4$. *c. Hint:* Look at the proof of the Conjugation Theorem 27.2.

## Chapter 28 — Sylow Theory: The Preliminaries

**a.** $\mathbb{Z}_9$ and $\mathbb{Z}_3 \times \mathbb{Z}_3$.

**b.** $G$ has an element of order 3 and one of order 5. Their product is of order 15.

**c.** $(123)$ and $(145)$, for example.

**d.** See Cayley's Theorem.

**e.** $H$ inherits the action of $G$ on the set $X$.

**f.** Here, $|X| = 4$, $|\text{Fix}(G)| = 1$, and $p = 3$.

**1.** *Hint:* This is similar to groups (or rings) of sequences we saw before – for instance in Exercise 6.19. Note the operation is coordinate-wise.

**3.** *Hint:* The base case follows from Theorem 28.2.

**5.** *Hint:* Need to show the relation is reflexive, symmetric, and transitive.

**7.** *Hint:* This action fails to be homomorphic.

**9.** *a. Hint:* Need to show if $g, h \in N(H)$ then so is $gh^{-1}$. *b.* Suppose $g \notin N(H)$ but $gHg^{-1} = H$; this is a contradiction. If $H$ normal in $G$, then $N(H) = G$. *c. Hint:* Need to show first that $C(H)$ is a subgroup of $N(H)$, then show it is normal in $N(H)$. *d. Hint:* $g \in C(H)$ implies $ghg^{-1} = h$ for all $h \in H$. Define a homomorphism from $N(H)$ to $\text{Aut}(H)$ with kernel $C(H)$.

## Chapter 29 — Sylow Theory: The Theorems

**a.** The First Sylow Theorem.

**b.** 6 is not a prime power.

**c.** The subgroups of order 16 are the Sylow 2-subgroups, and they must be conjugate.

**d.** Since all are of order $p^2$, they must be abelian: $\mathbb{Z}_5 \times \mathbb{Z}_5, \mathbb{Z}_{25}$. $\mathbb{Z}_7 \times \mathbb{Z}_7, \mathbb{Z}_{49}$.

**e.** If $m = 1$, $G = \mathbb{Z}_3$. If $m = 3$, $|G| = p^2$, and so is abelian. If $m = 5$, Theorem 29.8 applies.

**f.** 1, 2, 2, 1.

**g.** $\mathbb{Z}_2 \times \mathbb{Z}_4 \times \mathbb{Z}_3$. $\mathbb{Z}_2 \times \mathbb{Z}_{25} \times \mathbb{Z}_3$. $\mathbb{Z}_2 \times \mathbb{Z}_4$. $\mathbb{Z}_2 \times \mathbb{Z}_9$. $\mathbb{Z}_4 \times \mathbb{Z}_3$.

**h.** 1.

**i.** 2.

**j.** $4 \cdot 9 \cdot 5$. No.

**j.** $4 \cdot 9 \cdot 5$. Yes.

**1.** For example, $(14)(123)(14) = (234)$ and $(14)(132)(14) = (243)$, so $(14)\langle(123)\rangle(14) = \langle(234)\rangle$.

**3.** *Hint:* This follows in a similar manner to the proof of Theorem 29.8. Also note Theorem 29.9.

**5.** 1 or 7.

**7.** *Hint:* Use the Third Sylow Theorem to show there must be one Sylow 2-subgroup.

**9.** Sylow 2-subgroups: $\{\iota, (12)\}$, $\{\iota, (13)\}, \{\iota, (23)\}$. Sylow 3-subgroup: $\{\iota, (123), (132)\}$.

**11.** $|S_5| = 120 = 2^3 \cdot 3 \cdot 5$. There are either 1, 3, 5, or 15 Sylow 2-subgroups. Note that the cyclic group generated by a 4-cycle is a Sylow 2-subgroup and in each of these cyclic groups, there are exactly 2 4-cycles (the generator and its inverse). Now there are a total of 30 4-cycles in $S_5$ which would imply 15 different Sylow 2-subgroups – all isomorphic to $\mathbb{Z}_4$. In a similar manner, we see there are 1, 4, or 10 Sylow 3-subgroups. These are obviously isomorphic to $\mathbb{Z}_3$ and generated by 3-cycles, each Sylow 3-subgroup containing 2 3-cycles. There are 20 2-cycles in $S_5$ implying there are 10 Sylow 3-subgroups.

There are either 1 or 6 Sylow 5-subgroups, each isomorphic to $\mathbb{Z}_5$ and generated by a 5-cycle. There will be 4 5-cycles in each Sylow 5-subgroup. There are a total of 24 5-cycles and so 6 Sylow 5-subgroups.

**13.** *Hint:* For $(g, h) \in \mathbb{Z}_{12} \times \mathbb{Z}_{10}$ to be of order 8, $g$ would need to be of order 4 and $h$ of order 2. But then $(g, h)$ would have order 4.

**15.** $kj$.

**17.** Use the group $S$ of infinite sequences of real numbers, under addition (see Exercise 6.19).

**19.** *Hint:* The Third Sylow Theorem says that there is 1 Sylow $p$-subgroup, which would be isomorphic to $\mathbb{Z}_p$.

# Chapter 30 — Solvable Groups

**a.** Yes.

**b.** No. $S_3$.

**c.** Pick any proper subgroup $H$ of $Q_8$. Then $H$ and $Q_8/H$ are abelian. (Check these assertions.)

**d.** By Third Sylow Theorem, there is 1 Sylow 2-subgroup, $H$, which is abelian, being of order $2^2$. But $G/H$ is isomorphic to $\mathbb{Z}_3$.

**e.** By Third Sylow Theorem, there is 1 Sylow 3-subgroup, $H$, which is abelian, being of order $3^2$. But $G/H$ is isomorphic to $\mathbb{Z}_2$.

**1.** *Hint:* Both $G$ and $H$ have subnormal series with abelian quotients. Now construct one for $G \times H$.

**3.** An easy proof uses the Fundamental Theorem for Finite Abelian Groups 29.12.

**5.** *Hint:* The only subgroups of $\mathbb{Z}$ are of the form $n\mathbb{Z}$. So regardless of choice for $G_{n-1}$, it will not be simple. Thus the composition series can't even get started.

**7.** *Hint:* Proceed by induction on the order of the group. Use Exercise 6.

**9.** *Hint:* Note that $105 = 5 \cdot 21$. How many Sylow 5-subgroups are there?

# Chapter 31 — Quadratic Extensions of the Integers

**a.** $(3 + 2\sqrt{2})(1 + \sqrt{2})^k, (3 + 2\sqrt{2})(1 - \sqrt{2})^k$, any integer $k$.

**b.** $5 + i, -5 - i, -1 + 5i, 1 - 5i$.

**c.** $n = \sqrt{n} \cdot \sqrt{n}$.

**d.** No. 1 is not irreducible.

**e.** No. 3 and $3 + 3\sqrt{2}$ are associates.

**f.** $\{7a + b\sqrt{7} : a, b \in \mathbb{Z}\}$.

**g.** $3 + \sqrt{2}, 5 + \sqrt{2}, 5 + 2\sqrt{2}, 1 + 3\sqrt{2}$, etc.

**h.** No ($3$ in $\mathbb{Z}[\sqrt{2}]$).

**i.** Yes. No.

**1.** $c = \pm 1, d = 0$ or $c = 0, d = \pm 1$ are solutions to the Diophantine equations.

**3.** *Hint:* An element of $\mathbb{Z}[\sqrt{n}]$ is of the form $a + b\sqrt{n}$, for $a, b \in \mathbb{Z}$. $2\sqrt{3} = \sqrt{12} \in \mathbb{Z}[\sqrt{3}] \cap \mathbb{Z}[\sqrt{12}]$.

**5.** $(8 + 3\sqrt{7})^n$. $\sqrt{7}(8 + 2\sqrt{7})^n$.

**7.** $8 = 2 \cdot 4 = (1 + \sqrt{-7})(1 - \sqrt{-7})$.

**9.** *Hint:* There is no element in $\mathbb{Z}[\sqrt{n}]$ with norm 2.

**11.** *Hint:* $N(\sqrt{p}) = p$. $\sqrt{6} = (3 - \sqrt{6})(2 + \sqrt{6})$.

**13.** *Hint:* See Exercise 14.10 for a way to show the ideal is not principal. Example 14.6 shows how to find a homomorphism to a field for which the ideal is the kernel.

**15.** *Hint:* Suppose by way of contradiction that $\sqrt{n} = \frac{a}{b}$. Square both sides, clear the denominator, and then use the Fundamental Theorem of Arithmetic.

**17.** *a. Hint:* Need to show $\bar{I}$ is closed under subtraction and absorption. *b. Hint:* Consider Exercise 13.4a, and Example 14.6. *c. Hint:* If $I = \langle a + b\sqrt{n}\rangle$, then what should $\bar{I}$ be generated by?

# Chapter 32 — Factorization

**a.** None. 2 in $\mathbb{Z}[\sqrt{-5}]$. 6 in $\mathbb{Z}$.

**b.** This is similar to Example 32.3.

**c.** $a = uu^{-1}a$. No, primes must be non-units.

**d.** See Theorem 32.4.

**e.** $\langle x^2 - 1\rangle \subset \langle x + 1\rangle$. $\langle x + 1\rangle = \langle 2x + 2\rangle$. Strictly larger ideals correspond to strictly smaller degrees.

**1.** *Hint:* This is nearly identical to the proof for $\mathbb{Z}$ (Theorem 2.7).

**3.** Let $a = 3$. If $a = 3b$, then $a = 3 \cdot 3 \cdot b = \cdots$. Consider $(1, 0)$.

**5.** *a. Hint:* Consider $(1 + i)(a + bi)$. *b. Hint:* The norm preserves multiplication.

**7.** *Hint:* If $I_n$ proper, then $1 \notin I_n$.

**9.** Suppose $I$ an ideal of $R$ not finitely generated. Let $a_1 \in I$. Since $I$ not finitely generated, $\langle a_1\rangle \subset I$. Now pick $a_2 \in I$ but $a_2 \notin \langle a_1\rangle$. Then $\langle a_1\rangle \subset \langle a_1, a_2\rangle$. Continuing in this manner we get an infinite tower of ideals, all inside $I$.

Now suppose every ideal is finitely generated and $I_1 \subset I_2 \subset I_3 \subset \cdots$ is a tower of ideals. Let $I = \cup I_i$. $I$ is finitely generated, so $I = \langle a_1, a_2, \ldots, a_n\rangle$. Each $a_i$ must be in some $I_k$. Let $I'$ be the largest of these. Then $I = I'$ and so $I'$ is the largest of the ideals in the tower.

# Chapter 33 — Unique Factorization

**a.** PID implies UFD.

**b.** No, but soon.

**c.** $\mathbb{Z}$. $\mathbb{Z}$. $\mathbb{Z}[\sqrt{-5}]$. Any field.

**d.** It is a field.

**e.** $1 = \gcd(9, 50) = 9x + 50y \in \langle 9, 50 \rangle$.

**1.** *Hint:* Try 14. Then $\langle 2 \rangle$ will be maximal among principal ideals but is contained in $\cdots$

**3.** *a.* *Hint:* Both $a$ and $b$ are elements of $\langle d \rangle$. *b.* *Hint:* $d \in \langle a, b \rangle$. *c.* *Hint:* If $\langle d \rangle = \langle c \rangle$ then $\cdots$ *d.* *Hint:* All elements of $\langle a, b \rangle$ are linear combinations of $a$ and $b$.

## Chapter 34 — Polynomials with Integer Coefficients

**a.** GCD identity, PID.

**b.** 6, 7, 42.

**c.** $\mathbb{Q}$ is a field.

**d.** $2 \cdot 3 \cdot (x - 1)(x^2 + x + 1)$, no. $3x(x^3 - 2)$, no. $(x + 1)^2(5x - 1)(x - 2)$, yes.

**e.** None. $\mathbb{Z}[x]$.

**f.** $x^4 + 1$, none, 3, $-1$, none, $x^2 - 2$, $2x + 2$.

**1.** *a.* The content must be 1. *b.* $3x + 1$.

**3.** *Hint:* If $x^3 + 2x + 7$ factors one factor must either be constant or linear. But this polynomial is monic so no constant factors. If a linear factor, then the polynomial has a root of either $\pm 1$ or $\pm 7$.

**5.** *Hint:* Show a linear combination of $3x + 1$ and $x + 1$ is a linear combination of 2 and $x - 1$. Use an argument like Example 9.12 to show this ideal is not principal.

## Chapter 35 — Euclidean Domains

**a.** $q = 16$, $r = 4$. $q = (1/2)x^2 - 2x - (1/4)$, $r = 2x - (13/4)$. $q = 4 - i$, $r = -1$. $q = 7/\pi$, $r = 0$.

**b.** Find a non-unit in the ideal with smallest valuation.

**1.** *Hint:* Divide $m + n\sqrt{2}$ by $r + s\sqrt{2}$. Write $\dfrac{m + n\sqrt{2}}{r + s\sqrt{2}}$ is the form $u + v\sqrt{2}$, where $su, v \in \mathbb{Q}$. Pick integers $q_1$ and $q_2$ so that $q_1 + q_2\sqrt{2}$ is 'close' this number. Calculate the appropriate remainder and show its norm is small enough.

**3.** *Hint:* Let $\beta = 2 + \sqrt{2}$, $\alpha = 2$ and follow what you did in Exercises 1 and 2.

**5.** *a.* *Hint:* The valuation preserves multiplication, and so if $d$ is a divisor of $a$, $v(d) \leq v(a)$. *b.* *Hint:* Take your cue from the existential proof of Theorem 2.4. $d$ will be the element of the form $ax + by$ with smallest valuation. *c.* *Hint:* Recall $\langle a, b \rangle = \{ax + by\}$. *d.* *Hint:* $\langle a, b \rangle = \langle d \rangle$ if and only if $d = ax_0 + by_0$ for some $x_0, y_0 \in D$ and $d \in \langle a \rangle$ and $d \in \langle b \rangle$. Use part (b).

**7.** Let $d = \gcd(a, b)$ in $\mathbb{Z}$ and suppose $s + ti$ is a gcd of $a$ and $b$ in $\mathbb{Z}[i]$. Use Exercise 5 to argue that both $d$ and $s + ti$ are actually generators for the ideal $\langle a, b \rangle$ in $\mathbb{Z}[i]$.

**9.** For the Gaussian integers, we chose the quotient by looking at ordinary distance in the complex plane. For $\mathbb{Z}[\sqrt{2}]$, the 'distance' will involve hyperbolas!

## Chapter 36 — Constructions with Compass and Straightedge

**a.** Yes; construct a 30° angle (how?) and bisect it.

**b.** Yes.

**c.** No.

**d.** No; it is imaginary.

**e.** Because we could construct a square with area $\pi$.

**f.** Because $\sqrt[3]{2}$ is the length of the edge of a cube with volume 2.

**g.** An arbitrarily long straightedge.

**h.** An arbitrarily large compass.

**1.** *Hint:* Using labels as the construction in the proof of Theorem 36.2: Construct $A$ so $|\overline{VA}| = a$, $Q$ so $|\overline{VQ}| = b$ and $P$ so $|\overline{VP}| = 1$. Then $B$, so $\overline{QB}$ is parallel to $\overline{AP}$. Now show $|\overline{VB}| = b/a$.

**3.** To construct $\sqrt[4]{3}+1$, construct $3$, $\sqrt{3}$, $\sqrt[4]{3}$, then $\sqrt[4]{3}+1$. To construct $\sqrt{\frac{5}{2} + \sqrt{7}}$, construct $2$, $5$, $7$, $\frac{5}{2}$, $\sqrt{7}$, $\frac{5}{2} + \sqrt{7}$, then $\sqrt{\frac{5}{2} + \sqrt{7}}$.

**5.** *Hint:* Solve $(a + b\sqrt{2})^2 = 3 + 2\sqrt{2}$.

**7.** *a. Hint:* The interior angle for a pentagon is $108°$. *b. Hint:* Note that $|\overline{AE}| = |\overline{BE}| = 1$. *c. Hint:* We can construct $\rho$, so we can construct the $\triangle BCE$. Then construct $\triangle ABC$ and finally the pentagon.

## Chapter 37 — Constructibility and Quadratic Field Extensions

**a.**
$$\sqrt{6} \in \mathbb{Q}(\sqrt{6}).$$
$$\sqrt[8]{6} \in \mathbb{Q}(\sqrt{6}, \sqrt[4]{6}, \sqrt[8]{6}) \supseteq \mathbb{Q}(\sqrt{6}, \sqrt[4]{6}) \supseteq \mathbb{Q}(\sqrt{6}) \supseteq \mathbb{Q}.$$
$$\sqrt{2 + \sqrt[4]{5}} \in \mathbb{Q}\left(\sqrt{2 + \sqrt[4]{5}}\right) \supseteq \mathbb{Q}(\sqrt[4]{5}) \supseteq \mathbb{Q}(\sqrt{5}) \supseteq \mathbb{Q}.$$

**b.** $\mathbb{C}$.

**c.** No.

**d.** Yes: $\mathbb{K}(i)$; No.

**e.** No.

**f.** If $(a, b)$ is the point of intersection, then both $a$ and $b$ are elements of the field $F$.

**g.** Yes.

**1.** *Hint:* Attempt to solve $(a + b\sqrt{3})^2 = 5$.

**3.** *Hint:* Simplify $a + b\sqrt{p}$, where $a, b$ are of the form $c + d\sqrt{q}$. Now do same with $p$ and $q$ reversed.

**5.** To show that this set is a commutative ring with unity is straightforward.

**7.** *Hint:* Let $y = x^2$ and write the quartic given in terms of $y$.

**9.** The line in question has equation $y = mx + b$, passes through the point $(3, 1)$, and intersects the circle. Perform this intersection algebraically. When does a quadratic equation have a unique root?

## Chapter 38 — The Impossibility of Certain Constructions

**a.** $\sqrt[3]{2}$,   $\pi$.

**b.** We could construct a right triangle with hypotenuse $1$ and one side of length $\cos\theta$. The adjacent angle would be $\theta$.

**c.** Some irrational numbers are constructible, such as $\sqrt{2}$.

**d.** Double the length of a side of the original square, then construct the square root of this. Use this length for the sides of a new square.

**e.** Simply construct an edge of twice the original length.

**1.** *a. Hint:* Consider the proof of the appropriate lemma in this chapter. *b. Hint:* Start with $x = \sqrt{4 + \sqrt{7}}$ and remove the square roots. Your results should be a quartic.

**3.** Follow the hint given.

**5.** *Hint:* This is a slight generalization of Lemma 38.3; the same proof will work.

**7.** *Hint:* This is a slight generalization of Lemma 38.1; the same proof will work.

**9.** *a. Hint:* This is a bit messy. Need to integrate difference of line and parabola from $x_1$ to $x_2$ to get the area of the segment. Likewise to get area of the triangle, need to evaluate two integrals, or else use the cross product. *b.* Find the height (any height) of the triangle. Then construct a triangle with same base but $\frac{4}{3}$ the height. By Exercise 4a, we can square this triangle.

**11.** *a.* First construct an angle of $30°$. Now let $1/2^m < \epsilon\frac{\pi}{6}$. Bisect your $30°$ angle $m$ times. *b.* *Hint:* Limit as $n \to \infty$ of $n\psi$ is $\infty$. *c.* Take the closer of $n\psi$ and $(n-1)\psi$. *d.* *Hint:* Let $\epsilon$ be your tolerance. First perform parts (b) and (c) for $\epsilon/3$; let $\psi$ be the angle and $n$ be the multiple. Then divide $n$ by 3 (disregard the remainder).

# Chapter 39 — Vector Spaces I

**a.** $(4,2)$, $(0,-4)$, $(4,6)$, $(1,-1/2)$, $(-2,1)$.

**b.** 0 is a scalar, **0** is the zero vector.

**c.** Scalars are 0, 1, 2. Vectors are the polynomials $0, 1, 2, x, x+1, x+2, 2x, 2x+1, 2x+2, x^2, x^2+1, x^2+2, 2x^2, 2x^2+1, 2x^2+2, x^2+x, x^2+x+1, x^2+x+2, x^2+2x, x^2+2x+1, x^2+2x+2, 2x^2+x, 2x^2+x+1, 2x^2+x+2, 2x^2+2x, 2x^2+2x+1, 2x^2+2x+2$.

**d.** Yes, No.

**e.** $-\mathbf{v}$ is the vector added to $\mathbf{v}$ to get **0**. $(-1)\mathbf{v}$ is the vector $\mathbf{v}$ multiplied by the scalar $-1$. Both vectors are the inverse of $\mathbf{v}$.

**f.** Yes, No.

**g.** No.

**1.** *Hint:* The properties follow from properties of the polynomial ring $\mathbb{Q}[x]$.

**3.** *Hint:* Use the fact that operations are coordinate-wise, and $F$ is a field.

**5.** *Hint:* Use appropriate defining properties of vector spaces. To show that $\varphi_r$ is an injection, you must use that $r^{-1} \in F$.

**7.** *Hint:* The properties follow from properties of $M_2(\mathbb{R})$.

**9.** *Hint:* The properties easily follow from properties of fields. (Notice which way the inclusion goes. It is usually not the case that $F$ is a vector space over $E$ since closure under scalar multiplication is a problem.)

**11.** *Hint:* Note addition is defined point-wise.

**13.** Consider the scalar product $(1/2)1$.

# Chapter 40 — Vector Spaces II

**a.** $\{(1,0),(0,1),(1,1)\}$ in $\mathbb{R}^2$; can't happen; $\{(1,0),(0,1)\}$ and $\{(1,0),(1,1)\}$ in $\mathbb{R}^2$; can't happen; $\{(1,0),(0,1),(1,1)\}$ in $\mathbb{R}^2$; $\{(1,0,0),(0,0,1),(0,1,0),(1,1,1)\}$ in $\mathbb{R}^3$.

**b.** The set of vectors would not be linearly independent.

**c.** 2, 1.

**1.** *Hint:* Suppose $a\sqrt{2} + b\sqrt{3} = 0$ for $a, b \in \mathbb{Q}$.

**3.** *Hint:* To prove these are vector spaces is straightforward. For a basis: Take care that scalar multiplication yields a different set of vectors depending on the field you use.

**5.** *Hint:* You should find $mn$ matrices.

**7.** *Hint:* If $\mathcal{S}$ is a finite subset of $\mathbb{Q}[x]$, give a limit on the degree of a polynomial expressible as a linear combination of vectors from $\mathcal{S}$.

**9.** *a.* *Hint:* The field will be the same. The subset will involve vectors. *b.* *Hint:* Note that most of the properties of a vector space will be inherited. Closure is important.

**11.** Yes, Yes, Yes, Yes.

**13.** $\{(1,1,0),(2,3,-1),(1,4,2)\}$.

**15.** *Hint:* First write $\mathbf{v}$ as a linear combination of the $\mathbf{v}_i$. Now simply substitute the linear combination of all the $\mathbf{v}_i$'s in terms of the $\mathbf{w}_j$'s

**17.** *Hint:* If $\mathcal{B}$ is a maximal independent set in vector space $V$, and $\mathbf{v} \notin \mathcal{B}$, then $\mathcal{B} \cup \{\mathbf{v}\}$ is linearly dependent.

## Chapter 41 — Field Extensions and Kronecker's Theorem

**a.** $x^3 - 2$; $x^3 - 2$ or $x - \sqrt[3]{2}$; $x^2 - 2x - 1$; $x^4 - 2x^2 - 1$; $x^2 - 2x + 5$; $x^2 - 2\pi x + \pi^2 + 1$.
**b.** Yes, No.
**c.** The original field.
**d.** There are none.
**e.** None, $\pi$, $\sqrt[3]{2}$.
**f.** $0$, $\mathbb{Q}$; $0$, $\mathbb{Q}$; $0$, $\mathbb{Q}$; $11$, $\mathbb{Z}_{11}$; $0$, $\mathbb{Q}$; $3$, $\mathbb{Z}_3$.
**g.** See Kronecker's Theorem 41.1.
**h.** Yes, $\cos 20°$ is a root of $8x^3 - 6x - 1$.
**i.** $\sqrt[3]{2}$, $\cos 20°$.
**1.** *Hint:* Use Kronecker's Theorem 41.1, together with induction on the degree of $f$.
**3.** *a. Hint:* Find a polynomial in $\mathbb{Z}[x]$ with root $a/b$. *b. Hint:* Think Gauss's Lemma 5.5.
**5.** *Hint:* Think trig identities.
**7.** *a. Hint:* What elements must belong to any subfield of $F$? *b. Hint:* This proof will be similar to part (a).
**9.** *Hint:* First factor $f$.

## Chapter 42 — Algebraic Field Extensions

**a.** $1$, $\sqrt[3]{2}$, $\sqrt[3]{4}$; $1$; $1$, $1 + \sqrt{2}$; $1$, $\sqrt{1 + \sqrt{2}}$, $1 + \sqrt{2}$, $(1 + \sqrt{2})^{3/2}$; $1$, $1 + 2i$; $1$, $\pi + i$.
**b.** $3/2 \cdot 1 - 1/2(1 + \sqrt{2})$, $8/17 \cdot 1 + 1/17\sqrt[3]{2} - 2/17\sqrt[3]{4}$.
**c.** No, $\mathbb{Q}(\pi)$ over $\mathbb{Q}$; Yes; No, $\pi$ in $\mathbb{Q}(\pi)$ over $\mathbb{Q}$; Yes; Yes.
**d.** It divides every polynomial with that root, and it is monic.
**1.** *a.* See answer to Exercise 8.15. *b. Hint:* Here, $p = 2$.
**3.** *a.* Simply perform the (tedious) computation. *b.* Ditto. *c.* Factor the two quadratics by using the quadratic formula.
**5.** *Hint:* You will need to solve for the six rational number unknowns that will serve as coefficients on the powers of $\alpha$. Cross-multiply to eliminate the denominator, and then use the minimal polynomial to eliminate powers of $\alpha$ higher than five, as we did in the proof of Theorem 42.3.
**7.** *Hint:* The proof follows closely along the lines given in Exercise 6.
**9.** *Hint:* To show the other roots are not in the given extensions, follow the idea in Example 42.9 where it is shown that $\sqrt[3]{2} \notin \mathbb{Q}(\sqrt[3]{2}\zeta)$. Your isomorphisms should map a root to a root, and fix elements in $\mathbb{Q}$.
**11.** $x^4 - 2x^2 + 9$. Now show this polynomial cannot be factored in $\mathbb{Q}[x]$. For another root, consider the complex conjugate.

## Chapter 43 — Finite Extensions and Constructibility Revisited

**a.** They are equal.
**b.** $E$ is either $F$ or $K$.
**c.** They are equal.
**d.** See Theorem 43.1; $\mathbb{Q} \subset \mathbb{Q}(\pi)$ is not algebraic; $\mathbb{Q} \subset \mathbb{Q}(\pi)$ is not finite; consider the field of all constructible numbers as an extension field of $\mathbb{Q}$; see the previous field.
**e.** $7^6$.
**f.** $\mathbb{Q}(\sqrt{2})(\sqrt{3})$; There are none; $\mathbb{Z}_2(\alpha)$ where $\alpha$ is a root of $x^3 + x + 1$.

**1.** *Hint:* First show that $\sqrt{2}+\sqrt{3}$ is a root of $x^4-10x^2+1 \in \mathbb{Q}[x]$, and that this polynomial is irreducible in $\mathbb{Q}[x]$. Therefore, $[\mathbb{Q}(\sqrt{2}+\sqrt{3}) : \mathbb{Q}] = 4$, and consequently $\sqrt{2}+\sqrt{3} \notin \mathbb{Q}(\sqrt{2})$. Conclude that $\sqrt{3} \notin \mathbb{Q}(\sqrt{2})$.

**3.** *Hint:* For a computational proof, compute $(\sqrt{2}+\sqrt{3})^3$ and then subtract an appropriate multiple of $\sqrt{2}+\sqrt{3}$ from your result to get a multiple of $\sqrt{2}$. From this, argue that

$$\sqrt{2} \in \mathbb{Q}(\sqrt{2}+\sqrt{3}).$$

Then conclude that $\sqrt{3} \in \mathbb{Q}(\sqrt{2}+\sqrt{3})$. Thus, $\mathbb{Q}(\sqrt{2},\sqrt{3}) \subseteq \mathbb{Q}(\sqrt{2}+\sqrt{3})$. To show containment the other way is easy. A proof along the lines of Exercise 1 using Theorem 43.2 is also possible.

**5.** *Hint:* Suppose by way of contradiction that $[F : \mathbb{Q}] = n$.

**7.** *Hint:* Use Theorem 43.2, but be careful.

**9.** *a. Hint:* $\mathbb{Z}_2$ is the prime subfield here. *b. Hint:* In this finite case you can actually write down all possible irreducible polynomials in $\mathbb{Z}_2[x]$ of degree 2 or 3. Show by direct calculation that $\alpha + \beta$ is not a root of any of these. Conclude that $\mathbb{Z}_2(\alpha + \beta) = \mathbb{Z}_2(\alpha, \beta)$.

**11.** *Hint:* Pick up the calculations with the equation for $\sqrt{2}$ just before the Quick Exercise in Example 43.3.

**13.** *a. Hint:* Consider Exercise 5. *b. Hint:* Show that if $\alpha \in K\backslash\mathbb{A}$ then $\alpha$ is transcendental over $\mathbb{Q}$ and so transcendental over $\mathbb{A}$.

# Chapter 44 — The Splitting Field

**a.** True; True; False ($\mathbb{C}$ is a normal extension of $\mathbb{Q}$); True; True; False (let $F = \mathbb{Q}$ and $f = x^3 - 2$); False (let $F = \mathbb{Q}$ and $f = x^2 - 2$); True; True; False (consider $\mathbb{Q} \subset \mathbb{Q}(\sqrt[3]{2})$).
**b.** Let $\zeta$ be the primitive cube root of unity. Then the answers are: $\mathbb{Q}(\sqrt[3]{2}, \zeta)$; $\mathbb{R}(\zeta) = \mathbb{R}(i) = \mathbb{C}$; $\mathbb{Q}(i)(\sqrt[3]{2}, \sqrt{3})$; $\mathbb{C}$.
**c.** $F$.

**1.** *a. Hint:* Perform the multiplication, remembering that $\alpha = \langle f \rangle + x$ and the prime field is $\mathbb{Z}_2$. *b. Hint:* $\alpha^2 = \langle f \rangle + x^2$. Show you get all 8 elements of $\mathbb{Z}_2(\alpha)$ as a result of field operations starting with 1 and $\alpha^2$. *c. Hint:* $\alpha^2 + \alpha = \langle f \rangle + x^2 + x$. Show you get all 8 elements of $\mathbb{Z}_2(\alpha)$ as a result of field operations starting with 1 and $\alpha^2$.

**3.** *a. Hint:* This should be straightforward. *b. Hint:* This should be straightforward. *c. Hint:* Use induction on the degree of $g$. *d. Hint:* Use induction on $n$.

**5** *a. Hint:* If $f$ has a linear factor, $f$ has a root. If $f$ factors into two quadratics, consider the substitution $y = x^2$. *b.* Let $\alpha = \sqrt{2+\sqrt{2}}$ and $\beta = \sqrt{2-\sqrt{2}}$. Then $f = (x - \alpha)(x + \alpha)(x - \beta)(x + \beta)$. *c.* $\mathbb{Q}(\alpha, \beta)$. *d. Hint:* Show that $\beta$ can be expressed in terms of $\alpha$, and so the splitting field is $\mathbb{Q}(\alpha)$.

**7.** *Hint:* Show $K$ must contain *all* roots of $f$.

# Chapter 45 — Finite Fields

**a.** $\mathbb{Z}_5[x]/\langle p \rangle$ where $p = x^2 + 2$ would work, among others.
**b.** It has prime order.
**c.** $\mathbb{Z}_9$; $\mathbb{Z}_3 \times \mathbb{Z}_3$; $GF(9)$. The latter is the only field. The last two are isomorphic as additive groups.

**1.** *a. Hint:* Use Exercise 2.19 and the binomial theorem Exercise 6.17. *b. Hint:* Same as part (a), but use a slight variation on Exercise 2.19. *c. Hint:* There are two cases, depending on whether $p$ is even or odd.

**3.** *Hint:* Use Exercise 45.1a.

**5.** *Hint:* You will need to show that the polynomial $x^{m-1} - 1$ divides $x^{n-1} - 1$ if and only if $m$ divides $n$.

**7.** *Hint:* Look at the polynomial term by term. Use Exercise 45.1a and Exercise 45.6.

**9.** *a. Hint:* First factor $x^8 - x$ in $\mathbb{Z}_2[x]$ into two linear polynomials and one of degree 6, then show $f$ divides this degree 6 polynomial. *b. Hint:* Your work in part (a) should have this essentially done. *c. Hint:* What are the 8 elements of $\mathbb{Z}_2(\alpha)$? Show these are each roots of $f$. *d. Hint:* Think order of these fields. *e. Hint:* Write out the 8 elements in the field $\mathbb{Z}_2(\beta)$. Be careful your map preserves the operations.

**11.** Just compute.

# Chapter 46 — Galois Groups

**a.** True; True; True; False; False (it is the trivial group).

**b.** Cyclic group of order ten.

**c.** Cyclic group of order eleven.

**d.** Example 46.5; None; Examples 46.7, 46.8, 46.9.

**1.** Klein Four group; cyclic group of order two.

**3.** Cyclic group of order four; cyclic group of order two.

**5.** *a. Hint:* Use Gauss's Lemma 5.5. *b. Hint:* Use the argument from Section 38.2 first letting $\theta = 40°$ then $\theta = 10°$. *c. Hint:* First show $\gamma \in \mathbb{Q}(\beta)$ by noting that $-2\sin(10°) = -2\cos(80°)$ and then using the double angle formula. Now show $\alpha \in \mathbb{Q}(\gamma)$ using double angle formula and converting to sine. *d. Hint:* Consider Theorem 46.1. *e. Note:* If you use the discriminant of the conic (as in Exercise 10.12.d), the root will appear to be complex – in fact the imaginary part is zero.

**7.** *Hint:* $f$ is irreducible by Eisenstein's Criterion 5.7. Now to show the derivative has 2 real roots, look at the second derivative.

**9.** *Hint:* Follow the proof of Theorem 46.2, noting that Theorem 42.3 applies to simple extensions.

**11.** *a. Hint:* Need to show $\sqrt{3}i \in \mathbb{Q}(\zeta)$ and $\zeta \in \mathbb{Q}(\sqrt{3}i)$. *b. Hint:* Follow the technique of Example 43.3.

**13.** Note the group isomorphism described in the proof of Theorem 46.3.

# Chapter 47 — The Fundamental Theorem of Galois Theory

**a.** $\alpha_1\alpha_2 + \alpha_1\alpha_3 + \alpha_1\alpha_4 + \alpha_1\alpha_5 + \alpha_2\alpha_3 + \alpha_2\alpha_4 + \alpha_2\alpha_5 + \alpha_3\alpha_4 + \alpha_3\alpha_5 + \alpha_4\alpha_5$.

**b.** $\text{Gal}(K|E)$ is a subgroup of $\text{Gal}(K|F)$.

**c.** $\text{Gal}(E|F)$ is isomorphic to $\text{Gal}(K|F)/\text{Gal}(K|E)$.

**d.** $\text{Fix}(H_2) \subseteq \text{Fix}(H_1)$.

**e.** There are none.

**f.** Because any group of order 4 has a subgroup of order 2.

**1.** *Hint:* For the Galois group, see Theorem 46.4.

**3.** *Hint:* It is not easy to find all of these fields. Keep in mind that the splitting field includes such elements as $\sqrt{2}, i, \sqrt{2}i$, etc., and that every distinct subgroup must correspond to a distinct intermediate field.

**5.** *Hint:* All the field axioms should be easily verified.

**7.** *Hint:* You are doing induction on $n$ here. Try $n = 2$ and $n = 3$ to see the pattern for what needs to be done.

**9.** 8, 4, 4, 2, 4, 2, 4.

**11.** $[K : \text{Fix}(H)] = [S_5 : H] = 40$. *a.* Field is characteristic zero. *b.* 40. *c.* $K$ is normal over $\mathbb{Q}$. *d.* $H$ is not normal in $S_5$. *e.* No. *f.* Let $J = \{3, 4\}$. *g.* $\text{Fix}(H) \subset \mathbb{Q}(\alpha_3, \alpha_4)$. *h.* 2. *i.* $H$ is normal in the copy of $S_3$. *j.* $\mathbb{Z}_2$.

# Chapter 48 — Solving Polynomials by Radicals

**a.** $\mathbb{Q} \subset \mathbb{Q}(\sqrt[4]{2}) \subset \mathbb{Q}(\sqrt[4]{2}, i)$.

**b.** $\mathbb{Q} \subset \mathbb{Q}(\sqrt[3]{2}) \subset \mathbb{Q}(\sqrt[3]{2}, \zeta)$.

**c.** $\mathbb{Q} \subset \mathbb{Q}(\sqrt[3]{2})$.

**d.** See Theorem 46.4.

**e.** Yes.

**f.** Find an irreducible cubic with three real roots.

**g.** $f$ is not solvable by radicals.

**1.** Make a chart as in Example 46.9, considering that $\alpha = \sqrt[3]{1 + \sqrt{2}}$ can be mapped only to $\alpha, \alpha\zeta, \alpha\zeta^2$, $\zeta$ can be mapped only to $\zeta, \zeta^2$, and $\sqrt{2}$ can be mapped only to $\pm\sqrt{2}$.

**3.** *a. Hint:* This is a straightforward algebra exercise. *b. Hint:* You know that the $\alpha_i$ are distinct (why?). Use this to argue that the $\beta_i$ are distinct. Then show that the permutations in $H$ are the only elements of $S_4$ that leave the $\beta_i$ fixed. *c. Hint:* What do you know about conjugation and cycle structure? *d. Hint:* Find a group homomorphism from $\mathrm{Gal}(E/F)$ onto $G$ with kernel $H$.

**5.** $g = x^3 + 8x = x(x^2 + 8)$. Clearly the splitting field for $g$ is $\mathbb{Q}(2\sqrt{2}i)$, which as a quadratic extension has Galois group over $\mathbb{Q}$ isomorphic to $\mathbb{Z}_2$.

**7.** *a.* Eisenstein Criterion. *b. Hint:* Show the function has a global minimum, and that it has a negative value there. *c. Hint:* See Exercise 3. *c. Hint:* Use the Rational Root Theorem and Gauss's Lemma. *d. Hint:* Use the derivative. *e. Hint:* The argument of Example 46.8 applies. *f. Hint:* Consider Corollary 43.5 applied to the splitting field of $f$.

# Guide to Notation

In this appendix we provide a guide to the mathematical notation we use in this book. In many cases we provide a reference in the text where the given notation first occurs. Rather than attempting an alphabetical listing, we have grouped this list of notations conceptually.

## Set Theory

Modern mathematics is expressed in terms of set theory, and we use standard notation for these concepts. In what follows, assume that $A$ and $B$ are sets:

If $a$ is an element of $A$, then we write $a \in A$. If it isn't, we write $a \notin A$.

The *intersection* of $A$ and $B$ is

$$A \cap B = \{x : x \in A \text{ and } x \in B\}.$$

The *union* of $A$ and $B$ is

$$A \cup B = \{x : x \in A \text{ or } x \in B\}.$$

The *set difference* is

$$A \backslash B = \{x : x \in A \text{ and } x \notin B\}.$$

If $A$ is a subset of $B$, we write $A \subseteq B$. If it is a *proper* subset we write $A \subset B$. We can change emphasis and write these two as $B \supseteq A$ and $B \supset A$, respectively.

If the set $A$ is finite we denote by $|A|$ the number of elements in $A$.

The *set product* is the set of all ordered pairs, with first entry from $A$ and second entry from $B$:

$$A \times B = \{(a, b) : a \in A,\ b \in B\}.$$

If $A$ and $B$ are rings or groups, we can place a ring (Example 6.10) or group (Section 222.4) structure on $A \times B$.

## Numbers

$\mathbb{N}$ is the set of *positive integers* (or *counting numbers*), $\mathbb{Z}$ is the set of all *integers*, and $\mathbb{Q}$ is the set of *rational numbers* (Chapter 1).

For integers $a, b$ and positive integer $n$, $a \equiv b \pmod{n}$ means that $a$ and $b$ are *congruent modulo n* (Chapter 3).

$\mathbb{R}$ is the set of *real numbers*, and $\mathbb{C}$ is the set of *complex numbers* (Exercise 6.11 and Section 8.5). $\mathbb{S}$ is the *unit circle*, the set of complex numbers of length one (Example 19.14.)

$\mathbb{K}$ is the set of *constructible numbers* (Section 36.2), and $\mathbb{A}$ is the set of *algebraic numbers* (Exercise 41.4).

## Functions

$\det(A)$ is the *determinant*, which is a real-valued function defined on square matrices (Example 22.5).

$\deg(f)$ is the *degree* of the polynomial $f$ (Section 4.1). cont $f$ is the *content* of a polynomial $f \in \mathbb{Z}[x]$ (Section 34.3).

$\phi(n)$ is *Euler's phi function* (Exercise 24.15).

$\gcd(a, b)$ is the *greatest common divisor*, for integers, polynomials, or Euclidean domains (Section 2.2, Section 4.6, Exercise 35.4). Likewise, $\text{lcm}(a, b)$ is the *least common multiple* (Exercise 2.11).

$|\alpha|$ is the ordinary absolute value, if $\alpha$ is a real number; it is the *modulus* if $\alpha$ is complex (Section 8.5). $\arg(\alpha)$ is the *argument* of the complex number (Section 8.5).

We use several important functions from calculus: the *natural logarithm* $\log(x)$, the *exponential function* $\exp(x) = e^x$ and the *trigonometric functions* $\sin(x), \cos(x), \tan(x)$ etc.

# Rings

In what follows, assume that $R$ is a ring and $F$ is a field.

$\mathbb{Z}_n$ is the ring of integers, modulo $n$ (Chapter 3 and Example 6.4). Of course, we also can consider it as a cyclic group (Example 19.4 and Example 21.18). $R[x]$ is the ring of polynomials in indeterminate $x$ and coefficients from the ring $R$ (see Exercise 6.23); particularly important are $\mathbb{Q}[x]$ and $\mathbb{Z}[x]$ (Chapters 4 and 5), and $F[x]$, where $F$ is a field (Chapter 9). $\mathbb{Q}[x, y]$ is the ring of polynomials in two indeterminates $x, y$ (Exercise 9.31).

$\mathbb{Z}[i]$ is the ring of *Gaussian integers* (Exercise 6.12); more generally, $\mathbb{Z}[\sqrt{n}]$ is a *quadratic extension of the integers* (Section 31.1). These rings are equipped with a *norm function* $N(\alpha)$ (Section 31.2).

$M_2(R)$ is the ring of $2 \times 2$ *matrices*, with entries from the ring $R$; see Example 6.13 and Exercise 6.8.

For any set $X$, $P(X)$ is the set of all subsets of $X$; it is called the *power set* and can be made into a ring (Exercise 6.20).

$C[0, 1]$ is the ring of all real-valued continuous functions on the unit interval (Example 6.14).

If $I$ is an ideal of the ring $R$, then $R/I$ is the *ring of cosets, modulo I* (see Section 19.2).

$U(R)$ is the *group of units* of the ring $R$ (Section 8.2 and Section 19.3). For a field $F$, we denote $U(F)$ by $F^* = F \backslash \{0\}$ (Section 19.3).

If $\varphi$ is a (ring or group) homomorphism, then $\ker(\varphi)$ is its kernel (Section 12.1 and Section 26.1).

Given $a \in R$, $\langle a \rangle$ is the *principal ideal* generated by $a$ (Section 9.1). More generally, $\langle a_1, \cdots, a_n \rangle$ is the *ideal generated by the $a_i$* (Exercise 9.12).

$N(R)$ is the *nilradical* of $R$ (Exercise 7.15).

$Z(R)$ is the *center* of $R$ (Exercise 7.12).

# Groups

In what follows, assume that $G$ is a group.

$D_n$ is the $n$th *dihedral group*, the group of symmetries of the regular $n$-gon (Section 17.2 and Exercise 20.29).

$S_n$ is the $n$th *symmetric group*, the group of permutations of a set of $n$ elements (Section 18.2). $A_n$ is the *alternating group*, the subgroup of $S_n$ consisting of the even permutations (Chapter 27).

$Q_8$ is the group of *quaternions* (Example 19.16). See Exercise 21.12 for our usual notation.

$Ha$ and $aH$ are right and left cosets of the subgroup $H$ of the multiplicative group $G$ (Section 24.2 and Section 25.1). Additive cosets look like $H + a$ or $a + H$.

$G/H$ is the *group of cosets* for $G$ with normal subgroup $H$ (Section 25.2).

$[G : H]$ is the *index* of the subgroup $H$ in $G$; that is, it is the number of distinct cosets $H$ has (Section 24.2).

$\langle a \rangle$ is the *cyclic subgroup of G generated by g* (Section 21.3).

For $g \in G$, $o(g)$ is the *order* of the element $g$ (Section 21.1).

$\text{End}(G)$ is the *endomorphism ring* for the group $G$ (Exercise 22.14).

$Z(G)$ is the *center* of $G$ (Exercise 20.16).

$\mathrm{Orb}(x)$ is the *orbit* of $x$ (Section 28.2).
$\mathrm{Stab}(x)$ is the *stabilizer* of $x$ (Section 28.2).
$\mathrm{Fix}(g)$ is the set of fixed points of $g$ (Section 28.2).

## Fields

If $F$ is a field, $F(\alpha)$ is a *simple extension* — the smallest field containing $F$ and $\alpha$ (Sections 42.2 and 42.3).

For fields $F \subseteq E$, $[E : F]$ is the *degree* of the field extension: The dimension of $E$ as an $F$-vector space (Section 43.1).

$\mathrm{Aut}(R)$ is the *automorphism group* of the ring $R$ (Example 19.21). $\mathrm{Gal}(E|F)$ is the *Galois group of $E$ over $F$* (Section 46.1).

$GF(p^n)$ is the *Galois field* with $p^n$ elements (Section 45.1).

# Index